Kevin M. Quinley, CPCU, ARM Dı

BUSINESS AT RISK

How to Assess, Mitigate, and Respond to Terrorist Threats

The National Underwriter Company
PROFESSIONAL PUBLISHING GROUP

P.O. Box 14367 • Cincinnati, Ohio 45250-0367
1-800-543-0874 • www.nationalunderwriter.com

P.O. Box 14367
Cincinnati, OH 45250-0367

International Standard Book Number 0-87218-702-0

Printed in the United States of America

Dedications

Donald L. Schmidt

To Sherry, Danny, and Cari.
And to my 295 colleagues lost on September 11, 2001.

Kevin M. Quinley

To the fallen, and those who fight for freedom.

Table of Contents

Foreword

It is by now a truism that the terrorist attacks of September 11 were a wake up call for America and for our government. For the first time in almost 200 years America was attacked by foreign forces on its own soil. It was one of the bloodiest days in our history.

Less obviously, the events of September 11 should be a wake up call for American businesses. Over the past thirty years, 80 percent of all terrorist attacks on American targets have been against American businesses. This was also true in September. Over four hundred businesses were directly impacted by the attacks.

Responsible businesses owe it to their employees and shareholders to consider the implications of the new situation. In particular, managers must ask themselves whether they have adequate plans in place to address the new threats, which must today include dealing with the threat of chemical, biological or radiological attacks.

In this important book, Don Schmidt and Kevin Quinley have done an excellent job of portraying the main elements of the new terrorist threat. They offer the reader an extensive catalog of measures that prudent businesses should follow to mitigate these threats. As they make clear, these steps must cover everything from the immediate reaction of employees at the scene of the incident, to the way senior managers plan for business recovery after the event. Finally, they look carefully at the alternative means by which a company can assure its financial recovery and how the financial services industry has been impacted.

This book is an important contribution to new thinking about the new threat of terrorism. It should be must reading for businesspeople.

Ambassador L. Paul Bremer III
Chairman, Chief Executive Officer
Marsh Crisis Consulting
Chairman, National Commission on Terrorism

Acknowledgments

I would like to acknowledge and thank those who made it possible for me to co-author this book.

Vivian Menna and Glenn Buser, Managing Directors of Marsh, recognized the importance of this project and strongly promoted it. Thank you both.

Walter Wilk, Jr., of Marsh Risk Consulting allowed me the time and flexibility to focus on the book. Thanks Walter. I couldn't have done it without your support.

I would also like to thank the technical experts who peer reviewed my chapters of the book. Thanks to John A. Edgar, CBCP, CSP, and Al Berman, CBCP, both of Marsh Risk Consulting; Lloyd Bokman of the Ohio Emergency Management Agency and Chair of the NFPA 1600 Technical Committee; and Craig Garlock, PE, of Marsh Crisis Consulting.

I would also like to thank Ambassador Jerry Bremer. Your foreword validates the importance of this work to the business community.

Writing a book is challenging, but a great editor makes it a lot easier. Diana Reitz provided excellent direction and was always a pleasure to work with.

Last, but certainly not least, I would like to acknowledge the incredible support of my family. Writing a book while working a full-time job leaves little time for anything or anyone else. Thanks Sherry, Danny, and Cari for your encouragement, support, and understanding while I was sequestered away for days on end.

Donald L. Schmidt

Introduction

July 4, 1776. April 12, 1861. December 7, 1941. November 22, 1963. And now September 11, 2001.

Days that will be forever etched in the history of America. Some of us recall where we were that day in December when the Japanese attacked Pearl Harbor. More of us remember what we were doing when we heard the newscast from Dallas about President John F. Kennedy's assassination.

Each of us knows exactly where we were on September 11, 2001. Some were busily setting up meetings or making conference calls. Many of us were working on our computers. Others were at home, getting ready to go to work or sending the kids off to school.

But we stopped in mid-sentence when we heard that American Airlines Flight 11 had crashed into the North Tower of the World Trade Center. Minutes later we were stunned by the news that a second aircraft hit the South Tower. And then the Pentagon and Pennsylvania.

The fallout from September 11, 2001, will continue to be felt for years to come. The word *terrorism* is now commonplace in the vocabulary of average Americans. Business and risk managers are exhibiting an increased awareness of the need to professionally manage the risks of terrorism both on and off American soil. The insurance industry has been faced with the largest insured loss in its history.

On that day, many financial services professionals who previously had been able to maintain an objective stance in the midst of catastrophe found themselves the victims. They were coping with not only their clients' losses, but also with personal and professional losses that previously had been incomprehensible.

This book was developed—at least in part—in order to portray the effect of terrorist acts on the business community and financial services industry. It relays information about assessing, mitigating, responding to, recovering from, and financing the risk of terrorism, all now critical to business operations. It includes valuable pre- and post-loss information, offering guidance to those with an expanded awareness of just how unnatural a catastrophe can be.

Our emphasis is on the practical. In many cases the authors have distilled the discussion into handy checklists, suggestions, charts, dos and don'ts. Nevertheless, we realize that there can be no one perfect guidebook to managing the risk of terrorist threats. There is no "One-Minute Risk Manager for Terrorism" or "Ten Steps to Protect Against Terrorism." Each solution or set of suggestions must be adapted, customized, and tailored to each particular organization's needs.

Charts and checklists are starting points—not ending points—for building a strong risk management plan. The goal is to upgrade an organization's invulnerability to terrorism losses and to cushion financial blows caused by terrorism losses. Risk professionals and businesses grope for practical solutions more than a theoretical understanding of terrorism and its threats. The book is more practical than theoretical.

One of the first elements addressed is the challenge of *defining* terrorism. Chapter 1 discusses just what we mean by the *terrorism threat.* Subsequent chapters offer information on risk identification and vulnerability assessment; prevention and mitigation of terrorism; effective response and recovery from terrorist threats and attacks; and financial considerations in paying for and transferring the risk of terrorism to either an insurance or alternative financing mechanism.

Assessment, Mitigation, Response, Recovery, Financing

This book outlines five key aspects of managing the risk of terrorism: assessment, mitigation, response, recovery, and financing. After defining terrorism it moves from pre-loss risk, threat, and vulnerability assessment to business impact analysis in chapter 2. Chapter 3, Mitigation Strategies, offers powerful insight into site selection and business design, physical and operational security, and computer and cyber security.

The fourth chapter, Organizing and Planning to Respond, offers a systematic approach to establishing a plan that assigns responsibilities to trained team members in an effort to safeguard the health and safety of employees, as well as to protect business assets. Included is information on emergency response; crisis management; business recovery; command, control, and communications; pre-incident planning; and writing the plan.

The chapter on emergency response discusses both the threat of a possible terrorist attack as well as an actual terrorist event. It offers detailed information on attack indicators, how to determine protective actions for various types of hazards, and procedures for specific types of threats. Chapter 6, Recovery, provides specific information about activating the recovery plan, addressing personnel safety and health, and communicating with various company and public sectors.

The fall of 2001 illustrated just how crucial proper training and preparation is to survival after an incident. Chapter 7, Implementation and Training, summarizes these critical implementation steps and offers practical insight into staff and employee training, plan testing, and plan audits.

Donald L. Schmidt, ARM, senior vice president, managing consultant, and operations manager within the Risk Consulting Practice of Marsh in Boston, wrote the first seven chapters.

The second half discusses the financial impact of terrorist losses, especially the impact on the insurance marketplace and individual business policyholders. Chapter 8 begins by outlining the reasons that the incidents of September 11 had such a profound effect on the insurance and reinsurance marketplace and how this impact could create future solvency problems. Tips and guidelines on how insurance buyers and risk managers can best survive a hard insurance market are highlighted in chapter 9.

Chapter 10, Insurance Coverage Issues, offers insight into the coverage issues that arose in the aftermath of the 2001 terrorist incidents. It includes discussions of how the number of occurrences impact insurance recovery; whether the standard war exclusion might apply; the application of the debris removal clause on property policies; the pollution exclusion; and business interruption and workers compensation coverage implications. The chapter continues with a discussion of the response of primary insurers to the loss of reinsurance for terrorist threats.

The government and industry approach to reinsurance pressures caused by the September attacks is discussed in Chapter 11.

Chapter 12, Subrogation and Recovery Options against Terrorists, discusses why and how subrogation and civil recovery may impact the ultimate effect of terrorist incidents. The final chapter, Alternative Risk Techniques, points out the renewed interest in alternative risk transfer mechanisms as a result of insurance marketplace conditions following September 11. The use of surplus lines carriers, captives and risk retention groups, catastrophe bonds, and self-insurance and higher retentions are offered as potential offsets to difficult insurance marketplace conditions.

Kevin M. Quinley, CPCU, ARM, senior vice president, risk services, for MEDMARC Insurance Company in Chantilly, VA, wrote this second section.

It is important to note that these topics are dynamic, especially in regard to the state of the insurance market, legislative developments related to federal reinsurance initiatives, and regulatory treatment of terrorism exclusions. However, an effort has been made to present current information that in all likelihood will continue to affect the subject.

Terrorism in America

Recent terrorist events have clearly changed life in America. One challenge of managing the terrorist threat is maintaining a constant awareness that future terrorist losses may look significantly different from those of September 11. They could arise from nuclear, chemical, or biological agents, or they could involve cyber crime. They may incorporate conventional or unconventional weaponry.

But fortune still favors the prepared. Despite multibillion-dollar losses, the insurance industry is expected to absorb the September 11th blows and survive. Business owners, risk managers, safety and loss control specialists, facilities managers all continue with the business of assessing, mitigating, responding to, recovering from, and ascertaining the financial impact of what terrorist acts can do. This book offers tools to help in that process.

Chapter 1

The Terrorism Threat

Introduction

The September 11, 2001, attacks on the World Trade Center and the Pentagon, along with the downing of a fourth American flagged airliner ushered in a new paradigm in terrorism. The method of attack; the magnitude of death, injury, and destruction; and the perceived threat to the American population were inconceivable to most. Weeks later, a series of anthrax-laden letters were posted to government officials and the news media, causing additional deaths and illnesses and exacerbating the fear that Americans are at risk at home and in the office. Less publicized, a serious computer worm—Nimbda—infected many computer networks and forced the shutdown of essential connectivity while systems were cleansed.

Years ago, terrorists targeted United States interests abroad. Aircraft were skyjacked; embassies, military installations, or vessels were bombed; and government and military officials were abducted or assassinated. Today, terrorists have shown they will attack American interests on American soil on multiple fronts and in ways inconceivable a short time ago. The impact of terrorist attacks has reached new levels—and could escalate further—if terrorists further employ weapons of mass destruction.

> "Relatively small amounts of some chemical and biological agents can create mass casualties, potentially causing large numbers of fatalities and an overwhelming number of injuries. The consequences of a WMD [weapons of mass destruction] incident could also include economic damage, environmental contamination, international repercussions, increased internal police powers, and deleterious psychological effects on citizens."
>
> Executive Session on Domestic Preparedness, Harvard University

What is Terrorism?

The U.S. *Code of Federal Regulations* defines terrorism as "...the unlawful use of force and violence against persons or property to intimidate or coerce a government, the civilian population, or any segment thereof, in furtherance of political or social objectives."[1] The Federal Bureau of Investigation (FBI) further describes terrorism as either domestic or international, depending upon the origin, base, and objectives of the terrorists.

The FBI recorded 327 incidents or suspected incidents of terrorism in the United States between 1980 and 1999[2]. Of these, 239 were attributed to domestic terrorists and 88, to international terrorists. These incidents resulted in the deaths of 205 persons and more than 2,037 injuries. Since 1968, more than 14,000 terrorist attacks have taken place worldwide, resulting in more than 10,000 deaths[3].

Table 1. Terrorist Incidents

Date	Incident	Consequences
September 9-18, 1984 and September 19-October 10, 1984 (two waves)	Salmonella Typhimurium sprayed on salad bars in approximately 10 The Dalles, Oregon, restaurants to sicken potential voters to fix a local election; perpetrated by followers of *Bagwan Shree Rajneesh;* not identified as a biological weapons attack until one year later.	751 illnesses
December 21, 1988	Pan Am flight 103 exploded, and pieces of the plane fell onto the Scottish town of Lockerbie. Perpetrated by agents of the Libyan government.	259 people on board and 11 people on the ground were killed.
February 26, 1993	Bombing of the World Trade Center, New York	6 killed, 1,000+ injured
March 20, 1995	Sarin gas attack on the Tokyo, Japan, subway system by Aum Shinrikyo cult.	12 killed, several hun dred injured, thousands sought treatment
April 19, 1995	Bombing of the Alfred P. Murrah Federal Building, Oklahoma City, OK.	168 killed, 642 injured
October 1995	Amtrak's Sunset Limited derailed in Arizona. Rail spikes had been removed and a safety device was bypassed. A note signed by "Sons of Gestapo" criticizing the U.S. government was left at the scene.	1 killed, 78 injured
July 27, 1996	Bombing of Centennial Olympic Park, Atlanta, GA.	2 killed, 112 injured
May, July, and August 1998	19 women's clinics in Florida, New Orleans, and Houston were attacked with butyric acid; Army of God claimed responsibility.	14 injured
August 1998	Cancer patient who may have a grudge against doctors and police was caught stuffing 100 envelopes with cyanide packaged to look like free samples of a nutritional supplement (Marina del Rey, CA).[4]	Post offices nationwide were searched for additional mailings
1999	Homemade industrial chemical devices used to attack movie theatres.	Minor injuries and several evacuations
December 1999	Plot to infect wheat crops with destructive fungus.	Prevented by law enforcement
September 11, 2001	Two hijacked aircraft flown into the World Trade Center, New York; third aircraft hijacked and flown into the Pentagon; fourth aircraft hijacked but crashed in rural Pennsylvania.	More than 3,000 killed, more than 1,000 injured including hundreds of firefighters and police officers; twin towers and adjacent buildings destroyed; four airliners with passengers and crew lost[5]
October 2001	Anthrax mailed to offices in Florida and government offices in Washington, D.C., possibly from a postal facility in New Jersey.	5 deaths, 6 ill from inhalation anthrax; 7 suffer cutaneous anthrax and survive; mail delivery curtailed in affected areas

Domestic Terrorists

The domestic, or homegrown, terrorism threat includes right-wing, left-wing, and special interest orientations with causes related to American political and social concerns. Right-wing terrorists often embrace principles of racial supremacy or adhere to conspiracy-oriented philosophies; they often are anti-government and anti-religion.[6] Left-wing terrorists follow a revolutionary socialist doctrine against capitalism and imperialism. Special interest terrorist groups seek to influence people on specific issues rather than widespread political change. These issues include animal rights, pro-life or anti-abortion, the environment, and nuclear. For example, the Earth Liberation Front (ELF) claimed responsibility for the arson attack on a ski resort in Vail, Colorado, that caused $12 million in damage.

> "The United States and other nations face increasingly diffuse threats. Potential adversaries are more likely to strike vulnerable civilian or military targets in nontraditional ways to avoid direct confrontation with our military forces on the battlefield, to try to coerce our government to take some action terrorists desire, or simply to make a statement."
>
> U.S. General Accounting Office

Left-wing causes were predominant in the 1960s to 1980s, anti-government right-wing groups perpetrated many acts in the 1990s and special interest groups and international terrorists are the threat in the early years of the new millennium.

Law enforcement authorities often classify individual acts of terrorist groups as hate crimes, not terrorism, when they are perpetrated against individuals or small groups. Although not classified as terrorism, bombings, vandalism, arson, and other acts can cause significant damage to individual businesses and threaten the health and safety of employees, customers, and visitors.

International Terrorists

The FBI divides the international terrorist threat into three categories: state sponsors of international terrorism, formal terrorist organizations, and loosely affiliated extremists and rogue international terrorists. Many terrorist organizations are autonomous and transnational with their own infrastructure, training facilities, and financial support network. These groups have been known to include the Irish Republican Army, Palestinian HAMAS, The Egyptian *Al-Gama al-Islamiyya* (IG), Lebanese *Hizballah*, and the terrorist network of Shaykh Omar Abdel Rahman and Osama Bin Laden's *al Qaida*.

The FBI reports that *Hizballah* has carried out numerous attacks on American interests around the world, but not in the United States. Shaykh Omar Abdel Rahman was convicted of conspiracy in a plot to bomb several buildings in New York City and the Holland and Lincoln tunnels. The terrorist network of Osama Bin Laden has been named in the September 11, 2001 attacks on the World Trade Center and Pentagon.

Trends in Terrorism

In its 1999 report on terrorism, the FBI stated that the trend in terrorism is toward fewer but more destructive attacks. Contrary to the overall trend toward fewer attacks, cases involving the use, or threatened use, of weapons of mass destruction (nuclear weapons and biological or chemical agents) investigated by the FBI have shown a steady increase since 1995. Most were hoaxes perpetrated against office buildings, schools, federal government facilities, courthouses, and women's reproductive health centers.

The sarin gas attack in the Tokyo subway system in 1995 that killed twelve and injured hundreds ushered in a new era of chemical weapons of mass destruction (WMD). Members of the Aum Shinrikyo cult poured liquid sarin on the floors of subway cars and allowed it to evaporate. Several years later, anthrax, a bacterial agent that naturally exists in the soil and in some animals, was spread via the U. S. Postal system. This terrorist action claimed multiple lives and impacted mail service in many communities and the U.S. government during Fall 2001. Butyric acid was used to terrorize women's clinics in multiple Florida cities, New Orleans, and Houston in 1998. The "Army of God" claimed responsibility for the attacks.

> "It also appears likely that as governments 'harden' (or make more secure) official targets, terrorists will increasingly seek out more vulnerable 'softer' targets, such as high-profile offices of multinational firms and Americans traveling and working abroad."
>
> Terrorism in the United States, 1999
> FBI

Terrorists can be individuals like Theodore Kaczynski who acted alone for seventeen years or part of an international organization. Some are well funded—the Aum Shinrikyo cult in Japan reportedly had assets of one billion dollars—and some operate with little means.

Extremists such as Timothy McVeigh and Terry Nichols (convicted in the 1995 Oklahoma City bombing) and Eric Robert Rudolph (charged in the 1996 Centennial Olympic Park bombing) have shown that small groups with limited means can perpetrate destructive acts.

More formalized terrorist organizations with "cells" or ad hoc groups that form to achieve a specific objective are difficult to identify, infiltrate, and neutralize. Osama Bin Laden's network and those of Ramzi Yousef (the reported mastermind of the 1993 attack on the World Trade Center also connected to a 1995 plot to bomb U. S. flagged air carriers transiting the Far East) have proven they are capable of perpetrating catastrophic attacks.

Animal rights and environmental extremists were active in the late 1990s with acts of vandalism, destruction of property and other criminal acts. Anti-abortion groups have attacked women's clinics and abortion doctors.

Terrorism has become a means to an end rather than just a means to publicize a cause.

Terrorist Threats

There are many potential terrorist threats—many are well known and have been committed for decades, if not centuries. Bombings, arson fires, hijackings, kidnappings or hostage taking, and assassinations are common. Biological and chemical weapons have been used in wartime for centuries. The specter of Weapons of Mass Destruction (WMD) looms ever larger as experts and pundits debate whether nuclear weapons have been acquired or will be acquired by terrorists.

> "With advanced technology and a smaller world of porous borders, the ability to unleash mass sickness, death and destruction today has reached a far greater order of magnitude. A lone madman or a nest of fanatics with a bottle of chemicals, a batch of plague-inducing bacteria, or a crude nuclear bomb can threaten or kill tens of thousands of people in a single act of malevolence."
>
> Secretary of Defense William S. Cohen, November 1997

As terrorism continues to unfold around the world, there are emerging threats. Threats to agriculture are called "agroterrorism." Technological research into fields such as high-energy radio frequency and electromagnetic pulse weapons may present credible threats in the future.

Today, businesses must be prepared to deal with many different terrorist threats as well as the increasing frequency of workplace violence. This book focuses on major terrorist threats—including chemical, biological, and radiological weapons—as well as conventional bombs. It also addresses the threat of cyberterrorism.

Biological & Chemical Agents

Hostile nations, terrorist groups, and even individuals may target Americans, our institutions, and our infrastructure with biological and chemical agents. Even though there are significant technical and operational challenges in manufacturing and dispersal of lethal quantities of biological and chemical agents, a number of notable events have proven it is possible.

In 1984, followers of the Bagwan Shree Rajneesh in Oregon contaminated restaurant salad bars with salmonella bacteria to prevent people from voting in a local election. Although no one died, 751 people were sickened with a foodborne illness.[7]

The sarin release by the Aum Shinrikyo cult was preceded in June 1994 when the cult used sarin undetected in another Japanese city—an incident that produced seven deaths and 200 injuries. Furthermore, the group attempted to produce and disperse biological agents on nine occasions between 1990 and 1994 in Tokyo and other nearby areas—to no effect.[8]

More recently anthrax-laden letters were sent to news media offices and governmental officials.

The Chem-Bio Threat

- "Hoaxes and threats are more likely than use.
- Use of chemicals is more likely than biological substances.
- Small-scale attacks are more likely than large-scale attacks.
- Crude dispersal in an enclosed area is the most likely mode of attack."

Milton Leitenberg,
Center for International and Security Studies, University of Maryland

The RAND database of international terrorist incidents identified over 9,000 acts of terrorism, but only one hundred involved the use of weapons of mass destruction. From 1975 to August 2000, the Monterey Institute's Center for Nonproliferation Studies database includes 139 cases where chemical or biological substances were used. These statistics don't include the anthrax attacks that occurred in Fall 2001.

Chemical and biological agents pose different sets of problems. For example, most chemicals quickly affect individuals directly exposed to the agent within a given geographical area. In contrast, the release of a biological agent may not be known for several days or weeks, and both perpetrators and victims may travel miles away from the point of release before an incident is detected. In addition, some biological agents produce symptoms that can be easily confused with influenza or other less virulent illnesses. Finally, biological agents such as smallpox cause a communicable disease that can spread throughout the population.

Biological Agents

Biological agents or pathogens are weapons of mass destruction—or mass casualty weapons since they aren't usually associated with property damage. However, a former Soviet scientist testified before a Congressional committee that biological agents could be produced to corrode equipment, degrade plastics, or render fuels useless.[9]

Biological agents are based on bacteria, viruses, rickettsia, fungi, or toxins produced by these organisms. Dozens of different agents can be used to make a biological weapon, and each agent will produce a markedly different effect. The differences include length of time an agent can survive after release, dose required to infect a patient, contagiousness, and the type of disease that the agent produces.

> **Biological** agents are infectious microbes or toxins used to produce illness or death in people, animals, or plants. Biological agents can be dispersed as aerosols or airborne particles. Terrorists may use biological agents to contaminate food or water because they are extremely difficult to detect. Incubation periods range from hours to weeks.

There are two types of formulations, liquid and dry powder. The liquid is easier to produce, but the dry powder form stores longer and is easier to deploy. The equipment used to produce biological agents is almost entirely dual-use: stainless-steel fermenters suitable for growing anthrax are routinely used to produce legitimate products such as vaccines, vitamins, food supplements, biopesticides, and fermented beverages. Even freeze-drying (lyophilization) and milling machines, which can be used to convert bacterial or viral agents into a dry powder for optimal dissemination as a fine-particle aerosol, have become standard equipment in the pharmaceutical industry.[10]

Attacks can include contamination of food or water supplies, which are then ingested; release of infected vectors (e.g., fleas or mosquitoes), which then bite their victims; or creation of an aerosol cloud, which is inhaled by victims. This is the most effective terrorist method.

Contamination of Food or Water

Members of the Ragneesh cult sprayed salmonella on salad bars in Oregon in 1984, which resulted in 751 people becoming sick. This attack was not identified as terrorism until the following year because medical surveillance was not able to identify the outbreak. Medical surveillance systems today would be more effective as local, county, and state public health officials work with the U. S. Centers for Disease Control and Prevention (CDC) to establish better reporting systems to quickly identify an outbreak.

An attack on drinking water supplies has been feared, and security of our reservoirs and water distribution systems has been increased. However, poisoning of the public water drinking supplies to kill large numbers of persons is difficult. Public water supplies are highly filtered and chemically treated before water reaches the tap. Chlorine would kill all but the hardiest of biological agents. Small-scale contamination of a public water system would be easier and could be accomplished by backflow into public water mains downstream of any water treatment facility. A number of accidental incidents highlight this possibility. In 1997, firefighting foam backed up through hoses into the City of Charlotte's water system resulting in a ban on showering or drinking. In 1998, chemical additives to a fire sprinkler system in Bridgeport, Connecticut, backflowed into the public water system resulting in water use restrictions.

An aerosol cloud of microscopic particles could be created by exploding a bomb or spraying. The effectiveness of the attack would be determined by the amount of agent that survives the explosion or spraying, the effective particle size (particles must be small enough to be inhaled into the lungs), the wind speed and direction to move the cloud over a populated area, and prevailing weather conditions. Persons exposed to a sufficiently dense portion of cloud of small enough particles could be infected. In addition, some agents would settle to the ground potentially contaminating buildings, water, and food sources.

The prospect of all of these conditions being met to infect large numbers of people is remote. An aerosol release of suspected "weapons-grade" anthrax from Soviet Military Compound 19 near the city of Sverdlovsk (a city of over one million residents[11]) occurred on April 2, 1979. Of the ninety-six people infected, sixty-six died of pulmonary (inhalation) anthrax.

Dispersion of biological weapons by aerosol means can be accomplished with simple equipment available at local retailers. However, this equipment does not have the high capacity necessary to spray large quantities of small particles sized to infect large numbers of people. In addition, equipment must keep the organisms alive.

Unmodified agricultural sprayers (such as crop-dusters) are not well suited for disseminating biological agents, but these devices could be used to spread chemical agents over a fairly large area, provided that the perpetrator takes precautions when filling the sprayer tanks and is aware of meteorological dynamics. Some agents such as sarin do not burn readily and could be dispersed with an explosive charge.[12]

Once released into the environment, biological agents can travel great distances. However, they are sensitive to sunlight, humidity, pollutants, temperature, and exposure to oxygen. Agents lose virulence at the rates of 2 to 30 percent per minute. One means to overcome the sensitivity to environmental factors is to release an agent into a closed environment, such as a subway system or enclosed building. Releasing an agent into a building could contaminate multiple floors—more if the attacker knew the arrangement of the building's HVAC system.

Pathogenic microbes are widely available, either from areas where diseases such as anthrax or plague are endemic, or from the hundreds of culture collections scattered across the globe that provide seed stocks for biomedical researchers and commercial biotechnology firms. This situation is further complicated because some dangerous pathogens are not only studied by biomedical researchers but have also become commercial products. Pharmaceutical companies currently produce large quantities of botulinum toxin (trade name Botox) for medical and cosmetic purposes.

The former Soviet Union produced hundreds of tons of anthrax and dozens of tons of smallpox and plague. The collapse of the former Soviet Union has raised concerns that terrorists may be able to gain supplies of weaponized biological agents.

Scenarios of a Biological Attack

Dealing with a biological weapon attack would be very challenging. Persons exposed to a biological agent would not know for hours—or in some cases weeks—that they had been exposed and were contagious. During this incubation period, the period between infection and onset of symptoms, infected persons would be able to travel around the world spreading the disease, so the number of potentially infected persons could rise dramatically.

Healthcare providers would be the first to spot infections. However, since most doctors have never seen smallpox, plague, or anthrax infections, their diagnosis would be delayed. Quick and accurate diagnostic tests are not available for all pathogens. Individual healthcare providers and hospitals would report their findings to county and state public health officials, who in turn would notify the U.S. Centers for Disease Control and Prevention (CDC). Reporting systems developed by the CDC would help compile and detect clusters of cases.

Treatment of a large number of infectious patients would strain public healthcare resources. Most hospitals are not equipped to handle large numbers of infectious patients and could be overwhelmed with fearful individuals seeking treatment—even though they may not have been exposed. The Scud missile attacks in Israel in 1991 resulted in over 1,000 emergency room visits, but only 22 percent were for direct casualties; the balance were for psychiatric care.[13] One study estimates the ratio of psychiatric cases to medical casualties to be 4:1.[14]

News media reports would provide a valuable service to notify people that may have been exposed—assuming the time and place of the attack was determined—but the same news reports would also create a demand surge on healthcare providers.

An act of terrorism would likely cause widespread fear, so accurate and up-to-date information must be provided to calm the public and prevent an overwhelming demand for health care. Although panic is rare, if the public perceives there is a high degree of risk, small chance of escape and an inadequate response to the attack, then widespread panic becomes more probable.

Healthcare providers must also be protected from infected patients. Some may seek to take care of their own families; others may become infected before realizing what they have been treating. This could reduce the number of healthcare professionals available to deal with an emergency.

Types of Biological Agents

There are four common types of biological agents: bacteria, viruses, rickettsia, and toxins.

Bacteria and Rickettsia. Bacteria are single-celled organisms that multiply by cell division and can cause disease in humans, plants, or animals. Although true cells, rickettsia are smaller than bacteria and live inside individual host cells. Examples of bacteria include anthrax, cholera, plague, tularemia; an example of rickettsia is Q fever.

Viruses. Viruses are the simplest type of microorganisms. They lack a system for their own metabolism and therefore depend upon living cells to multiply. This means that a virus will not live long outside of a host. Viruses that could serve as biological agents include smallpox, Venezuelan equine encephalitis, and the viral hemorrhagic fevers such as the Ebola and Marburg viruses and Lassa fever.

Toxins. Toxins are toxic substances of natural origin produced by an animal, plant, or microbe. They differ from chemical agents in that they are not manmade and typically are much more complex. Toxins, in several cases, are easily extracted for use as a terrorist weapon and, by weight, usually are more toxic than many chemical agents.

The four common toxins thought of as potential biological agents are botulism (botulinum), SEB (staphylococcal enterotoxin B), ricin, and mycotoxins.

In Table 2 the CDC lists "Threat Agents of Concern."[15] Three categories or threat levels are listed, with Category A being the most significant.

Table 2. Biological Agents

Category	Biological Agent	Profile
Category A	Anthrax (B. anthracis) Smallpox (Variola virus) Plague (Y. pestis) Tularemia (F. tularensis) Botulism (Botulinum toxin) Viral Hemorrhagic fever viruses (Ebola, Marburg, Lassa)	Cause high mortality with potential for major public health impact Might cause public panic and social disruption Require special action for public health preparedness Easily transmitted person-to-person (other than anthrax & botulism)
Category B	Q fever (Coxiella burnetti) Brucellosis (Brucella species) Glanders (Burkholderia mallie) Alpha viruses (VEE, EEE, WEE) Ricin toxin (from castor beans) Epsilon toxin of Clostridium perfringens Staphylococcus enterotoxin B Some food/waterborne pathogens	Moderately easy to disseminate Cause moderate morbidity and low mortality Require specific enhancements of CDC's diagnostic capacity and enhanced disease surveillance
Category C	Nipah virus Hantavirus Tickborne hemorrhagic fever Tickborne encephalitis viruses Yellow fever Multi-drug-resistant tuberculosis	Includes pathogens that could be engineered for mass dissemination because of: • Availability • Ease of production and dissemination • Potential for high morbidity and mortality and major public health threat

Source: U.S. Centers for Disease Control and Prevention

Anthrax[16]

Bacillus anthracis, the etiologic agent of anthrax, is a large, gram-positive, nonmotile, spore-forming bacterial rod. Human anthrax has three major clinical forms: cutaneous (dermal or skin form), inhalation, and gastrointestinal. If left untreated, anthrax in all forms can lead to septicemia and death.

Symptoms of disease vary depending on how the disease was contracted, but symptoms usually occur within seven days. The incubation period for cutaneous anthrax ranges from one to twelve days, and inhalation anthrax, from one to seven but up to sixty days.

Most anthrax infections occur when the bacterium enters a cut of abrasion on the skin. Skin infection begins as a raised bump that resembles a spider bite, but it develops into a vesicle and then a painless ulcer with a characteristic black necrotic (dying) area in the center. Lymph glands in the adjacent area may swell. About 20 percent of untreated cases of cutaneous anthrax result in death. Deaths are rare if patients are given appropriate antimicrobial therapy.

Inhalation anthrax is the most lethal form of anthrax. Anthrax spores must be aerosolized in order to cause inhalation anthrax. The number of spores that cause human infection is unknown. It resembles a viral respiratory illness, and initial symptoms include sore throat, mild fever, muscle aches, and malaise. These symptoms may progress to respiratory failure and shock, with meningitis frequently developing.

Gastrointestinal anthrax usually follows the consumption of raw or undercooked contaminated meat; it has an incubation period of one to seven days. It is associated with severe abdominal distress followed by fever and signs of septicemia. Direct person-to-person spread of anthrax is extremely unlikely, and anthrax is not contagious. Therefore, there is no need to quarantine individuals suspected of being exposed to anthrax or to immunize or treat contacts of persons ill with anthrax, such as household contacts, friends, or coworkers unless they also were also exposed to the same source of infection.

Smallpox

The last naturally acquired case of smallpox occurred in 1977, and, in the United States, routine vaccination ended in 1972. In the majority of cases, smallpox is spread by infected saliva droplets that expose a susceptible person having face-to-face contact with the ill person. People with smallpox are most infectious during the first week of illness because that is when the largest amount of virus is present in saliva. However, some risk of transmission lasts until all scabs have fallen off.

The incubation period is about twelve days but ranges from seven to seventeen. Initial symptoms include high fever, fatigue, and head and backaches. A characteristic rash, most prominent on the face, arms, and legs, follows in 2-3 days. Lesions become pus-filled after a few days and then begin to crust early in the second week. Scabs develop, then separate, and fall off after about three to four weeks.

The majority of patients with smallpox recover, but death may occur in up to 30 percent of cases.

The CDC has guidelines to swiftly provide vaccine to people exposed to this disease. Vaccine is in secure storage, and the U.S. Department of Health and Human Services has accelerated production of a new smallpox vaccine. If the vaccine is given within four days after exposure to smallpox, it can lessen the severity of illness or even prevent it.

Much of the U.S. population may have partial immunity from prior immunization programs, but approximately half has never been vaccinated. Immunity can be boosted effectively with a single revaccination. Prior infection with the disease grants lifelong immunity.

There is no proven treatment for smallpox, but research to evaluate new antiviral agents is being done. Patients with smallpox can benefit from supportive therapy (e.g., intravenous fluids, medicine to control fever or pain) and antibiotics for any secondary bacterial infections that may occur.

Symptomatic patients with suspected or confirmed smallpox are capable of spreading the virus. Patients should be placed in medical isolation so that they will not continue to spread the virus. In addition, people who have come into close contact with smallpox patients should be vaccinated immediately and closely watched for symptoms.

Botulism

Botulism is a muscle-paralyzing disease caused by a toxin made by a bacterium called Clostridium botulinum. Foodborne botulism occurs when a person ingests preformed toxin that leads to illness within a few hours to days. With foodborne botulism, symptoms begin within six hours to two weeks after eating toxin-containing food.

Symptoms include double vision, blurred vision, drooping eyelids, slurred speech, difficulty swallowing, dry mouth, and muscle weakness that always descend through the body. Paralysis of breathing muscles can cause a person to stop breathing and die unless assistance with breathing (mechanical ventilation) is provided.

Botulism is not spread from person to person; foodborne botulism can occur in all age groups.

The CDC maintains a supply of antitoxin against botulism. The antitoxin is effective in reducing the severity of symptoms if administered early on. Most patients eventually recover after weeks to months of supportive care.

Pneumonic Plague

Plague is an infectious disease of animals and humans caused by the bacterium Yersinia pestis. Y. pestis is found in rodents and their fleas in many areas around the world.

Pneumonic plague occurs when Y. pestis infects the lungs. The first signs of illness are fever, headache, weakness, and cough that produces bloody or watery sputum. The pneumonia progresses over two to four days, may cause septic shock, and, without early treatment, death.

Person-to-person transmission of pneumonic plague occurs through respiratory droplets, which can only infect those having face-to-face contact with an ill patient. Early treatment of pneumonic plague with antibiotics is essential. There is no vaccine against plague, but prophylactic antibiotic treatment for seven days will protect persons who have been exposed.

Chemical Agents

Chemical agents kill or incapacitate people, destroy livestock or ravage crops. Some chemical agents are odorless and tasteless and are difficult to detect. They can have an immediate effect (a few seconds to a few minutes) or a delayed effect (several hours to several days).

The FBI defines chemical agents as "solids, liquids, or gases that have chemical properties that produce lethal or serious effects in plants and animals."[17]

The 1995 sarin gas attack in the Tokyo subway system resulted in twelve deaths and several hundred injuries.[18] A Pentagon study estimated that the open-air disbursement of twenty-two pounds of sarin agent would kill fifty people, and 2,200 pounds could kill 10,000 persons. The required quantity would be less if the gas were dispersed indoors where it would not dissipate as quickly.

Chemical agents fall into five classes:

- ❑ **Nerve agents**, which disrupt nerve impulse transmissions.
- ❑ **Blister agents**, also called vesicants, which cause severe burns to eyes, skin, and tissues of the respiratory tract.
- ❑ **Blood agents**, which interfere with the ability of blood to transport oxygen.
- ❑ **Choking agents**, which severely stress respiratory system tissues.
- ❑ **Irritating agents**, which cause respiratory distress and tearing designed to incapacitate. They also can cause intense pain to the skin, especially in moist areas of the body. They are often called riot control agents.

Most chemical agents are liquids and evaporate at different rates. Some, such as cyanide, are very volatile; blister agent mustard and nerve agent VX are similar to light motor oil. An "effective" attack would require aerosolizing the liquid into tiny droplets that could be inhaled. This can be done by spraying, such as the aerial spraying of pesticides on crops. Chemical agents can also be aerosolized by an explosion—a great concern during the Scud missiles attacks in Israel during the 1991 Gulf War.

The tiny droplets of chemical agent are dispersed by the wind, if outside, or by a building's heating, ventilation, and air conditioning system. Ideal targets for this type of terrorist attack would be events where large numbers of people congregate, such as for sporting events, concerts, conventions, or large meetings.

Chemical warfare agents can be produced using forty-year-old technology and readily available methods. Certain World War I-era chemical warfare agents, such as phosgene, hydrogen cyanide, and sulfur mustard, are relatively easy to manufacture. For example, sulfur mustard can be produced in small quantities in a crude facility such as a basement laboratory. Nerve agents require more technical sophistication.[19]

Many experts suggest that graduate-level chemistry skills are needed to produce chemical weapons. Experience as an organic chemist would be needed to make sarin or nerve agents in the "V" class.[20] There are many technical challenges, and the process is not without significant risk. However, experts told the U.S. General Accounting Office that terrorists do not need sophisticated knowledge or dissemination methods to use toxic *industrial* chemicals as weapons.[21] For example, the "Army of God" attacked nineteen women's clinics in Florida, New Orleans, and Houston with butyric acid in 1998.

Nerve Agents[22, 23]

Nerve agents are similar in nature to organophosphate pesticides but with a higher degree of toxicity. All are toxic at small concentrations (a small drop could be fatal). Agents include sarin (GB); Soman (GD); tabun (GA); and V agent (VX).

Nerve agents are identified with a letter system. In the case of GA, GB, and GD, the first letter "G" refers to the country (Germany) that developed the agent, and the second letter indicates the order of development. In the case of VX, the "V" stands for "venom," while the "X" represents one of the chemicals in the specific compound.

Nerve agents resemble water or light oil in pure form and have no odor. The most efficient distribution method is in the aerosol form. Small explosions and spray equipment can be used. Nerve agents also kill insects, birds, and other animals as well as humans. Many dead animals at the scene of an incident may be a warning sign of an attack.

Victims' symptoms will be an early outward warning sign of the use of nerve agents. There are various generic symptoms in humans that are similar to pesticide poisoning. The victims will salivate, lacrimate, urinate, and defecate without much control.

Other symptoms may include:

- ❑ Eyes: pinpointed pupils, dimmed and blurred vision, pain aggravated by sunlight;
- ❑ Skin: excessive sweating and fine muscle tremors;
- ❑ Muscles: involuntary twitching and contractions;
- ❑ Respiratory system: runny nose and nasal congestion, chest pressure and congestion, coughing and difficulty in breathing;
- ❑ Digestive system: excessive salivation, abdominal pain, nausea and vomiting, involuntary defecation and urination; and
- ❑ Nervous system: giddiness, anxiety, difficulty in thinking and sleeping (nightmares).

Blister Agents

Blister agents are also referred to as mustard agents due to their characteristic smell. In a pure state, they are nearly colorless and odorless, but slight impurities give them a dark color and an odor suggesting mustard, garlic, or onions.

They are heavy, oily liquids, dispersed by aerosol or vaporization, so small explosions or spray equipment may be used.

They readily penetrate layers of clothing and are quickly absorbed into the skin. Mustard (H, HD) and lewisite (L) are common blister agents. All are very toxic, although much less so than nerve agents. A few drops on the skin can cause severe injury, and three grams absorbed through the skin can be fatal. Clinical symptoms, which may not appear for hours or days, include:

- ❑ Eyes: reddening, congestion, tearing, burning, and a gritty feeling; in severe cases, swelling, severe pain, and spasm of the eyelids;

- Skin: within one to twelve hours, initial mild itching followed by redness, tenderness, and burning pain, followed by burns and fluid-filled blisters. The effects are enhanced in the warm, moist areas of the groin and armpits;

- Respiratory system: within two to twelve hours, burning sensation in the nose and throat, hoarseness, profusely running nose, severe cough, and shortness of breath; and

- Digestive system: within two to three hours, abdominal pain, nausea, bloodstained vomiting, and bloody diarrhea.

Blood Agents

Blood agents interfere with the ability of the blood to transport oxygen and result in asphyxiation. Under pressure, blood agents are liquids. In pure form, they are gases. Precursor chemicals are typically cyanide salts and acids—common industrial chemicals that are readily available. All have the aroma of bitter almonds or peach blossoms.

Common blood agents include hydrogen cyanide (AC) and cyanogen chloride (CK). CK can cause tearing of the eyes and irritate the lungs. All blood agents are toxic at high concentrations and lead to rapid death. Affected persons require removal to fresh air and respiratory therapy. Clinical symptoms of patients affected by blood agents include respiratory distress; vomiting and diarrhea; vertigo and headaches.

Pulmonary (Choking) Agents

Choking agents stress the respiratory tract. Severe distress causes edema (fluid in the lungs), which can result in asphyxiation resembling drowning. Chlorine and phosgene, common industrial chemicals, are choking agents. Clinical symptoms include severe eye irritation, coughing, and choking. Most people recognize the odor of chlorine. Phosgene has the odor of newly cut hay. Since both are gases, they must be stored and transported in bottles or cylinders.

Irritating Agents

Irritating agents, also known as riot control agents or tear gas, are designed to incapacitate. Generally they are nonlethal; however, they can result in asphyxiation under certain circumstances. Common irritating agents include chloropicrin, MACE (CN), tear gas (CS), capsicum/pepper spray, and dibenzoxazepine (CR). Clinical symptoms include:

- ❑ Eyes and throat: burning, irritation, tearing;
- ❑ Respiratory system: respiratory distress, coughing, choking, and difficulty breathing; and
- ❑ Digestive system: high concentrations may lead to nausea and vomiting.

These agents can cause pain, sometimes severe, on the skin, especially in moist areas. Most exposed persons report the odor of pepper or of tear gas. Outward warning signs include the odor of these agents and the presence of dispensing devices. Many are available over the counter.

Nuclear/Radiological Devices

There are two primary scenarios for a nuclear or radiological attack. One is the detonation of a nuclear bomb. The second is the detonation of a conventional explosive device incorporating nuclear materials—a so-called *dirty bomb* or radiological dispersal device (RDD).

The detonation of a low-yield, nuclear device in an urban center could kill tens of thousands of people. Although it is unlikely that terrorists would manufacture a nuclear warhead, it is technically possible. Another scenario would involve the theft of a nuclear weapon. For example, many news agencies have reported that nuclear weapons from the former Soviet Union are unaccounted for. A small tactical nuclear weapon could be smuggled into the United States via cargo containers off-loaded at the many seaports around the country.

A dirty bomb would consist of radioactive material wrapped around a core of conventional high explosives, which upon detonation would spew radioactive particles into the environment. There are many radiation sources, but the most power-

> "Large-scale, indiscriminate violence has become the reality of terrorism in the 1990s. At the same time, owing to the collapse of the Soviet Union and the proliferation of nuclear weapons development programs, the acquisition of nuclear materials, the biggest technical hurdle, may have become easier. How close we are to that theoretical point in time when capabilities meet intentions I cannot say, but we are closer. Of course, in focusing on the high end of the nuclear spectrum—a nuclear bomb—we should not ignore the possibility of lesser actions involving radioactive material."
>
> Brian Michael Jenkins, RAND Corporation, November 15, 2001

ful are also the most dangerous to handle. Spent radioactive waste from nuclear fuel rods would be extremely hazardous to handle, making it "proliferation-proof" according to some experts. Weaker sources of radiation are available from many sources.

Hospitals, some laboratories, and factories use low-grade radioactive materials. The Nuclear Regulatory Commission has recorded hundreds of cases in which radioactive material has been lost, stolen, or unaccounted for. The United Nations' International Atomic Energy Agency has reported hundreds of illicit sales of nuclear waste and radioactive materials, including many from the former Soviet Union.

The extent of bomb damage is dependent upon the yield of the explosives, but radioactivity could contaminate a much wider area. Contaminated debris would also complicate efforts to treat casualties, and public fear would be difficult to manage.

Other scenarios involve detonation of a large conventional bomb near a nuclear power plant or radiological cargo in transit; a portable stinger-type missile attack on a nuclear plant; or the crashing of a commercial jet liner into a nuclear power plant. Any catastrophic attack on a nuclear plant could result in the meltdown of the reactor core similar to Chernobyl or a dispersal of spent radioactive waste stockpiled on-site.

Incendiary Devices

An incendiary device is any "mechanical, electrical, or chemical device used intentionally to initiate combustion and start a fire."[24] Incendiary devices may be simple or elaborate and come in all shapes and sizes. An incendiary device can be a simple match applied to a piece of paper, a matchbook-and-cigarette arrangement, or a complicated self-igniting chemical device. Normally, an incendiary device is a material or mixture of materials designed to produce enough heat and flame to cause combustible material to burn once it reaches its ignition temperature.

Each device consists of three basic components: an igniter or fuse, a container or body, and an incendiary material or filler. The container can be glass, metal, plastic, or paper, depending on its desired use. A device containing chemical materials usually will be in a metal or other nonbreakable container. An incendiary device that uses a liquid accelerator usually will be in a breakable container (e.g., glass—a "Molotov cocktail.")

Once the incendiary device ignites a fire, the scope of damage would be limited only by installed fire protection systems, fire suppression activities, and the continuity of combustible construction or building contents.

Explosive Devices

Government sources estimate that 70 percent of all terrorist attacks worldwide involve explosives—bombs are a weapon of choice amongst terrorist groups. The FBI reports that of 3,163 bombing incidents in the U.S. in 1994, 77 percent were due to explosives. In these situations 78 percent of all bombs detonated or ignited. Another 22 percent failed to function as designed; only 4 percent were preceded by a warning or threat.

The FBI also noted three other interesting facts:

- ❑ When public safety agencies know of the presence of a device, they have only a 20 percent chance of finding it.
- ❑ Hundreds more "hoax" bomb incidents are reported each year.
- ❑ Residential properties are the most common targets for bombers.

Cyberterrorism

The U.S. *Code of Federal Regulations* defines terrorism as "...the unlawful use of force and violence against persons or property to intimidate or coerce a government, the civilian population, or any segment thereof, in furtherance of political or social objectives." Cyberterrorism is simply the use of computer systems to perpetrate the crime. Some cyber attacks are designed to intimidate or coerce, and many coincide with geopolitical events. The greatest threat is a well-coordinated attack on multiple fronts to disrupt critical infrastructure.

Cyberterrorism warrants special treatment not only because of its potential to cause significant direct damage but, more so, because of its potential for indirect damage or consequences. Cyberterrorism is also insidious—the undetected penetration of computer systems, the theft of sensitive or proprietary information, the corruption of critical data, or the use of the penetrated system to access other systems all pose enormous loss potential.

An analysis done at Dartmouth College highlighted several trends in cyberterrorism:[25]

- Cyber attacks are increasing in volume, sophistication, and coordination.
- Cyber attacks can be expected after political conflicts and could further complicate matters after a major terrorist attack.
- After the United States's accidental bombing of the Chinese embassy in Belgrade, Yugoslavia, in 1999, many U. S. Government Web sites were defaced in the name of China. The U. S. Departments of Energy and the Interior, The National Park Service, and the White House Web site were defaced or forced to shut down.
- Cyber attacks are directed at both government and private interests—in particular high value targets such as networks, servers, and routers with symbolic, financial, political, and tactical value.
- Palestinians have assaulted Israeli banking and financial institutions. The Code Red worm attacked the White House.

Another case shows the political connection as defined by the U. S. Code. In the summer of 1997, CNN included a link to a U.S.-based Internet Service Provider that hosted a Web page for the ETA, the Basque separatist group in Spain. The Spanish government asked CNN to remove the link. CNN refused. The California ISP was attacked for over a week, resulting in denial of service to the ISP's 13,000 subscribers.

The attacks stopped only after the ISP removed the material.

Critical Infrastructure

Information technology has changed the way business is transacted, government is operated, and national defense is conducted. These systems now rely on an interdependent network of critical information infrastructures. We depend upon this public/private infrastructure for water, power, and fuel; voice and data communications; business transactions; and movement of people and goods. We also rely upon public emergency services to respond promptly when called. Much of our infrastructure is controlled using computer systems that rely on the same telecommunications system that businesses rely upon.

The U. S. economy is increasingly dependent on electronic commerce. Business-to-business electronic transactions were $671 billion in 1998, with $92 billion over the Internet and 33 percent compounded annual growth expected.[26]

"Systems and services critical to the American economy and the health of our citizens—such as banking and finance, 'just-in-time' delivery system for goods, hospitals, and state and local emergency services—can all be shut down or severely handicapped by a cyber attack or a physical attack against computer hardware," according to Gov. James S. Gilmore III of Virginia.[27] An attack on the telecommunications system that provides connectivity for a computer-controlled electric power grid could black out an entire region of the country.

"Systems and services critical to the American economy and the health of our citizens—such as banking and finance, 'just-in-time' delivery system for goods, hospitals, and state and local emergency services—can all be shut down or severely handicapped by a cyber attack or a physical attack against computer hardware."

Gilmore Commission
November 2001

Critical infrastructure includes:

- ❑ Electric power transmission and distribution systems
- ❑ Gas and oil storage and distribution systems

- ❑ Telecommunications systems
- ❑ Banking and finance industries
- ❑ Public drinking water supplies
- ❑ Transportation systems
- ❑ Essential public emergency services (police, fire and emergency medical services)

Other attacks on critical infrastructure that did not result in catastrophic consequences include:

- ❑ In 1997, a juvenile in Massachusetts was able to hack into the former NYNEX (now Verizon) telecommunications system in Massachusetts. The attack impacted telecommunications services to the air traffic control system at the Worcester, Massachusetts, airport.
- ❑ A 1996 attack by a hacker in Sweden flooded a 911 emergency telephone reporting system in Florida, preventing legitimate calls from reaching the dispatch center.

These incidents may not be classified as cyberterrorism because the motive or intent was not to support a political or social agenda. However, the results illustrate what a true cyberterrorist could perpetrate.

The U.S. government recognizes the need to protect critical infrastructure. The National Infrastructure Protection Center (NIPC), an interagency warning and response center, was created in 1998 by presidential directive and is housed at FBI headquarters. The NIPC is tasked with deterring, detecting, and responding to unlawful acts, including terrorism, involving computer and information technologies. On October 16, 2001, President Bush issued Executive Order 13231. It created the "President's Critical Infrastructure Protection Board" to coordinate federal efforts and facilitate cooperation with the private sector, state and local governments, academic organizations, and federal agencies. In addition the order also established the National Infrastructure Advisory Council (NIAC) to advise the president on the security of information systems in banking and finance, transportation, energy, manufacturing, and emergency government services.

Computer Security

In testimony before a Senate subcommittee, Stephen E. Cross noted that inadequate computer security creates vulnerability to cyber attack and that the financial consequences are enormous. "Financial losses from network security breaches regularly exceed $100 million a year," he said. "Denial-of-service attacks on Web sites in February 2000 were estimated to cost more than $1.2 billion. On the days of the attacks, losses exceeded $100 million. Looking to the future, the FBI estimates that computer crime will cost companies $10 billion."[28]

The CERT® Coordination Center statistics for 1988 to 2001 show a substantial increase in the number of reported computer security incidents—from 3,734 in 1998 to 34,754 through the first three quarters of 2001. Reported vulnerabilities increased from 262 in 1998 to 1,820 in the first three quarters of 2001.[29] These statistics are probably understated because the Center estimates the majority of incidents go unreported.

The Computer Security Institute's sixth annual "2001 Computer Crime and Security Survey"[30] confirms that the threat from computer crime and information security breaches continues unabated and that the financial toll is mounting. Highlights of the survey include:

Eighty-five percent of respondents (primarily large corporations and government agencies) detected computer security breaches within the last twelve months of the report period.

Sixty-four percent acknowledged financial losses due to computer breaches. Thirty-five percent were willing and/or able to quantify their financial losses and reported $377,828,700 in financial losses, compared to only $265,589,940 in 2000. The most serious financial losses occurred through theft of proprietary information and financial fraud.

For the fourth year in a row, more respondents (70 percent) cited their Internet connection as a frequent point of attack than cited their internal systems (31 percent).

Cyberterrorism Methods

Terrorists can use various methods to pursue their political or social agendas. These include:

- Hacktivism (a combination of hacking and activism)
- Virtual sit-ins and blockades
- Email bombs
- Web site hacking and computer break-ins

Hacktivism

Hacktivism is the convergence of hacking with activism—or electronic civil disobedience.[31] It includes Web page defacements and denial-of-service attacks. Popular targets are the Web sites of government agencies, educational institutions, commercial interests, and cultural institutions. However, any site with an exploitable vulnerability is subject to attack.[32]

In late April and early May 2001 hacktivists and cyber protesters assaulted U. S. Web sites after a Chinese fighter jet was lost in a collision with a U.S. reconnaissance aircraft. This effort was led by pro-Chinese hackers and resulted in the defacement or crashing of 100 government and commercial Web sites. American hackers retaliated by damaging 300 Chinese Web sites.[33]

Virtual Sit-Ins and Blockades

Cyber protesters try to raise awareness of their cause by overwhelming a Web site with traffic. Automated software that accesses a targeted site repeatedly every few seconds is often used.

One such attack was coordinated by the Electronic Disturbance Theater (EDT) to dramatize the plight of the Mexican Zapatistas. It attacked the Web sites of the President of Mexico, the Pentagon, and the Frankfurt Stock Exchange, and delivering more than 600,000 hits per minute per site. The Pentagon site sensed the attack and countered by redirecting the offending browsers to an alternate site where an applet would be automatically downloaded, thereby tying up each machine.[34]

Email Bombs

Automated software can bombard an electronic mail "post office" with messages in a short period of time, overloading the post office and preventing receipt of legitimate messages. They are clearly used as a means of protest against the receiving site.

Web Site Hacks

The most frequent Web site hacks involve defacing the content of the Web site to display a message. Another means to modify what visitors will see when they visit a site is to tamper with the Domain Name Service so visitors are redirected to another site.

Computer Viruses & Worms

Malicious computer viruses and worms, including the Code Red virus, cost $12.3 billion[35] in 2001. The Melissa Macro Virus (1999) and Nimbda virus (2001) have shut down many corporate networks and destroyed thousands of computer files that had to be recovered or reconstructed. Both are malicious forms of computer code that infect computers and spread via local area and wide area networks. A worm is an autonomous software program that spreads on its own, whereas a virus is a program that infects computer files by inserting a copy of itself into the file. Unlike the computer worm, a virus requires human involvement (usually unwitting) to propagate.

Consequences of a Terrorist Attack

Estimates of the total direct cost of the attacks on the World Trade Center exceed $50 billion dollars. The indirect costs could easily add billions more to the total, but it will be many years before the final cost is known. The consequences of a terrorist event include more than just economic losses. The impact on the family members and coworkers of those lost was profound and long lasting. The impact on American society and, to some extent, the world community was also significant.

Direct Costs

Businesses face enormous potential losses from a significant terrorist attack. Much of the loss may be insurable, but much is not. The impact on profits can also be significant. Quarterly earning reports from major corporations with significant operations in the World Trade Center reflected charges in the tens of millions of dollars for uninsured losses. These charges paid life insurance benefits, extended medical coverage for surviving family members, provided support services for family members and surviving colleagues, relocated operations to alternate space, wrote-down intangible assets and noncash items such as prepaid rent and leases.[36]

The attacks on the World Trade Center and the Pentagon took place at a time when the U.S. economy was softening. However, most experts agree that the attacks helped push the economy along a path toward recession. Many segments of the economy were impacted, including the travel, hotel, and hospitality industries. Another terrorist attack that restricts the movement of people, such as an attack with an infectious biological agent could have a similar impact on these industries. Most businesses are impacted by declines in the overall U.S. economy and would see diminished earnings in case of a significant terrorist event.

Property Damage

Property claims from a major bombing event could include building, contents, and direct business interruption losses. These losses can extend well beyond the immediate building. For example, twenty-five buildings near the nine-story Alfred P. Murrah Federal Building in Oklahoma City were severely damaged or destroyed and another 300 were damaged.

A review of the World Trade Center disaster by Risk Management Solutions[37] identified some of the causes of property damage.

- ❑ **Fire.** Buildings in close proximity to the WTC were vulnerable to fire from the conflagration of the collapsed, burning towers. Firefighting efforts were focused on life safety, and evacuated buildings were largely left to burn.

- **Falling Debris**. The disintegrating building showered the surrounding area with debris and steel, and concrete building members were ejected well beyond the footprint of the buildings.

- **Airborne Dust.** Dust and debris were carried as far as two miles from the WTC site.

- **Pressure Wave.** Building collapse of the WTC towers created a pressure wave, similar to very strong wind gusts, that caused some damage to windows and cladding on buildings within, at most, 650 feet of the WTC complex.

- **Vibration and Ground Deformation.** Ground deformation associated with a sudden transfer of load would potentially damage the foundations of nonpiled buildings, roads, and underground infrastructure, including pipes and subways. Reports of gas leaks and road damage were probably caused by the collapse of underground structures. Some speculate that the collapses caused ground vibrations sufficient to damage some nearby properties.

Similar property damage and/or contamination can be expected in other terrorist incidents. Environmental contamination would result from the rupture of hazardous materials storage tanks or the release of hazardous materials into the atmosphere. A chemical, biological, or radiological incident would result in contamination of buildings and equipment surrounding the impact of any device or the release of any agent.

People

Direct casualties from a terrorist incident can be significant. Over 3,000 persons were lost in the World Trade Center and Pentagon attacks, including civilians, military personnel, police, and firefighters. More than 1,000 were injured, and others were affected in the following months by respiratory problems created by the lingering smoke and dust.

The costs for direct casualties include medical expenses, workers compensation indemnity payments, and administrative costs. In addition, life insurance payments and accidental death and dismemberment (AD&D) claims resulted. Disability insurers reported increased claims from individuals far from the site of the World Trade Center.

Thousands of people sought counseling to deal with the loss of their family members, friends, and colleagues. Employees heavily utilized Employee Assistance Programs (EAPs) following the attacks, and many companies reported heavy demand on their EAP providers for employees located far away from New York or Washington, D.C.

Lingering concerns about security continued to haunt many—months after the attack. Research reported in the *New England Journal of Medicine* showed that many Americans suffered from trauma-related stress. The nation's mental health must be considered another vulnerability and a cost to be born by businesses remote from the site of any attack.

Business Interruption

Business interruption is the loss of profits of a business that suffers an indemnified loss when it is unable to manufacture its product(s) or provide its service. Business interruption claims typically pay only after a defined waiting period, typically twenty-four to seventy-two hours, and only if the business can demonstrate a clear loss of profits. Depending upon the complexity of the business entity, significant interdependencies with other operations of the company could broaden the interruption. A loss at one critical facility could shut down the entire company.

Contingent business interruption claims arise when a company is unable to provide its service or manufacture its product because a critical supplier suffers an indemnified loss and is unable to provide its supplies or service. A major terrorist incident impacting many businesses or affecting the movement of goods and provision of services could generate contingent business interruption claims.

Just-in-time delivery systems are scheduled so a component part arrives within a few hours of the time it will be installed in a product. For example, wire harnesses may be shipped to arrive immediately prior to installation in new motor vehicles. If the wire harnesses do not arrive on time, the automotive production line shuts down—at great penalty to the wire harness manufacturer. Following the attacks of September 11, security along the U.S. Canada border was increased. Many automotive suppliers in Canada had difficulty getting their products into the U.S. because of traffic delays at the border crossings.

Another example of the potential business impact is the retail mall in the World Trade Center. These stores generated three times the sales per square foot than average stores.[38] After the attacks, these retailers were forced to relocate to other parts of the metropolitan area with less customer traffic and lower ability to generate the same high level of sales.

Indirect Costs

A major terrorist incident could present many different indirect consequences. If infrastructure is damaged, services could be disabled even to business without direct property damage. For example, Con Edison lost two electrical power substations in the World Trade Center area, and electricity was shut down in the area for over a week. Gas and steam lines were also impacted.

Transportation systems can also be greatly impacted by terrorist incidents. Subway stations beneath the site were damaged, affecting the commuting patterns of many people for many months. Surface roads in the area were impassible, and tightened security precautions limited vehicular traffic for many months following the attacks. Government prohibition or limitations on the movement of people into the World Trade Center area resulted in the vacancy of buildings that had not been heavily damaged.

Companies seeking to relocate from their lost offices faced the very difficult task of finding replacement space near their original sites and paying costs associated with relocation and access.

Companies that lost employees also lost the specialized knowledge and expertise of those who were lost—impacting projects in midstream. Firms also were faced with a challenging job market when hiring replacement employees. Salaries offered to applicants who know they are in demand can be much higher than those paid to their predecessors.

One type of loss that is difficult to quantify is the lost time and productivity associated with the mental anguish of a terrorist attack and concern about future attacks. Employees who may not have been physically injured may not be emotionally prepared to return to work. They may file stress claims or simply resign their positions.

Other Issues of Concern

A widespread health emergency resulting from a biological terrorist attack would most likely present many very challenging political, legal, moral, and ethical issues. Mandatory quarantines, enforced vaccinations, martial law, and other government-mandated actions would cause serious impact on everyday daily life and the ability to travel freely.

A major terrorist event raises stress in the workplace. Workers may refuse to work in a building they feel is not secure, may refuse to travel on aircraft, or may refuse to go to locations where they feel their security might be threatened. Post-traumatic stress claims could lead to provision of special accommodations for individual employees.

End Notes

[1] 28 C.F.R. Section 0.85.

[2] Federal Bureau of Investigation, *Terrorism in the United States 1999.*

[3] U. S. Department of State, Office of the Coordinator for Counterterrorism, *Patterns of Global Terrorism 1999.*

[4] "Post Offices Nationwide Search For Cyanide-Stuffed Mailings, " *The Topeka Capital-Journal via Associated Press*, 1998.

[5] As of December 31, 2001, there are 2,937 people dead or missing in the World Trade Center attacks, according to Mayor Rudolph Giuliani, including 157 passengers and crew and 10 hijackers on the two airplanes involved in the attacks. At the Pentagon, the U.S. Department of Defense said 125 people are missing or dead. There were 64 passengers and crew killed on the plane involved, including five hijackers. The plane that crashed in Pennsylvania carried 44 passengers and crew, including four hijackers.

[6] Federal Bureau of Investigation, *Terrorism in the United States 1999.*

[7] Senate Committee on Governmental Affairs, *Homeland Security A Risk Management Approach Can Guide Preparedness Efforts*, Testimony of Raymond J. Decker, Director, Defense Capabilities and Management, U.S. General Account Office, October 31, 2001, GAO-02-208T.

[8] Milton Leitenberg, Center for International and Security Studies, University of Maryland, *An Assessment of the Biological Weapons Threat to the United States*, A White Paper prepared for the Conference on Emerging Threats Assessment: Biological Terrorism, at the Institute for Security Technology Studies, Dartmouth College, July 7-9, 2000.

[9] U.S. Congress, Joint Economic Committee, *Terrorist and Intelligence Operations: Potential Impact on the U.S Economy*, Testimony of Dr. Kenneth Alibek, Program Manager, Battelle Memorial Institute, May 20, 1998.

[10] Senate Committee on Governmental Affairs, Subcommittee on International Security, Proliferation, and Federal Services, *The Proliferation of Chemical and Biological Weapons Materials and Technologies to State and Sub-State Actors*, Testimony by Jonathan B. Tucker, Ph.D., Director, Chemical & Biological Weapons Nonproliferation Program, Center for Nonproliferation Studies, Monterey Institute of International Studies, November 7, 2001.

[11] Sverdlovsk has been renamed Yekaterinburg.

[12] Senate Committee on Governmental Affairs, Subcommittee on International Security, Proliferation, and Federal Services, *The Proliferation of Chemical and Biological Weapons Materials and Technologies to State and Sub-State Actors*, Testimony by Jonathan B. Tucker, Ph.D., Director, Chemical & Biological Weapons Nonproliferation Program, Center for Nonproliferation Studies, Monterey Institute of International Studies, November 7, 2001.

[13] Bleich A, Kron S, Margalit C, Inbar G, Kaplan Z, Cooper S, Solomon Z., *Israeli Psychological Casualties Of The Persian Gulf War: Characteristics, Therapy, And Selected Issues*, Department of Mental Health, Israel Defense Forces Medical Corps, 1991.

[14] Presentation of Colonel Ann E. Norwood, *Psychiatric Aspects of Chem/Bioterrorism*, Department of Psychiatry, Center for the Study of Traumatic Cases, Uniformed Services University, Bethesda, MD.

[15] Centers for Disease Control and Prevention, "Bioterrorism Preparedness and Response Initiative 'A Strategy for Public Health'" (paper presented at the 2001 Mid-Year Conference of the National Emergency Management Association, Washington, D.C., February 12, 2001.)

[16] U.S. Centers for Disease Control and Prevention, *FAQ'S about Nerve Agents*, Internet: http://www.bt.cdc.gov.

[17] Federal Bureau of Investigation, *Weapons of Mass Destruction Incident Contingency Plan*.

[18] Thousands sought treatment but were not treated.

[19] Senate Committee on Governmental Affairs, Subcommittee on International Security, Proliferation, and Federal Services, *The Proliferation of Chemical and Biological Weapons Materials and Technologies to State and Sub-State Actors*, Testimony by Jonathan B. Tucker, Ph.D., Director, Chemical & Biological Weapons Nonproliferation Program, Center for Nonproliferation Studies, Monterey Institute of International Studies, November 7, 2001.

[20] Shackell, John M., *Terrorist Use of Weapons of Mass Destruction*, briefing from open sources, undated.

[21] U.S. General Accounting Office, *Combating Terrorism Need for Comprehensive Threat and Risk Assessments of Chemical and Biological Attacks*, GAO/NSIAD-99-163, September 1999.

[22] U.S. Centers for Disease Control and Prevention, *FAQ'S about Nerve Agents*, Internet: http://www.bt.cdc.gov.

[23] U.S. Department of Justice, Office of Justice Programs–Bureau of Justice Assistance, Federal Emergency Management Agency, and United States Fire Administration–National Fire Academy, *Emergency Response to Terrorism Self-Study*, FEMA/USFA/NFA-ERT:SS, June 1999.

[24] U.S. Department of Justice, Office of Justice Programs–Bureau of Justice Assistance, Federal Emergency Management Agency, and United States Fire Administration–National Fire Academy, *Emergency Response to Terrorism Self-Study*, FEMA/USFA/NFA-ERT:SS, June 1999.

[25] Michael A. Vatis, *Cyber Attacks During the War on Terrorism: A Predictive Analysis*, Institute for Security Technology Studies at Dartmouth College, September 22, 2001.

[26] Senate Committee on Armed Services, Subcommittee on Emerging Threats, *Cyber Threats and The U.S. Economy*: Testimony of Stephen E. Cross, Director, Software Engineering Institute, Carnegie Mellon University, February 23, 2000.
[27] House Science Committee, *Testimony of Governor James S. Gilmore III, Governor of the Commonwealth of Virginia and Chairman of the Advisory Panel to Assess the Capabilities for Domestic Response to Terrorism Involving Weapons of Mass Destruction*, October 17, 2001.
[28] Senate Committee on Armed Services, Subcommittee on Emerging Threats, *Cyber Threats and The U.S. Economy: Testimony of Stephen E. Cross*, Director, Software Engineering Institute, Carnegie Mellon University, February 23, 2000.
[29] The CERT® Coordination Center (CERT/CC) is a center of Internet security expertise, at the Software Engineering Institute, a federally funded research and development center operated by Carnegie Mellon University. It studies Internet security vulnerabilities, handles computer security incidents, publishes security alerts, researches long-term changes in networked systems, and develops information and training to help improve security.
[30] The "Computer Crime and Security Survey" is conducted by CSI with the participation of the San Francisco Federal Bureau of Investigation's (FBI) Computer Intrusion Squad. The aim of this effort is to raise the level of security awareness, as well as help determine the scope of computer crime in the United States.
[31] Denning, Dorothy. "Activism, Hacktivism, and Cyberterrorism: The Internet as a Tool for Influencing Foreign Policy," February 2001
[32] National Infrastructure Protection Center, "The Threat to the U.S. Information Infrastructure", October 2001.
33 National Infrastructure Protection Center, "The Threat to the U.S. Information Infrastructure," October 2001.
[34] Denning, Dorothy. "Activism, Hacktivism, and Cyberterrorism: The Internet as a Tool for Influencing Foreign Policy," February 2001
[35] Abreu, Elinor Mills, "Stand By for More Nasty Web Attacks in 2002," (estimates from Computer Economics, Carlsbad, CA), Reuters, December 27, 2001.
[36] Aaron Elstein, "Aon, Marsh & McLennan Tally 'Ground Zero' Costs", *The Wall Street Journal Online*, November 8, 2001.
[37] Risk Management Solutions, *World Trade Center Disaster RMS Special Report*, September 18, 2001.
[38] Westfield America Inc. as reported by Motoko Rich and Erin White, "High-Volume Trade Center Stores Seek New Spots with Similar Allure," *The Wall Street Journal*, November 2, 2001.

Chapter 2

Risk Assessment

Risk Assessment Process

Risk is a measure of the frequency or probability of a negative event and the severity or consequences of that negative event. You don't have to plan for events with zero probability or events that have no consequences. Unfortunately, terrorism in America is a real threat, and the consequences of a terrorist attack can be catastrophic. Chapter 1 highlighted many terrorism incidents that have occurred in the United States over the past decade; such events can impact anyone—while handling mail, eating from a restaurant salad bar, or working quietly in an office.

Statistically, the frequency of terrorist attacks is low. The risk of a catastrophic attack is very low, but the risk is still real, and the potential consequences warrant appropriate attention. Cyberterrorism, or attacks on computer systems, are increasing in frequency and the consequences are becoming more severe.

Figure 1. Risk Assessment Process

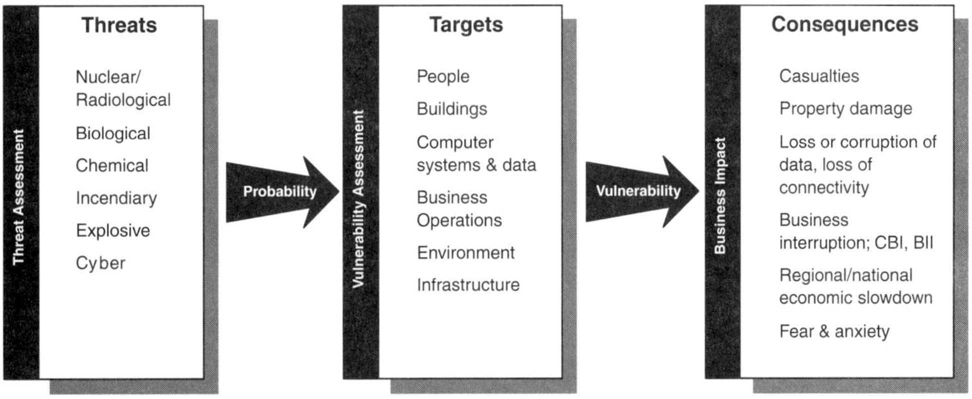

Events that occur frequently must be addressed through a combination of appropriate risk management techniques, including prevention, mitigation, response, recovery, risk financing, or risk transfer. All of these techniques are addressed in this book.

Terrorism risk assessment is a process not unlike the assessment of risk from other perils. The process is broken into three component parts and answers the following questions:

- **Threat assessment**: What threats are credible? What is the probability a threat will be perpetrated against a specific target?
- **Vulnerability assessment**: What are the physical weaknesses in site layout, buildings, utility systems, and computer networks that make people and business functions more vulnerable? What gaps exist in security policies, procedures, or other programs that would impact the ability to deter, detect, or effectively respond to a terrorist threat or attack?
- **Business impact analysis**: What would be the consequences of a terrorist attack and how would those consequences impact business operations?

Threat Assessment

Threat assessment identifies the types of terrorist attacks that could occur and quantifies the probability of occurrence. The following threats were presented in Chapter 1 as the primary terrorist threats:

- Nuclear or radiological
- Biological agents
- Chemical agents and industrial chemicals
- Incendiary devices
- Explosive devices
- Cyber attacks

The probability of a terrorist threat is a measure of the capabilities, impact, intentions, and past activities of potential terrorists. Capability to perpetrate a terrorist act is dependent upon the ability to manufacture or acquire a weapon and to carry out the terrorist act. Impact is the consequence of the act including casualties, property damage, and business interruption. Intentions are the motivations of a terrorist or terrorist organization to perpetrate acts of violence as explained in Chapter 1.

A nuclear or radiological incident could involve detonation of a thermonuclear device; explosion of a "dirty bomb" (radiological dispersion device); or the release of radioactive material from an attack on a facility that uses or stores radioactive materials (e.g., bomb, aircraft, or missile attack on a nuclear power plant).

An attack with biological agents could include the intentional dispersal or distribution of biological agents such as anthrax, smallpox, botulism, and the plague. Anthrax can be sent through the mail system, and food can be contaminated with salmonella. Smallpox and plague are infectious diseases that could spread widely.

A chemical attack could include use of the chemical warfare agents discussed in Chapter 1.

There are tens of thousands of hazardous industrial chemicals. Everyday these chemicals are used at industrial facilities, manufacturing plants, and research laboratories and transported throughout the United States via truck, rail, pipeline and along our navigable waterways. Release of, or accidents involving, hazardous materials are common, and the consequences of these incidents can be significant. Fires, large numbers of injuries, and fatalities have occurred in many large-scale incidents. Environmental contamination is also another major concern. Areas surrounding hazardous materials incidents are significantly impacted as roads are closed, rail transportation is shut down, and people are forced to shelter in place. Therefore, both the direct and indirect consequences of these types of events must be seriously evaluated.

An attack with an incendiary device would likely lead to a fire that could spread unless controlled by fire suppression systems, a private fire brigade, or the public fire department.

Bombings are historically the most likely terrorist events and have been the method of choice for most terrorists. Vehicle bombs can weigh thousands of pounds and cause catastrophic consequences, and a letter bomb can be lethal for anyone who opens the package.

Cyberterrorism includes a variety of attacks on networked computer systems.

Vulnerability Assessment

A vulnerability assessment is the process of identifying weaknesses in perimeter security, buildings, utility systems, personnel protection systems, or computer systems that may be exploited by terrorists. This process also identifies opportunities to mitigate vulnerabilities and reduce exposure to loss. Mitigation is addressed in Chapter 3.

The vulnerability assessment should also evaluate operational security and emergency response capabilities. The assessment can identify gaps that would limit an organization's ability to deter terrorists, detect threats quickly, and respond effectively to safeguard people and protect assets.

The vulnerability assessment should be done by qualified persons who understand the potential threats posed by terrorists and are able to identify vulnerabilities in building placement, layout, and construction; utility and computer systems; and physical security. The experts must also be able to identify gaps in operational security; computer security; and emergency response and recovery capabilities. The experts will require the assistance of persons within the organization who are knowledgeable about the organization's facilities, security, information technology, business operations, and other functional areas. Outside experts that may need to be engaged include structural engineers, fire protection engineers, security experts, law enforcement officials, terrorism experts, and computer security professionals.

Business Impact Analysis

The vulnerability assessment identifies weaknesses in physical assets and gaps in programs. A business impact analysis then quantifies the impact on business operations resulting from the varying consequences of a terrorist attack. The consequences of an attack could be localized within a building or widespread to affect a large region of the country. The resulting business impact could be minimal if damage to a building is minor, or it could be catastrophic if critical functions are lost and can't be easily replaced or the loss occurs at a critical time.

A cyber attack that compromises a critical Web server for an online retailer could prevent receipt of customer orders. The result would be a loss of sales that may never be recovered. If the cyber attack took place during a season when sales were slow, then the financial impact would be low. However, if the cyber attack prevented receipt of customer orders at a time when sales peak, the financial impact could be much greater.

Threat Assessment

The probability of a terrorist attack is determined by the capabilities and intentions of known and would-be terrorists. U.S. law enforcement and intelligence agencies including the FBI (domestic counter-terrorism), Central Intelligence Agency, Defense Intelligence Agency, and the State Department's Bureau of Intelligence and Research (foreign intelligence), and others continually gather intelligence on known and suspected terrorists to learn about their motives, intentions, and capabilities. They also investigate terrorist incidents to learn about terrorist methods, targets and the consequences of the attacks. This intelligence gathering allows the government to identify likely targets for future attacks and to warn us if there is a credible threat.

The U.S. Department of Justice developed a system that is used by cities and counties to rate potential terrorist targets within their jurisdictions. This rating system includes seven factors that make a site or building more or less desirable as a target for terrorists. These factors relate to the social or political motivations of domestic and international terrorists discussed in Chapter 1.

The seven factors include:

- ❑ Awareness of, or visibility of, a site
- ❑ Importance or criticality of a site to the community
- ❑ Value of the site as a target to further a terrorist's social or political agenda
- ❑ Access to the site
- ❑ Hazards on-site such as industrial chemicals, explosives, biological agents, or radioactive materials

- ❑ Site population

- ❑ Potential for collateral casualties

Each of the seven factors in Table 1 are scored with a rating of 0 to 5 with 0 for no exposure, and 5 for the highest exposure. A site with a rating of 35 would be a highly attractive target for terrorists. A site with a 0 rating would not be attractive to terrorists but still could be subject to collateral damage or the indirect consequences of an attack.

Visibility. A high profile building that is a landmark within a city or region would be a more probable target than a facility located in a remote area that is unknown to most people.

Usefulness to the Community. A building or facility that is critical to the community is a higher profile target than a building or facility that fulfills no governmental, public service, or economic role within the community.

Value. The "value" of the target relates to whether an attack on the site would serve the social or political interests of a terrorist.

- ❑ International terrorists target government or economic symbols of America. The World Trade Center was believed to be attacked because it is a center of domestic and international finance.

- ❑ Eco-terrorists in the U.S. attack businesses such as the pulp and paper industry that they perceive as detrimental to the environment.

- ❑ Terrorists that espouse the cause of animal rights have targeted businesses that they believe are harmful to animals. These include fur retailers, cosmetics manufacturers and others that use animals for research.

- ❑ Radical anti-abortionists have attacked many women's clinics.

- ❑ Religious extremists have attacked churches, synagogues and mosques.

Accessibility. Just as high security can deter terrorists, lack of security could entice terrorists to target a building or site. Security considerations include perimeter fencing, lighting, security guards, protected entryway, surveillance equipment, intrusion alarms, and other access control systems. Parking within fifty feet of a building is also a consideration for vehicle bombs. Not allowing vehicular parking close to a building would reduce the probability of a vehicle bomb attack.

Hazards. This factor was developed with both military targets and civilian targets in mind. Some military facilities stockpile various weapons of mass destruction. Many public and private facilities also house "weapons of mass destruction." Some hospitals or research laboratories contain biological agents; there are thousands of facilities nationwide that have hazardous industrial chemicals, incendiary, or explosive materials; and there are thousands of facilities that have radioactive materials.

Population and Collateral Casualties. These two categories refer to the number of persons that could be affected by a terrorist incident. Population at site refers to the people within a building or group of buildings on-site. Collateral casualties refer to the number of people within a one-mile radius who would be affected by a terrorist attack on a nearby site.

Table 1. Site Vulnerability Assessment Factors[1]

Location	Visibility	Usefulness	Value	Accessibility	Hazards	Population	Collateral Casualties

Visibility – Awareness of the Existence and Visibility Of The Target	
Invisible: Classified Location	0
Very Low Visibility: Probably not aware of existence	1
Low Visibility: Probably not well known existence	2
Medium Visibility: Existence is probably known	3
High Visibility: Existence well known	4
Very High: Visibility Existence is obvious	5

Usefulness To Community	
No Usefulness	0
Minor Usefulness	1
Moderate Usefulness	2
Significant Usefulness	3
Highly Useful	4
Critical	5

Value of Target to Serve Motivations	
None	0
Very Low	1
Low	2
Medium	3
High	4
Very High	5

Potential Collateral Casualties Within One-Mile Radius	
0 to 100	0
101 to 500	1
501 to 1,000	2
1,001 to 2,000	3
2,001 to 5,000	4
> 5,000	5

Accessibility of the Target to Ingress and Egress	
Fenced, Guarded, Protected Entry, Controlled Access by Pass Only, No Vehicle Parking within 50 Feet	0
Guarded, Protected Entry, Controlled Access of Visitors and Non-Staff Personnel, No Vehicle Parking within 50 Feet	1
Protected Entry, Controlled Access of Visitors and Non-Staff Personnel, No Unauthorized Vehicle Parking within 50 Feet	2
Controlled Access of Visitors, Unprotected Entry, No Unauthorized Vehicle Parking within 50 Feet	3
Open Access to all personnel, Unprotected Entry, No Unauthorized Vehicle Parking within 50 Feet	4
Open Access to all personnel, Unprotected Entry, Vehicle Parking within 50 feet	5

Target Threat of Hazard. Presence of WMD Materials	
No WMD materials present	0
WMD materials present in moderate quantities, under positive control, and in secured locations	1
WMD materials present in moderate quantities and controlled	2
Major concentrations of WMD materials that have established control features and are secured in the site	3
Major concentrations of WMD materials that have moderate control features	4
Major concentrations of WMD materials that are accessible to Non-staff personnel	5

WMD includes hazardous chemicals, biological agents, or radioactive materials.

Population Max. Number Of People At Site	
0	0
1 - 250	1
251 - 500	2
501 – 1,000	3
1001 - 5,000	4
> 5,000	5

History of Past Incidents

Past incidents are an indicator of potential future trouble. The most notorious case is the World Trade Center. It was a high profile target in 1993 and again in 2001. There have been many attacks on houses of worship, abortion clinics, and women's reproductive health centers; these attacks can be expected to continue. Law enforcement authorities—especially the FBI, which has responsibility for domestic terrorism—can provide information on actual or attempted terrorist incidents in a particular area or against a type of organization.

Potential Targets

Government buildings are clearly targets for both domestic and international terrorists with hostility towards the U.S. government. Terrorists may strike merely because the government building is the only federal presence in a community; the building is conspicuous; it is the biggest local symbol of America; or the building is perceived to be vulnerable. Timothy McVeigh, who bombed the Murrah Federal Building in Oklahoma City, commented that he thought the Murrah Federal Building was vulnerable to attack.

Buildings, industrial plants, and other structures can be targeted by terrorists for various reasons. The following is a list of potential targets that score high when evaluated using the seven vulnerability factors documented in Table 1. Site Vulnerability Assessment Factors:

- ❑ Any building, arena, stadium, theater, shopping mall, or venue where large numbers of people congregate.
- ❑ Theme parks or amusement parks.
- ❑ Government buildings that house law enforcement (e.g., Department of Justice, FBI, U.S. Treasury/Bureau of Alcohol Tobacco and Firearms, U. S. Marshals, state police, etc.)
- ❑ Government buildings that house other departments (e.g., Internal Revenue Service, U.S. State Department, foreign consulates or embassies, etc.)

- ❑ Military installations including armories, training facilities, and recruiting offices.

- ❑ Centers of finance or buildings with high economic impact. This could include major U.S. multi-national corporations that are perceived by terrorists to be capitalistic icons.

- ❑ Places of historical or symbolic significance. For example, the Statue of Liberty, Washington Memorial, and other landmarks are symbols of America and could be targeted.

- ❑ Houses of worship. Many churches were destroyed by arson in the late 1990s. Vigilantes targeted mosques in the U.S. after the September 11 attacks.

- ❑ Infrastructure such as bridges, subways, airports, electrical transmission lines, power plants, and natural gas transmission pipelines.

- ❑ Private or public organizations that meet the following criteria may be targets of letter or package bombs:[2]

    - Provide products, materials, technical assistance, technical training, or operate plants or facilities within the country involved or connected with current terrorist activity.

    - Contribute money to, provided support for, or been politically affiliated with any charity, aid program, cultural exchange, or educational program for which public recognition has been received that could in any way be construed as affiliated with current terrorist targets.

    - Support political or social causes that would target them for radical domestic hate groups.

    - Refuse to do business with, have withdrawn from, or failed to successfully negotiate business contracts with companies, organizations, or governments within the last two years that are affiliated with current terrorists.

- Manufacture or produce weapons of war or military support items (e.g., radios, vehicles, uniforms) for the international arms trade that would normally bear markings to identify the organization as the manufacturer.

- Made public statements or been quoted, interviewed, or authored papers on any facet of current terrorist activity or topics.

❑ Facilities that manufacture, store, treat, or dispose of hazardous materials are thought to be targets since an attack may result in an explosion, fire, uncontrolled release of hazardous materials, and/or environmental contamination. Although not a terrorist incident, the 1984 Union Carbide incident in Bhopal, India, demonstrated that a chemical release could kill thousands. Sources of chemicals in the communities include:[3]

- Chemical manufacturing plants (chlorine, peroxides, other industrial gases, plastics, and pesticides).

- Food processing and storage facilities with large ammonia tanks,

- chemical transportation (rail tank cars, tank trucks, pipelines, and river barges).

- Gasoline and jet fuel storage tanks at distribution centers, airports, and barge terminals.

- Compressed gases in tanks, pipelines, and pumping stations.

- Pesticide manufacturing and supply distributors.

- Educational, medical and research laboratories.

Some of the more common types of chemicals that could be used in improvised weapons in the communities include:

- Eye, skin and respiratory irritants (acids, ammonia, acrylates, aldehydes, and isocyanates).

- Choking agents (chlorine, hydrogen sulfide, and phosgene).
- Flammable chemical industry gases (acetone, alkenes, alkyl halides, amines).
- Aromatic hydrocarbons that could be used as water supply contaminants (benzene, etc.).
- Oxidizers for improvised explosives (oxygen, butadiene, and peroxides).
- Aniline, nitrile, and cyanide compounds that could be used as chemical asphyxiants.
- Compressed hydrocarbon fuel gases that could be used as incendiaries or simple asphyxiants (liquefied natural gas, propane, isobutane).
- Liquid hydrocarbon fuels that could be used as incendiaries or water supply contaminants (gasoline, jet fuel).
- Industrial compounds that could be used as blister agents (dimethyl sulfate).
- Organophosphate pesticides that could be used as low-grade nerve agents.

❑ Nuclear power plants are targets because an attack could release radioactivity into the atmosphere. After the September 11 attacks on the World Trade Center and Pentagon, the Nuclear Regulatory Commission ordered nuclear plants to adhere to the highest security level. The 104 operating nuclear reactors in the U.S. are built to the highest level of engineering design in the civil world, second only to the protection of U.S. nuclear armaments. However, they are not built to withstand a wide-bodied aircraft with thousands of pounds of jet fuel aboard. Besides the reactor core, spent fuel storage pools are not as well protected.[4]

It is important to emphasize that a facility does not have to be the intended target of a terrorist attack to suffer severe consequences. Buildings and people in proximity to the target can suffer severe effects. An attack with a "dirty bomb" (radiological dispersion device) would cause environmental contamination beyond the blast zone, possibly prohibiting occupancy of buildings not damaged by the blast.

An attack on a facility with hazardous materials could cause an uncontrolled release of chemicals that could be injurious to a wide area. Occupancy of the affected area could be restricted for a long period. The same could be true for abutters to roadways, rail lines or shipping channels heavily used for transportation of hazardous materials.

Since subway systems are a proven target (e.g., Tokyo, 1995), businesses in proximity to the entrances or exhaust ventilation shafts of rail tunnels and subways could be exposed.

Vulnerability Assessment

A vulnerability is a weakness in the siting of a building; the building and its utility and control systems; space planning within the building; or protection of targeted portions of a building. Gaps in management programs that deter, detect, respond to, or recover from terrorist attacks are also vulnerabilities. A vulnerability assessment is a process to identify these weaknesses or gaps that would increase the susceptibility of occupants, operations, and assets to a terrorist attack.

Persons who are knowledgeable in the threats posed by terrorists should complete a vulnerability assessment. This includes the threats defined in Chapter 1: chemical, biological, nuclear/radiological, incendiary, explosive, and cyber. These hazards can be grouped into three classes for the purposes of the vulnerability assessment: explosions, airborne hazards, and cyber attacks. The assessment should include a survey of the property; interviews with building management, engineering, and security personnel; reviews of construction plans and documentation; a review of emergency procedures; and an evaluation of critical life-safety systems.

The assessment will focus mostly on the physical aspects of a building and the security measures that deter or respond to terrorist activity. However, the assessment should always consider how people are exposed and how they might respond in an emergency.

People

Most terrorist attacks threaten the health and safety of people. Even cyberterrorism can jeopardize the safety of people if critical infrastructure such as the electric power grid, air traffic control system, or public safety communication systems are compromised. Whether office workers in a high-rise building, a receptionist at a women's health clinic, a passenger aboard a jetliner, or residents living near a chemical plant, all are at risk.

Occupants of buildings are subject to terrorist attacks in many ways. Flying glass from a bomb attack causes more injuries to people than the blast wave; and chemical or biological agents dispersed outside can infiltrate a building and be spread by ventilation systems.

Buildings or structures that house large numbers of people are targets because of the potential to inflict large numbers of casualties with a single strike. Companies that house a large percentage of their employees in a single high-rise building risk losing the bulk of their staff in a single event. Auditoriums, large conference, or meeting facilities within a building would also be an area to closely review security precautions because many persons are at risk from a small-scale inside attack.

The vulnerability of people within a building is clearly a product of the susceptibility to attack of the building that houses them. However, certain features of the building, including construction and protective systems—as well as response procedures—can heavily influence survivability. These include:

- ❑ Emergency occupant notification systems
- ❑ Adequacy of means of egress
- ❑ Emergency procedures to respond to terrorist threats and evacuate or shelter occupants in place

Occupant notification systems include fire alarm systems, emergency voice communication systems, and public address systems. Emergency alarm and communication systems would be used to notify occupants to evacuate if there were an attack within the building, or to shelter in place if there were a chemical or radiological incident outside the building.

Occupant notification systems should be designed, installed, tested, and maintained in accordance with national standards or local building and fire codes. Building operations personnel should know when and how to use these systems.

Means of egress is the system of corridors, exit signs, exits, stairwells, and exit discharges to facilitate prompt evacuation of building occupants. The *Life Safety Code®* published by the National Fire Protection Association (NFPA) or local building codes specify the number of exits required from each floor of a building and the arrangement of those exits. A professional competent in applying these codes and standards should evaluate whether the complement and arrangement of exits is sufficient for the number of occupants. Inadequate means of egress can result in a slow evacuation or trapped occupants.

Disabled, elderly, or very young occupants may be incapable of self-preservation in case of an emergency and may need assistance with evacuation. These persons should be identified so plans can be established to assist them. If large numbers of persons need assistance, specific plans and special arrangements may be needed. Evacuation planning is discussed in Chapter 4.

Site Location, Layout, & Protection

The building site can heavily determine its susceptibility to attack and its vulnerability to damage either directly or from an attack on a nearby target. The vulnerability assessment should include a review of the site including the positioning of buildings on the property; the nature of, and separation to, adjacent buildings; the location of parking areas and public streets; and vehicular traffic patterns.

Proximity to Potential Targets

Properties around the building site that are highly desirable targets for terrorists raise the susceptibility to collateral damage if the neighboring property is attacked. The vulnerability assessment should include a review of buildings or facilities that surround the property. Begin with those immediately within the building, if the building is multi-tenanted, and then look at adjacent buildings. Tenant directories provide a list of occupants, and building managers can confirm specific occupancies.

The "safe" distance from a potential target is difficult to determine since some terrorist acts affect a wide area. However, targeted buildings within the same building or within a one-hundred-foot radius are significant. Targeted buildings within a 500-foot radius present moderate hazard, and those within a 1,000-foot radius can't be ignored. High hazard chemical facilities and transportation lines (road, rail, pipeline, etc.) can pose a significant risk if there is an uncontrolled release of hazardous chemicals. The hazard is determined by the chemical release, the quantity released, and meteorological conditions at the time.

Figure 2. Potential Targets

Government Services

- Government offices — especially law enforcement, IRS
- Courthouses
- Military installations (includes reserves, armories, and recruiting offices)
- Diplomatic embassies and consulates

Emergency Services

- Police
- Fire
- Emergency medical services
- State and local emergency operations centers

Public Health

- Hospitals
- Emergency medical centers

Utilities

- Electric power generation & transmission
- Gas storage & distribution
- Petroleum storage & distribution
- Telecommunications

Water & Sewer Systems

- Water storage tanks
- Wells
- Pumping stations
- Water purification plants
- Wastewater treatment plants

Information & Communications

- Newspapers
- Radio stations
- Television broadcasting
- Radio communications, switching, and cable TV

Institutions

- Scientific research
- Academic institutions (colleges & universities)
- Museums
- Schools

Commercial & Industrial Facilities

- Chemical manufacturing
- Industrial plants with hazardous materials
- Petrochemical plants
- Business or corporate centers
- Malls/shopping centers
- Hotels/convention center

Transportation

- Railroads & rail yards
- Interstate highways
- Tunnels
- Subway system
- Airports
- Oil or gas pipelines
- Seaports or river ports
- Bus terminals
- Bridges
- Ferry terminals
- Truck terminals

Banking & Finance

- Banks
- Financial Institutions

Recreational Facilities

- Sports arenas or stadiums
- Auditoriums
- Theaters
- Parks
- Casinos
- Concert halls or pavilions
- Restaurants frequented by target population

Special Events

- Parades
- Religious services
- Festivals
- Celebrations
- Abortion clinics
- Agriculture

Placement of Buildings

The setback or distance of a building's walls to the unsecured perimeter of the property should be measured. The shorter the setback, the greater the exposure to attack. A 10,000-pound bomb concealed in a small box van can cause a lethal blast at 300 feet. A 1,000-pound bomb in a full-sized sedan is lethal at 125 feet. Even a hand carried bomb can cause injuries and property damage if exploded in close proximity to a building. Release of a chemical agent near a building may allow the agent to infiltrate the building or be sucked into the building's ventilation system. A setback of at least one-hundred feet is a minimum. Anything less raises the damage potential of an attack.

The arrangement of the building on the site should be out of alignment with the surrounding streets. Angling exterior walls so they are not parallel to street can deflect the blast effects of a bomb.

The shape of the building should be noted. Any building that has a shape that would contain an explosive blast (e.g., "U," "H," or "L") should be noted if the opening of the shape (the top of the "U") is pointed towards the street.

The occupancies of individual buildings on a campus should be noted. In particular, note whether critical functions, operations, and people are placed in buildings that are farthest from the perimeter.

Perimeter Security

The security of the perimeter of the property can prevent an attacker from penetrating into the building compound. A walking tour of the perimeter can identify physical barriers and other features that would make it difficult for a terrorist to attack. The following features should be evaluated:

- ❑ Distance from perimeter to important buildings.
- ❑ Distance from perimeter to roadways, streets, parking areas.
- ❑ Presence or absence of any walls, fencing, ditches, embankments, or other physical barriers that could

prevent unauthorized access of people or motor vehicle. Are the barriers continuous and in good condition? Are the barriers easily penetrated by scaling fences or climbing trees?

- ❑ Are there areas along the perimeter that could easily conceal a person?
- ❑ Is the entire perimeter well-illuminated at night to provide good visibility for security personnel?
- ❑ Are all sections of the perimeter monitored by security guards or with surveillance equipment?

Vehicular Traffic & Parking

The primary concern of vehicular traffic is the detonation of vehicle bombs. Of secondary concern are potential incidents involving the transportation of hazardous materials. The farther motor vehicles are away from the site the better.

- ❑ What is the distance from roadways, driveways, or parking areas to the sides of important buildings?
- ❑ What is the speed limit on roadways within one-hundred feet of the building? The faster the traffic moves, the less reaction time there is to a threat. Faster traffic can drive over curbs and other obstacles.
- ❑ What are the traffic patterns around important buildings? Do the patterns allow high-speed traffic towards, away from, or along a building? Traffic that is moving towards a building presents a greater hazard.
- ❑ Are there any speed bumps, changes in pavement, narrowing of lanes, or other traffic measures designed to slow traffic?
- ❑ Where motor vehicles are allowed to travel within one-hundred feet of an important building, are there barriers to prevent a vehicle from driving up to the front door? This could include curbs, guardrails, concrete "Jersey barriers," bollards, or concrete planters with shrubbery. Are the barriers spaced or placed to prevent a vehicle from driving between them?

- ❑ Are there separate parking areas for employees and visitors? Are the cars entering parking areas screened before entrance?

Vehicle Bomb Exposure

Bombs carried in motor vehicles present the most serious conventional bomb exposure. A bomb-laden van was used to attack the World Trade Center in 1993, and Timothy McVeigh attacked the Murrah Federal Building in Oklahoma City in 1995 with an ammonium nitrate fuel oil bomb hidden in a rented truck.

Bombs hidden in motor vehicles present a range of exposures depending upon the size of the vehicle, type of explosive, amount of explosives, and location of the vehicle relative to a building when the bomb explodes. Figure 3 from the U.S. Bureau of Alcohol, Tobacco and Firearms (ATF) depicts the potential capacity of explosives, lethal air blast range, evacuation zone, and falling glass hazard.[5] A compact car can hold 500 pounds of explosives, and a fourteen-foot box van can hold 10,000 pounds of explosives.

A vehicle bomb that explodes outside a building can kill or injure people from the air blast, flying shrapnel from the disintegrating vehicle, and falling glass or other hazards from buildings damaged by the explosion. Unless the bomb is designed to be incendiary, only combustibles in the vehicle or immediately adjacent to the vehicle will be ignited.[6]

Explosion of a vehicle bomb within a parking garage of a building can cause significant damage to the building and its occupants. The 1993 bombing of the World Trade Center resulted in six fatalities and over 1,000 injuries. Explosives packed in a van were detonated inside the parking garage, causing serious structural damage and a raging fire.

A detonation of an explosive within a confined space, such as an underground parking garage, prevents the hot gases of the explosion from expanding and mixing with cooler surrounding air. In addition, the explosion will overturn vehicles and rupture fuel tanks and gas lines within the building, adding fuel to the fire. A blast can also cause structural damage, impair fire protection sprinkler systems, and compromise critical utility systems. A raging fire—uncontrollable because fire protection systems are impaired—can then threaten occupants.

ATF research also provides the following guidance:

- The "Minimum Evacuation Distance" is the range at which a life-threatening injury from blast or fragment hazards is unlikely. However, non-life-threatening injury or temporary hearing loss may occur.
- Hazard ranges are based on open, level terrain.
- Minimum evacuation distance may be less when an explosion is confined within a structure.
- The hazard range of falling glass is dependent on line-of-sight from explosion source to window. The hazard is falling shards of broken glass.
- An explosion confined within a structure may cause structural collapse or building debris hazards.
- Additional hazards include vehicle debris.

Figure 3. Vehicle Bomb Evacuation Distance (ATF)

(www.atf.treas.gov/pub/fire-explo_pub/i54001.htm)

ATF	VEHICLE DESCRIPTION	MAXIMUM EXPLOSIVES CAPACITY	LETHAL AIR BLAST RANGE	MINIMUM EVACUATION DISTANCE	FALLING GLASS HAZARD
	COMPACT SEDAN	500 Pounds 227 Kilos *(In Trunk)*	100 Feet 30 Meters	1,500 Feet 457 Meters	1,250 Feet 381 Meters
	FULL SIZE SEDAN	1,000 Pounds 455 Kilos *(In Trunk)*	125 Feet 38 Meters	1,750 Feet 534 Meters	1,750 Feet 534 Meters
	PASSENGER VAN OR CARGO VAN	4,000 Pounds 1,818 Kilos	200 Feet 61 Meters	2,750 Feet 838 Meters	2,750 Feet 838 Meters
	SMALL BOX VAN *(14 FT BOX)*	10,000 Pounds 4,545 Kilos	300 Feet 91 Meters	3,750 Feet 1,143 Meters	3,750 Feet 1,143 Meters
	BOX VAN OR WATER/FUEL TRUCK	30,000 Pounds 13,636 Kilos	450 Feet 137 Meters	6,500 Feet 1,982 Meters	6,500 Feet 1,982 Meters
	SEMI-TRAILER	60,000 Pounds 27,273 Kilos	600 Feet 183 Meters	7,000 Feet 2,134 Meters	7,000 Feet 2,134 Meters

Computer software can be used to model the effects of explosive devices on individual structures taking into account many different site and building-specific factors. These models can estimate building damage, glass damage, and potential casualties.

Buildings

Terrorist threats to buildings can be perpetrated from outside and from within. A chemical or biological agent could be sprayed from an aircraft over the building or a letter containing anthrax spores could be opened inside. A vehicle bomb could be exploded outside the front door or a package bomb could be smuggled inside. An incendiary device could ignite a fire inside or outside.

A building vulnerability assessment includes evaluation of construction features, life safety systems, utilities, space planning, targeted areas, as well as physical and operational security. The assessment begins with a building survey.

Building Survey

A proper vulnerability assessment requires review of building plans and specifications, a walking survey of the building, and interviews of people who operate and provide security for the building.

The building survey should begin with a review of building plans and specifications. The type of construction and exterior glass; gross space per floor; occupant load per floor; number and location of exits; locations and features of critical utility systems; and the location of targeted or hazardous operations should be noted. The document review should be followed by a walking survey.

The building survey should identify the type of ventilation system (e.g., natural ventilation, individual through-the-wall units, or a duct system with air-handling units). The locations of mechanical rooms that house air-handling units should also be noted.

Space Planning

Critical assets should be placed in the most protected part of the building; this includes both people and critical functions or systems. To determine the vulnerability of people, evaluate the locations of the offices, meeting or conference rooms, auditoriums, and day care centers.

- ❑ Are offices, meeting or conference rooms, auditoriums, and day care centers placed away from vulnerable sides of a building? Vulnerable sides are adjacent to public streets or in the middle of the "L" or "U."

- ❑ Are areas where visitors have unrestricted access located so they are easily monitored and controlled by security? These can include lobbies, package reception areas, loading docks, and public restrooms or cafeterias.

- ❑ Are security zones established around critical functions like computer rooms and executive offices?

- ❑ Are there any areas easily accessible to visitors where bombs or other devices could be easily hidden? These include alcoves, unsecured equipment rooms, and especially parking areas.

Building Construction

The type of construction and structural design can greatly determine how well the building will be able to withstand the blast of an explosive device. All structures are vulnerable to bomb blasts, but some are more vulnerable than others. The most vulnerable structural types are wood frame buildings, followed in descending order by ordinary construction (joisted masonry), reinforced concrete, and steel frame.[7] A solid foundation system and strong, load-bearing walls are better able to withstand a blast than weaker structural systems.

Structural engineers are best qualified to evaluate how well a building could withstand a blast. Computer models are also available to calculate the effects of varying size bombs on the structural integrity and the number of potential casualties.

Life Safety/Means of Egress

Adequate means of egress and other life-safety systems are critically important when a threat is recognized and prompt evacuation is required.

- ❑ Are there an adequate number of exits? A minimum of two exits are required for each floor, although additional may be required depending upon the size and layout of the building.
- ❑ Do the exits have sufficient capacity to handle the maximum occupant load?
- ❑ Do any of the components of the means of egress (doors, stairwells, etc.) impede egress from a building?
- ❑ Is the travel distance to reach an exit excessive?
- ❑ Are exit components (exit access, exit stairwells, and exit discharge) properly illuminated?
- ❑ Are emergency lighting and exits signs adequate to ensure that evacuees can locate an exit?
- ❑ Is there an effective evacuation plan that is practiced regularly? Do occupants know the location of primary and secondary exits as well as where to assemble after exiting the building?
- ❑ Are there at least two assembly areas in opposite directions in case one area is not safe to approach?

Evacuation of a large or multistory building could take considerable time. Persons who are incapable of self-preservation (e.g., young children, the elderly, sick or injured) and those with disabilities are especially at risk.

- ❑ Are there provisions to identify and evacuate persons needing special assistance?
- ❑ Are refuge areas available for those who cannot evacuate because of airborne hazards in the egress path, such as fire or smoke, or who are incapable of negotiating stairs?

Refuge areas can be constructed to provide protection from fire, smoke, airborne hazards, or blast effects.

Exterior Glass

Exterior glass is highly susceptible to the blast of an explosion, and flying glass is the most serious hazard for building occupants.

- ❑ Is the exterior glass designed to resist breakage? This could include tempered glass, polycarbonates, laminated, and film-backed glass or blast resistant secondary glazing inside exterior glazing.

- ❑ Are bomb blast net curtains installed on the inside of the exterior glass?

Vestibules

Vestibules, airlocks, and revolving doors help control the intake of cold or warm air and could help contain the intake of contaminants. In a high-rise building this is most important on a cold winter day because the building tends to act as a chimney to suck in outside air. Any contaminants collecting on the ground outside the building could be drawn into the building when doors are opened.

- ❑ Are vestibules provided to restrict the infiltration of outside air?

Ornamentation & Signs

- ❑ Are there any signs or ornamentation that can be easily dislodged by an explosion?

- ❑ Do any of the signs on buildings provide information that could be helpful to terrorists? This could include information about the assets or hazards on-site.

- ❑ Are hazardous materials placards located so they are not visible to potential terrorists but are visible to responding firefighters?

Building Utility & Control Systems

Many utility systems are necessary for occupancy of a building and to support business functions. These include electricity, water, heating (gas, steam, and/or fuel oil), and telecommunications. Other systems are critical for the protection of occupants during an emergency. Command centers, if provided, are critically important to control utility and emergency systems during an emergency.

The vulnerability assessment should evaluate the location, arrangement, and vulnerability of critical utility systems and any command center.

Command Center

In many buildings, command centers allow building engineers to monitor and control all building systems and to operate emergency systems as well. In some buildings, these centers are separate; in others, control panels are scattered throughout the building. On large campuses, there may be a central control center or multiple control rooms scattered about. In smaller buildings, there may be nothing but a utility room and a fire alarm control panel.

❑ Is the command center located in a secure location? Command centers should not be located adjacent to lobbies, loading docks, package reception areas, or in parking garages unless they are properly protected from bomb blasts.

❑ Does the command center have full control over critical building utility systems and emergency systems? These include:

- ventilation systems
- smoke control systems
- emergency voice communication or fire alarm systems

Full control over these emergency systems will allow operators to promptly shut down ventilation to shelter in place or exhaust airborne hazards. It will also allow the command center to promptly communicate with building occupants.

Critical Utility Systems

There are many different utility systems supporting a building. These include:

- ❑ Electrical feeds from the public utility grid
- ❑ Water supplies for domestic and fire protection service
- ❑ Natural gas lines
- ❑ Fuel oil storage tanks
- ❑ Steam lines
- ❑ Voice/data lines

Typically these utilities are located in first floor or basement areas exposed by parking, loading docks, or other areas where an explosion is likely to occur.

- ❑ Are there multiple connections for critical utilities (e.g., water mains, electrical feeds and switches, gas lines, and telecommunications links)?
- ❑ Are all of these utilities exposed to parking areas, loading docks, or package reception areas?
- ❑ If they are exposed, are they protected against blast effects?
- ❑ Are these utilities and any mechanical equipment rooms clearly marked to make it easier for an intruder to target them?
- ❑ Are the automatic transfer switch and electrical conduits to emergency generators exposed the same as the primary electrical power supply?
- ❑ Are fire protection pumps and water mains exposed or protected in blast resistant enclosures?
- ❑ Are any utilities subject to flooding if water mains rupture during an explosion?
- ❑ Are utility areas protected by higher security including physical barriers, locks, closed-circuit television, and intrusion alarm systems?

- ❑ Has a qualified fire protection engineer evaluated the adequacy of the building's fire protection systems including sprinklers, water supplies, fire alarm systems, and communications systems?

- ❑ Are all emergency systems inspected, tested, and maintained in accordance with applicable codes and standards?

HVAC Systems

Ventilation systems take in fresh outside air and mix it with air recirculated within a building. Any release of a chemical or biological agent outside can be taken into a building by the ventilation system; once the contaminant enters the building it can be spread throughout by the ventilation system. Ongoing government-sponsored research will help determine how agents move throughout a building and are deposited by the HVAC systems, as well as how buildings can be more efficiently decontaminated.[8]

The vulnerability assessment should evaluate the arrangement of the ventilation system, including controls, fresh air intakes, mechanical rooms, and system operation.

- ❑ Can the ventilation system be controlled quickly using the standard complement of personnel on duty during normal operating hours? This includes shutting down any fans that take in outside air or switching all fans to exhaust mode. In a large building there are many controls and switches, and they could be located in many different locations.

- ❑ Are the controls for individual zones properly marked?

- ❑ Are on-site personnel well trained to control the air handling system if a threat is detected?

- ❑ Is the ventilation system equipped with automatic dampers on fresh air intakes and exhaust fans to help stop the migration of chemical or biological agents into the building?

- ❑ Where are the locations of outside air intakes? Are they accessible to potential terrorists?

- ❑ Are the intakes located close to ground level, in close proximity to adjacent buildings, or near other vulnerable areas? Intakes located close to the ground are vulnerable not only to intentional release into the intake, but also from agents that settle to ground level from a release in an adjacent area.
- ❑ Is access to rooftop air intakes secure?
- ❑ Are mechanical rooms housing air-handling units partitioned from adjacent areas with solid walls and locked?
- ❑ If mechanical rooms are accessible from the outside of the building, are they properly secured with fencing or other physical barriers?

Target Areas

Areas of a building that receive visitors, packages, and motor vehicles are more vulnerable than other areas that are secured for employee and authorized visitor access only. The "target" areas include lobbies, mailroom or package reception points, indoor parking or garages, and any loading dock. These areas should be constructed and secured to provide additional protection.

Lobbies

Effective screening at building entrances can help detect containers that could hold chemical and biological agents or incendiary or explosive devices. Screening should be done at both the primary visitor entrance and any other entrances used by contractors.

- ❑ Is the lobby area separated from the remaining portion of the building by substantial walls?
- ❑ Are all visitors directed to a main reception point where their access to the property is cleared?
- ❑ Are all packages, bags, and other articles searched?
- ❑ Is screening equipment used to detect hazardous containers or packages?

- ❑ Have security guards and receptionists been properly trained on what to look for? (aerosol cans, pressurized containers, spray devices, liquid containers, bottled gas, and personal-sized containers of irritating agents)

- ❑ Are all visitors issued badges that clearly show what part of the building they are visiting, and are they escorted while on-site?

- ❑ Are unexpected visitors—especially contractors—carefully screened to ensure their visit is properly authorized and that packages they bring into the building are safe?

Mail Rooms & Package Delivery

Areas that receive packages not screened for hazardous materials have a higher exposure. The building survey should include a review of the construction of these areas to determine the potential impact of a bomb explosion or the release of a chemical or biological agent. The survey should also review the operation of these rooms to determine whether adequate procedures are in place and personnel know what to look for and what to do with suspicious packages.

- ❑ Are these areas separated from surrounding areas by full height, fire resistive, and/or blast-resistant walls?

- ❑ Are individual air-handling units provided for these areas, or are they served by the building's air-handling system? If these areas do not have separate air handling systems, an airborne hazard could be spread over a wider area of the building.

- ❑ Is the ventilation system for these areas arranged to create a slight negative pressure within the enclosure to prevent migration of contaminants? An airlock or vestibule may be required to maintain the negative pressure if frequent access to the protected enclosure is necessary.

- ❑ Is there a designated receptacle to contain any suspicious package or letter that is received?

- ❑ Have procedures been established for identification and proper handling of suspicious packages?

- ❑ Are the personnel within the mailroom periodically trained on suspicious package recognition and handling?

Indoor Parking & Loading Docks

Indoor parking, loading docks, and garages present a serious hazard from vehicle-concealed explosives. Restrictions on the number or kinds of vehicles can reduce the hazard, and security practices can provide deterrence.

- ❑ What vehicles are permitted to park in the building? Preferably, parking should be restricted to known passenger vehicles. This limits the size of the vehicle and the capacity of explosives that could be hidden in a vehicle.

- ❑ Is parking restricted to vehicles with a valid parking permit?

- ❑ Must drivers entering the garage show identification or prove they are the valid user of a parking permit?

- ❑ Are all vehicles inspected prior to entering the garage or loading dock to detect telltale signs of explosives?

- ❑ For loading docks, are all vehicles allowed to enter and then cleared, or is access only allowed for vehicles that are expected and have been precleared?

- ❑ Are vehicles weighed as they enter the garage to determine if weights exceed expected values?

Physical Security

As part of the vulnerability assessment, a security expert should review building security. The overall level of security should be determined by the threat assessment, and the value of the building, its occupants, and operations to the organization. A highly valued facility that is a highly desirable target warrants a high level of security. The security expert can identify building zones (e.g., lobbies, mail rooms, executive offices, computer facilities, etc.) that require higher levels of security.

Adequate security of the building as a whole is necessary to deter would-be terrorists and to detect suspicious activity or actual threats. Physical security includes both barriers and electronic systems. Since no barrier or electronic system can stop a determined terrorist with time and expertise, qualified security personnel also may be needed.

Physical Barriers

- ❑ Are all exterior doors and windows in good condition and locked?
- ❑ Are windows breakage-resistant?
- ❑ Are visitors limited to one entrance that is properly staffed to screen the visitors and examine packages?
- ❑ Are all other entrances locked or provided with electronic card access or a similar system?
- ❑ Is the access to passenger and service elevators restricted to authorized employees and escorted visitors?
- ❑ Are elevators programmed to restrict access to authorized floors only?

Electronic Security Systems

- ❑ Are all exterior doors and interior doors that enclose high security zones electronically monitored to detect intrusion?
- ❑ Are vulnerable exterior windows equipped with breaking glass detection?
- ❑ Are other potential points of entry, such as roof hatches and skylights, equipped with intrusion detection?
- ❑ Are all electronic devices monitored at a constantly attended location, on-site if a highly valued building?
- ❑ Are security personnel who monitor the electronic systems properly trained and qualified?

Operational Security

Operational security includes the policies, procedures, staff, and training to deter terrorists, to detect threats, and to respond promptly and effectively to a threat or emergency situation. In its simplest form, operational security is locking of doors and windows and vigilance on the part of all building occupants. However, for high value buildings that are highly attractive targets for terrorist, the scope of operational security can be significant.

Security Policies, Procedures, Staffing, & Training

The vulnerability assessment of any building should include a review of security systems in place. This includes physical barriers (e.g., doors and locks), intrusion detection systems (e.g., contact switches, sound and motion detectors), but also the security organization's policies, procedures, quality of staff, and training. Keep in mind that even the best combination of physical barriers and intrusion detection systems can be beaten—if a terrorist has sufficient time and expertise.

Physical security should be evaluated from the outside in and from top to bottom. Approach the building from all sides to identify vulnerable points of access. Assess the accessibility of rooftop hatches and the penetration of exterior windows and doors. Verify that sufficient intrusion detectors have been provided for all openings in the building exterior, and test existing detectors to be sure they work.

Review the security staff organization for an effective chain of command to address security issues and proper supervision of individual guards. Assess the competence of security staff, especially their knowledge of the facility, ability to recognize threats, and ability to respond to emergencies.

Since any package or vehicle that enters the building is a potential terrorist threat, policies for access to buildings should be reviewed to determine whether they address the current security threat to the organization. Vehicles—in particular large delivery vehicles that are not expected—are perhaps the greatest threat. Employee and visitor access policies, systems, equipment, and procedures should be reviewed to reduce the probability that a weapon can be brought into the building.

Computer Security

The vulnerability assessment should address computer security, particularly security against cyber attacks. Computer security should be properly managed within the organization and policies, procedures, practices, training, detection, and response should be a part of a program to maximize the security of information technology.

Management Policies & Procedures

Qualified persons with defined responsibilities should manage computer security. The vulnerability assessment should review documentation and audit practices to verify that the following are properly addressed, including:

- ❑ Identification and assessment of computer security risks.
- ❑ Development and implementation of computer security policies and procedures.
- ❑ Provision of training.
- ❑ Auditing to identify computer vulnerabilities and the effectiveness of existing controls.
- ❑ Development of recovery plans in the event critical information technology is lost or compromised.
- ❑ Implementation of procedures for resolution of identified vulnerabilities and reporting of computer security incidents.

Hardware & Software Controls

- ❑ Access controls should be required for all administrators and systems professionals.

- ❑ Standards for assignment, review, and revocation of passwords or revision of access privileges should be established.
- ❑ Software development and modifications should be carefully controlled to ensure that any changes are properly authorized, tested, and safe prior to implementation.
- ❑ Responsibilities should be segmented so one individual does not have the ability to develop, test, and implement software applications or modifications without independent approval.
- ❑ Hardware and/or software firewalls should be provided wherever there is connectivity outside the network.
- ❑ Detection of unauthorized use or access should be monitored to identify violations before they can cause serious problems.
- ❑ Installation of any hardware that could open a hole in firewalls should require special review and authorization.
- ❑ Operating system and application software should be updated as security vulnerabilities are identified.

Known Vulnerabilities

Domain Name Servers

The domain name service architecture should be evaluated to avoid creating a single point of failure that could result in an extended loss of connectivity. The Domain Name System (DNS) is a distributed catalog that allows users to access Internet resources by using familiar text strings like www.nationalunderwriter.com instead of difficult numeric addresses like 65.171.14.3. An organization establishing an online presence will generally specify two or more name servers that provide authoritative DNS information. If DNS becomes unavailable, access to common resources such as Web browsing, email, remote login capability, and other fundamental Internet services can be totally disrupted. In this sense, DNS can be a single point of failure presenting a risk of total loss of electronic connectivity for a company.[9]

Cyber Vulnerabilities

[The following information is adapted from the SANS/NIPC *Twenty Most Critical Internet Security Vulnerabilities,* and is used with permission of the SANS Institute.]

Cyber attacks by definition strike computer systems that are connected via local and wide area networks to computer networks outside of the building—including and especially the Internet.

The SANS Institute and the National Infrastructure Protection Center (NIPC) have published a document on the Internet called *The Twenty Most Critical Internet Security Vulnerabilities.*[10] This document can be used to identify and prioritize vulnerabilities that expose computer systems to cyber attack. The list of vulnerabilities is segmented into three categories: General Vulnerabilities, Windows Vulnerabilities, and Unix Vulnerabilities.

The majority of successful attacks on computer systems via the Internet can be traced to exploitation of security flaws on this list. For instance, the Code Red and Nimbda worms can be traced to exploitation of unpatched vulnerabilities.

This short list of software vulnerabilities accounts for the majority of successful attacks, simply because attackers are opportunistic—they take the easiest and most convenient route. They exploit the best-known flaws with the most effective and widely available attack tools. They count on organizations not fixing the problems, and they often attack indiscriminately, scanning the Internet for vulnerable systems.

The SANS and NIPC paper includes Internet addresses for software that can automate finding or diagnosing many of the Internet vulnerabilities.

Default Installs of Operating Systems and Applications

Most software, including operating systems and applications, comes with installation scripts or programs. The goal of these installation programs is to get the systems installed as quickly as possible, with the most useful functions enabled and the least amount of work by the system administrator. To accomplish this goal, the scripts typically install more components than most users need. The vendor philosophy is that it is better to enable functions that are not needed than to make the user install additional functions when they are needed. This approach, although convenient for the user, creates many of the most dangerous security vulnerabilities because users do not actively maintain and patch software components they don't use. Furthermore, many users fail to realize what is actually installed, leaving dangerous samples on a system simply because they do not know they are there.

Run a port scanner and a vulnerability scanner against any system that is to be connected to the Internet. Analyze the results under the principle that the systems should run the smallest number of services and software packages needed to perform the tasks required. Every extra program or service provides a tool for attackers—especially because most system administrators do not patch services or programs they are not actively using.

Accounts with No Password or a Weak Password

Most systems are configured to use passwords as the first, and only, line of defense. User IDs are fairly easy to acquire, and most companies have dial-up access that bypasses the firewall. Therefore, if an attacker can determine an account name and password, he or she can log on to the network. Easy to guess passwords and default passwords are a big problem; but an even bigger one is accounts with no passwords at all. In practice all accounts with weak passwords, default passwords, and no passwords should be removed from your system.

In addition, many systems have built-in or default accounts. These accounts usually have the same password across installations of the software. Attackers commonly look for these accounts because they are well known to the attacker community. Therefore, any default or built-in accounts also need to be identified and removed from the system.

To determine if you are vulnerable, you need to know what accounts are on your system. The following steps should be performed:

- ❑ Audit the accounts on your systems and create a master list. Do not forget to check passwords on systems like routers and Internet-connected digital printers, copiers, and printer controllers.

- ❑ Develop procedures for adding authorized accounts to the list and for removing accounts when they are no longer used.

- ❑ Validate the list on a regular basis to make sure no new accounts have been added and that unused accounts have been removed.

- ❑ Run a password-cracking tool against the accounts looking for weak or no passwords. (Make sure you have official written permission before employing a password-cracking tool.)

- ❑ Have rigid procedures for removing accounts when employees or contractors leave and when the accounts are no longer required.

Nonexistent or Incomplete Backups

When an incident occurs, recovery from the incident requires up-to-date backups and proven methods of restoring the data. Some organizations make daily backups but never verify that the backups are actually working. Others construct backup policies and procedures but do not create restoration policies and procedures. Such errors are often discovered after a hacker has entered systems and destroyed or otherwise ruined data.

A second problem involving backups is insufficient physical protection of the backup medium. The backups contain the same sensitive information that is residing on the server, and should be protected in the same manner.

An inventory of all critical systems must be completed. Then a risk analysis should be performed identifying what the risk and corresponding threat is for each critical system. The backup policies and procedures should clearly map to these key servers. Once these systems have been verified, the following should be validated:

- ❑ Are there backup procedures for those systems?
- ❑ Is the backup interval acceptable?
- ❑ Are those systems being backed up according to the procedures?
- ❑ Has the backup media been verified to make sure the data is being backed up accurately?
- ❑ Is the backup media properly protected in-house and with off-site storage?
- ❑ Are there copies of the operating system and any restoration utilities stored off-site (including necessary license keys)?
- ❑ Have restoration procedures been validated and tested?

Large Number of Open Ports

Both legitimate users and attackers connect to systems via open ports. The more ports that are open the more possible ways that someone can connect to your system. Therefore, it is important to keep the least number of necessary ports open on a system. All other ports must be closed.

The netstat command can be run locally to determine which ports are open, but the best way to have confidence in the scans is to run an external port scanner against your systems. This will give you a list of all ports that are actually listening. If the results of netstat differ from the port scanning results, you should investigate. Once the two lists agree, go through the list and validate why each port is open and what is running on each port. Any port that cannot be validated or justified should be closed. The final list should be recorded and used to audit the ports on a regular basis to make sure no extraneous ports appear.

Not Filtering Packets for Correct Incoming and Outgoing Addresses

Spoofing IP addresses is a common method used by attackers to hide their tracks when they attack a victim. For example, the very popular smurf attack uses a feature of routers to send a stream of packets to thousands of machines. Each packet contains a spoofed source address of a victim. The computers to which the spoofed packets are sent flood the victim's computer, often shutting down the computer or the network. Filtering traffic coming into your network (ingress filtering) and going out (egress filtering) can help provide a high level of protection. The filtering rules are as follows:

- ❑ Any packet coming into your network must not have a source address of your internal network.
- ❑ Any packet coming into your network must have a destination address of your internal network.
- ❑ Any packet leaving your network must have a source address of your internal network.
- ❑ Any packet leaving your network must not have a destination address of your internal network.
- ❑ Any packet coming into your network or leaving your network must not have a source or destination address of a private address or an address listed in RFC1918 reserved space. These include 10.x.x.x/8, 172.16.x.x/12 or 192.168.x.x/16 and the loopback network 127.0.0.0/8.
- ❑ Block any source routed packets or any packets with the IP options field set.
- ❑ Reserved, DHCP auto-configuration, and Multicast addresses should also be blocked.

Try to send a spoofed packet and see if your external firewall or router blocks it. Not only should your device block the traffic, but it should also produce a record in the log showing that the spoofed packets have been dropped. Note, however, that this opens up the door to a new attack—flooding the log file. Make sure your logging system can handle a heavy load; otherwise it could be vulnerable to a denial of service (DOS) attack. Programs like nmap can be used to send decoy packets or spoofed packets to test this type of filtering. Once filtering is set up, don't assume that it is working effectively. Test it often.

Nonexistent or Incomplete Logging

One of the maxims of security is, "Prevention is ideal, but detection is a must." As long as you allow traffic to flow between your network and the Internet, the opportunity for an attacker to sneak in and penetrate the network is there. New vulnerabilities are discovered every week, and there are very few ways to defend against an attacker using a new vulnerability. Without logs you have little chance of discovering what the attackers did. Without that knowledge, your organization must choose between completely reloading the operating system from original media and then hoping the data back-ups were good, or taking the risk that you are running a system that a hacker still controls.

You cannot detect an attack if you do not know what is occurring on your network. Logs provide the details of what is occurring, what systems are being attacked, and what systems have been compromised.

Logging must be done on a regular basis on all key systems, and logs should be archived and backed up. Most experts recommend sending all of your logs to a central log server that writes the data to a write once media, so that the attacker cannot overwrite the logs and avoid detection.

Review the system logs for each major system. If you do not have logs, or if they are not centrally stored and backed-up, you are vulnerable.

Vulnerable CGI Programs

Most Web servers, including Microsoft's IIS and Apache, support Common Gateway Interface (CGI) programs to provide interactivity in Web pages enabling functions such as data collection and verification. In fact, most Web servers are delivered (and installed) with sample CGI programs. Unfortunately, too many CGI programmers fail to consider that their programs provide a direct link from any user anywhere on the Internet directly to the operating system of the computer running the Web server. Vulnerable CGI programs present a particularly attractive target to intruders because they are relatively easy to locate and operate with the privileges and power of the Web server software itself. Intruders are known to have exploited vulnerable CGI programs to vandalize Web pages, steal credit card information, and set up back doors to enable future intrusions. When the Department of Justice Web site was vandalized, an in-depth assessment concluded that a CGI hole was the most probable avenue of compromise. Web server applications are similarly vulnerable to threats created by uneducated or careless programmers. As a general rule, sample programs should always be removed from production systems.

If you have any sample code on your Web server, you are vulnerable. If you have legitimate CGI programs, ensure you are running the latest version, and then run a vulnerability-scanning tool against your site. By simulating what an attacker would do, you will be prepared to protect your systems.

Other Considerations

Business Operations

The vulnerability assessment should address whatever materials, equipment, technology, methods, processes, and records or information the organization uses to achieve its mission, goals, and objectives. The operations of a professional services business will be different from a manufacturer. However, you can categorize the critical business functions and needs in the following categories:

- ❑ Critical staff
- ❑ Operating hours, days, and seasonal peaks
- ❑ Custom machinery, equipment, and technologies
- ❑ Critical utilities
- ❑ Critical supplies and suppliers of both goods and services

Every organization, regardless of whether it is for-profit or not-for-profit, large or small, manufacturing or service, requires these same categories of resources. Specific assessment questions include:

- ❑ What personnel are absolutely necessary? How many do you need? Can they work from remote locations?
- ❑ What hours, days, or time of year must your business be available to customers or constituents?
- ❑ What machinery, equipment, or technology is unique to your business and must be replaced in order to service your constituents or produce your product? For information technology, this includes valuable paper, records, drawings, computer applications, and data.
- ❑ What custom utilities are required? What unique equipment is needed to support a critical data processing facility, process control, or other specialized machinery and equipment?
- ❑ Review critical raw materials, subassemblies, supplies, and single or sole source suppliers. Are they subject to the same level of threat? Would they be lost in the event of the same threat? Have you identified and qualified any backup suppliers?
- ❑ Are similar facilities or operations within your organization available to support or replace your local operations?

Environment & Critical Infrastructure

All businesses rely upon critical infrastructure and lifelines to operate. Lifelines include utilities such as electricity, natural gas, steam, water, and sewer systems. Critical infrastructure also includes telecommunications and transportation. Open public streets, tunnels, bridges, and highways are needed for employees, customers, and visitors to reach places of business and to receive and deliver goods and services. Rail transportation moves heavy goods and raw materials; aircraft transport passengers around the country and the world.

A major terrorist attack could disable many of these lifelines and critical infrastructure. At the very least, the movement of people may be restricted or security may slow the movement of people and vehicles into an affected area or at border crossings. For example, security precautions at the U.S.-Canada border following the World Trade Center attacks in 2001 resulted in traffic restrictions in lower Manhattan and slowed delivery of goods to U.S. manufacturing plants dependent upon just-in-time delivery systems.

Business Impact Analysis

A business impact analysis (BIA) calculates the financial impact of a negative event. Since terrorist attacks can result in injuries, fatalities, property damage, business interruption, contingent business interruption, and even economic decline, all of these factors should be considered when calculating the business impact.

A fairly simplistic and traditional approach assumes the loss of an entire facility. Experts within the company build a flowchart of the company showing manufacturing, distribution, or other important functions. This analytical process requires identification of critical supplies, suppliers, facilities, and primary and dependent functions that enable the organization to satisfy its mission.

The loss of production or service capability is business interruption. If the company is fortunate to have duplicate facilities that are able to manufacture the product or duplicate the services that were lost, there is no business interruption (property damage is assumed).

Today, most companies have eliminated redundant operations and processes to increase efficiency and lower costs. With just-in-time delivery systems that bring materials to the production line within hours of their need, there often is little tolerance for any downtime. Therefore, any business interruption—within the company or a critical supplier—can have an immediate and significant impact. Electronic commerce exacerbates the problem. Technology allows managers to control facilities, product movement, and even manufacturing from afar. Loss of these critical systems can affect many locations that have not suffered property damage. In addition, loss of facilities or critical functions within the facility can significantly affect the organization.

Many terrorist incidents are not like traditional natural disasters. Terrorist acts can damage many buildings and injure or kill many people at the same time. The region surrounding an attack can be affected. Buildings may not be damaged, but airborne hazards could render an area uninhabitable, or authorities could restrict access to an area for an extended period. Of course, the fear and anxiety of employees affected by an attack could also lower productivity and increase manpower needs. Nightmare scenarios of a biological attack even hypothesize global implications after the release of an infectious disease.[11]

The data collection and analysis required for completing a BIA can be done through a combination of questionnaires, interviews, and physical surveys. Staff familiar with, or responsible for, the various business functions at each location should complete the questionnaires. Follow-up interviews can delve further into the supply chains for each location. Physical surveys can validate the important site business functions and identify any production bottlenecks. A blended approach that incorporates interviews, questionnaires, and surveys produces an outcome that is comprehensive, timely, and specific.

Scenarios

Multiple hypothetical scenarios are presented in Table 2. The scenario of a smallpox attack was developed as a training exercise. Others are based upon government training programs for emergency responders. The best training scenario is one based upon the operations of an organization known to all of the exercise participants. It allows everyone to visualize physical assets, understand the mission, and comprehend many of the critical functions of the organization. They are then better able to estimate the implications of various terrorist scenarios.

The goal of exercise scenarios is to train people to think about what could happen and how best to deal with what does happen. Opportunities to mitigate the exposure should be identified and pursued. These include enhanced property protection, operational and physical security, and changes in policies and procedures. Chapter 3 discusses mitigation, and later chapters discuss response and recovery.

Table 2. Hypothetical Terrorist Attack Scenarios

Threat	Scenario	Consequences	Business Impact
Bomb	Briefcase bomb explodes in lobby area of a mid-rise building housing corporate offices of a Fortune 100 company. Receptionist and several visitors awaiting entry to the building are killed. Heavy damage to the immediate area, but no apparent major structural damage. Smoke from a small fire spreads to several floors.	Building would be shut down for a few days to check for damage. Damage to lobby and adjacent areas would take weeks to repair. Many employees would feel grief and anxiety, and absenteeism may be a problem.	Business impact dependent upon whether any critical functions or people were lost. Absenteeism of some who are concerned about security may result in loss of productivity.
	A 10,000-lb. bomb concealed in a 14-foot box van is detonated at the loading dock of a 10-story building. The building houses corporate offices, R&D, data center, and manufacturing control for high technology manufacturer. Over 100 persons are killed and the building is destroyed. Heavy damage to all surrounding buildings.	Loss of entire building; fatalities to many key employees and injuries to many more. Loss of senior executives could have an immediate impact on investors who feel the company will die without proper direction. If data center is backed up and able to recover manufacturing control, may not be catastrophic. Loss of R&D staff and ongoing research (if not backed up) could cripple future product development.	There are many possible outcomes. Possibly the total collapse of the company. In other cases, the results could be more favorable.
Incendiary Device	Incendiary device ignited at trash compactor in the rear of the building causes a fire that extends into the building. Automatic fire sprinkler system operates as designed and damage is very limited.	Minimal consequences.	No business impact.
	Incendiary device ignites a fire within a concealed area in a one story distribution center. Sprinkler heads are obstructed by storage, and fire spreads into a concealed space and then the roof level. Heavy fire damages the distribution center. Terrorist claim of responsibility.	Heavy damage to the distribution center and total damage to inventory.	Business impact dependent upon whether there are any other distribution centers in the company that could handle customer orders. Since loss deemed a result of terrorism, there is limited insurance coverage, and this creates an immediate financial crisis.
Nuclear Weapon	Tactical nuclear weapon exploded in central city of major metropolitan area by international terrorists. Tens of thousands are killed and the radioactive plume spreads for miles downwind.	A nightmare scenario. Besides the loss of life, property damage, and business interruption, the consequences could include severe restrictions on travel, economic recession, or worse.	Any company in the area would face dire consequences and business impact. Loss of facilities and personnel, and probable direct business interruption and contingent business interruption.
"Dirty Bomb"	Truck bomb explodes three blocks away from a downtown office building; one hour later officials detect radioactivity in the area. Extent of radioactivity covers 10 square blocks.	Although there is no damage to the office building, the radioactivity in the area essentially shuts down the area for an extended period.	The business impact could be significant if the office building could not be inhabited. The business impact would depend on the quality of business recovery plans and the ability to shift personnel to other locations.

Threat	Scenario	Consequences	Business Impact
Biological Agent	Suspicious outbreak of "adult chicken pox" in Oklahoma City and two other states. Initially 12 people are diagnosed with suspected smallpox. Meanwhile tensions are high in the Middle East, but no one takes "credit" for attack. Limited supply of vaccine given to public health, emergency responders, and the military. On day six, 300 have died, and at least 2,000 are infected. Cases are suspected in 15 states, Canada, Mexico, and England. Oklahoma quarantined—no movement into or out of the state. Vigilantes are reported shooting at people trying to flee into Texas. Vaccine supplies are inadequate and the pace of distribution is slower than the spread of cases. Foreign countries ban all flights from U.S. Epidemic continues to spread across the globe…([11])	This was an exercise scenario conducted on June 22-23, 2001, titled "Dark Winter." The consequences of this scenario would be catastrophic to the U.S. in many compelling ways, and to individual businesses impacted by the loss or illness of many employees, restrictions in travel, and fear and anxiety. Since this "exercise," the U.S. government has ordered production of tens of millions of doses of smallpox vaccine to be deployed in a public health emergency. An outbreak of an infectious biological agent such as smallpox or plague is a nightmare scenario because of the potential to spread the disease over a wide area before preventive actions can be taken.	
	A letter containing Anthrax spores is received in the mailroom. The mailroom does not suspect the letter is a problem. Employee opens the letter on his desk and a cloud of "dust" envelops the area. The ventilation system picks up the "cloud" and spreads to undetermined parts of the building.	Several employees are treated for inhalation anthrax; many others are given prophylactic treatment. Building shuts down for cleaning, which takes two months.	Business impact is significant since the building can't be occupied and no one can reach important records and documents.
Chemical Agent	Sarin is spilled on the ground floor of a high-rise building that houses dozens of offices. The ventilation system picks up the chemical agent and spreads it through the lowest air-handling zone. Hundreds are exposed and several dozen are killed.	Building shutdown for cleanup; expected duration unknown. Loss of valuable staff; emotional impact is severe.	Business impact is significant if critical employees are lost, if the building cannot be occupied, and if business recovery plans are not able to overcome the loss of critical functions.
	Chemical agent is dispersed outside the federal building, which is adjacent to a large office building, shopping mall, and transportation center.	If meteorological conditions are favorable and the plume quickly disperses or moves away from people and important buildings, the impact could be minor. If not, the impact could be significant for people exposed to high concentrations.	Business impact would depend mostly on the exposure of people and buildings.
Cyber Attack	A computer worm spreads throughout a corporate network. The extent of the infection is difficult to determine.	The worm destroys numerous computer files and shuts down the corporate network. The server must be cleansed, and all clients connected to the server must be cleansed before the network is restored.	If backup tapes are not current or logs are not able to confirm when the worm attacked, it may be difficult to determine which backup tapes should be used for the restore. The further back in time needed to ensure the backup tapes are not infected, the more lost data.
	A hacker gains access to a Web server for an online retailer.	The Web server is shutdown after detection, but the loss or corruption of data is not easily determined. Sales are lost while the system is down.	The immediate impact is the loss of income from lost sales while the server is down. However, the real loss could be much worse if customer billing records are corrupted or data is stolen and distributed to others.

People

When evaluating critical personnel, consider the loss or unavailability of experience and technical expertise. Think about the potential loss of executive management; scientists and engineers key to research and development of products and services; successful marketing and sales staff; proven managers; and others who support information technology and other critical functions.

Loss of Buildings, Equipment, & Critical Functions

The business impact analysis should address the impact on the organization both from the loss of a facility or from part of a facility's capabilities. The impact of the loss of multiple facilities—if they are located in the same region—should also be addressed. Consider the loss of critical personnel, records, drawings, data and information, customer service capabilities, manufacturing processes, information systems, and administrative/support functions. Don't forget the loss of critical vendors, suppliers, and contractors that could impact operations if critical supplies or critical services are not received on time.

Lifelines

Lifelines are critical to supporting a building. Transportation systems, water, electricity, gas, steam, telecommunications, sewerage, or wastewater treatment, are many of the lifelines that businesses can't operate without. Consider scenarios where critical lifelines are unavailable or unreliable, and identify the impact of their loss. You may not experience a physical loss at your facility, but off-site incidents may cause you to lose critical utilities.

Restricted Access or Movement

Large-scale terrorist attacks such as the attack on the World Trade Center resulted in government restrictions on the movement of people and automobiles, inspections of trucks, and the closing of many roadways for an extended period. Authorities temporarily seized buildings around the World Trade Center, and occupants were prohibited from entering to restore operations. These disruptions should be considered in the business impact analysis—especially for time critical functions.

Contingent Business Interruption

Identification of the supply chain will highlight those companies that are single or sole source suppliers. They may have been selected for quality, price, or other considerations, but they present contingent business interruption exposures if they are unable to supply their goods or services. Identification of the supply chain will also enable you to make arrangements for backups.

Interdependencies

Many companies are like large puzzles with many interconnected pieces. Another analogy is a chain with multiple links. If one link is broken, the entire chain is broken. If one production facility goes down, it may not be able to provide materials used by another facility. These interdependencies must be identified to calculate the overall business interruption loss potential.

End Notes

[1]U.S. Department of Justice, Office of Justice Programs, Office for State and Local Domestic Preparedness Support, Assessment and Strategy Development Tool Kit, NCJ181200, 1999.

[2] United States Postal Inspection Service, *Security Plan for Suspected Letter and Parcel Bombs*, Internet http://www.usps.gov/websites/depart/inspect/

[3] Agency for Toxic Substances and Disease Registry (ATSDR), Department of Health and Human Services, "Industrial Chemicals and Terrorism: Human Health Threat Analysis," 1999.

[4] Dr. Nicholas Berry, *Keeping Nuclear Power Plants Safe from Terrorists,* Center for Defense Information, October 1, 2001.

[5] U.S. Bureau of Alcohol Tobacco and Firearms, *Vehicle Bomb Explosion Hazards and Evacuation Distance Tables,* www.atf.treas.gov/pub/fire-explo_pub/i54001.htm.

[6] Ronald J. Massa, "Vulnerability of Buildings to Blast Damage and Blast-Induced Fire Damage," *The World Trade Center Bombing Report and Analysis,* United States Fire Administration Technical Report Series, Report 076.

[7] Ronald J. Massa, "Vulnerability of Buildings to Blast Damage and Blast-Induced Fire Damage," *The World Trade Center Bombing Report and Analysis",* United States Fire Administration Technical Report Series, Report 076.

[8] A research team from the Department of Energy's Sandia National Laboratories has developed modeling and simulation tools for assessing the threat and vulnerability of buildings to chemical and biological attacks. This includes looking at how agents move and are deposited inside a building, developing and assessing mitigation strategies, guiding the use of detection methods, and examining the effectiveness of cleanup and decontamination efforts.

[9] National Infrastructure Protection Center, "Highlights," Issue 11-01, December 7, 2001.

[10] SANS Institute, *The Twenty Most Critical Internet Security Vulnerabilities (Updated) The Experts' Consensus,* Version 2.501, Internet [http://www.sans.org/top20.htm], November 15, 2001.

[11] This scenario was taken from the "Dark Winter" exercise conducted June 22-23, 2001, at Andrews Air Force Base, Washington, D.C. It was produced by the Center for Strategic and International Studies, Washington, D.C.; Johns Hopkins Center for Civilian Biodefense Studies; and ANSER Institute for Homeland Security .

Chapter 3

Mitigation Strategies

Site Selection & Building Design

The security of a building or group of buildings should begin with the selection of the site, followed by design of the building, then space planning, and finally operation of the completed building. Site selection can be one of the most important decisions followed by placement of the building on the site. The location of hazardous or target areas within a building should also be carefully considered; provision of additional protection may be warranted. Critical utility systems should be properly located to minimize vulnerability and to allow for prompt and effective control in an emergency. Life safety should be enhanced with an effective communication system and adequate means of egress.

Site Selection

If security were a primary factor in site selection, access control would be a primary consideration. A site within the center of a major city poses more risk than a building located in a suburban campus. The distance between a building and adjacent public streets, parking areas, and other public space is limited in the urban center, and it is very costly to acquire additional land for a buffer zone. Government Services Administration standards call for a one-hundred-foot setback from the perimeter.

All buildings on-site should be placed out of alignment with public streets and high-speed vehicle approaches. Angling of a building so exterior walls are not parallel to neighboring streets can help deflect some of the blast wave of an explosive device. Buildings with "U" or "L" shape can be more vulnerable if the blast occurs within the "U" or "L."

Surrounding buildings and their occupants should be surveyed to identify attractive targets for terrorism as described in Chapter 2.

Multitenanted Buildings

Tenants are faced with several challenges. First, tenants don't have authority to exclude other tenants that may be targeted by terrorists. A single tenant targeted by terrorists increases the threat to all building occupants. In addition, tenants probably have little or no control over building security, mitigation strategies, or emergency procedures. Tenants, however, are able to lease space in buildings that are not visible targets and to move out of buildings where the perceived threat is too high or the protection is unacceptable.

Building Design & Space Planning

There are many methods for designing a building to withstand potential terrorist acts. Qualified engineers with knowledge of the hazards and effects of bomb detonations and airborne hazards such as chemical and biological weapons can provide design input. A building structural frame can be designed to better withstand blast effects, and the exterior of the building can be designed to contain flying glass hazards. Blast-resistant, reinforced concrete walls can be constructed to protect critical areas or systems. Utility systems can be located in areas where they are less vulnerable and they can be better protected; redundancies can help ensure systems are available in case of a limited incident. Hazardous operations can be located and protected to contain the spread of airborne hazards or blast effects. Space planning can reduce exposure by placing people and critical business functions out of harms way.

Space Planning

Executive offices; meeting, conference, or public-assembly type occupancies; on-site day care centers; or other areas designed to house large numbers of or critically important people should be placed in the most highly protected portion of a building. These spaces should also be situated to facilitate effective security controls. Such spaces can be located on the inside of buildings, away from exterior glass facing public streets or parking areas. Conversely, corridors, storage, or other sparsely populated areas could be located on the outside or in more vulnerable areas of a building.

Security zones should be established for each category of risk. Public areas such as lobbies and waiting areas, package delivery, and loading docks are more vulnerable and require higher levels of security and protection. Critical equipment areas, including computer rooms and critical building utility systems, should be situated behind substantial barriers and secured.

Space planning should also minimize potential hiding places for bombs and incendiary devices. This could include alcoves, equipment rooms, and parking areas. These areas should be secured to prevent unauthorized entry.

Package & Visitor Reception Areas

Some spaces within a building are more susceptible than others. These include:

- Lobby
- Mail room
- Receiving, truck, or loading dock

If people, mail, or equipment and supplies enter the building prior to security screening, hazardous packages may enter as well. These spaces should be isolated from the remainder of the building to prevent the spread of hazardous materials. Isolation can be accomplished by providing:

- Separate air-handling units for each area.
- Exhaust fan(s) to create a slight negative pressure within the enclosure to prevent migration of contaminants. An airlock or vestibule may be required to maintain the negative pressure to allow frequent access to the enclosure.
- Full height, fire resistive, and/or blast-resistant walls around each area.

Mail Center

In addition to physical improvements, a container should be provided within the mail center to hold suspicious packages. The container should allow easy placement of the package inside and offer visibility without further handling. A Plexiglas type box with lid is an ideal container. A more substantial container should be used for suspected bombs.

Personnel who handle incoming mail may be provided with impermeable gloves such as nitrile or vinyl. At a minimum, employees who have cuts or abrasions on their hands should wear gloves while these injuries heal. Masks or respirators are not recommended for low risk facilities.

In addition to the physical arrangement and personal protective equipment, operational procedures should be implemented as follows.

- ❑ Nonessential personnel (e.g., contractors, visitors, etc.) should be restricted from mailrooms where airborne hazards may exist.

- ❑ Practices that generate dust, such as dry sweeping, dusting, and use of compressed air to clean machinery, should be avoided. Areas should be vacuumed with an industrial vacuum cleaner equipped with a high-efficiency particulate air (HEPA) filter. Conventional home or industrial vacuums should not be used since these vacuums may further disperse powdered agents.

- ❑ Employees should be instructed to wash their hands regularly with soap and water—whenever gloves are removed, before eating, and at the end of a shift.

- ❑ Mail center staff should be trained on:

  - Modes of anthrax transmission
  - Signs and symptoms of anthrax infection
  - Emergency procedures to deal with possible contamination
  - Protective clothing to minimize skin exposure
  - Care for abrasions that might provide an infection route

Horizontal Exits & Areas of Refuge

Full evacuation of complex and multistory buildings takes time—possibly an hour or more. Evacuation of persons who are incapable of self-preservation (e.g., young children, elderly, sick, or injured), and those with disabilities is extremely difficult. However, building design can provide two methods for safeguarding these persons.

The first concept is the design of horizontal exits. Horizontal exits pass from one fire area of a building to another. The *Life*

Safety Code® defines a horizontal exit as "A way of passage from one building to an area of refuge in another building on approximately the same level, or a way of passage through or around a fire barrier to an area of refuge on approximately the same level in the same building that affords safety from fire and smoke originating from the area of incidence and areas communicating therewith."[1] Horizontal exits eliminate the need to use exit stairs to evacuate a fire area; they also provide protection against fire spread and smoke.

Refuge areas can provide temporary protection for occupants who are unable to easily evacuate. Refuge areas can be constructed to provide protection from fire, smoke, airborne hazards, or blast effects.

Vestibules

Vestibules, airlocks, and revolving doors help control the intake of cold or warm air and could help contain the intake of contaminants. In a high-rise building this is most important on a cold winter day because the building tends to act as a chimney sucking in outside air. Any contaminants collecting on the ground outside the building could be drawn into the building when doors are opened.

Exterior Glass

Exterior glass is highly susceptible to impact from the blast of an explosion, and flying glass is the most serious hazard for building occupants. There are three methods for improving protection of exterior glass.[2] They include applying a transparent polyester anti-shatter film in conjunction with bomb blast net curtains; replacement with blast-resistant (e.g., laminated) glass; or installing blast-resistant secondary glazing inside exterior glazing. Window assemblies can be replaced with exterior glass, glazing, and net curtains that can contain glass fragments. Tempered glass, polycarbonates, laminated, and film-backed glass are currently in use in coastal areas to withstand hurricane-force winds.

Ornamentation & Signs

Signs and ornamentation that can be dislodged by an explosion should be removed or limited to reduce the hazard from flying debris.

Signs on buildings that house operations that are highly desirable targets for terrorist attack should not provide any information about the assets or hazards. Hazardous materials placards should be located so they are visible to responding firefighters but not easily visible to potential terrorists.

Critical Building Systems

Many utility systems must be operational to allow occupancy of a building and to support business functions. These include electricity, water, heating (gas, steam, and/or fuel oil), and telecommunications. Systems critical for protection of occupants in an emergency include:

- ❑ Ventilation systems
- ❑ Smoke control systems
- ❑ Emergency power supplies (generators and automatic transfer switches)
- ❑ Emergency voice communication or fire alarm systems
- ❑ Fire protection systems
- ❑ Means of egress (stairwells, exit doors, corridors, exit discharges, etc.)

Electrical feeds from the public utility grid, water supplies for domestic and fire protection service, natural gas lines, fuel oil storage tanks, steam lines, and voice/data lines typically enter a building at one or just a few points in the basement or first floor. The locations of these utility entrances typically are obvious; many pipes, cable trays, and mechanical rooms are clearly marked. A relatively small bomb explosion in the basement of a building adjacent to these critical utilities could damage them all.

Redundant feeds or connections should be provided for critical systems. They should be located in remote areas of the building not subject to the same threat. This could include redundant connections for telecommunications and electrical power supplies (e.g., primary feeds, switchgear, motor control centers, etc.)

The automatic transfer switch (ATS) that transfers power from the primary electrical supply to the emergency generator should be protected to the same level as the generator. Conduits should be protected against thermal and blast damage. The location of the ATS and generator should also consider the possibility of flooding caused by rupture of water supply lines and fire protection systems.

The locations of these critical systems can be concealed so their location or functions are not obvious to potential attackers. Additional security, such as physical barriers, controlled access, and surveillance (electronic and/or physical) can also be employed to better protect them.

Emergency systems should be closely evaluated to ensure they are properly designed for the occupancy. Ventilation systems must be arranged for quick shutdown or for full exhaust. Smoke control systems must be properly arranged to exhaust smoke or airborne hazards.

Emergency power supplies (generators) should be tested to ensure they have sufficient capacity to power emergency lighting—critical to illuminate exit travel paths, emergency voice communication systems, and smoke exhaust systems. The location of the automatic transfer switch should also be determined, and it should be offered the same protection as other critical utilities.

Verify that the emergency voice communication or fire alarm systems are properly designed and in reliable condition. The vulnerability of the "command center" that houses these systems should be reviewed. Often it is located adjacent to an unsecured lobby that is vulnerable to blast effects or would have to be evacuated if the lobby were contaminated with an airborne hazard. If located in the basement, is the command center vulnerable to a vehicle bomb? Scripts and procedures for operators should be reviewed to ensure clear messages are communicated to building occupants in case of an emergency. Scripts should be prepared for evacuation, sheltering in place, and other actions for various threats.

Means of egress (corridors, exit signs, exit doors, stairwells, exit discharges, etc.) should be inspected to ensure they are clear and unobstructed. It is critical that all occupants can find the primary and alternate exits and move unimpeded along corridors and stairwells to the exit discharge.

Physical Security

Physical security includes protection of the site and all buildings on it. Of primary concern is motor vehicle access to the site, so vehicle travel patterns and parking should be addressed. The security of the perimeter can also prevent access to intruders. Building security includes protection of the exterior to prevent unauthorized access; controlled access within the building; and protection of ventilation systems that could spread airborne hazards.

Site Security

Motor Vehicle Approach & Access

Circulation

Vehicle bombs are an important concern when establishing traffic patterns around a building. The goal is to restrict motor vehicles to roadways and driveways that are away from buildings or their vulnerable sides. In an urban setting public officials must be asked to adjust traffic patterns or restrict motor vehicle access.

The speed limit on adjacent and intersecting streets should be limited to thirty miles per hour through signs, speed bumps, changes in pavement, narrowing of lanes, or other means. Changes in traffic direction or flow can also restrict movement around a building.

High curbs, median strips, bollards, planters, or a combination of all four can be used to block vehicles from driving off the roadway toward a building. Physical barriers should have not more than four feet of space between them to block vehicle passage. Breaks in the curb, median, or barriers should only be provided at controlled entry points.

Site Access

Vehicle access to a site can be restricted to one or two points—one for service vehicles and the other for customers and visitors. These access points should be located away from busier streets or side streets where vehicles are traveling at high speeds.

Vehicle entrances located opposite intersecting streets, alleys, or curb cuts are vulnerable to high-speed approaches. Medians, bollards, or other barriers can be installed to prevent a vehicle traveling at a high rate of speed from breaking through. Barriers can include guardrails, concrete "Jersey barriers," large earth-filled planters, excavations, or ditches.

Vehicle entrances can be supervised with guard stations to screen vehicles and allow access to authorized vehicles only. Appropriate barriers can be placed at the security checkpoint and opened when the vehicle has been cleared. The checkpoint should be properly lighted for clear visibility of approaching vehicles at night.

Parking

On-site parking lots should be situated away from vulnerable buildings or vulnerable sides of buildings. Parking should be screened for employees or authorized visitors only, and adequate lighting should be provided to monitor parking lots. Surveillance cameras (with time-lapse recording) can be provided to allow continuous monitoring, especially in remote areas.

Parking should be prohibited along streets, especially if the street is within one-hundred feet of a building.

Perimeter

Ideally, the perimeter of the site should be as far away from important or vulnerable buildings as possible to reduce the damage potential from vehicle bombs. The perimeter should be configured so that security personnel can monitor it and identify potential threats.

Physical barriers, including fencing and walls, can enhance security along perimeter walls. Walls should be at least nine feet high to prevent a person from easily climbing over them. Proper perimeter illumination will help security staff detect potential intruders. Trees should not be placed in areas that would allow intruders to scale them to climb over a fence or wall, and shrubs should not provide hiding places for people or packages.

Lighting

Lighting is essential to deter would-be intruders, allow for proper observation, enable effective inspection of approaching and arriving vehicles, and detect other potential problems. It provides safety and enhances the feeling of security for personnel moving around the facility. Lighting for high-security areas, guard posts, and points of access should be connected to an emergency power supply.

Building Security

Building security is best provided by a combination of qualified personnel, physical barriers, and intrusion detection and alarm systems. Physical barriers such as walls, locks, and electronic access controls help keep people out of unauthorized areas. Intrusion detection systems— including surveillance cameras (closed-circuit television with or without recording capability)—and detectors sound an alarm when an intrusion is detected. However, a sufficient number of qualified security personnel is critical because any physical security measures can be compromised if an experienced terrorist has enough time.

A qualified security professional should review the proposed layout of a building, identify the zones that require higher levels of security, and then specify the type and location of barriers and electronic security equipment. The assessment begins outside, then moves inside, from top to bottom. All areas should be assessed and assigned a priority level (from high to low).

This security survey should evaluate the vulnerability of entrances and exits to and from the building. Particular attention should be paid to trees, ledges, balconies, or storage structures that could be used to scale fences or other perimeter barriers to access windows or the roof top. Roof top hatches and skylights should be inspected along with exterior doors and windows.

Electronic Security Systems

Electronic systems can be employed to detect unauthorized entry. Some systems, like closed-circuit television, require constant monitoring to detect suspicious activity, while others sound an alarm when the detector has been activated. The threat assessment can help determine the level of security protection desired and the types of electronic devices that are needed.

Electronic systems should be placed around the building perimeter, including doors and windows. They can also be provided for vulnerable roof hatches, skylights, and other potential points of entry. Systems can also be employed to restrict access within a building.

All electronic systems should be monitored at a constantly attended location such as a security command center or remote location such as a private alarm company. Electronic systems should utilize equipment that has been tested and approved or certified by a nationally recognized testing laboratory and installed in accordance with applicable standards. Certified installation helps to assure that the system meets recognized standards.

Security or facilities staff should be familiar with the proper operation of security systems so they are properly armed and not compromised. Periodic inspections should verify operational readiness.

Doors, Stairs, Elevators, & Windows

Entrances & Doors

The greater the number of entrances or doors, the more difficult it is to control access to a building. Therefore, access points should be restricted. Visitors and package delivery should be controlled (see "Visitor & Package Screening" later in this chapter).

Elevators & Stairwells

Access to floors can be restricted by controlling access to elevators and stairwells. Visitors should be screened at a reception point prior to accessing an elevator bank. This is especially critical for package delivery personnel and contractors who may enter via a service entrance and then use a service elevator or stairwell that can reach all floors in a building. Lock stairwell doors (on the stairwell side) to prevent unauthorized entry to floors.

Windows

Exterior windows that can be opened should be locked to prevent entry. Breakage resistant windows should be considered for high security areas.

Ventilation Systems

Control of ventilation systems could be critically important to prevent the intake of hazardous materials from an external attack or to exhaust chemical or biological agents that may have been dispersed inside. In large buildings, these systems can be extensive with many air-handling units and controls located in many different parts of the building. Controlling systems from a secured, central location would allow prompt supervision of the system in case of an attack. Arrangement of ventilation systems for full exhaust, either for the entire building or for individual zones, can help to purge contaminants that may enter the building.

Fresh Air Intakes

Fresh air intakes should be located well above grade away from potential attackers. In addition, elevated air intakes provide passive protection against chemical or biological agents that naturally collect at ground level.

Installation of automatic dampers on the fresh air intakes and on exhaust fans can help prevent the migration of chemical or biological agents into a building.

Security

Air intakes located at or near grade should be protected against malicious acts. This can include fencing, surveillance cameras with time-lapse recording capability, and motion detectors to detect unauthorized access and alert security personnel.

Mechanical, fan, and air-handling unit rooms should be secured with solid walls and locked, alarmed doors at all times to prevent the introduction of chemical or biological agents into the air-handling system.

Operational Security

The simplest form of operational security is to keep doors closed and locked to prevent unauthorized access. If a terrorist can't access a building, it is impossible to place an explosive or other device. This includes storage lockers, manholes, and other areas where a device could be easily hidden. Hiding areas that are easy to access and in close proximity to targets (e.g., people, structural members, critical utilities and systems, etc.) should be secured.

Maintenance hatches that require periodic access and can't be locked can still be equipped with a seal that, when broken, indicates unauthorized access.

Good housekeeping both inside and outside reduces the opportunity for placement of an explosive or dispersal device. Trash receptacles have been used to hide explosive devices, so they should be located away from critical areas.

Exterior furniture should not provide a hiding place. Shrubbery should not obscure visibility.

Visitor & Package Screening

Effective screening at building entrances can help detect containers that could contain chemical and biological agents as well as incendiary or explosive devices. Access control requires physical barriers to direct all entrants to a security checkpoint, manual package searches, or screening equipment (e.g., x-ray or metal detectors) to detect any hazardous containers or packages. Screening must be done by qualified personnel who operate within legal requirements.

Package Screening

The following containers should be investigated and prohibited in the building if deemed a security threat:

- ❑ Aerosol cans or other pressurized containers
- ❑ Manual or electric spray devices
- ❑ Containers of liquids or powders
- ❑ Bottled gases for repair or maintenance of the building
- ❑ Personal protective dispensers of irritating agents (e.g., pepper sprays, mace, etc.)

Examine the label to ensure it matches the contents. Missing or altered labels could be a legitimate reason to exclude the container from the building.

Visitor & Employee Access

All visitors should be directed to a main entrance, and all side or rear entrances should be locked and provided with keyed or electronic access to authorized persons. Electronic surveillance of rear and side doors can help increase security. All personnel should be required to check in at a staffed checkpoint during off-hours.

Visitors should be required to wear badges after entering, and they should be escorted in secure areas at all times. Employees should wear photo identification badges that are visible, so visitors without a badge are easily identified.

Contractors should be screened to verify they are authorized to enter the building, and they should be escorted while on-site. Unknown contractors who arrive on-site unexpectedly should be scrutinized. Toolboxes, equipment, and supplies should be carefully examined to detect items that could pose a threat. Contractor badges should define the area they are authorized to visit.

Electronic card access systems provide multiple benefits. First, they can be programmed to allow access to selected areas only during specified hours. In addition, electronic systems can provide a record of personnel who have entered a building or area. If employees are also required to electronically “sign out” of a building or area, and the electronic records of their entrance and exit are available when evacuation is ordered, it may be easier to account for personnel. There are many potential flaws in this system, so it should only be considered a tool when accounting for personnel evacuated from a building.

For example, the electronic access control system at the World Trade Center included electronic cards for all employees and visitors to the site. However, the electronic records were not backed up off-site in real time, so there was no available record of who was in the building at the time of the disaster.

Motor Vehicle Access to Garages, Inside Parking, or Loading Docks

Below grade parking, loading docks, and garages present a hazard to the building from vehicle concealed explosives. Restrictions on the number or kinds of vehicles can reduce the hazard, and security practices can provide deterrence.

- ❑ Restrict parking to passenger vehicles only. This limits the size of the vehicle and the capacity of explosives that could be hidden inside a vehicle.
- ❑ Restrict parking to vehicles with valid parking permits.

- ❑ Inspect all vehicles entering a garage to detect telltale signs of explosives and to match driver with parking permit.

- ❑ Screen all vehicles that enter loading docks and allow only vehicles that have been cleared.

- ❑ Weigh vehicles entering a garage. Vehicles with weights exceeding expected values can be prohibited or further inspected to verify the contents are not hazardous.

- ❑ Spot check all vehicles entering the garage.

Security Policies & Procedures

Security Policy

A security policy should be developed for and signed by a senior manager to define the levels of security within the facility. The policy should define any controversial access control procedures (e.g., restrictions on personnel or vehicle access) and the organizational structure, staffing, equipment, and training required to maintain the desired security level. High valued or highly vulnerable properties should have high security, and the policy statement should "codify" the authority of the security organization and provide the necessary support to maintain the required level of security.

Threat levels and the corresponding response procedures should be written into the policy. For example, authority to evacuate a building when a specified threat level is reached should be included. This will ensure prompt action during an emergency.

The use of firearms by security personnel should be defined in the policy. This includes the carrying or use of firearms by off-duty law enforcement officers working part-time for the company.

Threat Response Procedures

Threat response procedures are covered in Chapter 5.

Security Organization & Staffing

Organization

A head of security should be named and vested with the authority to hire, organize, and supervise staff. The head of security should be responsible for training and equipping security personnel.

The security organization should develop a close working relationship with local law enforcement personnel. This may enable the security chief to obtain information about potential but unannounced threats. Procedures and means to promptly receive terrorist threat information should be developed.

Staffing

The staffing of the security organization should be based upon the security review, facility size, and functions.

Computer & Cyber Security

Computer Security Management

Cyberterrorism is an attack from outside the organization through electronic connectivity with the outside world. To combat cyber attacks, every organization needs a computer security program to safeguard information technology. A qualified person should be assigned this responsibility, which includes:

- ❑ Identification and assessment of computer security risks
- ❑ Development and implementation of computer security policies and procedures
- ❑ Provision of training

- ❑ Auditing to identify computer vulnerabilities and the effectiveness of existing controls
- ❑ Development of recovery plans in the event critical information technology is lost or compromised
- ❑ Implementation of procedures for resolution of identified vulnerabilities and reporting of computer security incidents

Hardware & Software Controls

Access Controls

Access controls should be implemented to limit or detect unauthorized or inappropriate access to computer resources. Computer facilities should be safeguarded through physical barriers (walls, doors, and locks), and intrusion alarms. Computer applications and data should be protected with passwords, digital certificates, and other means commensurate with the risk to the organization and the user's authority and need to access the resource. High-level "administrator" privileges with unrestricted access to all resources should be carefully assigned.

Standards should be assigned for legal passwords; passwords should expire periodically; access privileges should be reviewed and/or revised when a user changes positions within the organization; and access should be promptly terminated when a user leaves the organization.

Software Development

Software development and modifications should be carefully controlled to ensure that changes are properly authorized. New and modified software should be properly tested before implementation.

Duties should be segmented so one individual does not have the ability to develop, test, and implement software applications or modifications. Independent approval should be required at defined stages after validation from appropriate reviewers.

Firewalls

Computer networks should be protected from attack by hardware and software firewalls. Even a single computer connected to the Internet is vulnerable to attack.

Detection

Systems personnel should continuously monitor systems and network performance to identify suspicious activity. Log files produced by firewall systems or Web servers should be reviewed frequently to identify possible attacks. System administrators should respond promptly to users reporting a potential problem.

Computer system files and directories should be inspected periodically, and unauthorized hardware such as modems should be investigated to determine if it has opened a hole in the firewall.

Security alerts that are available automatically via the mailing lists of multiple computer security organizations (see Appendix A for listings) should be reviewed to learn about recent vulnerabilities and software patches.

An intrusion detection system is needed to monitor systems activity and to alert systems administrators of potential problems.

Known Security Issues

The following are security-related issues identified by the Systems Administration, Networking and Security Institute (SANS) in conjunction with the National Infrastructure Protection Center, which is part of the FBI. Permission to use the information was granted by the SANS Institute. Additional details are available at the Institute's Web site, http://www.sans.org/top20.htm.

Domain Name Servers

Any organization that is dependent upon Internet access needs to review its Domain Name Server (DNS) architecture to ensure its reliability. Pay particular attention to adequate redundancy and physical dispersion of the organization's name servers to avoid a single point of failure. Both of these issues can be resolved in a variety of ways, including dispersing name servers across geographic locations, arranging for mutual backup DNS service with another company, or contracting with a third party to provide additional name servers.[3]

Internet Security Vulnerabilities

Default Installs of Operating Systems and Applications

Remove unnecessary software, turn off unneeded services, and close extraneous ports. This can be a tedious and time-consuming task. For this reason, many large organizations have developed standard installation guidelines for all operating systems and applications. These guidelines include installation of only the minimal features needed for the system to function effectively.

Accounts with No Passwords or Weak Passwords

Give all accounts with no password a new password or remove the account. When users are asked to change and strengthen their passwords, they often pick another one that is easy to guess. User passwords should also be validated when they change their passwords. Computer programs are available to reject any password change that does not meet security policy.

Nonexistent or Incomplete Backups

Backups must be made daily. The minimum requirement in most organizations is to perform a full backup weekly and incremental backups every day. The backup media should be verified at least once a month with a restore to a test server to see that the data is actually being backed up accurately. This is the minimum requirement. Some companies perform full backups every day or several times a day. The ultimate backup solution is a fully redundant network with fail-safe capability—a solution required for critical real-time financial and e-commerce systems.

Large Number of Open Ports

Identify the minimal subset of ports that must remain open for the system to function effectively—then close all other ports. To close a port, find the corresponding service and turn it off or remove it.

Not Filtering Packets for Correct Incoming and Outgoing Addresses

Rules for this should be set up on the external router or firewall.

Nonexistent or Incomplete Logging

Set up all systems to log information locally and to send the log files to a remote system. This provides redundancy and an extra layer of security; the two logs can be compared against one another. Any differences could indicate suspicious activity. In addition, this allows cross checking of log files. One line in a log file on a single server may not be suspicious, but the same entry on fifty servers across an organization within a minute of each other may be a sign of a major problem.

Wherever possible, send logging information to a device that uses write-once media.

Vulnerable CGI Programs

The following are essential tasks that need to be done to protect against vulnerable CGI programs:

- ❑ Remove all sample CGI programs from any production Web server.
- ❑ Audit the remaining CGI scripts and remove unsafe CGI scripts from all Web servers.
- ❑ Ensure all CGI programmers adhere to a strict policy of input buffer length checking in CGI programs.
- ❑ Apply patches for known vulnerabilities that cannot be removed.

- ❑ Make sure that the CGI bin directory does not include any compilers or interpreters.
- ❑ Remove the "view-source" script from the CGI bin directory.
- ❑ Do not run Web servers with administrator or root privileges. Most Web servers can be configured to run with a less privileged account such as "nobody."
- ❑ Do not configure CGI support on Web servers that do not need it.

Antivirus Software

Antivirus software should be installed on all computers and configured to provide maximum protection. If the organization's mail server does not scan mail messages automatically, the antivirus software should be configured to scan electronic mail messages` as they are downloaded from the mail server. Antivirus programs should be run periodically to scan for viruses on local drives.

Some antivirus applications rely upon virus "definitions" or signatures to detect many viruses, and these signatures must be updated frequently. Some antivirus software vendors provide automatic configuration that will download and install these files. This provides maximum protection.

Electronic Mail

Electronic mail is often the source of virus attacks on personal computers and local area networks. Viruses, worms, and other destructive or disruptive computer code are downloaded and unknowingly run by unsuspecting users. A number of steps can be taken to reduce this vulnerability.

The preview pane of email software should be closed, since it opens a message automatically before the user has the opportunity to decide whether the message is expected and probably safe.

Computer users should be instructed to be careful when opening email attachments from unknown senders. These attachments could contain viruses, worms, or other malicious or disruptive code.

Other Mitigation Strategies

Employee Training

Employees should be taught to be security conscious. This is true for the entire gambit of exposures, from helping to identify suspicious packages and suspicious activities to diligently following company policies for the use of passwords on their computers. Suspicious activities would include anyone placing a package in an unusual place rather than accidentally dropping the package in a very public area as well as suspicious email messages. Reporting procedures should be clearly defined, so suspicions can be promptly communicated to security personnel or directly to the police.

Emergency Response & Business Recovery Programs

Effective response to and recovery from a terrorist attack are discussed in Chapters 4, 5, and 6.

End Notes

[1]NFPA 101, *Life Safety Code®*, 2000 Edition, National Fire Protection Association, Quincy, MA, 2000

[2] *Bombs: Protecting People and Property, A Handbook for Managers*, 4th edition, Home Office, United Kingdom, 1994.

[3] National Infrastructure Protection Center, "Highlights," Issue 11-01, December 7, 2001.

Chapter 4

Organizing & Planning to Respond

Plan Components, Objectives, & Standards

A terrorist incident could occur at any time. Most attacks have been localized with minor casualties and property damage, although recent history has proven that catastrophic events can occur. The actions taken during the critical initial minutes of the attack can often determine its ultimate impact. A well thought out plan that assigns responsibilities to trained members of a team can safeguard the health and safety of employees and protect company assets.

Sole dependence upon public emergency services is not sufficient, since a major terrorist attack could overwhelm police, fire, and emergency medical services. They also cannot be expected to respond as effectively in unfamiliar surroundings. Therefore, every facility should have a site-level emergency response plan that is coordinated with local police, fire, and public health services.

A comprehensive plan to respond to and recover from terrorist incidents includes emergency response, crisis management, and business recovery components.

- ❑ Emergency Response is the site-level actions taken to stabilize an incident.
- ❑ Crisis management is an executive or corporate-level function that includes decision-making and communication components.
- ❑ Business recovery includes the actions taken to recover critical functions within predetermined timeframes.

Small companies or organizations with a single facility can develop a single, integrated plan that incorporates all component parts. Larger companies require more extensive planning at the local, division, and/or corporate levels to address both site-specific emergency response needs; recovery of interdependent operations; and crisis communications. Terrorist events are headline news, and a carefully crafted communications strategy is needed to inform important stakeholders.

The three components of effective planning are shown in Figure 1. The planning and involvement of ascending levels of a company vary significantly, depending on loss potential. A minor incident that causes little damage may only involve site-level emergency response. A major terrorist attack that causes loss of life, significant property damage, and business interruption requires effort at all levels of a company. Site-level emergency response efforts stabilize the situation, and resources throughout the company are needed to restore critical functions until the facility is restored or relocated. A major event would be well publicized, and stakeholders would be concerned about the company for differing reasons. The executive level crisis management team would direct overall relief efforts and communicate internally and externally as needed.

Figure 1. Plan Components & Levels of Involvement

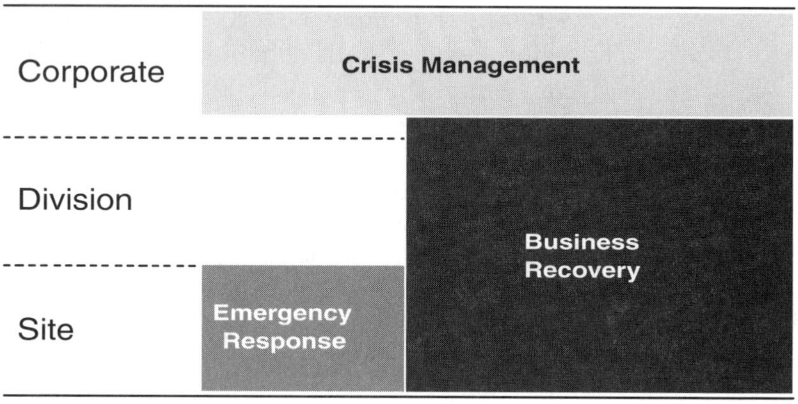

The three components of effective planning can respond individually, simultaneously, or in succession. It all depends upon the scope of the incident. The duration of each phase of response can also vary considerably. For example:

Scenario 1: A large bomb concealed in a van is detonated in the public street adjacent to a mid-rise corporate office building and data center. The explosion results in many casualties, the building is seriously damaged, and all business activity will have to be relocated.

This scenario would require all components of the plan to respond in rapid succession. Emergency response would tend to the casualties; crisis management would address the corporation's overall response and communication with stakeholders; and business recovery plans would address the restoration of critical functions.

Scenario 2: A "dirty bomb" is detonated late at night blocks away from a corporate office building. There is no damage to the building, and no employees were in the building at the time of the explosion. Later that night, officials announce that they have detected radioactivity in the area, but they are not sure of the extent. It could take days to assess the extent of contamination, so all buildings are ordered closed for at least a week.

In this scenario, the emergency response plan would not be needed. However, the unavailability of the corporate data center could have a major impact on operations, so the business recovery plan is activated—data processing is relocated to a hot site facility. The corporate crisis management team also meets off-site to plan overall strategies and to communicate with stakeholders.

Scenario 3: Over the weekend, the news media report a terrorist claim of responsibility for poisoning sugar supplies in the region. Further reports say it was confectioners' sugar. The exact location of the poisoning is unknown. Officials, who initially tried to quiet the reports, announce they are investigating. Public concern quickly grows. The operators of a large regional bakery are concerned that customers will not buy their products because of the threat to confectioners' sugar—they are one of the biggest users of sugar in the area.

This claim of an attack could be real or false. There are no reported casualties and no damage to the bakery's facilities. However, the potential drop in demand for their products could be huge. This scenario requires a senior level crisis management team to decide what to do and to communicate very effectively. Even if the claim proved to be a hoax, there could be lingering doubts about the safety of the sugar. Continued efforts would be required to reassure the public that the bakery's products are safe.

Figure 2 depicts a timeline with a parallel line for each of the three components of effective planning. Typically, the emergency response plan responds as soon as a threat is detected or an incident occurs. The duration of the emergency response phase could be minutes or hours. In extreme cases such as the September 11 attacks on the World Trade Center, the emergency response phase lasted for days. If the threat or potential consequences of a threat or incident are significant, the crisis management plan should be activated to address the company-wide response. The "crisis" could last for days, weeks, or longer in extreme cases. If there is actual damage or damage is imminent, the business recovery plan is activated. Catastrophic damage to a facility could require an extended recovery period, although critical functions should be restored using alternate strategies. Delayed restoration of critical functions could threaten the company.

Figure 2. The Response Timeline

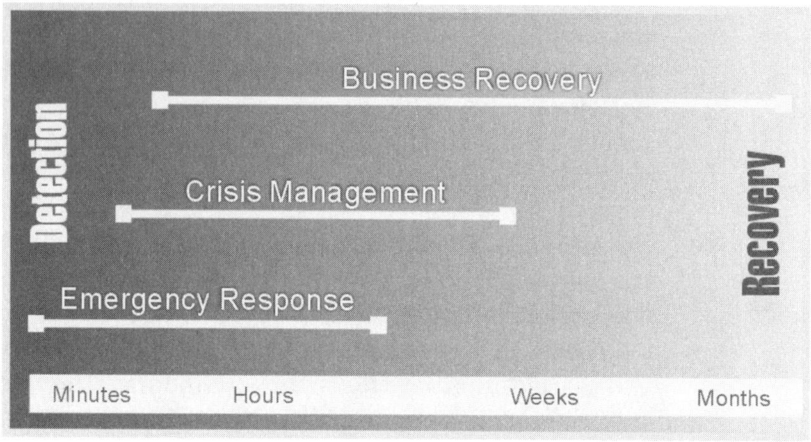

National Standards & Regulations

There are many best practices, standards, and regulations for emergency response, crisis management, and business recovery planning. Best practices are developed by professional organizations. Standards define performance and practice in a document that can be adopted and enforced by political jurisdictions. Regulations are mandatory requirements promulgated by government agencies.

One standard that addresses risk assessment, mitigation, response, and recovery is NFPA 1600 *Standard on Disaster/Emergency Management and Business Continuity Programs*. It was developed by a technical committee of the National Fire Protection Association (NFPA) first as a recommended practice in 1995, and then reissued as a formal standard in 2000. It has been approved as an American National Standard, and it has been endorsed by the Federal Emergency Management Agency (FEMA), the National Emergency

Management Association (NEMA), and the International Association of Emergency Managers (IAEM). In addition to the U.S. representation, international committee members also participated so the document reflects best practices from around the world.

NFPA 1600 establishes a common set of criteria for emergency management and business continuity programs. It provides criteria to assess current programs, or to develop, implement, and maintain programs to mitigate, prepare for, respond to, and recover from disasters and emergencies. It is not a "how to" guide. However, plans and programs that meet the performance standard outlined in NFPA 1600 should be comprehensive and stand up to internal and external scrutiny.

Business recovery planning is required for many firms in the financial services industry and to protect healthcare records and information. The Federal Financial Institutions Examination Council (FFIEC) mandates planning for the banking industry. Organizational structure, data backup, and recovery requirements are specified. The federal Health Insurance Portability and Accountability Act (HIPAA) requires continuity planning, testing, and maintenance to ensure protection of healthcare records. The Security and Exchange Commission (SEC) and National Association of Securities Dealers (NASD) require business continuity planning for broker dealers, investment managers, and other financial services firms that handle securities.

There are many other standards that apply to building construction; the design, installation, and testing of utility systems; and life-safety systems, computer security, operations within a building, and threat-specific response and recovery procedures. Plan developers should be mindful of regulations such as OSHA standards and fire codes that require planning as well as best practices that define what should be done. This will help ensure that a facility is compliant with the law and that plans are well founded.

The Planning Process

Development of effective plans to deal with threatened or actual terrorist events is best done with a systematic approach. The approach must be founded on management support; the participation of managers and staff on planning committees; and the development of organized teams with defined procedures commensurate with the potential threat, site vulnerability, and impact on the company.

The multi-step process includes:

1. Writing a management policy
2. Organizing planning committees
3. Conducting a risk assessment as described in Chapter 2
4. Assessing the availability and capabilities of public emergency services, company personnel, and equipment resources
5. Determining the most appropriate level of response based upon company needs and regulatory requirements
6. Organizing emergency response, crisis management, and/or business recovery teams
7. Writing the plans
8. Training personnel
9. Exercising each plan
10. Maintaining the plan

Management Policy

A policy statement is a relatively short document, but it is very important. The policy statement defines in broad terms the commitment of the organization to safeguarding people and protecting assets through risk assessment, mitigation, response, and recovery. Management must be informed of specific requirements before they approve the document. This includes applicable regulatory requirements; risk assessment and mitigation; development of plans at varying levels of the organization; ongoing training; provision of equipment; and time to do it properly. A signed management policy statement can codify the desired level of response and ensure continued support at all levels of a company. Conversely, a weak or missing policy statement ensures that individual facilities or departments within a facility will have to continually battle for the attention and support of management.

Management must also direct managers, supervisors, and employees to actively participate in the planning process. Financial support for purchasing equipment, providing training, or simply allowing personnel to attend training and participate in drills is critical. Management must also hold those involved in the planning process accountable for implementing and maintaining the appropriate level of preparedness.

Planning Committees

As noted in the introduction to this chapter, comprehensive planning is required at the local, division, and/or corporate levels of an organization. It includes emergency response at the site-level, crisis management at the corporate-level, and business recovery at local, division, and/or executive/corporate levels. Development of these plans requires committees at all levels.

The involvement of personnel from many departments provides broad expertise that will enhance the planning process. In addition, the planning process itself is educational for those who participate.

Planning committees should be organized for each plan component. Table 1 shows suggested participation by department.

Table 1. Department Participation

Emergency Response	Crisis Management	Business Recovery
• Site or Facility Management	• CEO, COO, & CFO	• Manufacturing, Production, or Operations
• Manufacturing or Operations Management	• Executive VPs or heads of affected operating units	• Engineering or Facilities Management
• Engineering or Facilities Management	• Public or Media Relations	• Administrative Support
• Environmental Health & Safety	• Government Affairs	• Information Technology Telecommunications
• Security	• Investor Relations	• Sales & Marketing
• Medical Department or Nurse	• Human Resources	• Purchasing
• Media Relations	• Legal Counsel	• Logistics
• Human Resources	• Administrative Support	• Finance
• Maintenance or Housekeeping	• Ad HOC Experts (within & outside the organization)	• Legal/Contracts
		• Customer Service

A terrorist incident could require significant internal resources and involve many different outside resources. Government agencies including law enforcement, fire, hazardous materials, emergency medical services (EMS), and public health officials would respond. A significant incident would involve a multitude of state and federal agencies as well.

Recovery from an incident could involve many different suppliers or contractors to assist affected employees and to recover critical functions. Healthcare and employee assistance program (EAP) providers may be needed. Many companies rely upon outside public relations firms to craft and deliver messages to external stakeholders.

Effective planning requires contact with many of the public agencies that may respond. These include not only police, fire, and EMS, but also local emergency planning committees. This is discussed further in the *Pre-Incident Planning* section of this chapter.

Assessment of Available Resources

The planning process should inventory available resources that could be deployed to respond to or recover from a terrorist incident. These resources include personnel, equipment, and systems such as ventilation, fire suppression, or life-safety systems that are designed to prevent or control an emergency. Personnel who would be capable of responding to emergencies and their level of education and training should be cataloged. Facilities that operate beyond weekday business hours must determine the number of personnel available evenings, overnight, and on weekends.

Organizations

The three different plan components described in Figure 1—emergency response, crisis management, and business recovery—require different teams operating at different levels within the organization. The emergency response team works at the site level, whereas the business recovery team(s) works at the local level, division level, and in conjunction with the top or corporate level.

There may be multiple business recovery teams, each designed to restore a different critical function. For example, one might work to restore information technology at a contracted hot-site facility. Another team might work to transfer manufacturing to an alternate facility.

The emergency response team assesses damage incurred and communicates with the business recovery team leader, who has to decide whether recovery strategies should be employed. The crisis management team monitors all activities, adjusts direction as needed, allocates resources, and communicates with important stakeholders. Figure 3 highlights the relationships.

Figure 3. Organizational Relationships

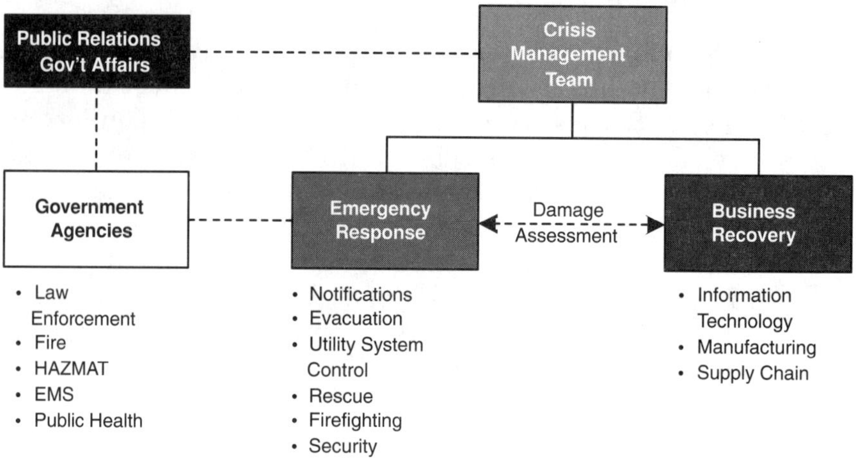

Emergency Response Team

The emergency response team (ERT) is responsible for stabilizing an incident. Primary activities are to detect a threat or an actual incident, determine the best protective actions, and follow predetermined procedures for the type and scope of incident. Detonation of a bomb would involve notifying public emergency services, alerting occupants to evacuate or shelter in place, and assessing casualties and property damage. Specific emergency procedures are discussed in Chapter 5.

The emergency response team may be very small if a limited number of personnel is available, the facility is limited in size, and the capabilities of the public emergency services are adequate. A large facility that is a highly desirable terrorist target should organize a very capable team that involves many departments and personnel.

Organization

An organizational structure, which defines responsibilities and lines of authority, is critical to the success of the ERT. During an emergency, accurate information must be gathered quickly; decisions must be made by a knowledgeable leader; and orders must be given to members. The ERT commander (the incident commander) must be known to all employees and responding public emergency services. The incident commander's authority also must be understood and agreed upon prior to an incident. An example ERT organization chart is shown in Figure 4.

The ERT should be staffed with sufficient members to cover the facility when it is occupied. Staffing should be large enough to overcome vacations, travel, and other work commitments. In addition, staffing must recognize that during a terrorist incident, many members may not be able to reach the facility or may want to leave to tend to family members. These situations are common and can deplete staffing. The ERT should also identify ad hoc participants such as contractors that could address specific needs. Identification badges should be issued so members will be able to reach the facility if there are governmental restrictions on travel. In addition, identification badges will allow ERT members to move through the building if security restrictions are in place.

Figure 4. Example Emergency Response Team

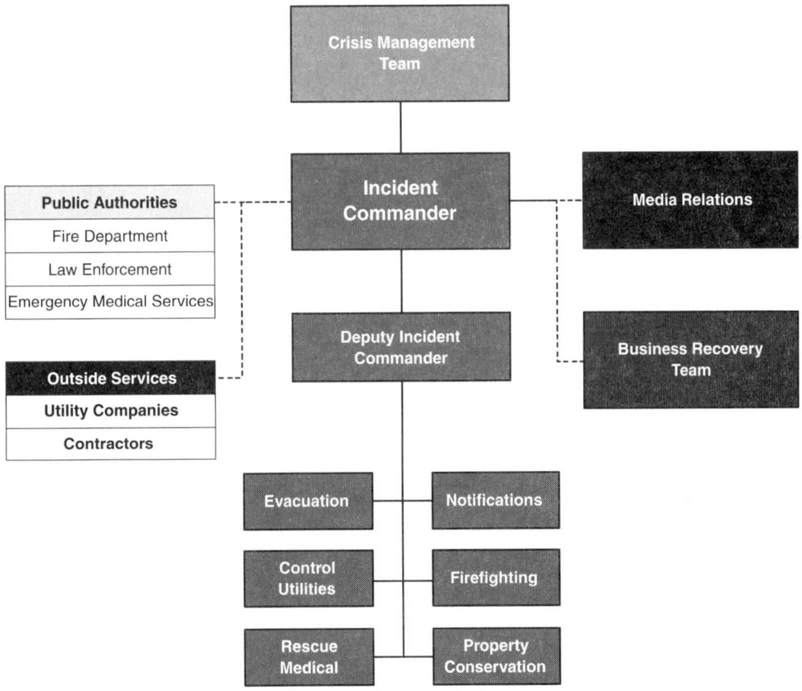

Incident Commander

The incident commander is the most critical member of the emergency response team. The IC must be capable of handling the stress incumbent in directing an emergency operation. The IC must also be very familiar with the facility's construction, hazards, fire protection systems, utility systems, emergency response plan, and the persons or agencies that will respond to an incident.

Duties of the incident commander include:

- ❑ Determining the types of terrorist incidents that may occur and developing procedures to respond to them.
- ❑ Coordinating pre-incident planning and emergency response procedures with outside agencies.
- ❑ Ensuring that public emergency services are called when necessary.
- ❑ Directing all emergency activities, including evacuation or sheltering of personnel.
- ❑ Directing the shutdown of operations when necessary.

Emergency Response Team Members

The emergency response team may be called upon to carry out many critical tasks to safeguard the health and safety of building occupants and to protect company assets. Initially, the team has to detect a threat or actual event. Detonation of a bomb would be immediately apparent, but dispersal of a chemical agent outside the building probably would not be immediately discovered. Once the threat is identified, the ERT must carry out defined emergency response procedures including:

- ❑ Assessment of the threat or incident
- ❑ Notification of public emergency services
- ❑ Alerting building occupants
- ❑ Evacuation or sheltering of occupants
- ❑ Supervision or control of building utility systems (HVAC, life safety, and fire protection)
- ❑ Provision of first aid

- ❑ Security of buildings and grounds
- ❑ Rescue of trapped occupants
- ❑ Firefighting (if trained)
- ❑ Containment and clean up of hazardous materials (if trained and equipped)
- ❑ Assessment of damage

The emergency response team is in the best position to assess damage from the terrorist incident. This information is essential for the crisis management and business recovery teams to decide the most effective strategy for recovery.

If plans call for personnel to perform firefighting or hazardous materials (HAZMAT) operations, they must be physically capable. OSHA regulations require medical evaluations. Personnel who respond to medical emergencies should be trained in first aid, universal precautions against bloodborne pathogens, and cardiopulmonary resuscitation (CPR).

Other members of the ERT most likely will include staff from engineering, maintenance, or facilities departments. Members of these departments can be assigned to the ERT or support the ERT by supervising mechanical and utility systems. Security personnel can be assigned to evacuate building occupants, direct public emergency services to the scene, and control access to the site. Housekeeping is typically assigned to salvage and clean up operations.

A sufficient number of personnel should be trained in basic first aid and CPR, both of which are taught by the American Red Cross and other organizations. This applies to facilities located in urban areas where ambulances or paramedics can be delayed by traffic congestion or high demand, as well as in rural areas where equipment and staff may be scarce. Prompt first aid treatment can often stabilize victims until ambulances arrive.

In addition, a suitable isolation area should be identified to house victims contaminated with chemical or biological agents. This area should be easily quarantined and accessible from the outside. Ventilation should be separate from the building's air-handling system. A makeshift shower facility with a containment area should be available to prevent waste from entering public sewers or the environment.

Crisis Management Team

What is a crisis? For the purpose of this book, a crisis is a terrorist attack or threat of a terrorist attack that could or does jeopardize the survival of an organization. A crisis may result from a direct attack on a facility within the organization, or it could result from the indirect consequences of an attack. A terrorist attack that sickens employees in their homes; poisons raw materials in a product manufactured by the company; or prevents access to buildings, supplies, or markets critical to the survival of the organization can be a crisis, too. A terrorist incident may also result in a crisis because critical suppliers or customers are seriously impacted by an event that does not directly impact a facility within the organization. In each of these cases, the crisis must be managed by a group of individuals that can assess the situation and direct efforts to mitigate the consequences.

A crisis management team (CMT) is separate and distinct from the site emergency response team and business recovery teams. The CMT is organized at the senior levels of an organization, and it can operate from any location where the team can assemble and properly communicate. It is responsible for overseeing the entire organization's response to a terrorist incident—not just the individual facility or facilities that are directly affected.

The crisis management team must assess the impact of the incident on the organization as a whole and must anticipate how stakeholders will react. The CMT then develops communication strategies to inform stakeholders and allay fears. The CMT fulfills a damage control role of sorts since its communications strategy can prevent or soften negative reaction by financial stakeholders.

The CMT should set up an emergency operations center and media center as described later in this chapter.

Organization

The Crisis Management Team is usually comprised of senior management staff. Permanent members may include managers from operations, legal, public affairs, governmental relations, human resources, security, safety, and administrative support departments. Ad hoc members may include internal and external technical experts required to provide advice on the hazards and possible consequences of an attack and remediation strategies that could be employed. Experts in chemical, biological, and radiological incidents as well as structural engineers could be included with the list of experts. The CMT would evaluate the experts' advice and formulate the organization's strategy.

The crisis management team must establish the protocols and procedures to swiftly approve decisions. These protocols and procedures must be consistent with the company's rules of governance. A decision to spend a million dollars would normally require approval by multiple layers of management before reaching the chairman of the board. However, during a crisis, the approval could go from the business continuity leader directly to the chairman of the board. The rules of governance are not violated in this example, but the time necessary to make the decision is reduced.

Communication with Target Audiences

Prompt and efficient communication within the organization and to external audiences is critical in a crisis. The responsibilities and lines of communications to and between the crisis management team, emergency response teams, and business recovery teams should be well established. All other departments must understand with whom they are to communicate to gain approvals, provide information to, and receive information from. If responsibilities are known and lines of communication are well established, communication will be enhanced.

Procedures and protocols must also be consistent with the organization's disclosure policy. This ensures that information is not leaked prematurely to investors before a wider disclosure to the media.

The crisis management team must identify and prioritize the audiences that must be reached after a terrorist attack. The prioritization of target audiences is based upon the importance of the audience to the continued success or survival of the organization. Target audiences include:

Employees and their Families. In the immediate aftermath of a major terrorist incident that injures or kills employees, family members and coworkers are the most important audiences. The organization has an obligation to the family members of affected employees to provide accurate information on the status of their loved ones. Physical outreach is necessary.

Key Customers. If customers begin to abandon a company because they think the organization may not be able to meet their needs, effective communication and reassurances may dissuade them from making a change.

Investment Community. The value of a public company's stock is determined by many factors, including the assessment of analysts and reaction in the financial markets. Effective public statements that a company is able to overcome the incident can calm the fears of investors and analysts. After September 11, many corporations took out full-page ads in national newspapers to explain that they would be able to meet their customers' needs despite the loss of many employees and important facilities.

Regulators. Highly regulated industries such as financial services, food processing, and pharmaceuticals must reach out to regulators to ensure that they approve of changes in operations.

Critical Suppliers. Suppliers that are critical to a company must be embraced so the supply of goods or resources is not interrupted.

News Media. The local, regional, and national media must be addressed since they will convey a message to the world—good or bad. Specific instructions for dealing with the news media are provided later in this chapter.

Community. The affected community may need to be addressed, especially if a terrorist attack on the organization results in consequences to the community at large. Examples could include an attack on a chemical plant that releases hazardous chemicals into the environment. An attack on a major employer in an area would also significantly interest the community, even those not employed by the firm.

Table 2. Working Groups & Target Audiences

Group	Audiences	Activities
Public Relations or Communications	• News Media • Community	• Local, regional, & national print and electronic • Employee communication • Community relations • Call centers
Investor Relations	• Partners • Investors • Analysts	• Impact on operations, short and long- range financial outlook
Government Affairs	• Government regulators	• Regulatory impact • Reporting requirements
Marketing & Sales	• Key Customers	• Communicate any impact on orders or delivery schedules
Finance	• Banks and lenders	• Emergency financing
Risk Management	• Insurance Broker • Underwriters • Claims adjusters	• File insurance claims • Coordinate with adjuster • Settle claims
Human Resources	• Family members • Employees	• Account for missing staff • Family outreach & support • Employee records & information • Employee and family counseling assistance (EAP)
Benefits Managers	• Healthcare Providers • AD&D Insurers • Life Insurers	• Expedite provision of benefits to affected employees and their families
Purchasing	• Critical Suppliers (raw materials or services)	• Schedule • Alternate suppliers • Redirect supplies
Others	• Contractors • Suppliers	• Restoration of damaged or affected facilities

Once these audiences are identified, determine the best means to communicate with them. Backup channels of communications should also be identified because primary channels may be unavailable or overloaded. The protocols or procedures for scripting and delivering the message should be decided in advance. This ensures that the most appropriate person delivers the correct message.

The means of communicating with the various audiences varies. Telephone contact with most important stakeholders, such as important customers or suppliers, is best. Contact with the news media can be done face-to-face or via public statements, press releases, Web sites, fax, or electronic mail. Even toll-free calling centers can help spread the message to employee stakeholders.

The text of press releases, fact sheets or other written or verbal statements should be scripted in advance—as much as possible. Space should be left to answer expected questions.

The scripts should include the name, address, and types of operations at each facility. In addition, these scripts should highlight positive factors of the facility. This can include building construction, fire protection, or life-safety systems, as well as the emergency response and business recovery plans in place. Additional details about each facility within the organization should also be prepared for reference or distribution as needed.

Business Recovery Team

The business recovery team is responsible for identifying functions that are critical to the organization's ability to fulfill its mission. These critical functions must be prioritized based upon the length of time they can be lost before significant harm is done to the organization. This time frame is called the *recovery time objective*. Strategies should be developed to restore or replace critical functions within the recovery time frame.

The business recovery team can be organized at the local or a higher level. The business recovery leader is responsible for directing business recovery efforts in conjunction with the crisis management team.

The business recovery organization should be led by a senior level manager who is authorized to organize a steering committee and develop recovery plans. A planning committee is comprised of members from important departments such as manufacturing, production, operations, engineering, facilities, human resources, finance, and administrative support (see Table 1.) Individual department or functional teams may also be necessary to develop specific recovery plans for individual departments or functions.

Critical Functions & Recovery Times

Critical functions or processes should be identified from the top down. Senior managers can help identify the most critical functions. They can also provide perspective on recovery time objectives. Questionnaires and interviews with department heads and others can then be used to capture information such as interdependencies, supplies, resources, and other needs required to support the critical functions.

Critical functions are those that generate the most revenue; are essential to meet required regulatory requirements or quality standards; are essential to maintain market share; or are important to maintaining relationships that are critical to success. A long list of criteria can be applied to determine what is critical. A flow chart of production methods or service delivery can help to identify the critical path and functions.

Once critical functions have been identified, the resources necessary to support the function must be identified. These resources fall into several broad categories:

- ❑ Facilities and supporting infrastructure
- ❑ Equipment (office equipment, computers, networks, telecommunications, or production equipment)
- ❑ Supplies (raw materials, subassemblies, or critical services from within the organization or outside)
- ❑ Vital records (software applications and electronic data files, paper records, engineering drawings, product specifications, etc.)
- ❑ Personnel and expertise

Resource needs must address the lead-time necessary to obtain the needed equipment, raw materials, component, or support service. Lead times are typical for products or equipment that are custom manufactured, in short supply, or made at a distant location. These can include confirmation of quality standards or restrictions (quality, regulatory, availability, etc.) that can impact the delivery of replacements.

Any peaks in production or service demand (at quarter or year-end, prior to holiday season, etc.) should be addressed.

Figure 5. Example Production Flowchart

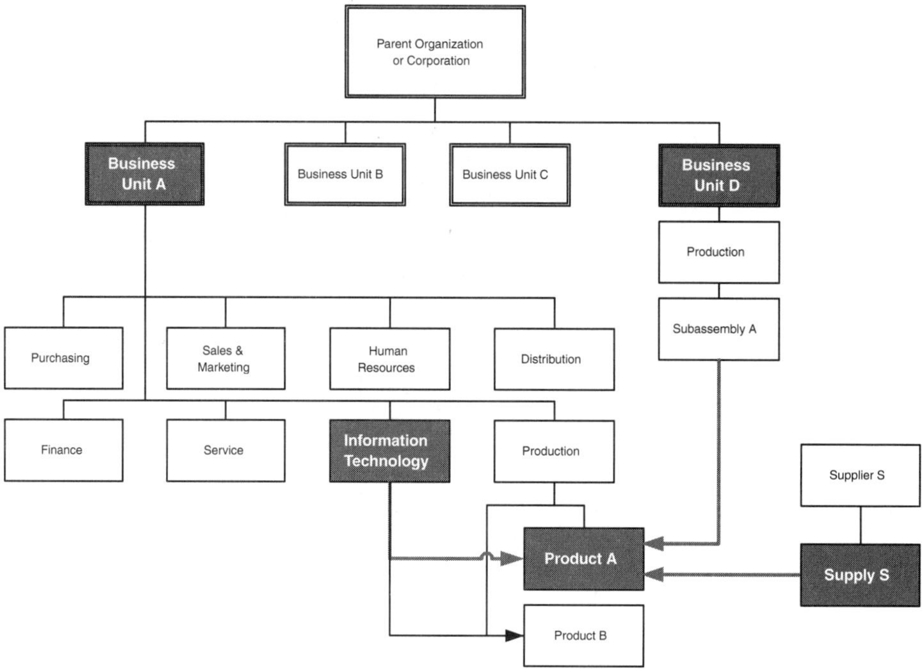

Figure 5 shows a hypothetical manufacturing corporation that has four business units, each producing different products. Business Unit "A" manufactures the corporation's most lucrative product. This business unit also produces a secondary product, "B." Product "A" requires "Supply A" from outside "Supplier S" and a subassembly from Business Unit D. The production of Product "A" also requires computerized production control. The IT department provides this.

Product "A" is the lifeblood of the corporation, so an initial business recovery plan would focus on the restoration of Product "A" production. The business impact analysis determined that recovery within thirty days is required before there is irreparable harm to the corporation. Table 3 shows an analysis of the resource requirements, sources, and recovery time for producing Product "A." The computerized control systems required to manufacture Product "A" becomes a critical function that must be restored before manufacturing can resume. Other internal or external resources may also become critical to support the manufacturing of Product "A."

Table 3. Example Critical Processes or Function

Process or Function	RTO (days)	Resources Required	Internal/External Source
Manufacture of Product "A"	30	Supply S	Supplier S
		Subassembly D	Business Unit D
		Computerized manufacturing control	Internal IT systems
Computerized Manufacturing Control systems	30	Electricity	Public utility with specialized UPS and power conditioners
		Climate controlled environment	Self-contained HVAC unit
Continue Listing			

Recovery Strategies

Recovery strategies should be developed to work around or replace critical functions or processes. Low priority operations would probably be discontinued until functions that are more important are restored. Operations could be shifted to another owned facility, or work could be contracted out to others. Production methods could be changed, or work could be done manually with additional staffing. There are many possible strategies that can provide temporary functionality.

Strategies must be developed for all supporting functions or processes and the resource supplier. This includes raw materials, components, support services, equipment, people, expertise, knowledge, records, drawings, specifications, etc. All must be identified, and backups should be in place—if possible and practical.

Recovery strategies should be documented in the plan. The plan should address what needs to be done; what business recovery team or group is responsible for the restoration; procedures for restoration; and step-by-step procedures with defined tasks. Supporting documentation such as the names, telephone numbers, and email addresses of persons, suppliers, and contractors should be incorporated into the plan. Lists of equipment and supplies should be compiled along with the names of primary and alternate suppliers.

Command, Control, & Communications

Terrorist threats or actual incidents require immediate and coordinated action. Effective command, control, and communications are essential to achieve the best outcome. This includes a system to control internal teams and resources following defined procedures, as well as coordination with external agencies and resources. Systems and procedures are needed to communicate with groups of people as well as individual members of a team. In addition, compatible equipment and common terminology are needed to ensure the accurate exchange of information.

Incident Command System

An incident command system that identifies the person in charge and defines the responsibilities of each team and external agency is the best means to effectively control an incident.

The Incident Command System (ICS) was developed in the public sector to command the extensive resources of public emergency services that may respond to an incident. Companies with extensive operations, emergency response teams, crisis management teams, and/or recovery teams should implement an incident command system. The ICS should address:

- ❑ Person in charge (name and title)
- ❑ Chain of command
- ❑ Advisory roles (technical recommendations)
- ❑ Veto powers (who has the power to overrule a decision)

- ❑ Command post and emergency operations center activation and operation
- ❑ Common terminology (organizational functions, resources, and facilities)
- ❑ Resource management

The ICS used by public authorities and easily adapted for private use includes a chain of command and four sectors for planning, logistics (supply), finance, and operations. A typical public sector incident command chart is shown in Figure 6.

Figure 6. Public Sector Incident Command System

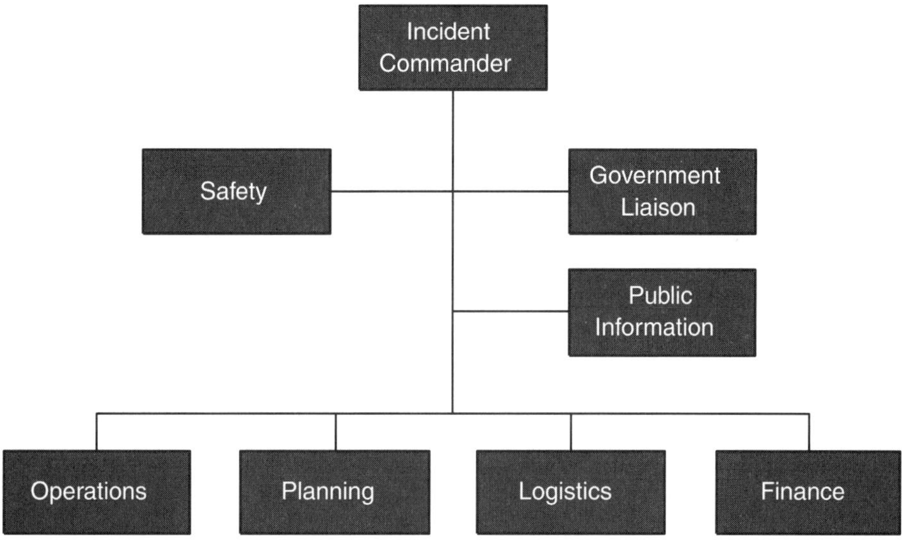

Large-scale incidents that involve multiple agencies utilize a unified command to ensure effective decision-making. The unified command is the assembly of top commanders from many of the agencies that respond. The commanders then make joint decisions to prevent contradictory actions that could affect incident control. Terrorist incidents typically involve many public agencies in addition to private sector organizations that may be affected by the incident. An understanding of the unified command concept by private organization will enable a more effective public/private response.

An example incident command system for a private organization is shown in Figure 7. This system includes an incident commander who has the responsibility and authority for directing all of the company's resources to control an incident.

The ICS has four sectors responsible for related groups of tasks. The sectors are operations, planning, logistics, and finance. Individual activities and responsibilities include the following:

Operations. The operations sector could include multiple teams working to contain the impact of a terrorist incident. Their activities could include notification of building occupants to evacuate or shelter in place; control of building utilities including ventilation, life-safety, and fire protection systems; security; first aid; firefighting; and property conservation.

Planning. The planning sector is responsible for assessing the situation and determining what action should be taken. This includes assessing damage so the business recovery team can determine whether to activate the recovery plan.

Logistics. The logistics sector provides resources to assist in the response and recovery from a terrorist incident. Manpower, medical support, facilities, equipment, supplies, and transportation are just a few examples of the resources that need to be managed.

Finance. The finance sector is responsible for managing the money. An emergency requires unusual and extraordinary spending, and procurement must be done in accordance with an organization's rules of governance. This sector is also responsible for maintaining records, documenting extra expenses, filing insurance claims, and providing compensation.

A qualified person should be assigned to oversee health and safety. Since this is such an important role, it reports directly to the Incident Commander (IC). A scribe should also be assigned to assist the IC to log a chronology of events, commands issues, action items, etc.

The organization's spokesperson should be assigned to media relations in accordance with established policies and procedures. Governmental affairs representative(s) should be assigned to liaise with government agencies that may respond or impact an organization's ability to recover from a terrorist incident.

> Use of the incident command system doesn't mean an organization has to name dozens of people to different tasks. Rather it is a methodology to assign responsibilities in a coordinated fashion and control them to maximize the benefit.

Use of the incident command system doesn't mean an organization has to name dozens of people to different tasks. Rather it is a methodology to assign responsibilities in a coordinated fashion and control them to maximize the benefit. Many existing departments can easily be inserted into the ICS chart.

Figure 7. Private Sector Incident Command System

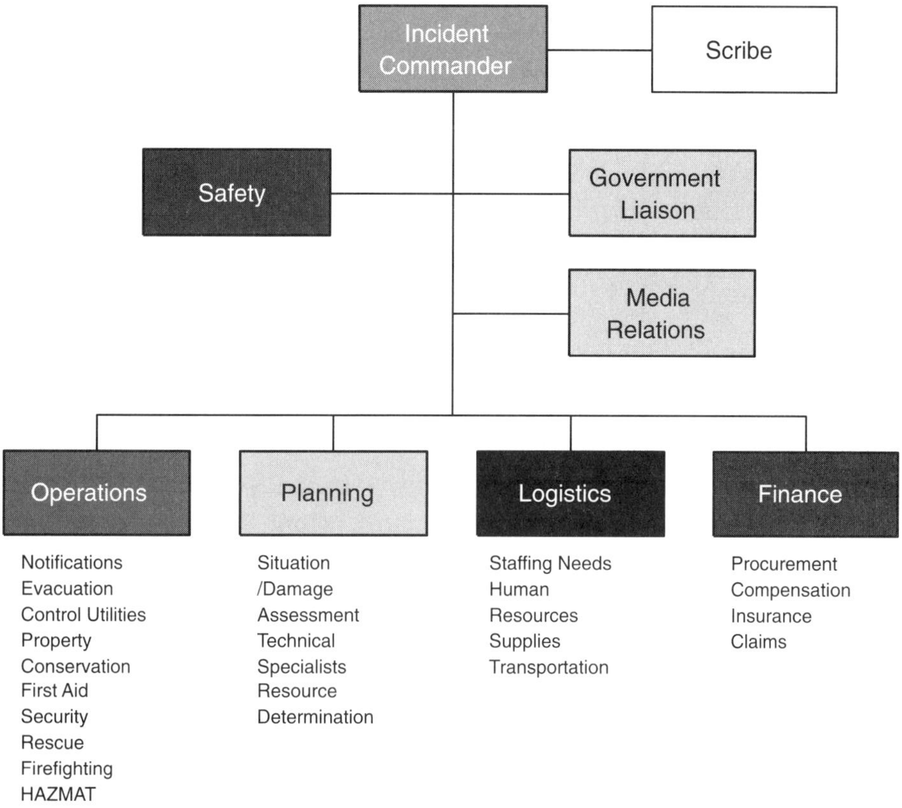

Emergency Operations Center

An emergency operations center (EOC) is a facility where members of the emergency response, crisis management, and/or business recovery teams can assemble together or separately to deal with a terrorist incident. Those assembled monitor the situation and direct response and recovery operations. The EOC is different from a command post, which is the forward location where the incident commander or deputy directs the emergency response team. In some situations, the incident commander may direct all activities from the EOC.

A minimum of two emergency operations centers should be established in case one EOC is uninhabitable due to the emergency. For example, one EOC can be located on-site, and a second should be located at a distance so that it is not subject to the same incident. Since terrorist attacks can cause widespread destruction or disruption, a location ten miles or more away is appropriate.

The EOC should be located in a part of the building that is least vulnerable to attack. Interior locations that are away from any windows and constructed with substantial walls are best. Vapor clouds settle to the ground making first floor areas more vulnerable. Basements are subject to flooding when water mains break. Locations near target hazards such as loading docks, mail centers, lobbies, and public streets should be avoided.

The EOC should be sized and equipped to support all teams for extended operations lasting many days. Separate or adjoining facilities should be considered for teams that have to address different objectives. For example, the crisis management team should have its own facilities to address issues that concern the entire organization. The business recovery team may require separate facilities to support its activities. Depending upon the organization, this may not be an issue if the incident affects a location far removed from the crisis management team and business recovery efforts are managed elsewhere. Figure 8 shows an EOC with a main conference room and separate rooms for communications and side meetings.

The primary and secondary EOCs should be accessible 24/7 and equipped as follows:

• 12 Telephones with direct incoming and outgoing lines (recording capability is helpful, but callers must be informed they are being recorded and have right of refusal.) • Speakerphone for conference room table • One dedicated incoming fax machine; one for outgoing faxes • Computer with printer and access to corporate databases, intranet, Internet. • Electronic mail with dedicated "EOC" email address • Television set with off air antenna to receive news reports • AC and battery powered AM/FM radio for radio news • Portable two-way radios with spare batteries and chargers to operate on all company radio channels • Photocopy machine	• Reams of paper for printers and lined paper for handwriting, pens • Flip charts with markers (Dry erase boards with markers and erasers but not in place of flip charts.) • Extra copies of emergency response, crisis management, and business continuity plans for all members plus extra copies • Employee lists by department with work, home, and cellular telephone numbers; and work and home email addresses • Contractor lists • Telephone directories • Copies of drawings for all building utilities including ventilation, electrical, gas, water, and sewage. • Maps of the area surrounding the building plus wider area maps of the city and regional area. • Video and still cameras with spare videotapes and film for documenting damage and cleanup activities.

The EOC must have communications links to the corporate crisis management team as well as the person in charge of business recovery. In addition, there should be links to public agencies to provide and receive official information. The EOC is not the site for handling the news media. This is addressed later in this chapter.

Figure 8. Example Emergency Operations Center

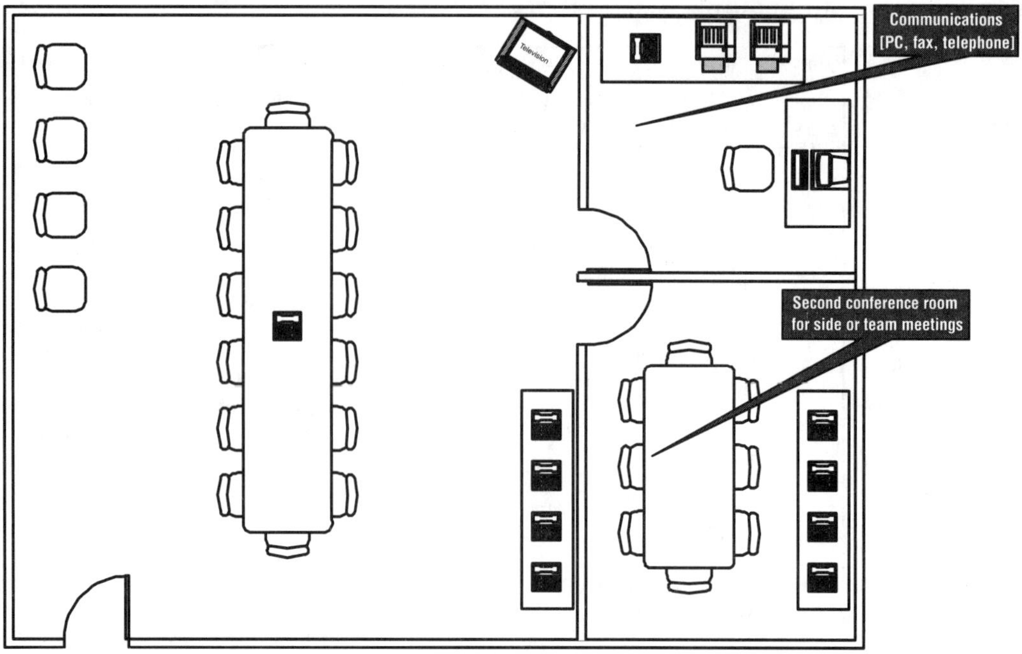

Communications

Effective communication is critical during emergencies. This includes effectively relaying information from one person to another as well as the use of systems and equipment to transmit or receive the information. An emergency by definition is something that happens unexpectedly, and it creates considerable confusion and emotional reaction. Therefore, effective communication needs to be addressed as part of plan development.

There are many different groups to communicate within during or after a terrorist incident. These include:

- ❑ Emergency Response Team
- ❑ Crisis Management Team
- ❑ Business Recovery Team
- ❑ Public emergency services (e.g., local police, fire, EMS, and public health)
- ❑ State & federal agencies (FBI and possibly many others)

- ❑ Public utilities (electric, gas, water, sewer, telecommunications)
- ❑ Contractors
- ❑ Suppliers
- ❑ Customers
- ❑ Regulators
- ❑ Management
- ❑ News Media
- ❑ Employees

The emergency response plan should include the circumstances that require communication with these groups and the procedures or scripts to be used.

The emergency response plan should include telephone numbers for each of the groups that need to be reached. Telephone numbers for nonemergency use during normal business hours should be included as well as emergency, off-hours telephone numbers. Include landline, cellular, pager numbers, and email addresses if available.

Communication with employees can be a daunting task due to the sheer numbers involved, and accounting for a large number of employees who have been evacuated from a building is very difficult. Inevitably many scatter and don't wait to be counted at an assembly area. Therefore, employers need to reach out to their employees by any means necessary to verify that they are safe. Arrangements can be made with local television or radio stations to broadcast official company information instructing employees to report to work or to check in with their supervisors. Telephone calling trees are popular but inherently flawed because the loss of one branch of the tree may mean that many people are not promptly called.

Additional means to notify employees include use of voice mail systems to post official information at predetermined times (e.g., 5 a.m. before employees leave their homes). Since telephone systems may be overloaded or unavailable, alternate means of communication should be established. Remote call centers that receive toll-free calls can act as reception points for callers requesting information when they are unable to connect with their normal points of contact. Call-in procedures can be more effective when large numbers of people are involved.

Web sites are another means of communicating with employees and other constituents. Many corporations impacted by the World Trade Center attacks in 2001 posted daily messages on their Web sites. Some even provided bulletin boards to allow employees, friends, and others to post and read messages.

Employee information should be kept up to date. This includes home address, telephone numbers, spouse or emergency contact, and family members as well as beneficiary information for life insurance, accidental death & dismemberment (AD&D), and other insurance policies. All information should be verified periodically to ensure it is current. This information should also be available to the primary and alternate emergency operations centers.

Scripts should be established as part of the emergency planning process to reduce the time necessary to communicate official information and to ensure consistency in the message. Messages should be crafted for use by the different communication methods (e.g., telephone operators or call centers that receive incoming calls, broadcast emails, Web site messages, or radio/television broadcasts).

Communications Equipment

A lot of different equipment may be used for communications. This includes building voice communication systems, public address systems, fire alarm systems, stairwell communication systems, two-way radios, pagers, and other means. Response to a serious terrorist attack will be extended, taxing communications capabilities. The susceptibility of systems to damage should be checked as noted in Chapter 2, and backup systems should be available if primary systems fail. Provide spare batteries, chargers, and cases for two-way radios. Ensure that other essential communications (e.g., base radio stations, telephone systems, computer terminals used for email, etc.) have backup power supplies.

Keep in mind that telephone communication systems may be unavailable during a serious terrorist attack. Telephone central offices or telephone lines may be damaged or disrupted, systems will probably be overtaxed, and wireless services may be purposely restricted to official users only. On-site telephone services that rely upon a PBX or central system may be disabled if the central system fails. Direct telephone lines should be provided to the EOC for this reason. Be prepared to use runners or messengers to communicate point-to-point within a building or campus.

News Media Relations

A terrorist attack will undoubtedly receive tremendous media attention. If the attack occurs at a company-owned facility or nearby, the press will descend upon the site to gather information and report their stories. Organizations that are well prepared to handle the media can benefit from positive news reports. Conversely, those that are unprepared often are viewed negatively in the press and have to spend time trying to correct misstatements. Therefore, organizations need to prepare in advance for the onslaught of reporters, the questions they are likely to ask, and the appropriate responses to give.

Primary and secondary briefing areas should be determined in advance, but not too close to the emergency operations center or command post where the incident commander is directing operations. In addition, the briefing area should be located away from any sensitive areas of the facility. Ideal locations are cafeterias or off-site hotel meeting rooms. A room that can provide a separate entrance and exit for the speaker will prevent the speaker from being bombarded with further questions when passing by a gauntlet of reporters. Escort reporters to ensure they do not enter unauthorized areas. If possible, the briefing area should be equipped with telephones for use by the media.

Media Relations Coordinator

Designate a spokesperson to speak to the news media. This can be a company employee or an outside public relations firm. If an outside firm is used, multiple means of contact 24/7 should be available. A primary spokesperson should be named as well as an alternate.

The designated spokesperson should be someone who is articulate, presents a calm, authoritative presence, and is knowledgeable about the organization and its activities. The spokesperson should receive professional training on how to handle the news media.

At least initially, the spokesperson should not be the senior executive in charge of the facility. Early on in an emergency, information is often incomplete and/or inaccurate. If an executive relays what turns out to be incorrect information, his or her credibility will be damaged.

The spokesperson should be briefed by the incident commander about the scope of the incident, extent of casualties, and other pertinent information. The spokesperson then must craft an authoritative message. The spokesperson should carefully assess the types of questions that could be expected and must be prepared to respond to the questions. The spokesperson should use scripts to communicate what is known but should avoid speculation. Another person should transcribe the spokesperson's address to the media, if possible. This will enable tracking of public statements, the type of questions that are raised, and the answers provided. Misstatements can be corrected, and the types of questions asked will enable preparation for future briefings.

Briefings should be scheduled frequently in the initial hours of the incident and then less often as the incident is brought under control. The briefing schedule should be arranged with the deadlines of the news media in mind. All media representatives should be asked to sign in with their name, media outlet, telephone, fax, and email addresses. This will allow for easier follow-up communications.

Procedures should also be established to respond to inquiries from the news media and others before or in between regularly scheduled briefings. Decide who should receive information requests or how they should be funneled. All communication should be logged, and all public statements or responses (paper or electronic) should be recorded so a consistent message is conveyed. As conditions change or updated information becomes available, all members of the media should be informed.

Identify the news media outlets that service the local area in advance. The news director's telephone, fax, and email addresses should be compiled with the procedures. This will facilitate communication.

General Rules: What to Do & Not to Do

There are some basic rules to follow when dealing with the news media:

- ❑ Remember comments are always on the record.
- ❑ Be calm, clear, concise, caring, honest, fair, responsive, and cooperative.
- ❑ Stick to the facts and avoid speculation.
- ❑ Don't mislead the media by covering up the facts.
- ❑ Convey information about casualties, property damage, and business impact only if the executive crisis management team authorizes disclosure.
- ❑ If you don't have the answer to a question, don't answer "no comment." Instead, take the reporter's name and inform the reporter you will research the question and provide an answer within a specified time frame.
- ❑ Follow a schedule (announce the location and times).
- ❑ Provide equal access to all media (local, regional, and national).
- ❑ Record what you say and when you said it.

Pre-Incident Planning

Pre-incident planning is the process of evaluating the construction, occupancy, protection, and other features of a building and compiling the information into a document that can be used to effectively deal with the consequences of a terrorist incident. The pre-incident plan is used by the site emergency response team and responding public emergency services to manage the terrorist incident. Information gleaned from the building survey described in Chapter 2 can provide much of the information that is needed.

The pre-incident plan should be compiled in conjunction with the public emergency services (e.g., fire, police, EMS) that will respond to the site. NFPA 1620, *Recommended Practice for Pre-Incident Planning*, 1998 Edition is a great resource for developing a pre-incident plan. An example form from NFPA 1620 is shown in Figure 9.

Community Hazards & Warning Systems

Hazardous Chemicals

Identify hazardous chemicals in the surrounding community that are manufactured or stored in significant enough quantities that they could be a threat. Local Emergency Planning Committees (LEPCs) can identify chemical plants or other facilities that submit regulatory required reports on hazardous chemicals[1]. Note the location of each facility, distance to your site, and types of chemicals that could be released.

Public Warning Systems

Identify any public warning system(s) that would be utilized to warn of an emergency. Some jurisdictions (cities, towns, or counties) have outdoor sirens to warn citizens of emergencies that could seriously affect them. These systems are activated by public emergency managers and can be programmed to sound in the affected areas only. When sirens sound, tune to radio or television stations for official announcements or information concerning the emergency. One radio station will probably be designated as the primary Emergency Alert System (EAS) activation station.

The EAS has replaced the Emergency Broadcasting System (EBS). This new technology is designed to allow statewide alerts as well as single or multiple county alerts. Police, fire, National Weather Service, and emergency management officials are all potential users of the EAS in case of an emergency that threatens life and property. Radio and television stations must conduct weekly EAS tests, and the full capability of the EAS must be tested once each month. Subordinate stations in the network must react to such tests within fifteen minutes of reception by carrying the voice and text messages.

Warnings could be issued for releases of hazardous chemicals from a manufacturing plant; transportation accidents involving hazardous materials; an emergency at a nuclear power plant; or suspected chemical, biological, or radiological incident.

Figure 9. Example Pre-Incident Planning Form (NFPA 1620)
(Permission to reprint from NFPA)

Pre-Incident Planning Form

Company Name: **Date**:

Street Address: **Telephone**:

Site Access & Restrictions: **Fences:** height and construction **Security**: Guard Service: Guard House location: Number on duty, knowledge, areas where guards do NOT have access. **Guard dogs**:

Lockbox: **Fire Command Center**: **Emergency Operations Center**: **Remote Annunciator**: **Fire Alarm Panel**:

Life Safety & Occupant Considerations

Number of Occupants: Days [8 - 4]: Evenings [4 - 12]: Nights [12 - 8]:

Handicapped/Special Needs:

Areas of Refuge:

Site Emergency Coordinator

Emergency Coordinator: **Telephone**:

Occupancy or Special Hazards

Type of Hazard: **Location**:

Special shutdown procedures (complex or extended operations):

Controlled environments:

Building Construction

Building Access: Doorways, Locking devices, Accessible windows, Fire escapes, Tunnels, Breechable walls

Length: **Width**: **Height**: **Number of Stories**:

Walls: Interior finish materials: **Floors**: [raised floors]: **Ceilings** [multiple/suspended ceilings]: **Roof**: [concealed spaces/multiple levels] **Roof Covering**:

Vertical and horizontal openings: Large undivided areas, Unprotected openings between floors, Stairwells, Elevator shafts, Utility shafts, Escalators; Type of Fire Doors:

Ventilation System: [type of system] [locations of controls & dampers]

Shelter-In-Place Controls:

Smoke Management: [manual] [automatic] Control locations:

Atriums: Location in the building, No. stories connected, No. stories open to atrium

Building Utilities

Electricity: enters property [overhead] [underground] at; Disconnects located at;Transformers: [PCB]

Emergency Power Supplies:

LPG or Natural Gas: for [building heat] [processes]; storage tank(s) located at: shut-off located at

Fuel Oil: for [building heat] [processes]; storage tank(s) located at; shutoff located at

Domestic Water:

Steam:

Elevators: number, floors served, type, restrictions and location, Fire Service Override, location of keys

Fire Alarm & Communication Systems

Detection System: Method of system activation (manual, automatic), Area of coverage, Fire alarm control panel (FACP) and remote annunciator panels, Type of automatic detectors provided (smoke, heat, other), Off-site alarm transmission, Name of and contact number for off-site monitoring agency:

Voice Alarm or Public Address System: Method and extent of occupant notification

Stairwell Telephone System:

Contact Person or Company Responsible for System Maintenance: [telephone number]

Contact information for person(s) responsible for maintaining the system:

Protection Systems

Automatic Sprinklers: Ceiling/Roof: In-Racks: Other: System Demand: gpm @ psi

Water Supplies: Type of supply: Fire pump: [diesel] [electric] Rating: gpm @ psi

Available Flow: Static: Residual: Flow: Available @ psi residual: **Required Flow**:

Hydrants: **Fire Department Connection**: Location: FDC supplies:

Standpipes & Inside Hose Connections: Type of system, hose outlets, Pressure available, control and sectional valves, FDC, risers:

Fire Extinguishers:

Special Protection Systems: [CO_2, Gaseous Extinguishing Agents, Foam-Water, Dry Chemical and Wet Chemical]: Type of system, Hazard or area protected, Location of control panels, Location of agent supply and reserve, Activation method, Hazards of protective agent

Exposures

North: **South**: **East**: **West**:

Environmental Concerns:

Building Ventilation System

Emergency response may require prompt control of the ventilation system. This could include shutting down fans that take in outside air or switching all fans to exhaust mode. In a large building there are many controls and switches, and they could be located in many different locations. The building survey should identify the arrangement of the system and the controls that must be operated to shut down air intake or to switch to full exhaust. Since prompt action would be required, the survey should estimate the amount of time necessary to accomplish the control of the system using the standard complement of personnel on duty during normal operating hours.

The following should be documented in the pre-incident plan:

- ❑ Identify the type of ventilation system (natural ventilation, individual through-the-wall ventilation units, or duct system with air-handling units).
- ❑ If the building has a ducted ventilation system, record the number of zones, air-handling units, and the locations of controls or switches for each.
- ❑ Document the procedures for purging after an internal release. This could include opening windows, doors, or turning on smoke exhaust systems, air handlers, and fans. If the building has smoke purge fans, determine if the intakes are located at ground level or elevated. Record the location of controls.
- ❑ Identify safe rooms or interior rooms with substantial construction, protected environment, or lower air exchange rate (e.g., rooms with dedicated air conditioning systems) that can be used to provide shelter.
- ❑ Note stairwells that are pressurized to keep out smoke.

Smoke Purge or Exhaust Mode

Compile a list of switches that control smoke exhaust, and mark them on a floor plan labeled "Smoke Exhaust." Note whether the smoke exhaust system is connected to the fire alarm system and is automatically operated when the fire alarm system is activated. The exact operation of the system should be verified to ensure proper operating modes are understood and that the system operates reliably.

Evacuation Plan

Every building should have an evacuation plan. This is not only best practice; it's also the law in most cases. OSHA standard 29 C.F.R 1910.38, *Employee Emergency Plans and Fire Prevention Plans,* (for employers of ten or more persons) and local, county, or state fire codes require evacuation plans.

Developing a good evacuation plan involves assessing building construction, occupancy, staffing, and other factors. These include:

- ❑ Construction and layout of a building
- ❑ Identification of hazards within a building
- ❑ Number and concentrations of occupants, including number and location of those with special needs
- ❑ Occupant notification systems
- ❑ Building utility systems (especially ventilation)
- ❑ Fire protection systems (e.g., sprinklers, smoke control systems, etc.)
- ❑ Means of egress system

The plan development should include floor plans with clearly marked exit travel paths—to primary exits. Also mark secondary exits.

Identify hazards within the building. "Hazards" include those parts of the building that have a higher than average likelihood of terrorist incident. These include the mail center, loading dock, and visitor reception areas. If these areas are targeted, evacuation routes must avoid them.

Concentrations of occupants into auditoriums, cafeterias, and conference facilities require extra attention to ensure there is adequate means of egress to handle the maximum number of occupants. These areas are also potential targets because of the concentration of people.

Special Considerations

Occupants with special needs should be identified in advance. This must be done voluntarily in most cases to protect the privacy of the individual. However, someone with an obvious disability will require assistance, and plans should address necessary aid. Anticipate that there will be others who will not identify themselves as having special needs but will nevertheless need assistance. These could include persons with arthritis or temporary conditions such as a sprained ankle or a broken leg. Heart disease, emphysema, asthma, or pregnancy can reduce stamina to the point where assistance will be needed to descend stairs. This is especially true if there are any airborne contaminants.

High-rise buildings present two challenges—the number of occupants and the time necessary to evacuate all occupants safely. Evacuation plans for high-rise buildings require an assessment of the building and its means of egress, as well as close coordination with the building's emergency response team and the local fire department. Full evacuation of high-rise buildings is not typically ordered when a fire alarm system is activated. Initially, the floor where the alarm sounds and two floors above that floor are evacuated down to the fire floor. The next zone to be evacuated would include the two floors below the fire floor because they would be used as staging areas by firefighters. The number and location of floors to be evacuated could change, depending upon the arrangement of the building's ventilation system and any vertical penetrations in the floors. The evacuation plan may call for all floors within the air-handling zone to be evacuated. Severe fires would require full evacuation. Occupants should not be directed to the roof.

Evacuation plans should identify areas of refuge. These include oversized landings of a stairwell or fire and smoke compartments within a floor. They should be equipped with a means of communication so those awaiting rescue can communicate with the evacuation leader or firefighters. The number of persons that can safely occupy the areas of refuge, and the circumstances in which they would be used should be addressed in the evacuation plan.

Elevators should not be used when evacuating due to fire, smoke, or airborne hazards. During a fire, elevators may be recalled automatically to the first floor. Older elevators may stop at the fire floor, jeopardizing the safety of elevator occupants. Elevators can help spread an airborne hazard.

The evacuation plan should be designed to overcome any limitations in the building's means of egress. Limitations could include limited audibility of the building's fire alarm or occupant notification systems, inadequate complement of exits, poor marking of exits, obstructions or impediments, long travel distances, or dead-ends. Every effort should be expended to resolve these problems, so evacuation is not hindered. However, if limitations are identified, evacuation plans can emphasize alerting in areas of high noise (or limited audibility of the alarm system); use of extra floor marshals to move occupants along; or the redirecting of occupants to overcome bottlenecks.

Multitenanted buildings require close coordination with the building's manager to ensure building and tenant plans work seamlessly. Close coordination and effective real-time communication during an evacuation is essential.

Evacuation plans must also address the shutdown of any processes or equipment that would create a hazard if they were left running after operators evacuated. The time necessary to shut down this equipment should be determined, and safety of operators should be addressed while they are shutting down the process or equipment.

Evacuation Team

Authority to order an evacuation should be vested in the leader of the emergency response team, the head of security, or other individuals who are able to quickly assess a potential threat and determine that it requires evacuation. This will speed the evacuation process.

A team of individuals is necessary to properly coordinate the evacuation of a building with a sizable number of occupants. The following are suggested roles for the evacuation team:

- ❑ **Evacuation Leader.** The leader could be the head of security, engineering, facilities, or another department. The leader will probably not be the same as the incident commander who is responsible for directing multiple teams during an emergency. The evacuation leader reports to the incident commander. An effective means of communication such as two-way radios should be provided for evacuation leaders to communicate with the incident commander at the command post.

- ❑ **Floor Wardens.** Individuals should be assigned to coordinate the evacuation of a defined part of a floor to ensure everyone within the sector evacuates. Wardens must ensure that all areas have been evacuated—including restrooms, storage rooms, and any areas where the occupants might not hear the evacuation order or alarm system.

- ❑ **Stairway Monitor.** The stairwell monitor should hold open the door while remaining out of the path of oncoming evacuees. The monitor should inform evacuees where to move to, the level of exit discharge (ground floor) or to a predetermined intermediate floor.

- ❑ **Elevator Monitor.** The elevator monitor prohibits evacuees from using elevators and directs them to the nearest stairwell. Both passenger and service elevator lobbies should be monitored.

- ❑ **Aides for Evacuees with Special Needs.** Pairing a capable individual with each evacuee needing special assistance will help speed evacuation. These aides or "buddies" should be assigned to assist their charge to an area of refuge where they will await rescue. Buddies should be located in close proximity to the person they will assist so they can establish contact quickly. Alternate buddies should be considered to ensure all persons with special needs are assisted.

- ❑ **Assembly Monitor(s).** A person or persons should be assigned to the assembly area to record the names of persons who arrive at the assembly point. Their job is to confirm that everyone who was in the building has evacuated. These monitor(s) should have access to the employee roster and visitor logs to allow easy check-off accounting of persons as they arrive at the assembly point.

Evacuation Scenarios

The decision to evacuate part or all of the facility depends upon the nature of the emergency and the area affected. A release of chemical or biological agents or a bombing inside a building requires evacuation. However, an incident that occurs outside the building and does not threaten the building or its occupants would not require evacuation. In fact, evacuation of a building could place evacuees into harm's way if the terrorist attack disperses a chemical, biological, or radiological hazard.

Bomb threats are another special case. There are multiple options for handling bomb threats. If a suspicious device is found or there is credible information that an explosive device is in a building, then the building should be evacuated. However, a telephoned bomb threat that includes no specific information may not require evacuation. Evacuation scenarios should be discussed with law enforcement authorities, and they should approve procedures. Specific procedures for handling bomb threats are covered in Chapter 5.

> Keep in mind that no evacuation plan is perfect, and it can't anticipate every possibility—especially every possible terrorism scenario. A well thought-out evacuation plan that is well communicated and executed should work well.

Keep in mind that no evacuation plan is perfect, and it can't anticipate every possibility—especially every possible terrorism scenario. Therefore, the incident commander has to assess the threat to building occupants and make the best decision possible. Keep in mind that occupants will tend to make their own decisions if they have had prior experience in an emergency or if there is a lack of leadership. A well thought-out evacuation plan that is well communicated and executed should work well.

Evacuation Routes & Assembly Areas

Evacuation routes should be established for every floor of each building. Each map should also indicate the location of the primary and secondary assembly areas.

Primary and secondary assembly areas should be separated from each other by direction (e.g., north and south) so both are not subject to the same terrorist event. They should be located away from the staging area for emergency vehicles and a minimum of a quarter mile apart; both should not be in the same path of prevailing winds.

Each assembly area should be given an easily remembered name—such as the name of a local landmark located adjacent to the spot. Maps showing assembly areas should be included with emergency information wallet cards given to employees.

Occupant Notification System

The type and operating instructions for any occupant notification system should be noted in the plan. This could include an emergency voice communication system that is designed for higher reliability and use during an emergency. It could also be a less reliable public address system.

All building occupants should be familiar with the sound of the evacuation alarm system so they will respond immediately.

Shelter-in-Place

An airborne hazard (e.g., chemical, biological, or radiological) outside of a building may require sheltering-in-place. For example, sheltering may be required if a hazardous plume is moving towards an area and there is insufficient time to evacuate the area before the plume arrives. In a center city, large numbers of people in mid- and high-rise buildings can't be evacuated in short order. Even if all people could be evacuated from buildings quickly, traffic congestion would make it impossible to move everyone outside the hazard zone.

An uncontrolled release of hazardous industrial chemicals from an industrial facility or chemicals in transport could result in a plume of hazardous chemicals moving across an area. Explosion of a dirty weapon could also result in airborne hazards outside but adjacent to a building.

Shelter-in-place procedures should be developed for these types of possibility. Inside rooms, such as conference rooms or break rooms that are large enough to allow ten square feet per person, are good choices. The rooms should be above the ground floor, windowless, and have a minimal number of vents and doors that have to be sealed to keep out infiltrating airborne hazards.

Compile a list of switches that control air-handling units, outside air fans, exhaust fans, and unit ventilators or single room air conditioners. Mark the location of each switch on a floor plan, and label each switch "Shelter-in-Place." The exact operation of the system should be verified to ensure proper operating modes are understood and that the system operates reliably.

Supplies should be kept on hand to equip the shelters. These should include:

- ❑ plastic sheeting to fit over windows and ventilation openings
- ❑ duct tape to seal cracks around windows, doors, and vents
- ❑ flashlights and batteries
- ❑ first aid kit
- ❑ cellular telephone or landline telephone dedicated for emergency use
- ❑ battery-powered radio to receive official information

The location of all controls to shut down the building's heating, ventilation, and air conditioning system should be identified and marked "Shelter-in-Place Shutoff."

Mail Center/Package Screening

A postal bomb plan coordinator and alternate should be appointed for the mail center. The function of the coordinator is to assume command when a suspicious postal item is encountered during the screening process. The coordinator is initially responsible for ensuring that personnel who have detected the item cordon off the area to limit the exposure of others. A direct line of communication should be provided between the mail center and security or other emergency contact.

The mail center should be equipped with a hazardous waste container for suspicious letters or packages. A holding/carrying container should be available when no loading dock or outside covered area is available to hold a suspicious item until police or postal inspectors arrive. The container should be placed away from high traffic areas where other employees are working, constructed of heavy wood or exceptionally strong plastic (not metal), and light enough for one or two individuals to carry. It is not intended to contain an explosion. It is extremely important that the container be well ventilated. A second container should be used for small items that are suspected of containing a powder or airborne hazard such as anthrax. This container should be able to contain any powder.

Isolation Area for Suspicious Postal Items

When the mail screening process identifies a suspicious item, it should be promptly removed to an isolation area. The area should offer a degree of isolation where a suspicious letter or package can be placed pending identification and/or the arrival of the police. The following attributes should be considered when selecting the isolation area:

- ❑ The isolation area should be easily accessible from the mail screening area.
- ❑ The hand transportation of a suspicious postal item from the screening area to the isolation area should not pass near areas of high employee population or heavy traffic.
- ❑ If possible, access to the isolation area should not involve the opening of doors, climbing of stairs, or passage through areas of clutter or poor illumination.
- ❑ The total distance from the mail screening area to the isolation area should not exceed fifty yards if possible.
- ❑ The isolation area should, whenever possible, be located outdoors and sheltered from the elements (a covered truck loading dock or an open shed area).

Screening Procedures for Postal Items

Incoming mail should be screened to identify one or more of the following indicators of a letter or parcel bomb:

- ❑ Foreign mail, airmail, and special delivery
- ❑ Hand delivered
- ❑ Unexpected or from someone unfamiliar to the addressee

- ❑ Receipt is followed by anonymous caller asking if item was received
- ❑ Marked with restrictive endorsements, such as "Personal", "Confidential", "Fragile—Handle with Care", "Rush–Do Not Delay", "To be Opened in the Privacy of…", "Your Lucky Day is Here", or "Prize Enclosed"
- ❑ No or excessive postage (usually stamps—not meter strips)
- ❑ Handwritten or poorly typed address, incorrect titles or titles with no name, or misspellings of common words
- ❑ Addressed to someone no longer with the organization or otherwise outdated
- ❑ No return address or one that can't be verified as legitimate
- ❑ Misspellings of common words
- ❑ Unusual or excessive weight for the size, lopsided, or oddly shaped
- ❑ Unusual shape, soft spots, or bulges
- ❑ Rigid envelope
- ❑ Lopsided or uneven envelope
- ❑ Oily stains or discoloration
- ❑ Strange odors
- ❑ Powdery substance on the outside
- ❑ Protruding wires or tinfoil
- ❑ "Sloshing sound." Mail bombs typically do not "tick"
- ❑ Pressure or resistance when opening the package—may indicate a mail bomb
- ❑ Excessive security materials, such as masking tape, string, etc.
- ❑ Visual distractions

Writing the Plan

The emergency response plan should be written as succinctly as possible. It should be well organized with a table of contents and tabs or page dividers so information can be found quickly. Supporting information should be included in the appendix or references should be made to other documents. Figure 10 depicts the outline of a suggested written plan.

The introductory section should include the management policy statement and an organizational statement. Definitions and examples of potential terrorist incidents should be provided where not obvious. In most cases, it is apparent to everyone that there is an emergency—a bomb explodes, for example. However, a strange odor, or one or more persons unexpectedly becoming dizzy or ill may be an emergency because of a release of an airborne hazard. If scenarios that constitute an emergency are identified in advance, there will be less confusion and indecision if an incident does occur.

Emergency Organization

A detailed description of the emergency organization should be included in the written plan. Include the names, titles, incident command structure, and responsibilities in this section. This section should not duplicate the organizational statement but should provide enough information to eliminate confusion and disagreements during an emergency. Coordination with public agencies that may respond to the facility should be defined in this section.

Protection of the safety and health of all members of the ERT is an important responsibility of the incident commander, and it is the responsibility of each team member as well. The use of personal protective equipment (where required by specific procedures), supervision of activities to ensure no one is left unaccounted, and other precautions should be included in the written plan. OSHA standards also require this.[2]

There may be a number of actual plan documents, and they may vary significantly, but they all need to address essential components.

Figure 10. Example Emergency Response Plan Contents

1. Introduction
- 1.1 Policy statement and objectives of the plan
- 1.2 Organizational statement
- 1.3 Definitions of potential terrorist threats

2. Emergency Response Team
- 2.1 Organization of the team
- 2.2 Command structure (Incident Command System)
- 2.3 Responsibilities of members
- 2.4 Safety and health protection of ERT members
- 2.5 Emergency operations center
- 2.6 Coordination with outside authorities

3. Communications
- 3.1 Emergency alarm and occupant notification systems
- 3.2 Radio communications
- 3.3 Notification of personnel on site and off site
- 3.4 News media relations

4. Emergency Procedures
- 4.1 Activation of the plan
- 4.2 Threat assessment
- 4.3 Evacuation
- 4.4 Sheltering-in-place
- 4.5 Combination evacuation and sheltering-in-place
- 4.6 Medical triage, treatment, and transportation
- 4.7 Isolation and decontamination
- 4.8 Fire
- 4.9 Suspicious package
- 4.10 Bomb threats (telephone, letter, or verbal)
- 4.11 Explosion
- 4.12 Potential airborne contaminants
- 4.13 Other security threats
- 4.14 Utility outages (water, gas, electricity, steam)
- 4.15 Property conservation

5. Training Requirements

6. Distribution of and Updating the Plan
- 6.1 Recordkeeping requirements
- 6.2 Plan distribution
- 6.3 Revision history

7. Appendix
- 7.1 Definitions
- 7.2 Pre-Incident planning forms
- 7.3 Diagrams and instructions for fire protection and life safety systems
- 7.4 Diagrams and controls of building ventilation, mechanical, utility, waste treatment and drainage systems
- 7.5 Site diagrams
- 7.6 Material Safety Data Sheets (MSDS's)
- 7.7 Other equipment referenced in the plan

Communications

Effective emergency communications are vital. The emergency coordinator and public agencies responding to the scene require accurate information to make decisions. Inaccurate information or information that is incorrectly relayed from the scene to the emergency operations center can prove disastrous as the emergency coordinator directs personnel and equipment. The plan document should address:

Emergency Alarm Systems. Employees must be trained on procedures to report emergencies and evacuate in the event the alarm sounds. In buildings that have complex fire detection, alarm, and communication systems, personnel who operate the systems must be adequately trained on how to use them, especially when coordinating evacuation. Occupant notification systems to shelter in place should also be devised.

Radio Communications. Two-way radios are invaluable tools for effective communications. Where portable radios have multiples channels, the emergency plan should specify the proper emergency radio frequency or channel to be used. For large-scale incidents, multiple channels may be used for command, evacuation, and support functions.

Notification of the Emergency Response Team. The emergency plan must address procedures for notifying members of the emergency response team both on- and off-site, during business hours and after. A telephone operator or receptionist for small facilities can handle notification. Where there is a constantly attended security office, responsibility for notification can be assigned to security personnel.

Notification of Emergency Services & Agencies. Procedures for notifying the appropriate public agencies must be specified in the emergency plan. The following personnel, agencies, and contractors should be included on notification lists:

- ❑ Members of the emergency response, crisis management, and business recovery teams
- ❑ Department managers (engineering, safety, security, maintenance, housekeeping, operations, transportation, etc.)
- ❑ Executive management (facility, division, corporate)

- ❑ Public relations, public affairs, or media relations department or outside consultants
- ❑ Risk management and insurance (insurance company claims departments, brokers)
- ❑ Fire department
- ❑ Law enforcement (local, county, and state police; FBI)
- ❑ Medical services (on-site medical staff, consulting physician, local hospitals or outpatient clinics, paramedics, ambulance services, and employee assistance plan providers)
- ❑ Utility companies (electricity, water, gas, steam, telephone)
- ❑ Contractors (elevator, plumbing, sprinkler, hazardous material clean-up, smoke/fire restoration, electrical)

Figure 11. Emergency Telephone List

		Organization or Person's Name	Emergency Telephone & Pager	Non-Emergency
Fire				
Police				
FBI				
Medical	Ambulance			
	Physician			
Hospital & Public Health	Hospital Poison Control			
	Public Health			
Government	Public Works			
	Environmental			
	Emergency Management			
Corporate or Division Management				
	Public Relations			
Insurance & Risk Management	Risk Management Insurance broker			
	Property Underwriter			
	Workers Compensation			
News Media	Radio Station			
	Television			
	Newspaper			
Utilities	Electric Co.			
	Voice Communications			
	Data Communications			
	Gas Co.			
	Steam			
	Water			
Contractors	Plumbing			
	Heating			
	Refrigeration			
	Electrical			
	Fire Alarm			
	Maintenance			
	Fire Sprinkler			
	Hazardous Materials			
	Smoke/Fire Restoration			
	Waste removal			
Transportation	Personnel			

Display of Information

Emergency response, crisis management, and business recovery plans can be compiled into paper or electronic documents, or both. Paper documents are easy to work with while sitting around a conference table in an emergency operations center. Electronic documents are easily accessible and updated. The best approach is a combination of the two. All documentation should be maintained in electronic format for all authorized users. Paper copies should be provided at selected locations, including the primary and backup EOCs as well as any existing security or other operations center. Plan documents must also be immediately accessible wherever the emergency operations center is located and wherever the incident commander establishes the command post.

Many formats are used for compiling required information. Some companies use sophisticated relational databases or special software. Others use common word processing, spreadsheet, and presentation graphics applications with which many people are familiar. Use of these applications with their linking capabilities allow compilation of information in a series of connected computer files.

There are other means of displaying information to different audiences:

- ❑ Flowcharts can be used to show a sequence of events—especially when there may be multiple criteria that must be evaluated to determine the best course of action.
- ❑ Wallet-sized cards are used to inform employees what to do for different types of emergencies.
- ❑ Flip charts with visible tabs for various types of emergencies can be posted in each department or work areas. These allow users to quickly retrieve information specific to the designated area or department.
- ❑ Sophisticated software applications are available to allow an entire building and its systems to be mapped to a computer system. When an alarm sounds or a threat is received, operators retrieve real-time information about the area of the alarm and the status of sys-

tems. Real-time monitoring of airborne plumes of chemical hazards can augment these systems. This capability is available from third party companies.

- Personal Digital Assistants (e.g., the Palm Pilot®) are very popular and can store information for relatively easy retrieval.

> No matter what method is used to compile plans and other documentation, everyone needs to have a basic understanding of what to do and how to do it.

No matter what method is used to compile plans and other documentation, everyone needs to have a basic understanding of what to do and how to do it.

Updates, Revision History, & Distribution

A plan is only good if it meets current needs. As conditions change— whether changes in facilities or personnel or the threat profile changes— plans should be reviewed and updated. As the plan is updated, the date and scope of revision(s) should be noted in a defined section of the document. This enables everyone at quick glance to determine if they are working with the latest edition.

A list of persons and locations (e.g., EOC) that receive a copy of the plan should be included with the appendix. The names or job titles of persons or departments who can be contacted for further information or explanation of duties should be included as well. It is important that all persons who have been assigned tasks within the plan receive a copy of the plan. They must know their assignments.

Record-keeping requirements (e.g., OSHA mandated training) should also be listed here, although the actual records should be filed elsewhere.

End Notes

[1] Facilities that manufacture, store, generate, treat, or dispose of hazardous materials in large enough quantities to warrant attention probably must report to government officials in accordance with one or more of the following regulations: *Emergency Planning & Community Right-to-Know Act (EPCRA), Risk Management Plan, Comprehensive Environmental Response, Compensation, and Liability Act (CERCLA), nonreporting Resource Conservation and Recovery Act (RCRA), Toxic Release Inventory System (TRIS)*. Other facilities that should be identified include utilities (e.g., electric power generation, gas, water treatment), nuclear facilities, chemical stockpile, and/or manufacturing sites.

[2] OSHA Standards on Fire Brigades, Hazardous Waste Operations and Emergency Response (1910.120), and Personal Protective Equipment (Subpart L), as well as NFPA 1500, *Standard on Fire Department Occupational Safety and Health Program*, specify requirements and provide guidance on these topics.

Chapter 5

Emergency Response

Response to Terrorist Threats & Attacks

Emergency response plans must address both the threat of possible terrorist activity and actual terrorist events. Public officials could announce there is credible evidence of a possible attack; a suspicious package could be received in the mailroom; or a bomb threat could be called into the switchboard. An attack could be immediately apparent—a vehicle bomb explodes in the public street. However, an outbreak of disease may not be identified as terrorism for hours or days.

The response to a terrorist threat or attack is determined by the credibility of the threat or the type and location of the attack. A bomb threat that includes very accurate information about the facility and a detonation time should result in at least a search of the building and possibly evacuation. Evacuation of an undamaged building after a nearby explosion could expose occupants to more harm than remaining in place—if the bomb dispersed radioactive material or other harmful agents.

Response plans must be specific to the type of attack, the potential hazards associated with the attack, and the proximity of the attack to the building. An outside attack could cause property damage and casualties or could disperse radioactivity, chemical, or biological agents into the air. These could infiltrate surrounding buildings. An inside threat must be contained and personnel moved to safety. Explosions can cause casualties requiring rescue and medical attention, building damage, and fires that could threaten the safety of building occupants. Incendiary devices could cause fires. In addition, after any attack—regardless of location—vigilance must be maintained because other attacks could follow.

Threat Levels

Threat levels are a means of defining the state of preparedness to respond to terrorist events. The FBI operates with a four-tier system ranging from four (normal) to one (a weapons of mass destruction incident has occurred).

Private organizations can benefit from defining similar levels of preparedness. When information or conditions warrant, the threat level can be increased. Communication of the change in threat level also calls for increased preparedness. The promptness of a change in threat levels could be critical if there were a short warning period before an actual incident occurs. The following are suggested threat levels for private organizations:

Threat Level 4 Normal conditions. Routine security and preparedness activities.

Threat Level 3 An increased threat of terrorist activity exists, but there are no known credible threats targeting a facility or industry of this kind. Security staff should be instructed to closely scrutinize all visitors and packages entering the building. Exterior surveillance should be increased. There should be constant monitoring of warning radios and other means of notification. An adequate complement of security personnel should be maintained during all operating hours. Emergency response plans should be reviewed and updated; personnel should be contacted to ensure they are available to respond if needed.

Threat Level 2 Public officials announce a credible threat that targets the organization, its industry, or its type of facility; or the facility is in proximity to other businesses or facilities that have been targeted. Provide enhanced security for the perimeter and screen all access to public areas. Provide additional staffing to maintain coverage. ERT members should be placed on alert to respond immediately if called. Plans should be up to date, and staff should be briefed daily. Drills should be conducted as needed. Constantly monitor public warning systems.

Threat Level 1 A terrorist incident has occurred or is believed to have occurred. Full security precautions and procedures should be implemented immediately. ERT members should be called in if it is safe for them to travel. Nonessential personnel should be evacuated to safety or sheltered, whichever is the safest option. The CMT and BRT should be placed on notice.

Threat Detection

When a terrorist attack occurs, the emergency response plan should be promptly activated. The emergency response team should be notified, and procedures specific to the threat should be carried out.

Notification of a threat may come from public officials if radio, television, or emergency alert system broadcasts are monitored or warning sirens sound. A threat may also be detected by reports of a malicious act or reports of people exhibiting various symptoms. The sound of an explosion may also indicate a terrorist attack. Other potentially serious events may not be easily detected.

Physical Evidence

Security personnel should be alert to potential signs of terrorist attack, as well as equipment, packages, or telephone calls that communicate a threat. In particular, security should look for any spray device (e.g., pressurized cylinder, batteries with a pump and nozzle, container of liquid, gas, or powder) that is out of place with its surroundings. Suspicious parcels that are left unattended in the building should also be closely scrutinized. The following are signs of a possible attack:

- ❑ Unscheduled spraying or unusual application of spray.
- ❑ Abandoned spray devices, such as chemical sprayers used by landscaping crews.
- ❑ Smoke or fog inside the building.
- ❑ Presence of low-lying clouds or fog-like condition not compatible with the weather.
- ❑ Explosions that disperse or dispense liquids, mists, vapors, or gas.
- ❑ Presence of unusual metal debris—unexplained bomb or munitions material, particularly if it contains a liquid.
- ❑ Explosions that seem to destroy only a package or bomb device.

- ❑ Presence of unusual liquid droplets. Surfaces may exhibit oily droplets or film, and water surfaces may have an oily film.

- ❑ Civilian panic in potential high-profile target areas (e.g., government buildings, mass transit systems, sports arenas, etc.).

- ❑ An unusually large or noticeable number of sick or dead wildlife that may range from pigeons in parks to rodents near trash containers.

- ❑ Lack of insect life. Standing water can be checked for the presence of dead insects.

In an explosion, the fact that radioactive material was involved may or may not be obvious, depending upon the nature of the explosive device used. Unless confirmed by radiological detection equipment, the presence of a radiation hazard is difficult to ascertain. Although many detection devices exist, most are designed to detect specific types and levels of radiation and may not be appropriate for measuring or ruling out the presence of radiological hazards.

Symptoms or Sensory Indications

Signs of a possible chemical attack could include multiple people who are nauseous, choking, complain of irritation of the eyes or throat, or collapse. This is especially serious if the personnel are outside a building or have just entered a building. One or more of the following may indicate a potential chemical agent has been released:

- ❑ Presence of unexplained or unusual odors (where that particular scent or smell is not normally noted).

- ❑ Unusual noises, such as the release of gas under pressure in or near a building.

- ❑ Multiple individuals exhibiting serious heath problems, ranging from nausea, excessive secretions (saliva, diarrhea, vomiting), disorientation, and difficulty breathing to convulsions and death.

- ❑ Considerable number of persons experiencing water-like blisters, weals (like bee-stings), and/or rashes.
- ❑ Discernable pattern to the casualties. This may be "aligned" with the wind direction or related to where the weapon was released (indoors or outdoors).
- ❑ Mass casualties without obvious trauma.

Most industrial chemicals and chemical warfare agents are readily detectable by smell with the notable exception of sarin, a nerve agent. Biological agents are imperceptible, and there are no real-time detection systems. Therefore, response procedures can't be based on detection. A biological hazard may be recognized only after a credible threat is received; bioterrorism devices, agents, or a clandestine lab is discovered; or there is a diagnosis of a disease caused by a bioterrorist agent. Public health surveillance will more than likely confirm the existence of a biological attack.

The most probable scenario of a chemical attack on a single building would involve release of the agent at or near ground level. When gases or aerosols are released at ground level, they tend to remain there under stable conditions. Plumes originating at ground level will be diverted upward as they travel over buildings. This is especially true when the ground is hotter than the air above it. If the width of the building is more than twice its height, the shortest path will be *over* the building.

The most practical method of initiating widespread infection using biological agents is through aerosolization, where fine particles are sprayed over or upwind of a target where the particles may be inhaled. An aerosol may be effective for some time after delivery since it will be deposited on clothing, equipment, and soil. When the clothing is used later, or dust is stirred up, responding personnel may be subject to "secondary" contamination.

Indicators of an Attack

Conventional Explosive Devices

A conventional explosive device is still an easy weapon to obtain and use. Components are readily available, and detailed instructions to build a bomb are easy to find. An explosive device can be used to cause massive localized destruction or disperse chemical, biological, or radiological agents.

Improvised explosive devices are categorized as explosive or incendiary, employing high or low filler explosive materials to explode and or cause fires. Large, powerful devices can be outfitted with timed or remotely triggered detonators and can be designed to be activated by light, pressure, movement, or radio transmission.

A bombing may be limited to a single explosion, or one or more explosions could follow in the local or a widespread area. Secondary explosive devices may be targeted against responders. Warnings will probably not precede an explosion. FEMA reports that less than 5 percent of actual or attempted bombings were preceded by a threat.[1]

Chemical Agents

Hazardous chemicals, including industrial chemicals and agents, can be dispersed via aerosol devices (e.g., munitions, sprayers, or aerosol generators), breaking containers, or covert dissemination. An attack might involve the release of a chemical warfare agent, such as a nerve or blister agent or an industrial chemical. Most chemical attacks will be localized, and their effects will be evident within a few minutes.

There are both persistent and nonpersistent chemical agents. Persistent agents remain in the affected area for hours, days, or weeks. Nonpersistent agents have high evaporation rates, are lighter than air, and disperse rapidly (except in small, unventilated areas), thereby losing their ability to cause casualties after ten to fifteen minutes.

It may not be possible to determine which chemical agent was used from the symptoms experienced. Chemical agents may be combined; therefore recognition of agents becomes more difficult.

Biological

Biological agents may be able to use portals of entry into the body other than the respiratory tract. Individuals may be infected by ingestion of contaminated food and water, or even by direct contact with the skin or mucous membranes through abraded or broken skin.

Exposure to biological agents may not be immediately apparent. Casualties may occur minutes, hours, days, or weeks after exposure. The time required before signs and symptoms are observed is dependent on the agent used. While symptoms will be evident, often the first confirmation will come from blood tests or through other medical diagnostic means.

When people are exposed to a pathogen such as anthrax or smallpox, they may not know that they have been exposed; those who are infected, or subsequently become infected, may not feel sick for some time. This delay between exposure and onset of illness, or incubation period, is characteristic of infectious diseases. The incubation period may range from several hours to a few weeks, depending on the exposure and pathogen. Unlike an incident involving explosives or some hazardous chemicals, the initial response to a biological attack on civilians is likely to be made by healthcare providers and public health officials.

Indicators that a biological incident has taken place may take days or weeks to manifest themselves, depending on the biological toxin or pathogen involved. The CDC developed the following list of epidemiological clues that may signal a bioterrorist event:

- ❑ Large number of ill persons with a similar disease or syndrome
- ❑ Large numbers of unexplained disease, syndrome, or deaths
- ❑ Unusual illness in a population

- ❑ Higher morbidity and mortality than expected with a common disease or syndrome
- ❑ Failure of a common disease to respond to usual therapy
- ❑ Single case of disease caused by an uncommon agent
- ❑ Multiple unusual or unexplained disease entities coexisting in the same patient without other explanation
- ❑ Disease with an unusual geographic or seasonal distribution
- ❑ Multiple atypical presentations of disease agents
- ❑ Similar genetic type among agents isolated from temporally or spatially distinct sources
- ❑ Unusual, atypical, genetically engineered, or antiquated strain of agent
- ❑ Endemic disease with unexplained increase in incidence
- ❑ Simultaneous clusters of similar illness in noncontiguous areas, domestic or foreign
- ❑ Atypical aerosol, food, or water transmission
- ❑ Ill people presenting near the same time
- ❑ Deaths or illness among animals that precedes or accompanies human illness or death
- ❑ No illness in people not exposed to common ventilation systems, but illness among those people in proximity to the systems

Nuclear/Radiological

Radiation is an invisible hazard. There are no initial characteristics or properties of radiation itself. Unless the nuclear/radiological material is marked to identify it as such, it may be some time before the hazard is identified. The following signs may be the only indication of a nuclear or radiological incident prior to deployment of monitoring equipment:

- A stated threat to deploy a nuclear or radiological device
- The presence of nuclear or radiological equipment (e.g., spent fuel canisters or nuclear transport vehicles)
- Nuclear placards or warning materials along with otherwise unexplained casualties

An intentional nuclear/radiological emergency could result from the use of different devices:

- **Improvised Nuclear Device.** This includes any explosive device designed to cause a nuclear yield. Either uranium or plutonium isotopes can fuel these devices, depending upon the type of trigger. While "weapons-grade" material increases the efficiency of a given device, materials of less than weapons grade can still be used.
- **Radiological Dispersal Device.** An RDD includes any explosive device utilized to spread radioactive material upon detonation. Any improvised explosive device could be used by placing it in close proximity to radioactive material.
- **Simple RDD.** A simple RDD is any device that can spread radiological material (including medical isotopes or waste) without an explosive.

Combined Hazards

Agents can be combined to achieve a synergistic effect—greater in total effect than the sum of their individual effects. Agents may be combined to achieve both immediate and delayed consequences. Mixed infections or intoxications may occur, thereby complicating or delaying diagnosis. Casualties of multiple agents may exist; casualties may also suffer from multiple effects, such as trauma and burns from an explosion, which exacerbate the likelihood of agent contamination. Be alert for the possibility of multiple incidents perpetrated in a single or in multiple municipalities.

Determination of Protective Actions

Protective actions are determined by the location of the terrorist incident; known and potential hazards; and obvious property damage and casualties. The response must also anticipate the possibility of further terrorist attacks.

The attack may be within the building, immediately outside, a few blocks or even miles away. An attack that occurs miles away could still pose a threat to the site from airborne hazards such as chemical or biological agents or radioactivity.

The scope of any hazards may not be known for hours or longer, so the initial threat assessment should anticipate potential hazards. Any explosion could release airborne hazards including chemical, biological, or radiological material. Symptoms or signs of chemical or biological agents should result in declaring an emergency, assessing what is known, and implementing protective actions.

Protective actions must address all the potential hazards posed by the terrorist act, damage resulting from the attack, or the threat of additional attacks. These include:

- ❑ Fire
- ❑ Building damage or collapse
- ❑ Airborne hazards
- ❑ Casualties
- ❑ Risk of secondary attacks of similar or different kind

A major explosion could cause significant property damage or casualties and could spread airborne hazards. The target building could collapse onto adjacent buildings or people in the street. Airborne hazards within a building could spread throughout the building to affect other occupants. An airborne hazard released outside could infiltrate a building.

Emergency procedures must first stress the protection of people. This could include evacuation from an unsafe building or environment or sheltering-in-place to protect people from airborne hazards outside the building. Any sick or injured people should be administered first aid by qualified members of the emergency response team or triaged for medical treatment by public emergency medical services. If the safety of building occupants is properly addressed and there are sufficient trained and equipped personnel, then other efforts could be undertaken. These could include firefighting, containment of hazardous materials that may have been released, salvage and property conservation. A large-scale incident that over-taxes public emergency services may require extraordinary efforts by the emergency response team and others to assist victims.

Emergency procedures must first stress the protection of people. If the safety of building occupants is properly addressed and there are sufficient trained and equipped personnel, then other efforts could be undertaken. A large-scale incident that over-taxes public emergency services may require extraordinary efforts by the emergency response team and others to assist victims.

The initial size-up of an incident requires determination of the type of hazards, the source of the hazard and the extent of any spread. Sophisticated emergency plans could include software that can estimate the effects of a nuclear, chemical, or biological release, including the area affected and consequences to population, resources, and infrastructure.[2]

Airborne Hazards

If the initial appraisal determines there is an actual, suspected, or potential airborne hazard, the source of the hazard should be determined. If an explosion or vapor cloud is reported or observed outside the building, protective actions should be based on *outside hazard*. If an airborne hazard could have occurred within the building, protective actions should be based upon *inside hazard*. If an outside airborne hazard has possibly infiltrated the building, protective actions should be based on *inside hazard*.

Inside Hazard

Once it is determined that an airborne hazard exists inside the building, protective actions must be based on whether the source originated inside or outside. Since it may not be possible to determine the source, protective actions should be taken based on the most likely source of the hazard.

If the source of the hazard originates within the building, the following action should be taken:

- ❑ Shut down all air-handling units until the type of hazard and extent of its spread can be determined.
- ❑ Evacuate the affected portions of the building. [See Evacuation Procedures later in this chapter.]

If the source of the hazard is inside and spread is localized or contained (the container or package is known):

- ❑ Shut down air handling units for the affected floor(s).
- ❑ Isolate the affected area by shutting doors.
- ❑ Evacuate the affected floors using egress routes that take evacuees away from the affected area.

Outside Hazard

If the source of the airborne hazard is clearly outside and there is no indication that it has infiltrated the building, the following actions should be taken:

- ❑ Shelter occupants. Communicate with building occupants via the emergency voice communication or public address systems to initiate sheltering procedures.
- ❑ Shut down all air-handling units and close all dampers on fresh air intakes.
- ❑ Close all doors and windows to prevent the infiltration of airborne hazards.
- ❑ Contact public emergency managers (e.g., police, fire, emergency management, or other governmental offi-

cials) to obtain official information about the emergency and suggested actions.

- ❑ Monitor television and radio reports to identify the source of the hazard, its location, the wind direction and speed, and estimated time the plume could reach the building.

Hazard Source Unknown

If the source of the airborne hazard is unknown, the following actions should be taken:

- ❑ If there is a perceptible odor, irritation to the eyes, or other obvious signs that the hazard has spread within the building (see previous sections), quickly determine if it is clear outside. If the outside is clear, evacuate the building.

- ❑ Evacuate if there are symptoms (see previous sections) but no obvious signs of a hazard within the building.

- ❑ Check for other possible sources:

  - Are there obvious signs of hazards on adjacent floors? If obvious signs are limited to one floor or part of it, evacuate that floor and isolate the area. Evacuate adjacent areas that are served by the same air-handling zone. Prepare to evacuate the building if signs spread to other areas.

  - If there are visible signs outside the building that indicate an airborne hazard (people are fleeing the area), it is likely an external release.

Emergency Response Procedures

Notifications

A terrorist attack may require notification of numerous public emergency service and law enforcement officials, including local and state police, FBI, fire department, emergency medical services, and public health officials. The emergency response team must be alerted as well. Assuming the incident is significant and will have a major impact on the organization, the crisis management and business recovery teams should be notified. Defined call lists and notification procedures as described in Chapter 4 should be followed.

The forward command post from which the incident commander will direct emergency operations should be identified and communicated to the ERT. The primary or secondary emergency operations center should be opened and equipped for use. The location of the EOC should be communicated to members of the ERT who are to report.

Evacuation

Members of the emergency response team should coordinate evacuation according to established evacuation plans (see Chapter 4 for plan development). Instructions should be broadcast to evacuate via established exit routes (or alternates in case the primary route is affected by a known hazard) and proceed to a specified assembly area. Selection of the primary or secondary assembly areas should be based on the location of the hazards.

Since no evacuation plan can address every contingency, there may be unanticipated scenarios. In these cases, react to known indicators, evaluate the potential threat, and decide whether to call for an evacuation, partial evacuation, or sheltering in place. The public fire department or other emergency services should be informed of the decision to remain in place if they have not ordered sheltering in place. Procedures for shelter-in-place follow.

Reevaluate the situation continuously and base decisions on sufficient, credible information. Logically evaluate what is known and what is not known, but don't speculate on the unknown. Use whatever information is available (e.g., AM/FM radios, television, eyewitness reports, visual observations, and other credible sources).

Reevaluate the situation continuously and base decisions on sufficient, credible information. Logically evaluate what is known and what is not known, but don't speculate on the unknown.

Sheltering-In-Place

Sheltering-in-place involves moving personnel to safe locations within the building. It is not a perfect solution because every building is subject to infiltration of outside air, and the protection that the building affords will diminish over time. Therefore, it is only a temporary measure—for no more than two hours.

Sheltering requires shutting down air-handling systems before the building is enveloped in the hazardous plume of chemicals or other hazard; all windows and doors also must be closed.

Once the hazardous plume passes and authorities advise it is safe to move outside, all windows and doors should be opened, and the building's air-handling systems should be arranged to maximize air exchanges. This will help cleanse the building's atmosphere.

In the event that sheltering-in-place is ordered by public authorities or deemed the best means of protection against external airborne hazards, follow these procedures:

- ❑ Use the building's public address or emergency voice communication system to direct occupants to predetermined shelter areas. If shelter-in-place areas have not been identified, interior rooms such as conference or break rooms—or others that are windowless—should be selected. Inside rooms on the side of the building facing away from the approaching plume are preferred.

- ❑ Equip the shelter areas with a telephone, AM/FM radio, and flashlights if possible. There must be some means to receive official information and to seek help.

- ❑ Take attendance sheets, visitor logs, etc. to the shelter area to account for all expected occupants.

- ❑ Shut down the heating, ventilation, and air conditioning system. Close dampers to prevent intake of airborne hazards and the spread of hazards that may infiltrate the building.

- ❑ Close all windows and doors; lock them.

- ❑ Seal all vent openings and cracks around doors and windows.
- ❑ After everyone arrives in the shelter area, lock the outside doors and place a sign that everyone inside is sheltering in place.
- ❑ Account for everyone who should be in the shelter.
- ❑ Monitor the AM/FM radio for official information.
- ❑ Once the "All Clear" message is received from public authorities, evacuate the building to predetermined assembly areas. Account for all evacuees.
- ❑ Open all doors and windows; arrange ventilation systems to purge inside area (switch to exhaust mode).
- ❑ Return to building when informed it is safe.

Purging

Once officials confirm that the hazardous plume has passed and it is safe to evacuate the building, the building should be purged of contaminates. Air-handling systems should be arranged for full exhaust. Purging should not be employed if there is any chance that there are residual contaminants outside the building that could be taken into the building by air-handling systems.

Suspicious Package or Letter

Suspicious Package Indicators

Some characteristics of incoming letters or parcels that should arouse suspicion include:

- ❑ Foreign mail, airmail, and special delivery
- ❑ Hand delivered
- ❑ Unexpected or from someone unfamiliar to the addressee

- ❑ Receipt is followed by anonymous caller asking if item was received
- ❑ Marked with restrictive endorsements, such as "Personal," "Confidential," "Fragile—Handle with Care," "Rush–Do Not Delay," "To be Opened in the Privacy of...," "Your Lucky Day is Here," or "Prize Enclosed."
- ❑ No or excessive postage (usually stamps—not meter strips)
- ❑ Handwritten or poorly typed address, incorrect titles or titles with no name, or misspellings of common words
- ❑ Addressed to someone no longer with the organization or otherwise outdated
- ❑ No return address or one that can't be verified as legitimate
- ❑ Misspellings of common words
- ❑ Unusual or excessive weight for the size, lopsided, or oddly shaped
- ❑ Unusual shape, soft spots, or bulges
- ❑ Rigid envelope
- ❑ Lopsided or uneven envelope
- ❑ Oily stains or discoloration
- ❑ Strange odors
- ❑ Powdery substance on the outside
- ❑ Protruding wires or tinfoil
- ❑ "Sloshing sound." Mail bombs typically do not "tick"
- ❑ Pressure or resistance when opening the package—may indicate a mail bomb
- ❑ Excessive security materials, such as masking tape, string, etc.
- ❑ Visual distractions

Suspicious Letter or Package Marked "Anthrax"

- ❑ If any contents have spilled, immediately notify security and the emergency response team. The ERT should shut down the building's ventilation system to prevent the spread of powder.
- ❑ Do not shake, open, or empty the contents of any suspicious envelope or package.
- ❑ Do not sniff, touch, taste, or look closely at it or at any contents that may have spilled.
- ❑ Do not carry the package or envelope, show it to others, or allow others to examine it.
- ❑ Place the envelope or package in a plastic bag or other type of container to prevent leakage of contents. Within the mail center, the package should be placed in the holding/transport container.
- ❑ If there is no container, cover the envelope or package with anything (e.g., clothing, paper, trashcan, etc.) and do not remove this cover.
- ❑ Leave the area and section it off to prevent others from entering (i.e., keep others away).
- ❑ Wash hands with soap and water to prevent spreading powder to the face.
- ❑ Follow procedures titled "When a Suspicious Item Is Found."

Figure 1. Suspicious Package Indicators

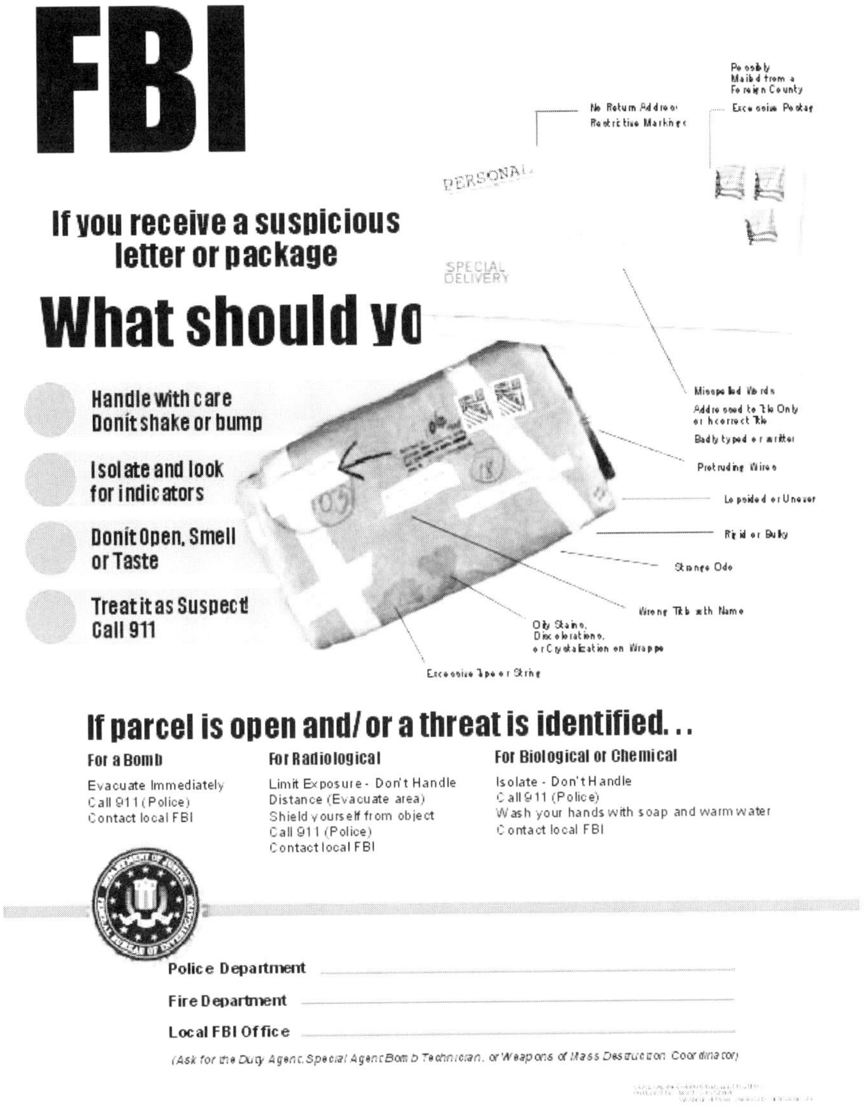

FBI
If you receive a suspicious letter or package
What should yo
Handle with care
Donit shake or bump
Isolate and look for indicators
Donit Open, Smell or Taste
Treat it as Suspect! Call 911
PERSONAL
SPECIAL DELIVERY
If parcel is open and/or a threat is identified...
For a Bomb
Evacuate Immediately
Call 911 (Police)
Contact local FBI
For Radiological
Limit Exposure - Don't Handle
Distance (Evacuate area)
Shield yourself from object
Call 911 (Police)
Contact local FBI
For Biological or Chemical
Isolate - Don't Handle
Call 911 (Police)
Wash your hands with soap and warm water
Contact local FBI
Police Department
Fire Department
Local FBI Office

When a Suspicious Item Is Found

When a suspicious item is found, take the following actions:

- ❑ Notify the emergency response team and/or security.
- ❑ Alert employees in the mail center or exposed adjacent areas of a suspicious letter or package. Inform them of the points of recognition, and tell them to remain clear of the isolation area.

- ❑ If the suspicious item is believed to be a bomb, evacuate the surrounding area, then the floors immediately above and below. Follow procedures for suspected bombs.

- ❑ Place the suspicious item in a sealed container and move it to the isolation area. If the package is believed to be a bomb, leave it alone.

- ❑ Have the employee write down any reasons for identifying the package as suspicious (e.g., excessive postage; no return address; rigid envelope; feel; etc.).

- ❑ Without making direct contact with the suspicious postal item, record all available information from all its sides:

  - Name and address of addressee
  - Name and return address of sender
  - Postmark cancellation date
  - Post office codes
  - Types of stamps
  - Any other markings or labels found on the item
  - Any other peculiarities associated with the item such as oil stains, tears, flap sealed with tape, flaps not glued down, etc.

- ❑ Record all information about the suspicious letter or package in an incident log. If possible, photograph the package from all sides without moving it. The photographs may be used to verify the package as legitimate or to show responding emergency personnel.

- ❑ Contact the addressee of the suspicious postal item to ask for identification or verification of the letter or package. (See following section.)

- ❑ If the package can't be verified as legitimate, or the package can't be verified as legitimate within a reasonable period, notify police. Inform them that a suspicious postal item has been detected and placed in a holding container in the isolation area. Provide the location of the holding area and the name of a contact person for the police to contact when they arrive.

- ❑ Stand by to offer in-house assistance to the police.

Identification or Verification of Suspicious Letter or Package by Addressee and/or Sender

Before calling the bomb squad, security or the emergency response team should attempt to find out if the sender or addressee knows of the item or its contents. If the addressee can positively identify the suspicious item, it may be opened by security with relative safety. If the addressee can't verify the letter or package and only the sender can be contacted, then management must determine whether the letter or package still represents a threat. Sample questions to ask the addressee and sender include:

- ❑ Is the addressee familiar with the name and address of the sender?
- ❑ Is the addressee expecting correspondence from the sender? If so, what is the nature of the correspondence?
- ❑ If correspondence is expected, what are its expected contents and approximate size?
- ❑ If the sender is unknown, is the addressee expecting any other business correspondence from the city, state, or country of origin?
- ❑ Is the addressee aware of any friends, relatives, service personnel, or business acquaintances currently on vacation or on business trips in the area of origin?
- ❑ Has the addressee purchased or ordered anything from a business concern whose parent organization might be located in the city, state, or country of origin?
- ❑ If the suspicious item has overseas markings (e.g., stamps, postmark, or address):
  - Has the addressee recently purchased an item that may have originally been manufactured in a foreign

country and for which a warranty, registration, or guarantee card was mailed?

- Has the addressee returned an item manufactured in a foreign country for repair or replacement that may have been transshipped for the required work?
- Can the addressee think of any reason for receiving mail from an overseas area?
- Has the addressee recently joined or contributed to any charitable, civil, religious, or international organization that might originate correspondence from the city, state, or country in the sender's address?

Bomb Threats

Bombs are the weapon of choice for terrorists, so every facility must be prepared to deal with the threat. Persons carrying explosive devices should be aware of the danger of premature explosion. There is a natural desire to set the explosive and leave the area. Quick in and quick out. Areas near a lobby and every unlocked door are spots that could be used to hide an explosive device. Other possible areas are stairwells, rest-rooms, janitor closets, unused offices, or display areas.

When a bomb threat is received, the predetermined chain of command must be triggered. A senior manager or the incident commander should be assigned as the person in charge. The police and fire departments should be notified. Based upon the available information, the credibility of the threat must be assessed, and the person in charge must decide:

- ❑ whether the building should or should not be evacuated,
- ❑ the scope of any evacuation—full or partial, and
- ❑ the duration of any evacuation.

Current policy in most law enforcement agencies forbids police officials from making decisions on whether to evacuate or not. In the event an explosive device is discovered, however, police or fire officials can order evacuation and other steps to protect the public. Therefore, it is essential that the details of the bomb threat be recorded and communicated to public authorities in order to make the necessary evacuation decision.

Instruct all personnel, especially receptionists and those at the telephone switchboard, what to do if a bomb threat call is received.

A calm response to the caller could result in obtaining additional information. This is especially true if the caller wishes to avoid injuries or deaths. If told that the building is occupied or cannot be evacuated in time, the bomber may be willing to give more information, such as the bomb's location, components, or activation.

The caller is the best source of information. When a bomb threat is called in:

- ❑ Keep the caller on the line as long as possible. Ask the caller to repeat the message. Record every word spoken.
- ❑ If the caller does not indicate the location of the bomb or the time of detonation, ask.
- ❑ Inform the caller that the building is occupied and detonation could result in death or serious injury to many innocent people.
- ❑ Pay particular attention to background noises, such as motors running, music playing, and any other noise that may provide clues about the caller's location.
- ❑ Listen closely to the voice (male, female), voice quality (calm, excited), accents, and speech impediments. Immediately after the caller hangs up, notify security and or the emergency response team. Fill out a checklist such as the one shown in Figure 2.
- ❑ Remain available, because law enforcement personnel will want to interview you.

When a written threat is received, save all materials, including envelopes or containers. Once the message is recognized as a bomb threat, avoid further unnecessary handling. Every possible effort must be made to retain evidence such as fingerprints, handwriting or typewriting, paper, and postal marks.

These will prove essential in tracing the threat and identifying the writer.

While written messages are usually associated with generalized threats and extortion attempts, a written warning of a specific device may occasionally be received.

Figure 2. Bomb Threat Checklist

Department of the Treasury
Bureau of Alcohol, Tobacco & Firearms
BOMB THREAT CHECKLIST

1. When is the bomb going to explode?
2. Where is the bomb right now?
3. What does the bomb look like?
4. What kind of bomb is it?
5. What will cause the bomb to explode?
6. Did you place the bomb?
7. Why?
8. What is address?
9. What is your name?

EXACT WORDING OF BOMB THREAT:

____

____

____

____

____

____

Sex of caller: _____ Race: _____

Age: _____ Length of call: _____

Telephone number at which call is received: _______

Time call received: _____

Date call received: _________

CALLER'S VOICE

☐ Calm	☐ Nasal
☐ Soft	☐ Angry
☐ Stutter	☐ Loud
☐ Excited	☐ Lisp
☐ Laughter	☐ Slow
☐ Rasp	☐ Crying
☐ Rapid	☐ Deep
☐ Normal	☐ Distinct

ATF F 1613.1 (Formerly ATF F 1730.1, which still may be used) (6-97)

☐ Slurred	☐ Whispered
☐ Ragged	☐ Clearing Throat
☐ Deep Breathing	☐ Cracking Voice
☐ Disguised	☐ Accent

☐ Familiar *(If voice is familiar, who did it sound like?)* ____________________

BACKGROUND SOUNDS:

☐ Street noises	☐ Factory machinery
☐ Voices	☐ Crockery
☐ Animal noises	☐ Clear
☐ PA System	☐ Static
☐ Music	☐ House noises
☐ Long distance	☐ Local
☐ Motor	☐ Office machinery
☐ Booth	☐ Other *(Please specify)*

____

BOMB THREAT LANGUAGE:

☐ Well spoken (education)	☐ Incoherent
☐ Foul	☐ Message read by threat maker
☐ Taped	☐ Irrational

REMARKS: ____________________

____

Your name: ____________________

Your position: ____________________

Your telephone number: ____________________

Date checklist completed: ________

ATF F 1613.1 (Formerly ATF F 1730.1) (6-97)

Search Techniques

If there is credible information to believe a device is hidden in the building, the police and fire department should be notified and the building should be evacuated. A designated search team should search the area where the device is believed to be or the entire building if necessary. Priority areas may also include critical business areas; areas subject to visitor or public access; or areas with concentrations of people.

Search personnel must be thoroughly familiar with all hallways, rest rooms, false ceiling areas, and other locations in the building where an explosive or incendiary device may be concealed. They should receive appropriate training on search techniques and the type of device to look for.

Two-Person Search Technique

When the search team enters a room to be searched, they should move to the center of the room and stand quietly. Instruct them to close their eyes and listen for a clockwork device. Frequently, a clockwork mechanism can be quickly detected without use of special equipment. Even if no clockwork mechanism is detected, the team is now aware of the background noise level within the room.

The leader of the search team should look around the room and decide how to divide it for searching. This includes dividing the floor space and determining the height above the floor for the first sweep. The first searching sweep will cover all items resting on the floor up to the selected height. The division of the floor space should be based on the number and type of objects in the room to be searched and not on the size of the room.

First Sweep

Look at the furniture or objects in the room and determine the average height of the majority of items resting on the floor. In an average room, this height usually includes table or desktops and chair backs. The first searching height usually covers the items that are to hip height.

After the room has been divided and a searching height has been selected, both individuals go to one end of the room division line and start from a back-to-back position. This is the starting point; the same point will be used on each successive sweep. Each person searches the room, working toward the other person, checking all items resting on the floor around the wall area of the room. When the two individuals meet, they will have completed a "wall sweep." They should then work together and check all items in the middle of the room up to the selected hip height, including the floor under the rugs. This first searching sweep should also include those items that may be mounted on or in the walls, such as air conditioning ducts, baseboard heaters, and built-in wall cupboards if they are below hip height.

Second Sweep

The search leader determines the height of the second searching sweep. This height is usually from the hip to the chin or top of the head. The two persons return to the starting point and repeat the searching technique at the second selected searching height. This sweep usually covers pictures hanging on the walls, built-in bookcases, and tall table lamps.

Third Sweep

When the second searching sweep is completed, the person in charge again determines the next searching height, usually from the chin or the top of the head up to the ceiling. The third sweep is then made. This sweep usually covers high mounted air conditioning ducts and hanging light fixtures.

Fourth Sweep

If the room has a false or suspended ceiling, the fourth sweep involves investigation of this area. Check flush or ceiling-mounted light fixtures, air conditioning or ventilation ducts, sound or speaker systems, electrical wiring, and structural framing members.

In conclusion, take the following steps when searching a room:

1. Divide the area and select a search height.
2. Start from the bottom and work up.
3. Start back-to-back and work toward each other.
4. Go around the walls and proceed toward the center of the room.

Suspicious Device Found

If a suspected explosive device is found, the building should be evacuated beginning with the floor where the device was found, followed by the floors above and below the danger area. Then all floors of the building should be evacuated as determined by the police or fire department officer in command.

Property Conservation

Salvage efforts designed to minimize further property damage should begin as soon as conditions permit. Valuable or critical property such as custom production machinery and equipment, computer systems and utilities, and valuable papers and records that were identified during the vulnerability assessment process should be targeted for salvage efforts.

Salvage equipment including fans, wet/dry vacuum cleaners, dehumidifiers, absorbent material, brooms, ropes, ladders, hand tools, hammers, staple guns, cleaning agents, plastic sheeting, and tarpaulins should be assembled. Outside contractors who can control damage or begin the repair of damaged buildings or equipment should be notified as soon as possible.

The following actions should be taken to minimize further property damage:

- ❑ Relocate storage or equipment that is exposed to water, smoke, or other damage if it is safe to do so.
- ❑ Cover valuable machinery and equipment with plastic sheeting. Make temporary repairs to broken windows, doors, or roof openings.

- ❑ Shut off broken piping systems. Keep floor drains unclogged to remove water. If there are no drains, use pumps, wet vacuums, brooms, mops, or squeegees to remove water or direct the flow outside the building.

- ❑ Close doors and windows to reduce smoke spread. Shut down ventilation systems to prevent the intake or spread of smoke, or use the system to exhaust smoke within the building.

- ❑ Ventilate all affected areas; clean up debris and damaged materials that smell or are water-soaked; and dry damaged equipment.

First Aid

First aid must be provided for any persons injured or sickened by a terrorist attack. However, several significant issues must be addressed depending upon the nature of the incident. If the attack involves chemical or biological hazards, the agents could contaminate rescuers providing first aid. This requires specialized personal protective equipment (PPE) that is probably only available from emergency services, healthcare professionals, or the military. Second, provision of medical care for victims of attack requires knowledge of what has caused the sickness or injuries and the proper methods of treatment. Third, rendering aid to a contaminated victim could contaminate the treatment area.

Organizations without trained personnel, equipment, and operating procedures to treat victims of chemical agents should:

- ❑ Access the victims only if it is safe to do so.

- ❑ Notify public emergency medical services (EMS) and public health department.

- ❑ Provide care for life-threatening conditions, if it does not expose the caregiver to the same hazard.

- ❑ Coordinate with the local EMS providers to ensure they are aware of the types of exposure that may require special equipment (e.g., decontamination) or treatment protocols.

- ❑ Universal precautions should be taken to prevent exposure to bloodborne pathogens.

Firefighting

When a fire or explosion is reported, the emergency response team must be alerted along with the municipal fire department. The fire department should be told the type of fire, its location, and the nature of any fire alarm zones that have been activated. For larger facilities, the fire department should be directed to respond to a specific gate or building entrance closest to the scene.

The incident commander should order evacuation of the building if there is any doubt that the fire cannot be quickly controlled. The incident commander should also set up a forward command post in front of the building or adjacent to the fire location. The incident commander must direct ERT members to their assigned areas based upon the conditions reported. When the public fire department arrives, the incident commander should inform the fire department incident commander of the nature of the emergency and action being taken. Building plans, a two-way radio (if used by the ERT), master keys, and other information should be provided as requested.

The ERT should supervise fire protection systems (e.g., sprinklers, fire pump, water supply control valves, etc.) and utility systems (generators, electrical disconnects, HVAC, gas, etc.) Duties include:

- ❑ Ensuring that all sprinkler systems and fire pumps are operating properly. The pump operator will keep the pumps operating until the incident commander orders them to shut down. After the incident is under control all sprinkler systems must be restored to full operation.

- ❑ Monitoring the emergency generator to ensure it has started and is running smoothly (if there is a power outage or potential outage).

- ❑ Disconnecting power to part or all of the building at the direction of the fire chief or incident commander.

- ❑ Operating the building's HVAC system under the direction of the municipal fire department incident commander.

Trained ERT members should use portable fire extinguishers if the fire is small and can be attacked with an unobstructed exit nearby.

Security should keep the building entrance and driveway areas free for fire department access. Occupants should be directed to emergency exits and assembly areas away from the building. Unauthorized persons should be kept off the property.

Rescue

Emergency procedures should identify the potential locations and types of specialized rescues that may be required. This includes high elevations, excavations, confined spaces, and other situations. Document this information in the emergency plan and communicate it to the arriving fire chief if there is a need for rescue.

If injured or trapped workers need to be rescued from confined spaces, the rescuers must be properly trained and equipped in accordance with OSHA Standard 1910.146 *Permit-Required Confined Spaces*. Rescuers must be provided with personal protective equipment including respirators and body harnesses; trained to perform the assigned rescue duties; be informed of the hazards of the spaces they enter; and have authorized entrants training as specified in the standard. Rescuers must be trained in first aid and CPR. Practice drills must be conducted annually.

Hazardous Materials

A terrorist attack that releases hazardous materials will require that public emergency services respond. The emergency response team and fire department (or other public agency responsible for HAZMAT incidents) should be notified immediately. The following available information on the nature of the attack and potential release of hazardous materials should be conveyed:

- ❑ Leak, spill, derailment, fire, or explosion

- Location; area affected (area of the building, the entire building, or threat to surrounding properties)
- Name of the material(s) released (material safety data sheet information, shipper, rail car 4-digit UN identification number, placard, or label)
- Quantity
- Type and hazards of the chemical
- Injuries
- Property damage
- Environmental impact

Material Safety Data Sheets (MSDS's) should be used to identify the chemical manufacturer, trade name, and chemical type. The emergency response instructions on the MSDS should be relayed to the incident commander.

Notify occupants in the immediate vicinity of the spill or leak immediately. If it is serious, sound the emergency alarm system for building evacuation.

The emergency response team should be assigned the following tasks depending upon the nature of the incident:

- Shut down air-handling system(s) if there are gases or vapors that spread through the system.
- Shut down energized electrical equipment if fire or explosion is possible. Sump pumps or electrical equipment downstream or at a lower level than the spill or leak should not be overlooked.
- Control the flow of liquid if it is safe to do so. First, stop the source, and then prevent liquids from flowing into sumps, drains, stairwells, shafts, or other low points. Sand or other inert material can be used to dike liquid flow.

Follow these precautions:

- ❑ No one should enter an area where they could be exposed to toxic chemicals unless they have been trained, are outfitted with personal protective equipment, and are paired with a buddy who can effect rescue.
- ❑ If vapors or gases in an enclosed area overcome someone, rescuers should not enter the area unless protected with self-contained breathing apparatus.
- ❑ Manholes or other confined space should not be entered unless personnel are trained, properly equipped, and supervised. Follow confined space entry procedures.
- ❑ If exposed to a chemical, either on clothing, skin, or by inhalation, provide decontamination and medical treatment immediately.
- ❑ Chemicals should not be cleaned up unless personnel are trained in the proper method; have been equipped with personal protective equipment; and have the necessary cleanup and containment equipment. It may be best to use outside contractors.

Government Response to a Terrorist Incident

Although situations may vary, government plans call for response to terrorist incidents and weapons of mass destruction (WMD) incidents as follows:

1. The first responder (e.g., local emergency or law enforcement personnel) or health and medical personnel will in most cases initially detect and evaluate the potential or actual incident, assess casualties (if any), and determine whether assistance is required. If so, state support will be requested and provided. This assessment will be based on warning or notification of a WMD incident that may be received from law enforcement, emergency response agencies, or the public.

2. The incident may require federal support. To ensure that there is one overall Lead Federal Agency (LFA), the Federal Emergency Management Agency (FEMA) is authorized to support the Department of Justice (DOJ) (as delegated to the Federal Bureau of Investigation [FBI]) until the Attorney General transfers the overall LFA role to FEMA. (*Federal Response Plan (FRP), Terrorism Incident Annex*) In addition, FEMA is designated as the lead agency for consequence management within the United States and its territories. In this capacity, FEMA will coordinate federal assistance requested through State authorities using normal FRP mechanisms.

3. Federal response will include experts in the identification, containment, and recovery of WMD (chemical, biological, or nuclear/radiological).

4. Federal consequence management response will entail the involvement of FEMA, additional FRP departments and agencies, and the American Red Cross as required.

End Notes

1 Federal Emergency Management Agency, "Guide for All-Hazard Emergency Operations Planning, State and Local Guide (101), Chapter 6, Attachment G – Terrorism," April 2001.

2 The "Directory of Atmospheric Transport and Diffusion Consequence Assessment Models," published by the Office of the Federal Coordinator for Meteorology (OFCM) is available both in print and online, http://www.ofcm.gov. The directory includes information on the capabilities and limitations of each model, technical requirements, and points of contact.

Chapter 6

Recovery

The Recovery Process

The potential impact of a terrorist attack or series of attacks can be catastrophic, and the recovery process can be enormous. However, a well-managed recovery plan can speed the process and reduce the business impact. Recovery plans must address the needs of employees, their families, and coworkers, as well as the physical restoration of facilities, equipment, and critical functions. The crisis management team must address the firm's reputation so important stakeholders do not harm it.

The recovery process begins as soon as it is safe. Each of the three organizations defined in this book must carry out their respective tasks under the unified command of senior management and supporting departments.

The process should include:

- ❑ Declaration of an emergency and activation of emergency response, crisis management, and business recovery plans
- ❑ Assessment of the situation
- ❑ Determination of critical needs
- ❑ Identification of available resources
- ❑ Application of available resources to meet critical needs

Effective recovery is dependent upon the ability of the organization to act promptly and decisively. Organizations with well-thought-out plans that assign responsibilities to trained personnel will be well prepared to meet the challenge.

Plan Activation

The Emergency Response Plan should be activated if it has not already been so. A command post should be set up for control of on-site operations. The primary on-site or secondary off-site emergency operations center should be opened. The Crisis Management and Business Recovery teams should be notified and assemble at their predetermined emergency operations center(s) if there is significant damage or it appears that operations will be impacted. Rosters of team members that report for duty should be compiled, and the command assignments should be incorporated into the incident command system. Distribute the name of the incident commander and the leaders of all other teams. Recruit replacements for missing members and plan for some turnover. Rotate staff to provide rest.

Incident Command System

The recovery process will involve many people within and outside the company as well as many different departments and teams of employees. The emergency response team will be actively involved in stabilizing the impact of the attack; the crisis management team is charged with managing the organization's response; and business recovery teams will begin to restore critical functions. A command structure and incident management system should be employed to ensure effective and distributed decision-making.

Figure 1. Incident Command Structure for Recovery

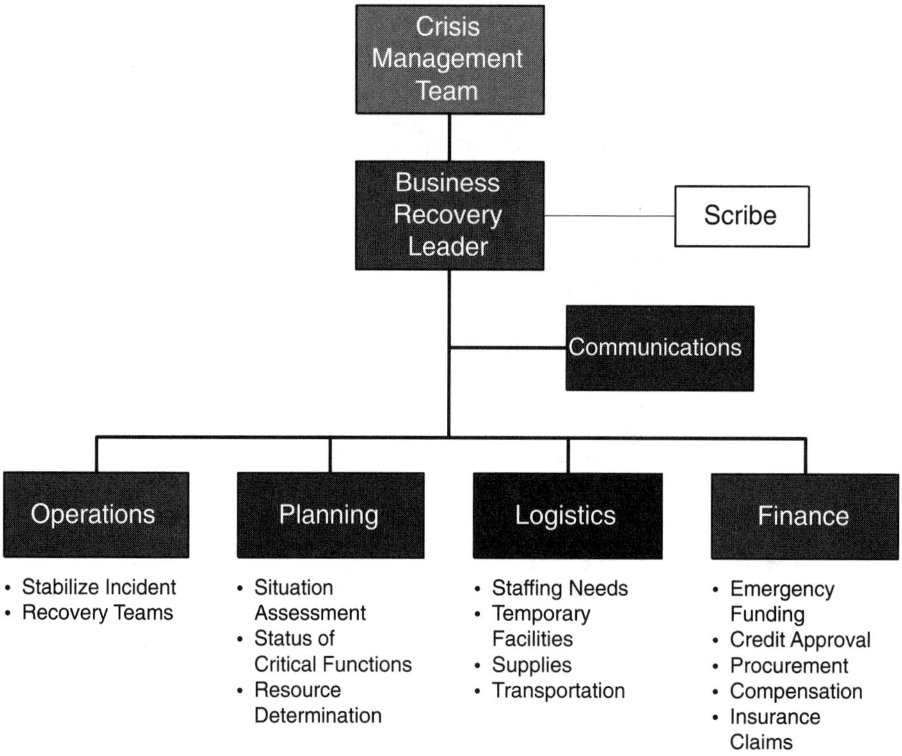

Delegate responsibilities according to response and recovery plans, and establish lines of communications in accordance with the incident command system to facilitate tracking of assignments. Additional resources can be assigned or reallocated to ensure priority functions are addressed first, with less important tasks following.

Situation Assessment

Initial activities will focus on assessment of the situation to determine the impact on people and business operations. The assessment could be very difficult because paths of communications will be hampered, and government authorities may prohibit access to individual sites. The assessment may also be more difficult because the extent of contamination from some terrorist incidents may not be readily known. Sampling and investigation may take considerable time. Therefore, the extent of any problem may not be known for hours or days. Plans should be developed for possible scenarios, and communication and recovery strategies should be readied for deployment as soon as the situation is understood.

Situation assessment includes evaluation of the impact of the terrorist event on the people, facilities, operations, and infrastructure that supports the organization. A terrorist event such as a bombing may impact only a single site or facility. However, an attack with chemical, biological, or radiological weapons of mass destruction could impact many facilities. A serious terrorist attack could also have significant consequences, such as travel bans or quarantines in particular areas; security measures that will limit or delay travel; or multiple effects on customers, suppliers, or raw materials.

Telecommunications and computer networks should be closely reviewed since cyber attacks can be expected. Networks and Web sites should be closely monitored for possible intrusion or hacking. Virus software should be updated to intercept viruses or worms. Telecommunications systems may not be able to handle the increased demands of users, and these systems could be subject to crashes.

ERT members can do preliminary on-site damage assessment; however, operations people familiar with utility systems and equipment may need to do a more thorough assessment.

The damage assessment process should identify damage by building, then department or area, then specific equipment.

Additional forms should be used to gather information on the utility systems that support the building. These include water, electricity, gas, steam, communications, sewerage, and transportation. Damage assessment should identify the scope of damage and the estimated time it will take to repair it after considering the impact of the terrorist attack on the surrounding area, employees, contractors, and suppliers.

Recovery Priorities

Recovery from a terrorist incident should be focused on the organization's critical resources, functions, and assets. These critical resources include people, critical functions necessary for the organization to satisfy its mission, and the reputation or standing of the organization in the marketplace. These resources should be identified during the plan development process, and recovery strategies should be developed to address their loss. If so, the recovery process is easier to execute. If not, the organization will be challenged to survive.

The loss or unavailability of employees with considerable experience and technical expertise can be a crippling blow to any organization. Promoting or reassigning others within the organization; hiring replacements; working overtime; and outsourcing are methods to address the loss of people. However, those who survive a terrorist attack may also need attention. The emotional toll of an incident could have a serious impact on survivors.

The loss or unavailability of employees with considerable experience and technical expertise can be a crippling blow to any organization. The emotional toll of an incident could have a serious impact on survivors.

Critical functions must be restored or replaced within an acceptable time frame before irreparable harm is done. Alternative operating strategies must be employed to temporarily replace functions such as customer service capabilities, manufacturing processes, information technology, and support.

A carefully crafted communications strategy must reach out to important stakeholders who may be concerned about the ability of the organization to survive or remain profitable in the wake of the terrorist attack. Stock prices on Wall Street are affected by many factors, including the perceptions of analysts and investors. A terrorist attack that kills high-level employees and damages or disables a critical facility could be perceived as a serious threat to earnings if not the survival of a company. A carefully crafted communications strategy could quickly dispel misconceptions or unfounded rumors and reassure important stakeholders.

Personnel Safety & Health

After an attack, priorities include provision of medical care for anyone injured in the attack and accounting for everyone who occupied the affected building or facility. Continued activities include coordination of medical care and family assistance. Significant events will also require assistance for coworkers.

Evacuation & Accountability

The first priority in any emergency is the health and safety of people—employees, visitors, contractors, and anybody on-site during an attack. Evacuation or sheltering procedures during the attack are designed to protect them. Once evacuated, all personnel must be accounted for to confirm their safety. Evacuation procedures discussed in Chapter 4 call for identification of primary and secondary assembly areas. Many employees may not reach the assembly area because they don't know the location, it isn't accessible, they are scared and want to go home as quickly as possible, they are injured, or worse.

The first priority in any emergency is the health and safety of people—employees, visitors, contractors, and anybody on-site during an attack. Evacuation or sheltering procedures during the attack are designed to protect them. Once evacuated, all personnel must be accounted for to confirm their safety.

Managers and supervisors are in the best position to account for staff. They can use employee rosters and visitor sign-in logs to account for personnel. Absent any rosters, any available list such as an office telephone directory can be used to systematically account for evacuees. This process should commence as quickly as possible to identify anyone who might be missing.

Anyone unaccounted for must be found, spoken to, or listed as "missing." Other employees might be able to account for a missing employee if they saw the person leave the building or knew he or she was not in the building at the time of the attack. Regardless of the method, every person should be accounted for.

Names of persons who remain unaccounted for after reasonable efforts to find or contact them should be reported to authorities. Concurrently, hospitals should be contacted to determine if the missing employees have been admitted.

A serious terrorist attack that results in substantial numbers of casualties may overwhelm hospitals, so victims could be transported to distant institutions. The names and telephone numbers of area hospitals should be included in the emergency response plan (see Chapter 4). If there are many deaths and bodies are accessible and recoverable (not contaminated), public officials will set up temporary morgue facilities for family members to identify their loved ones.

If bodies are not accessible or recoverable, facilities will be established for family members to formerly report the names of missing persons. Employers should be prepared to assist family members who will need to file missing persons reports.

Medical Care

Persons injured in a terrorist attack will probably be triaged on-site and may be treated at the scene and released. Injuries that are more serious will require transportation to a medical facility; this may not be the closest hospital because local hospitals may not be able to handle the number of patients. In addition, some patients may be transported to hospitals that can provide specialized care. Others may be quarantined if exposed to suspected biological agents.

The employer should determine the location of injured employees and assist in the coordination of medical care in accordance with the wishes of the employee's family. Information should be gathered for filing workers compensation claims.

Counseling

A terrorist attack is a violent, criminal act. The consequences of an attack can be catastrophic—deaths and injuries, property damage, and more. The loss of loved ones, family members, colleagues, and coworkers has a profound effect on survivors. An act of violence also threatens our sense of security and well-being. Those who escape an act of terrorism question why they survived and why others did not.

Reactions to an Attack

There are many reactions to a terrorist attack, and organizations must be prepared to deal with them. The health and welfare of an organization's most valuable resource—its employees—must be paramount in the recovery process. Organizations must recognize the potential impact on everyone affected by an incident—whether the incident directly or indirectly affected the organization's facility.

A RAND study[1] of Americans after the September 11, 2001, attacks reported that 44 percent had profound reactions to them. People were very upset by the events; had disturbing memories, thoughts, or dreams; had difficulty sleeping; were irritable and prone to angry outbursts; or had multiple reactions. Children were also seriously impacted.

The aftermath of an attack can result in many different feelings:[2]

- ❑ Shock and numbness
- ❑ Intense emotion
- ❑ Fear and insecurity
- ❑ Panic
- ❑ Guilt
- ❑ Distraction
- ❑ Anger and resentment
- ❑ Depression and loneliness
- ❑ Isolation
- ❑ Physical symptoms and distress
- ❑ Inability to resume normal activity
- ❑ Delayed reaction—Post Traumatic Stress Syndrome

Some people are more affected than others. These include survivors of past traumatic events, witnesses to the terrorist event or those who watched lengthy coverage of it, and those who were close to victims who were injured or killed.

Employers can expect a loss of productivity as employees attempt to cope with the events. Some may feel insecure about working in a building, such as a high-rise, that they feel is not secure or is a potential terrorist target. After September 11, 2001, many people avoided flying. However, proper support for employees can help avoid delayed reactions, absenteeism, and other problems resulting from repression of natural feelings.

Employer Response

Employers need to respond promptly and compassionately to the needs of employees and their family members. Acknowledge their feelings and reassure them that they are appropriate. Be flexible and offer time for employees to deal with their feelings and to honor the loss of coworkers, friends, and others. Harsh, unsympathetic treatment not only doesn't solve the problem but can result in feelings that the "company doesn't care about me."

Employers should be proactive in their response. If an employee assistance program is available, counselors should be brought in to teach managers how to reach out to staff and to recognize individuals who need attention. Counselors should also be made available to meet individually or in groups on-site. For those who seek privacy, provide information on confidential access to the EAP, such as through crisis hotlines or other qualified resources. Ongoing meetings may be necessary to deal most effectively with the trauma.

Employers can facilitate communication by providing regular briefings on what has happened, who has been affected, and what the company is doing to assist those directly and indirectly affected. Electronic or wall-mounted bulletin boards can be used to exchange messages of sympathy and to communicate information. They also provide a means for coworkers, friends, and colleagues in other companies to express condolences and offer support and assistance.

Managers should seek out staff members and employees having difficulty coping, and they should make referrals if requested. Managers should also allow employees time to congregate, discuss what has happened, and discuss coping mechanisms. Employees who continue to have difficulty should be referred for professional help. This difficulty could be manifested in poor job performance or problematic interaction with coworkers.

Family Support

If an employee is injured or killed, the employer should promptly reach out to the family and provide as much support as possible. A supervisor or coworker who is known and liked by the family can initially fulfill this important role. This family "relationship manager" can provide emotional support, relay information about the status of the loved one, and communicate plans the company may have for dealing with the incident. Company management should also reach out to the family to communicate the company's concern. The manager can also open lines of communication to support other family needs.

Supervisors in conjunction with human resources professionals should carefully collect belongings of deceased employees and package them for personal delivery to the family. This should be done with the greatest respect and in person at a time that is convenient to the family. The family might welcome work-related photographs, videotapes, or other remembrances of the employee.

Human resources staff should also reach out to the family when members are prepared to discuss benefits or other support. Counseling support that may be available from an employee assistance program should be offered to family members if it is within company policy. In addition, families may want assistance with legal matters and financial counseling to determine how best to utilize death benefits.

Funerals & Memorial Services

When lives are lost in a terrorist attack, funerals or memorial services provide an important means to express appreciation for the life of the deceased. They also allow survivors to come together and share common feelings of grief.

If the attack results in the loss of an employee or employees, an ecumenical service may be held at a local religious institution. Employees can participate in planning and delivering the service. This provides an opportunity for employees to come together and share grief in a structured and spiritual manner. It allows family members to join with the coworkers. Such organizational leadership can inspire and carry the organization through a very difficult time.

Critical Functions & Processes

Critical functions are functions or processes that are critical for the organization to fulfill its mission. The manufacturing company described in Chapter 4 had multiple business units producing multiple products. The flagship product was identified as critical in the business impact analysis. The company determined that the manufacturing of this product could be shut down for no more than thirty days before irreparable harm would be done to the company. Therefore, thirty days is the *recovery time objective.*

The situation assessment will determine the direct and indirect impact of the terrorist attack on the organization. The assessment will determine whether critical functions or processes were damaged or will be unable to resume. If so, the recovery strategies outlined in the business recovery plan must be employed.

The business recovery organization and individual teams with their defined roles, responsibilities, and prioritized list of tasks must be executed. Many problems will have to be overcome. Missing information, inadequate resources, and procurement problems are examples. The charge of the business recovery and crisis management teams is to quickly identify problems and determine the best solution.

Individual sectors of the organization must address their assigned responsibilities. If there is a problem with procurement due to purchase authority, the finance sector may have to step in. If other supplies or transportation are needed to obtain temporary staffing or to move people to an alternate site, the logistics sector would provide the solution.

Strong and efficient leadership facilitated by the incident command system and defined organizational structure will help alleviate disputes. Problems should be elevated to the highest level necessary to resolve them.

Roles & Responsibilities

Consistent with the suggested incident command system, the following are various responsibilities for departments aligned under the ICS.

Operations Sector

The operations sector includes the emergency response team(s) responsible for stabilizing the incident. Property conservation efforts should focus on salvage, cleanup, and minimizing further damage. Security of the site should be maintained if safe to remain on-site. Important records, drawings, and information can be recovered for use by the planning sector. The operations sector also includes teams for recovery of critical functions such as information technology and others.

Planning Sector

The planning sector is responsible for assessing the situation and determining what action should be taken. This is done under the command of the crisis management and business recovery leader. The status of critical functions and dependent resources must be assessed.

Initial plans must be established to recover critical functions—what needs to be done first, second, and so on. Can a damaged facility or function be restored or is relocation necessary? Long- and short-term infrastructure needs must be identified. Automated functions may have to be done manually, requiring additional manpower or extended work hours. Outsourcing may be necessary to replace a disabled function, requiring the qualifying and contracting of third parties.

Logistics Sector

If the scope of the loss is significant there will be many logistical needs. The purchasing department should set up open purchase orders for emergency purchases with specified suppliers or vendors. Specific types of goods and services include:

- ❑ Space for temporary or relocated workers or volunteers

- ❑ Temporary badges or means of official identification
- ❑ Food, water, clothing, toiletries, and shelter
- ❑ Office supplies and equipment
- ❑ Additional staffing for telephones, fax, messenger, manual tasks
- ❑ Transportation to and from the facility, victim assistance centers, healthcare providers, etc.
- ❑ Information to allow efficient use of public transportation (obtain schedules, routes, etc.)
- ❑ Lists of suppliers and emergency suppliers (include in contact lists and provide credentials or identification so they can access restricted areas)
- ❑ Contact information for professional resources (EAP, counselors, attorneys, interpreters, etc.)
- ❑ Communications facilities (set up toll-free telephone numbers, call centers, alternate telephone points of contact, additional or alternate email addresses)
- ❑ Fuel supplies for generators, etc.

Finance Sector

Financial management addresses the impact on cash flow and revenue streams. Credit facilities may need to be tapped for emergency cash during the recovery stage. The finance department may also have to set up methods for expedited approval of required payments in line with risk management and the organization's rules of governance.

Accounting procedures must be set up to classify expenditures as reimbursable under insurance policies or operating expenses.

Risk management probably would be responsible for submitting insurance claims for property damage, business interruption, and workers compensation. Expedited or advance payments should be pursued to fund recovery efforts. Guidelines on filing claims are included at the end of this chapter.

Internal & External Communications

One of the important responsibilities of the Crisis Management Team is communication with important internal and external stakeholders. These stakeholders can largely impact the success of the recovery effort and quite possibly the survival of the organization. Nervous analysts that perceive a company will be unable to meet earnings estimates could influence investors to sell stock. Internal stakeholders—including employees—must be informed of the impact of the terrorist attack on the company and, most importantly, their fellow employees. Before management has a chance to react, the news media may be knocking at the door asking what has happened, what is its impact, and what the organization is doing to rebuild.

A prioritized list of stakeholders such as those listed in Table 1 should be compiled or reviewed, and customized scripts should be prepared for each. This will enable the organization to respond more quickly and effectively. The trained media spokesperson can direct the media to a predetermined briefing area and provide regular updates.

As soon as possible the Crisis Management Team should meet and activate the crisis management plan. The availability of essential members, including the media spokesperson or outside firm, should be confirmed or a replacement should be named. Any scripts or templates prepared in advance should be reviewed and customized to reflect what has happened and the impact on the company.

The Crisis Management Team must assess the situation using the Planning Sector of the incident command system, such as the on-scene emergency response team commander, business recovery leader, or other qualified individual. The assessment includes the impact of the attack on people, facilities, operations, supporting infrastructure, the industry, and the economy as a whole. The assessment should be done in stages with the priority on people, followed by facilities and operations. Damage to the infrastructure may not be immediately known depending upon the type of attack. An attack with a chemical or radiological weapon could cause contamination that takes days to assess.

The Crisis Management Team must assess how important stakeholders will perceive the terrorist attack, and all messages must be crafted to allay their concerns as much as possible. Since information will be limited at first, the message must be redrafted as additional information is received. Public statements must be updated and reissued on schedule. The CMT must also assess how messages are received, with misinformation quickly corrected. Communication strategies are identified in Chapter 4.

Sensitive information should be withheld—especially the names of anyone killed or injured. Public authorities should provide identity confirmation, and family members must first be notified.

Table 1. Important Stakeholders

ICS Sector	Dept. or Group	Audiences	Activities
Communications	Public Relations or Communications	• News Media • Community • Employee communication • Community relations • Call centers	• Local, regional, & national print and electronic
Communications	Investor Relations	• Partners • Investors • Analysts	• Impact on operations, short and long-range financial outlook
Communications	Government Affairs	• Government Regulators	• Regulatory impact • Reporting requirements
Communications	Marketing & Sales	• Key Customers	• Communicate any impact on orders or delivery schedules
Finance	Finance	• Banks and Lenders	• Emergency financing
Finance	Risk Management	• Insurance Broker • Underwriters • Claims Adjusters	• File insurance claims • Coordinate with adjuster • Settle claim
Logistics	Human Resources	• Family Members • Employees	• Account for missing staff • Family outreach & support • Employee records & information • Employee and family counseling assistance (EAP)
Logistics	Benefits Managers	• Healthcare Providers • AD&D Insurers • Life Insurers	• Expedite provision of benefits to affected employees and their families
Logistics	Purchasing	• Critical Suppliers (raw materials or services)	• Schedule • Alternate suppliers • Redirect supplies
Logistics	Others	• Contractors • Suppliers	• Restoration of damaged or affected facilities

Example Timeline

The start and end points of the activities that take place after an attack are very hard to determine. Activities more often overlap than follow a chronological order. Figure 2 is presented for illustrative purposes only.

Figure 2. Example Recovery Timeline

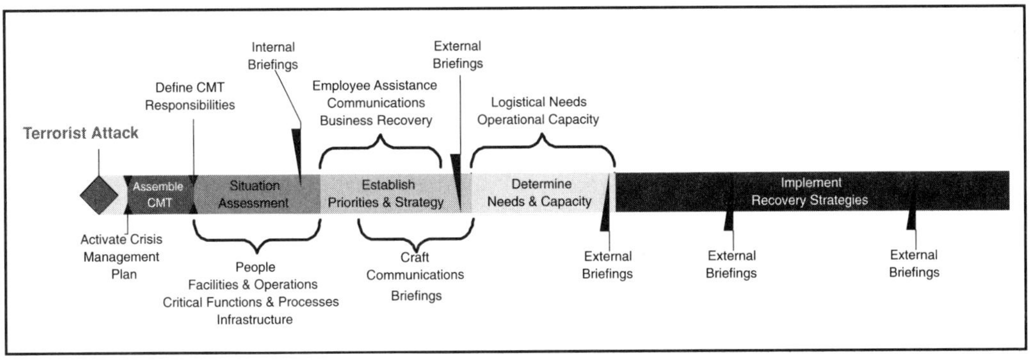

Post-Incident Debriefing

The response to any incident should be critiqued as soon as possible after the incident is stabilized—while the incident is still fresh in everyone's minds. The purpose of the debriefing is to identify changes in procedures, staffing, training, or physical improvements needed to improve response to a future incident. Members of the emergency response, crisis management, and business recovery teams that participated in the response should be assembled. The incident commanders should review the:

- ❑ Initial situation assessment
- ❑ Development of strategy and tactics
- ❑ Execution of policies and procedures
- ❑ Supervision, use, or performance of systems and equipment

The incident commander should solicit honest feedback on what parts of the operation succeeded and what did not. Why were some procedures executed well, and why did some actions not go well? Address the needs for increased staffing, better or more specific training, improved communication, enhanced coordination between public and private resources, physical deficiencies, and other components of the plan.

The results of the debriefing should be documented—especially actions needed to improve future responses. Recommendations should be forwarded to senior management and followed until either completed or dismissed. Plans and procedures should be updated.

Insurance Recovery

A large part of a business's recovery after a loss from terrorism may involve financial settlements from insurance policies. These range from property insurance on the company's physical plant to workers compensation coverage, disability policies, and life insurance for employees. Company officials must be prepared to begin to quantify the financial impact of the event and start the financial recovery process as quickly as possible.

Part of any risk manager's recovery plan should be an understanding of the postloss claim process. Depending on the scope of the terrorist incident, there could be massive injury and destruction, unusual contamination, or failure of public systems. Both the business and its insurance companies may be hampered by the situation itself, in addition to the increased level of frustration and stress that accompanies most large losses.

Because of this, businesses should incorporate a large loss procedure or protocol into their recovery plans. Risk management probably will be responsible for handling insurance claims, as noted in Chapter 4; guidelines on what must be done to pursue insurance proceeds should be incorporated into the recovery process.

The Process

Reporting to the Insurer(s)

Virtually all insurance policies require the insured to promptly—if not immediately—report losses. This requirement appears in the CONDITIONS section of the policy. After a loss from terrorism, the insured should immediately notify any and all insurance companies, keeping in mind that many types of coverage may be triggered.

Do not refrain from reporting a loss or claim to the insurance company because of (a) a belief that the loss will not be covered or (b) concern that insurance premiums will rise. Better to report the loss and learn later that it is not covered than to report it late and have the insurer deny coverage due to a breach of the policy condition. Let the insurer make a determination as to coverage.

Notify not only all primary insurers who might conceivably cover the loss but excess insurance carriers as well. Your insurance broker should assist with the notification process. Take the reporting requirement seriously because failure to do so may cause the insurer to deny coverage on an otherwise covered loss for late reporting.

Use phone, fax, e-mail, overnight delivery, certified mail, or a combination of methods to deliver prompt and verifiable notice to all carriers. Redundant back-up methods help assure that nothing slips through the cracks.

Gathering Information

The first rule is to notify all possible insurers immediately after a catastrophic terrorist event and urge them to send claims adjusters to the location. Many insurance companies will establish field claim posts in the vicinity. The adjusters will conduct their own investigation, but there is nothing to prevent an insured from gathering information proactively.

An insured's or risk manager's initial priority should be to preserve any and all physical evidence that might pertain to the cause of loss or its extent. This might involve not only preserving physical evidence but also taking quality color photographs of the damage. It might include identifying employees who are key witnesses so adjusters can interview them. Document any directives from adjusters to dispose of damaged property.

Often, the insured will be on its own until insurance adjusters arrive. Since a single loss from terrorism may trigger multiple policies, it is important to gather all information related to the event. This may be key to proving that a loss is or is not due to terrorism. (In the current insurance marketplace replete with terrorism exclusions, it may be critical for an insured to be able to prove that a given loss was *not* due to terrorism.) Financial recovery may be available from different lines of coverage—from life and disability to property insurance—and the business may have to prepare information for different insurers and multiple claims adjusters or teams of adjusters. Working with these adjusters to facilitate the information-gathering process will make their jobs easier; in turn, this will expedite the business's insurance recovery.

Occasionally, an insured may retain a public adjuster to conduct an independent loss investigation. This decision may depend on the amount and availability of internal resources schooled in insurance recovery. Typically, insurance adjusters represent the interests of the insurer, especially when the adjuster is an insurance company employee or an independent adjusting service that the insurer hired. Some policyholders are not totally comfortable with this, realizing that adjusting a large claim may involve gray areas of coverage and valuation. As a result, a business may hire a public adjuster to solely represent its interests.

Public adjusters are familiar with insurance policies, conversant with coverage interpretation, and focused to work with an insured to maximize insurance recovery. They usually work on either an hourly or fee-based compensation plan, or they might receive a percentage of the recovery that they can win for the policyholder.

Documentation

The postloss response and recovery plan should include a protocol for gathering such documentation, including monetary damage estimates and expenses, medical bills, extra expenses incurred to clean up the damage, and bills presented by emergency service personnel. Although it is not certain that all expenses will be covered, businesses should meticulously track them. Even temporary repairs to minimize the damage and prevent further loss may be covered under the applicable property insurance policies, and invoices or receipts documenting the amounts spent will be essential for later reimbursement by the insurance company. No documentation often means no reimbursement, even if coverage applies.

The insured should expect to spend considerable time and labor compiling this documentation to support insurance recovery. The insurance policy is a contract that places the burden of proof on the insured. Documentation might involve producing sales records, tax returns, copies of receipts, and statements from accountants or auditors.

Valuation Techniques

It is important to become familiar with the basis for valuing losses under insurance policies before the losses occur. This is particularly crucial when dealing with first-party property insurance coverages. Know, for example, whether the insurance policy pays on an *actual cash value* (ACV) or a *replacement cost* basis. In general terms, actual cash value is replacement cost less depreciation. It is best to review this before a loss—or ideally before buying insurance. The distinction makes a big difference on both the level of recovery and the pricing of coverage.

Astute risk managers will proactively convey information on valuation provisions to senior management quickly after a loss, especially if key limitations may constrain insurance recovery. For example, companies may select ACV coverage when designing their insurance programs because it is less expensive than replacement cost coverage. However, in the event of a catastrophic loss, the recovery also will be substantially lower.

With ACV, an insurer can take deep discounts off payment for old and worn-out property, as well as items that are still vitally useful to the company but not the most modern. Replacement cost—as the name implies—pays for what it would cost to replace property. There is no discount for depreciation.

Similarly, loss valuation methods can differ in time element or business interruption coverage. Does the insurance policy pay for lost revenue or for lost profits? The standard Insurance Services Offices (ISO) business income coverage computes recovery as net profit plus continuing expenses. There are a number of other methods available, and each could result in a different settlement amount. This can make a huge difference in the recovery amount, or even determine whether the insured enjoys a recovery.

These are simply two examples from common property coverages. The point is that an insured should be aware of the basis of valuation provided by the insurance contract. Prior to a loss, such a concern might seem theoretical, but the time to pay attention comes long before a loss arises. It is an example of how risk managers and insureds can avoid friction after a loss, making the stressful claims process more harmonious.

Dealing with Insurance Adjusters

A close working relationship with claims administrators can facilitate recovery after a disaster. A successful relationship is based on many factors.

First, there should be an orchestrated response from both the insured and the insurer, characterized by a high degree of cooperation among all parties. Further, companies that self-insure or have sizable self-insured retentions and face a large loss—such as the World Trade Center attack—must inventory and harness their strengths, preferably well in advance of a major calamity. Strengths can include access to a national claims network, experience in large loss settlement, and well-thought-out pre-incident planning. The company's business recovery teams should prepare lists of insurance company contacts, brokerage claims personnel, and qualified vendors as part of its recovery plan checklists. Guidelines for this are included in Chapter 4.

Corporate Large Loss Plan

There are other fine points that can help prepare a company to face a terrorist loss. One of these is to prepare a corporate large loss plan that could be used in the event of any large loss.

The first step is to determine management objectives. These may vary according to location. Second, identify potential coverage issues before a catastrophic claim occurs. For example, it is important to clarify the scope and any exclusions related to terrorism before a loss occurs. Senior management should be apprised of coverage limitations so that alternative financial recovery plans can be considered.

After the event, risk management staff should meet with insurance claim representatives as soon as feasible to confirm objectives, determine what part the insurance company staff can play in the plan, and agree on division of labor. Develop a detailed scope of loss at the first meeting, and then the recovery team can meet to revise the scope of loss periodically.

It is key to involve the insurer and make its representatives part of the team. Strive to communicate the story of the loss so insurance claims people can relate. The insurance company should be a resource, not an adversary. Involve the insurer in business meetings, not just in insurance meetings, so that its representatives are aware of business objectives. If insurers know the insured's business aims, they are much more likely to work with the insured to achieve them.

Discuss advances and payment requests up front with the insurer. These may be necessary in order to resume business as quickly as possible after the event and to forestall litigation later on. Risk managers should meticulously document all meetings, agreements, and commitments.

Business interruption losses pose significant challenges to companies in large loss negotiations. One of the first steps when proving a business interruption loss should be a meeting with the insurer(s), their consultants, any broker representatives, and the business's accounting and management personnel. Defining how income and revenue are earned at the stricken facility is an important priority. Other steps include:

- ❑ Agree on the information needed and acceptable method of documenting the loss—for example, do you need a copy of every lease, or just a sample or master copy? Photocopies or originals?

- ❑ Establish time frames for documenting loss and processing insurance payment: that is, assuming the documentation is provided, how soon can the insurer start making payments?

- ❑ Identify open issues prior to final negotiations with the insurer. Ask the adjuster or insurer representative, "Is there anything further you will need from us in order to process and pay this claim?" This avoids nasty surprises down the road.

End Notes

[1] RAND, "Research Highlights After 9/11: Stress and Coping Across America," RB-4546, 2001, Internet: http://www.rand.org.

[2] Office for Victims of Crime, U.S. Department of Justice, Office of Justice Programs, "OVC Handbook for Coping After Terrorism," NCJ 190249, September 2001.

Chapter 7

Implementation & Training

Summary of Important Steps

Chapter 2 describes a methodology for assessing an organization's risk to terrorist attack. Opportunities to prevent or mitigate the threat are presented in Chapter 3, and the development of plans to respond to and recover from an attack are described in Chapters 4 through 6. Preparedness for a possible terrorist threat or attack requires:

- ❑ Risk Assessment
- ❑ Mitigation
- ❑ Response plans
- ❑ Recovery plans

Senior management must endorse a clear policy requiring effective planning. Sufficient details about the required work and necessary long-term support should be provided so management can make an informed decision. Plans should be sustainable—whether you're planning for a terrorist attack or a more probable accidental event. Management must understand that there are initial and continuing commitments.

Plans are only as good as the work that goes into them and the knowledge and ability of those who have to carry them out. A plan that exists only on paper may be ineffective, whereas highly trained teams should be able to handle many different situations quickly and efficiently. Training, education, practice drills, and exercises are critically important.

Implementation of effective plans requires promulgation of policies by management and the writing of detailed procedures to respond to and recover from an attack. Systems, equipment, or facilities must be evaluated or provided. Everyone within an organization should be trained—basic training for most, but detailed training for those who must carry out the plans. Leaders must thoroughly understand the organization, its facilities, the potential threat, and effective response procedures.

The following is a brief summary of important steps in the development of response and recovery plans.

Policies & Procedures

Management policies should be promulgated to specify the level of preparedness and response organizations necessary to maintain it.

Since parking garages and loading docks are vulnerable to vehicle bombs, policies restricting access to these areas should be considered. Access to the mail center should also be restricted.

Visitor and contractor screening and access policies should be reviewed to specify who has access to buildings and at what times. Escort policies should be reviewed, especially as they apply to unfamiliar contractors who bring packages and equipment into a building.

Evacuation plans should be developed, posted, and practiced. Evacuation routes should be posted on every floor. Occupants with special needs should be identified and "buddies" assigned to assist them. Decisions on when to evacuate after a bomb threat is received should be determined in advance.

Pre-incident plans should be developed in concert with police, fire, and emergency medical services personnel to document information about the facility in a format that will be useful to these agencies during an emergency.

Organizations & Assignments

A planning committee should be established to develop organizations and plans. They should include emergency response, crisis management, and business recovery teams. Each organization may be subdivided into different teams or units for specific purposes. Existing department heads may be assigned tasks under the Incident Command System. All personnel need to understand their roles and responsibilities.

A media spokesperson should be identified, trained, and prepared with scripts to respond to the news media.

Equipment, Systems, & Facilities

Effective control of ventilation systems is critical to control the spread of airborne hazards into or throughout a building. Controls necessary to shut down air-handling systems as well as to purge a building should be identified and marked.

Communications equipment to receive early warning of threats or actual attacks should be stationed at a constantly attended location. This could include AM/FM radios tuned to Emergency Alert System broadcasters, television, or the Internet.

Life-safety systems and equipment, including adequate means of egress, and public address, emergency occupant notification systems, or other means to notify building occupants should be evaluated. Procedures for effective use should be documented. Consideration should be given to resolving deficiencies.

Equipment should be provided for the mail center to safely quarantine suspicious packages.

Security systems should be reviewed and updated to prevent access to vulnerable building areas, including ventilation systems.

Consideration should be given to providing enhanced photographic identification badges for employees and badges for visitors.

Primary and alternate emergency operations centers should be established and equipped for immediate use.

Compilation of Information

Employee information should be updated periodically and made accessible during an emergency. Contact lists for contractors, vendors, critical suppliers, important customers, government officials, regulators, and others should be compiled for emergency use.

Training

An emergency response plan is only as good as the ability of the ERT to evacuate occupants, control building systems, notify public agencies, and perform the other tasks that may be necessary. A crisis management plan is only as good as the ability of senior managers to assess a situation and its impact on the organization, and to make effective decisions promptly—including communication with critical stakeholders. A recovery plan is only as good as the strategies developed in advance and the ability of the business recovery team to execute them.

Training and education are important because they teach participants at all levels what they must do and when and how they must do it. Practice drills can be used to hone specific skills, such as locating and shutting down an air-handling unit or using a fire extinguisher. Exercises that test the knowledge of team members and the adequacy of the written plan can identify deficiencies. They also build confidence in team members and their ability to do the job.

Levels of Training

Training is required for everyone. Members of emergency response, crisis management, and business recovery teams require substantially more training than other employees.

All Employees

All employees need to understand evacuation procedures and the importance of reporting to the defined assembly area. Their safety—and possibly their survival—may depend on it. OSHA requires that all employees be trained in the facility's Emergency Action Plan (EAP)—when the plan is developed, whenever their duties under the plan change, and whenever the plan is changed. The *Life Safety Code*® requires periodic drills on varying schedules based upon the occupancy of a building.

Employees also should be trained to recognize a suspicious package and to challenge unescorted visitors that have entered restricted areas.

Emergency Response Team

The incident commander requires a high level of training. The IC must be able to assess a threat and determine protective actions. This requires a good understanding of nuclear, biological, and chemical weapons as well as conventional explosives and incendiary devices. The IC must be very familiar with buildings, systems, the ERT members, and how to deploy them effectively. The incident commander must also establish a good working relationship with public emergency services and senior management.

Evacuation leaders; floor wardens; "buddies;" stairwell, elevator, and assembly monitors must understand their individual roles. The evacuation team must memorize the location of both the primary and secondary assembly points. The assembly monitors must be able to use rosters and visitor logs to account for evacuees.

Training for ERT members is dependent upon the duties they are expected to perform. OSHA mandates training for fire brigades and hazardous materials response teams if they are organized. Training must be commensurate with duties and functions, including use of portable fire extinguishers.

Members responsible for supervision or operation of fire protection, alarm, communication, or ventilation systems must understand how these systems operate; the locations of controls; and what to do if a system malfunctions. Likewise, ERT members responsible for shutting down building utility and process systems—such as electrical distribution, emergency generators, HVAC systems, water, natural gas, steam, and mechanical or chemical process systems—must be trained on how to do so.

The training requirements for members of HAZMAT teams, if organized, varies according to the four different functions defined in the OSHA *HAZWOPER* standard—first responder awareness, first responder operations, hazardous materials technician, and hazardous materials specialist. Training requirements spelled out in the standard specify hours of training and demonstrated competency in the tasks specific to each assignment. Refresher training or verification of proficiency is required annually.

Personnel who respond to medical emergencies should be trained in first aid and CPR; it is preferable that they be certified by the American Red Cross or another qualified instructor. They should be trained to exercise universal precautions to prevent exposure to bloodborne pathogens such as Hepatitis B or HIV. This requires training upon initial assignment and annually thereafter. Specific training requirements are outlined in OSHA 1910.1030, *Bloodborne Pathogens*.

Instructors

Instructors should be able to demonstrate that they have a higher level of knowledge and training than those they train. Training assistance can be provided by a variety of sources including private contractors, local fire departments, insurance company loss control or safety engineers, or state fire training schools. First aid training is available through local chapters of the American Red Cross, hospitals, or other medical providers. Specialized terrorism training is also available through many public emergency management agencies at the local, county, or state levels. Online training programs are available through the Federal Emergency Management Agency at http://www.fema.gov.

Frequency of Training

The frequency of training varies according to the duties of the emergency response, crisis management, and business recovery teams. Training should be provided at least initially upon plan development or assignment to a team, then annually thereafter. Leaders should provide training to their respective teams at least quarterly.

Record Keeping

Training should be documented in human resource files. A master training record that describes the scope of training, list of attendees, and date of training should be maintained as well. Records should be accessible for review by regulatory or other interested authorities.

Drills & Exercises

A minimum level of training is required by regulations, but much more training and practice is required to develop and maintain a response *capability.* A response capability is an assurance that people will be able to respond promptly and effectively when asked to execute established plans and procedures on short notice. Well-trained members should be able to respond almost instinctively when notified. They must also be able to react to whatever happens because no emergency follows a script. Development of a response capability requires practice drills to master basic skills and exercises to ensure the pieces fit together. As the complexity of plans increases, so should the level of training, drills, and exercises. The need for drills and exercises is beyond the basic training and drills that are required by regulations. Figure 1 highlights this fact.

A response capability is an assurance that people will be able to respond promptly and effectively when asked to execute established plans and procedures on short notice. Well-trained members should be able to respond almost instinctively when notified. They must also be able to react to whatever happens because no emergency follows a script.

Figure 1. Training Requirements

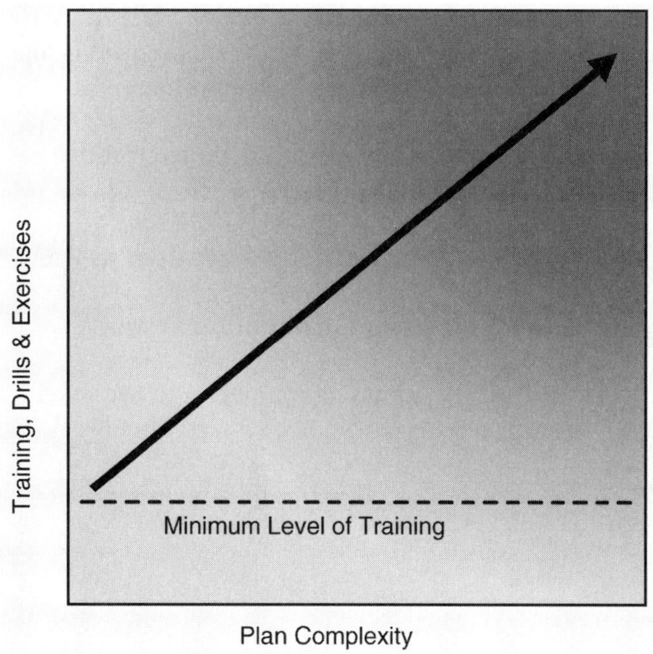

Practice Drills

Training and education teaches employees and team members their assigned tasks and roles. However, classroom instruction or one-on-one teaching provides only limited feedback on how well students comprehend the material. Basic instruction also doesn't evaluate how well plans can be executed. Therefore, drills and exercises are needed.

Drills are functional; they practice a specific part of a plan. They are important because people learn from doing. They also can identify inadequacies in procedures, as well as the ability of persons to execute them. Drills afford an opportunity to learn from mistakes, reassign tasks, or change a plan if certain procedures take too much time or are unrealistic.

Evacuation drills are a common example. Occupants of a building are notified to evacuate the building, and monitors evaluate where they go and how quickly they reach the assembly area. The monitors check whether anyone attempts to use an elevator, which is prohibited, and they direct these people to the nearest stairwell. Drills can test the knowledge of occupants on whether they can find the secondary exit by blocking the primary exit as if it were obstructed by the emergency. Timed drills can determine how long it takes to evacuate a building and reach assembly areas. This is valuable information when determining whether to shelter in place or evacuate if there is an approaching airborne hazard.

Many other drills may be required to practice specific components of the emergency response plan:

- Use of the public address system to warn occupants
- Shutting down air-handling systems
- Handling a suspicious package received in the mail room
- Operation of portable fire extinguishers

Exercises

Exercises afford an opportunity to learn by doing, work together as a team, validate the logic or decisions that were made when a plan was developed, and identify weaknesses before a real event occurs. Exercises can involve interaction with other components of response and recovery plans as well as external agencies and resources. Notification procedures can be tested, and the ability to communicate effectively can be practiced. Effective communications are crucial to a successful response.

Four levels of exercises can be run depending on objectives, available time, and effort. They include orientation, tabletop, functional, and full-scale exercises. Orientation exercises walk members of emergency response, crisis management, and business recovery teams through established plans. Roles and responsibilities are reviewed as well as specific procedures.

The next level exercise, and the most common, is called a "tabletop." Members of the emergency response or crisis management team assemble in a conference room—or preferably a properly configured emergency operations center—to deal with a simulated terrorist attack. The simulated attack is scripted in advance to provide the context. Initially, team members evaluate the situation, determine what needs to be done, and then execute procedures. This walk-through familiarizes team members with their roles and responsibilities.

Once the team becomes comfortable with its roles and responsibilities, "news bulletins" or "reports" can be announced. The new information tests their ability to evaluate a new situation, understand implications, and determine how best to respond. A trained facilitator can develop a meaningful scenario with many twists and turns to challenge the group and expose members to potential problems of a real crisis.

Exercises can be conducted with increasing levels of detail and participation. They can include an entire facility or multiple locations within an organization. Functional exercises involve more hands-on field simulation in addition to a meeting room session. Full-scale exercises can test many aspects of a plan with realistic actors and scenes, but they take considerable time and effort. Plans and procedures can be tested along with systems and equipment, such as communication systems. A simulated explosion could be staged that will require evacuation, rescue, medical treatment, firefighting, coordination with public emergency services, and activation of crisis management and business recovery teams and plans. Large exercises require a considerable amount of time and resources so they are not done as often.

An observer should document the exercise and record gaps or inadequacies in plans, procedures, or execution. A checklist that identifies critical steps or activities can be used to evaluate specific actions, such as prompt notifications or consideration of specific criteria before making a decision.

At the conclusion of an exercise, participants should critique it. They should compare the exercise results with the intended outcome and identify needed improvements. The facilitator can then focus on any problems or unresolved issues.

Business Recovery Plan Testing

The information technology component of business recovery plans requires testing to validate the compatibility of equipment that may be located at an alternate company location or at a vendor's cold or "hot site." Applications should be run on the off-site equipment to ensure compatibility. Testing should be done at least annually, as required by regulations, or after significant changes are made. More frequent testing may be required when a plan is new or if a plan has failed a prior test.

Partial tests can be conducted more frequently for critical components of the technology infrastructure. Vital records, data, and other information should be checked to ensure it is up to date and backed up securely off-site. Contact lists can be verified or test pages can be sent to members of recovery teams.

Audits

Audits are another means to validate that response and recovery plans are adequate. Internal or external audit personnel, qualified outside consultants, or members of the various response and recovery teams can conduct the audits. A successful audit requires a combination of document reviews, visual observations, and personal interviews.

The criteria for the audit should be based on the critical components of the emergency response, crisis management, and business recovery plans such as:

- ❑ Are organizations in place?
- ❑ Are membership rosters current (e.g., names, titles, telephone numbers, etc.)?
- ❑ Do interviews validate that members of each team are well aware of their roles and responsibilities?
- ❑ Are documents up to date?
- ❑ Is required equipment in place and in good working order?
- ❑ Does the schedule of training classes and exercises follow regulations and company policy?
- ❑ Have issues identified during exercises been properly addressed?

The results of the audit should be presented to management in a manner that will result in positive action and enhanced preparedness.

Plan Distribution & Updates

Distribution of the Plan & Updates

A list of plan recipients and the locations where copies of the plan are maintained should be documented. Copies may be kept in places such as the emergency operations center, in online folders, or on a dedicated Web site. Include procedures for updating the plan when physical conditions or personnel change. The names or job titles of persons or departments who can be contacted for further information should be noted as well. It is important that all persons who have been assigned tasks within the plan receive a copy of it. They must know their assignments.

An annual schedule should be established for reviewing and updating all plans. However, they also must be reviewed whenever there is a substantive change in personnel, facilities, or operations.

Record-keeping requirements should be specified in each plan document, but actual records should be filed elsewhere.

Other Considerations

Other programs are important to ensure preparedness for a terrorist attack or other emergency. These include programs to maintain critical systems and equipment, as well as loss prevention and security programs to minimize the probability of an incident. The programs listed below should be part of every facility's loss prevention program:

- ❑ Inspection, testing, and maintenance of fire protection and life-safety systems
- ❑ Impairments to fire protection and life-safety systems
- ❑ Loss prevention inspections

Chapter 8

Impact on the Insurance Market

Market Impact

The September 11th terrorist attacks have had an oceanic impact on the property and casualty insurance markets. This chapter provides a succinct overview of the contours of those changes.

Prior to September 11, 2001, a number of prominent features were becoming visible in the property and casualty marketplace, including:

- declining profits
- increasing insured losses
- rising premiums
- restrictive terms and conditions
- stable reinsurance capacity

After September 11, these signs became impossible to overlook. Further driven by terrorism losses, the property and casualty insurance industry suffered overall deterioration in its financial performance in 2001. The Insurance Services Office (ISO) predicted an industry-wide year-end combined ratio of 119.9 percent. ISO projected that losses from the events of September 11 alone degraded the insurance industry's combined ratio by 6.3 percentage points for the calendar year.[1]

Another large change triggered by the terrorist attacks was the hardening of the reinsurance market. In turn, this had a domino effect on the primary insurance market and accentuated the overall market hardening. Pricing policies of primary insurers are governed, at least in part, by the reinsurance market.

A special benchmarking survey conducted by the Council of Insurance Agents + Brokers (CIAB) revealed various market changes. They included more restrictions on coverage, higher deductibles, and elimination of blanket limits for some lines of coverage. Further, the study showed that insurers had become more exacting in terms of the information they demanded from insureds or prospective insureds, and that terms and conditions were being tightened, a situation that often was not negotiable.[2]

Is Terrorist Threat Uninsurable?

One reason that terrorism poses such a threat to the insurance industry is that many consider it uninsurable. Robert Hartwig, vice president and chief economist of the Insurance Information Institute in New York, states that, "Terrorism risk, for the moment, appears uninsurable because it's not possible to measure the risk."[3]

Insurance is predicated on *the law of large numbers*. In simplified form, this states that the ability to credibly predict future events rises with the increase in the universe of past events from which to extrapolate. If an event has occurred, say, 10,000 times in the past, the ability to predict it happening again is more accurate than if it has occurred ten times in the past.

Insurance is predicated on the law of large numbers. In simplified form, this states that the ability to credibly predict future events rises with the increase in the universe of past events from which to extrapolate.

Credible insurance pricing, which undergirds insurer solvency, is predicated on the law of large numbers. For fire, flood, and even hurricanes, large numbers exist. For terrorist attacks, though, an historical storehouse of data is lacking. This significantly impacts the insurance industry's concern over the insurability (or uninsurabuility) of the terrorism peril.

We should note, however, that others dispute the notion that terrorism is uninsurable. Some emphatically state that insurers are in the business of assuming risk and that they have the ability to insure terrorism losses. If they—the presumed experts and professionals—can't underwrite this risk, who can?

Editorialist Homan W. Jenkins, Jr., writing in the *Wall Street Journal* (Oct. 31, 2001), pointed out

> "Car accidents don't cause insurance companies to stop covering car crashes. Ditto hurricanes. Earthquakes are usually seen as an opportunity to sell a lot more earthquake insurance. Why not the same with terrorism insurance?"

Why not indeed? Insurance commentators may counter, however, that car crashes, hurricanes, and earthquakes are common. They have existed in sufficient numbers and over time to allow insurers some semblance of statistical credibility—and predictive capability—in pricing coverage and setting rates. Their universe of data and storehouse of historical experience is vaster, enabling them to translate risk into somewhat credible pricing. But there does not appear to be comparable information about terrorism.

Impact on Reinsurance Market

No discussion of the insurance market would be complete without covering the effect of terrorist acts on the reinsurance market. Before September 11, likely only those familiar with the insurance industry truly understood the term "reinsurance." After that date, however, many people outside the industry became familiar with it.

One reason is that the availability and price of reinsurance can dictate whether a risk manager can or cannot obtain primary coverage for various perils and at what prices. Part of the issue is capacity, that is, how much reinsurance will be available to carriers writing new policies? In some cases, the answer is, "Very little."

The January 2002 reinsurance treaties showed signs of tightened capacity for terrorism cover. Reinsurers imposed global terrorism exclusions on new and renewal treaties. Further, reinsurers have become more exacting in the underwriting process, demanding greater information and disclosure of underlying exposures. It also appears that reinsurance brokers have had a harder time getting manuscripted policies accepted.

Instead, reinsurers may increasingly insist on standard, non customized forms. In the aftermath of September 11 and the tightening reinsurance market, some large brokers and insurers have rolled out new programs to fill the void and provide more reinsurance capacity. They include Aon, Marsh, Inc., AIG, and Chubb. One reason that some brokers have been able to do this is that they do not have to recover as great of sums for past losses as do most insurers. This provides brokers with a window of opportunity to nimbly introduce new sources of reinsurance capacity.

Reflecting trends in the reinsurance marketplace after September 11, an "Insurance Market Overview" conducted by Willis (www.willis.com) reported:

> "To control these previously unforeseen correlations [between different policy types] reinsurers are likely to seek additional restrictions in coverage moving towards named perils and in some cases limited territorial scope. Reinsurers will be studying and may no longer be prepared to cover some of the widely drafted business interruption and suppliers' extension wordings currently in use."

Property Insurance Market Overview

The property insurance market was particularly hard hit. New reinsurance treaty quotations to primary carriers increasingly contain terrorism exclusions. Primary insurers are often compelled to pass along such coverage exclusions to existing and prospective policyholders. As noted in the previously mentioned CIAB study, deductibles and other terms and conditions on property insurance policies are cresting.

This situation may evolve into even more difficult problems. Developers that do not have coverage against terrorist acts may not be able to obtain loans for new construction, which could contribute to a downward spiral in the economy. One real estate consultant reported that a client—a shopping center—received notice from its lender. The lender required certification that the shopping center had insurance coverage for war risks. This was a question that lenders had not asked for perhaps a decade or more.[4]

Craig Thummel, an independent risk management consultant in Houston, Texas, described the situation in this way:

> "I just completed two property insurance renewals and thought the results might be of interest. Both suffered doubling of rates and higher deductibles, from $25,000 to $100,000 in both cases. The wind deductible increased to 2 percent for any storm (as opposed to a named storm). On one, the underwriter would not use a six-month period of (business) interruption, and

> would only quote using 365 days. No reason, just mandated.
>
> "A terrorism exclusion was mandated, and the language was non-negotiable. The exclusion is so broad as to possibly eliminate coverage for a situation where a terminated employee sets fire to an ex-employer's buildings, or eliminate coverage for computer or Web site hacker attacks. The language is so vague that I'm certain the insurer will invoke it at their [sic] pleasure and at the insured's expense either through denial of coverage or legal costs to fight it.
>
> "Renewal terms/conditions in both cases were available less than forty-eight hours prior to expiration, and some were not fully developed despite the broker's efforts. During the last three years, no insurer cared to visit the properties, and in both cases three insurers would consider the risk only if they sent a representative prior to working on a proposal.
>
> "Both property risks are very clean and loss free. One has been with the same underwriter for three years, and, when asked about the (over) $500,000 claim-free premium collected during that time, stated that they do not look at individual risks, only their book of business results. It went downhill from there.
>
> "There was no meaningful competition for either risk. Some underwriters just threw out some big numbers in the beginning and essentially said if the price was of interest, get back to them.
>
> "The best advice: warn clients (internal or external) early and often, and educate them about the reasons the market is in this chaotic (one word for it) state."

In early February, *Reuters Newswire* quoted Harvey Pitt, chairman of the U.S. Securities and Exchange Commission, as saying that the SEC is considering whether to require businesses that do not have terrorism insurance to disclose that fact to investors.

Property in transit will also be difficult to insure. The railroad industry is particularly hard hit by the lack of insurance coverage for terrorism. Some rail carriers transport hazardous shipments, and federal law does not allow rail carriers to refuse such shipments. These carriers have a huge liability exposure from terrorist acts. Obie O'Bannon, vice president of the Association of American Railroads says that rail carriers will have the Hobson's choice of either violating the law or being financially irresponsible by having no insurance coverage for loss due to terrorism.[5] Dissenting views over insurers' plights arise from some outside the insurance industry, however.

Some insurance buyers have expressed skepticism over the reasons for the rate increases; they question whether insurers are seizing upon the terrorist attacks as a pretext for raising rates across the board, including coverage lines not clearly impacted by the hijackings. Some may believe that insurers are using these threats as a way to exit the soft insurance cycle that has existed for the last ten-plus years. These same observers note that many insurance stocks have enjoyed rising valuations of as much as 30 percent since the attack.

The Insurance Services Office (ISO) filed several terrorism exclusions, which had been accepted by a majority of state insurance regulators. Notable exceptions were New York and California. Carrier-specific exclusions also are possible. Such exclusions are not limited to property insurance; they are being drafted for both property and casualty lines, with even some treading into personal lines coverage.

Casualty Insurance Market Overview

Workers compensation losses arising from the September 2001 terrorism attacks will be in the billions of dollars. The workers compensation insurance market was already hardening prior to these events, but the preexisting market trend is exacerbated by the disaster. Underwriters and insurers are much more cognizant of aggregate risk, i.e., having a large number of employees concentrated at one or more locations. Some workers compensation insurers will not renew accounts if, for instance, employers have 500-1,000 employees at one site.

Clearly, the September attacks have prompted workers compensation underwriters to pose questions rarely, if ever, asked before. For example, the Trammell Crow Company risk manager reported that his workers compensation insurer wanted to study the company's concentrations of employees at specific locales, something in which the underwriter never seemed previously interested. Clearly, insureds with high concentrations of employees at defined locations may need to brace for higher premiums or, in some cases, prepare for coverage being tough to find at all.

The issue is correlation of various exposures in one location. The September terrorist attacks resulted in insured losses across all lines of coverage. Workers compensation, life, health, disability, property, liability, aircraft, and excess coverages all were triggered by the attacks. Insurers and reinsurers that wrote multiple lines at the sites were affected simultaneously in a number of coverage areas. This appears to be one of the main reasons that insurers now are reviewing the correlation of risks, as opposed to confining their scrutiny to a particular line of exposure and coverage.

The umbrella insurance market also has been impacted. While capacity is available, some insurers are inserting terrorism exclusions on these policies as well.

Liability rates may also firm up and trend higher. The extent depends on the type of risk. Due to potential liability and legal exposure concerns, price hikes may rock anything having to do with airlines and aviation.

In general, risk managers can expect all insurers to ratchet up the rigor of their underwriting. The process of getting insurance quotes will take more time. Buyers may have less time available between receiving their insurance proposals and the date they need coverage.

Market Security Issues

Many buyers and brokers worry that the drain of financial resources will imperil the solvency of certain insurers. Actuaries at Tillinghast and Milliman estimate losses from the September 11, 2001, attacks to be in the range of $50 to $63 billion. Insureds should review the financial ratings of their insurance carriers, and agents and brokers should make sure that they do not place coverage with insurers that are financially unsound.

As a result of such concern, several major insurance brokers have been conducting their own in-house financial analysis and "stress tests" to gauge insurer financial health. Virtually all major insurance brokers have internal security committees to evaluate carrier financial strength. Although the insurance industry itself seems adequately capitalized to absorb the blow delivered on September 11, some individual companies may be threatened.

Insurers face a double whammy from terrorist threats. First are the additional claims and expenses flowing from the losses themselves. Second is the reduction in value of investments resulting from the market uncertainty that occurred post-attack.

The financial strength of carriers could be the weakest link in a business's terrorism risk management program. The penalty for this goes beyond any upbraiding by media commentators. After assessing price and service, insurance buyers must consider security a prime factor when purchasing coverage. If all other factors are equal, insureds should choose the insurance carrier with the higher financial rating.

To help with this, independent insurance company rating agencies grade insurers on their financial strength. Let's look at some of the prime resources available to risk managers who, in the wake of terrorism threats and solvency concerns, want to ratchet up the vigilance of their financial monitoring:

A.M. Best and Better

James Brittle is a broker and former director of risk management for LaRoche Industries in Atlanta, Georgia. He notes that many insurance brokers feel almost a compulsion to use A.M. Best ratings. As a risk manager in a former life, though, he utilized Weiss ratings (West Palm Beach, Florida). He feels they tend to provide a better picture of carrier strength and financial fortitude. Brittle feels that, "Best is prone to believe presentations more than hard data and that has been its downfall in the past." Brittle is partial to the Weiss ratings. Though he concedes that Weiss is an expensive service to maintain, the quality is strong.

Dave Parker, risk manager for Pima County, Arizona, likes a combination of the raters, usually relying heaviest upon

Standard & Poor's (S&P). Parker saw several insurers that had decent Best's ratings in the 1980s suffer problems. In hindsight, he feels that S&P was a better indicator at the time. Parker notes, "The lesson I've learned is to look at Best's, S&P, Moody's, and anything else I can get." If he sees any inconsistency, he then taps his insurance brokers for their recommendations.

Other Weathervanes

R. Scott Ecker, CPCU, AIAF, is a consultant with Atlas Risk Decisions, Brant, Michigan. He candidly admits that he hates accounting but studied insurance accounting with the purpose of being able to read a carrier's financial statement. "As flawed as they may be," Ecker says, "the IRIS tests may be one of the best early warning signs of trouble." While they take time to run and must be read as a whole since no one sign is a sure-fire indicator of trouble, Ecker believes they function as good weather vanes for risk managers.

Ecker believes that an integral part of a risk manager's job is assessing the capability of the transferee. "My crystal ball is in the shop, and the Psychic Hotline isn't returning my calls," Ecker notes, "but I've steered clients away from every major insurer debacle since Mission a year or longer before they hit the press."

Resources for Checking Fiscal Fitness

Risk managers can also contact the following insurance company rating services for financial health information:

A.M. Best (908) 420-0400 http://www.ambest.com/. Risk managers can get current information on important insurers throughout the country and the world with Best's Company Reports. Each Report contains the same in-depth information for the companies reported on in a companion publication, Best's Insurance Reports. This information is updated throughout the year.

Duff & Phelps (312) 368-3157 (http://www.duffllc.com/index.html).

Moody's (212) 553-0377 (www.moodys.com). Moody's provides credit ratings on roughly 700 insurance companies worldwide, with coverage growing rapidly in Europe and Asia. Moody's also rates the major reinsurers and financial guarantors. In addition, it provides performance ratings and supporting research on the Lloyd's syndicates through its Lloyd's Market Service. All Moody's insurance research services feature analyst access, credit opinions, in-depth analysis on individual companies, and industry-wide outlooks and commentary.

Moody's also provides both debt ratings and financial strength ratings for life and property and casualty insurance companies. Debt ratings reflect the companies' abilities to meet their obligations to repay interest and principal on outstanding obligations to fixed-income investors, including guaranteed investment contract (GIC) investors. Financial strength ratings are opinions of the ability of companies to repay punctually their senior policyholder claims and obligations.

Moody's Life Insurance Credit Research Service provides ratings and research on over 260 life insurers. Its Property and Casualty Insurance and Reinsurance Research Service provides ratings and comprehensive research on over 400 property and casualty, mortgage, and title insurers and reinsurers, as well as financial guarantors. Both services offer clients regularly updated rating lists and credit opinions; annual in-depth analysis on companies, industry outlooks, and statistical handbooks; special comments; and consultation with a staff of trained credit professionals.

Moody's also provides performance ratings and published commentary on close to sixty Lloyd's managing agencies and over ninety syndicates, as well as general commentary on trends affecting the Lloyd's insurance market. Its Lloyd's Service provides clients with Lloyd's Syndicate Rating Guide, Special Comments, syndicate and market forecasts, weekly market commentary, and analyst access and consulting services tailored to specific client needs.

Demotech, Inc. (614) 761-8602 www.demotech.com. Based in Columbus, Ohio, Demotech has developed a system of assigning financial stability ratings to insurers. Its proprietary Financial Stability Ratings® are assigned based upon quantitative models. It developed this model in 1988 as the first risk-based capital or dynamic financial analysis model universally applied to the insurance industry.

Demotech's highest Financial Stability Rating® is A?? (Unsurpassed), followed by A? (Unsurpassed), A (Exceptional), S (Substantial) and several levels of Financial Stability Ratings® that are average or below average.

Demotech claims that studies show that its Financial Stability Ratings® are generally more conservative than the ratings assigned by other firms. This might explain differences in the relative level of ratings for the same insurance company from one financial rating service to another.

To start its review, Demotech generally requests of an insurer:

- Five years of statutory annual statements
- Year-to-date information for the current year
- Latest Management's Discussion and Analysis
- Copy of latest actuarial report—not just the opinion letter, but the report in its entirety•
- Latest independent audit
- Latest insurance department examination
- Company's agreement to send them its quarterly financial information.

Standard & Poor's (212) 208-1527 http://www.standardandpoors.com/RatingsActions/RatingsLists/Insurance/InsuranceStrengthRatings.html. Standard & Poor's (S&P) offers a free "Insurer Financial Strength Ratings Search" in its Web site database. Type in the name of the insurance company, its country and state, and can see if it has received any type of S&P credit rating. A marginal or downgraded credit rating may portend collection problems in getting a claim resolved or paid.

Weiss Research (800) 289-9222 http://www.weissratings.com/products_pc.asp. Weiss Ratings tracks the financial safety of approximately 2,300 U.S. property and casualty insurance companies each and every quarter. It then issues "Weiss Safety Ratings" based on its analysts' reviews of publicly available information collected by the NAIC and supplemented by data it collects directly from the companies themselves. Unlike the other major rating agencies, Weiss does not accept compensation from the companies it rates. It trades on this to tout greater objectivity.

Grading Levels

Many risk managers limit their insurance placements to companies with the highest grade of financial rating, usually expressed by letters such as A+++, A+, etc. They avoid "grade-B" or lower insurers and sleep better knowing that their insurers have rock-solid financial stability, even if the premium is higher. If a company's coverage later evaporates due to insurer insolvency, the CEO will find small consolation in the fact that at least the risk manager saved money on the premium.

By contrast, some companies and risk managers may find it tempting to trade off price with security. Rating services are not infallible. Some highly rated insurance companies flounder while other companies with "B" ratings pay claims faithfully. As mutual fund and investment companies are fond of saying, "Past track record is no guarantee of future performance" and a company may pay more premium for the bank-vault security of an A-rated company. If risk managers strive to conserve dollars on the insurance budget in order to deploy them better elsewhere, though, they may find that the tradeoff is sensible.

In the face of terrorism loss and heightened concern over insurer solvency, risk managers should make this potentially weakest link one of the strongest parts of their risk management programs by vigilantly monitoring the financial health of their insurers.

Avoiding Insurance Scams

Tightening insurance capacity has a negative impact beyond high prices and availability of coverage. It can also attract the entry of less savory operators—scam artists—into the insurance arena. Some precedent exists for this threat to unwary buyers.

The liability insurance crunch of the 1980s spawned not only very reputable sources of insurance capacity—such as ACE and XL—but also some shady characters. These included some thinly capitalized insurers operating with Caribbean or

European domiciles. They offered deals that seemed too good to be true.

In the long run, they were. In still others, financial imprimaturs from accounting firms were on forged letterheads. In others, the new entrants were later found to be under capitalized. In other situations, the owners and managers of these enterprises were found to be engaging in fraud.

History may yet repeat itself. The phrase *caveat emptor*—"let the buyer beware"—has rarely been more apt. It is not far-fetched to envision that some policyholders, desperate to find terrorism insurance coverage or protection on other exposures, might gravitate to a new insurance player seeming to offer fantastic deals.

Red Flags

Certain red flags should arise when approached by an insurer having any of the following characteristics:

- ❑ New entrant to a particular market or to the insurance industry
- ❑ Domiciled in overseas locale that is lightly regulated
- ❑ Capitalization with unusual or illiquid assets rather than with widely traded stocks or bonds
- ❑ Management has been associated with prior insurance failures or insolvencies
- ❑ Deal seems too good to be true – offers materially lower rates for the same or better coverage terms[6]

Normal State of Affairs?

The soft insurance market of the past ten-plus years may have lulled some risk managers into a diminished state of vigilance regarding insurance scam artists. Further, some current risk managers may have been students during the time of the last hard market.

They may have begun to view a soft market as the normal state of affairs. Amidst the market aftershocks caused by past terrorism and its future threat, the best remedy here is not to be an unsuspecting buyer, but a shrewd customer with a keen eye toward financial rating and solvency.

> Tips for surviving in the hard insurance market following recent terrorist attacks:
>
> - Be prepared to provide insurance renewal information further in advance of renewal time than you might in normal markets or circumstances.
> - Express a willingness to meet with insurance underwriters as required. Better still, be proactive and seek out opportunities to meet with them.

Future acts of terrorism might be more limited in scope or more extensive than those already witnessed. This is a story in progress, not one that the authors can survey from the perspective of broad historical perspective. One does not need the vast expanse of time, however, to safely say that aftershocks from terrorist losses will continue to be felt in the insurance marketplace for years.

How lasting these changes will be is difficult to measure and involves an educated guess. Risk managers realize that, to a great degree, the insurance market has changed in ways that make it an even more challenging business environment. It will be tougher to procure financial protection and to do so at reasonable costs.

For this reason, the best-equipped risk managers are those armed with specific tools that will enable them to survive—and even thrive—in a hardening insurance market.

End Notes

1 "Terrorism Losses to Hike Combined Ratio Up to 120," *National Underwriter*, Nov. 26, 2001, p. 5.

2 CIAB 2001 Fourth Quarter Commercial Insurance Market Index, Council of Insurance Agents + Brokers, Jan. 16, 2002

3 "Sept. 11 Attacks Creating Coverage Challenges," *Business Insurance*, Nov. 5, 2001, p. 60.

4 "Firms, Insurers in Turmoil Over Terrorism Policies," *Washington Post*, Nov. 21, 2001, p. E2.

5 "Terrorism Insurance Effort Fails," *Washington Post*, Dec. 21, 2001, p. E4.

6 "Scam: Spotting Fraud," *Business Insurance*, Oct. 29, 2001, p. 30.

Chapter 9

Hard Market Survival Tips

Hard Cycle Tips

No one knows how long this insurance market cycle will last. Many insurance buyers and risk managers may find themselves ill equipped to adjust to its demands. One reason is that the insurance market has been soft for so many years. Some risk managers may never have experienced a hard market, and others may have gotten accustomed to thinking that a soft market—with low prices and abundant capacity—was the permanent and natural order of the universe. The change could induce a state of culture shock and unpreparedness in the face of the terrorism-induced hard market.

If American firms are relatively unprepared for risk managing terrorism, the picture is not much brighter across the Atlantic. A recent study by Marsh Ltd. in London revealed that only about half of the 600 European companies surveyed have a plan in place to address the high impact risk of loss from terrorist acts. Thirty-one percent of the European companies surveyed also said that management was unlikely to review risk more often than twice a year. Almost 20 percent confessed that such risk reviews occurred on only a sporadic or ad hoc basis.[1]

Risk managers who will survive and thrive are those armed with techniques for addressing high insurance prices and the relative unavailability of coverage. Suggested tools and techniques to use to build a strong risk management program in the face of terrorist threats include:

Hard and Soft Insurance Markets Defined

Hard market – insurance coverage is expensive and/or difficult to obtain. . . A seller's market where demand often outstrips supply.

Soft market – much price competition abounds as insurers cut premiums to gain market share. . . A buyer's market where supply often exceeds the demand.

Get out of the starting blocks early. Be proactive. Start the renewal process early. For example, in October or November it may be too late to begin working on a December 31 renewal. That is the time to start July 1 renewals. Start strategic planning at least six months in advance and be ready to go to market 120 days before the current policy expires.

Retool your insurance renewal dates. This relates to the preceding tip. Get away from the December 31 or July 1 renewal dates. Due to their popularity, insurance underwriters are often deluged at these times of year. Thus, they will not have the luxury of time to give individual renewals much consideration. A time-pressed underwriter is often an unyielding underwriter.

If possible, get current policies extended to less popular renewal dates such as April 30 or October 31. This also moves renewals away from the common reinsurance treaty date of January 1.

Some risk managers like the last day of the month because underwriters tend to look at their renewals one month at a time (as in "I can't talk about your August renewal now because I haven't finished my Julys").

Tune-up your loss runs. Get your loss runs in first class condition. Get dormant claims closed and have realistic reserves put on those that remain open. Include complete details on larger claims, since underwriters tend to look closely at loss runs to get a handle on past loss history and future loss potential.

Survival Tips in the Post-Terrorism Market:
Ten Ways To Tune-Up Your Loss Runs

1. Look for claims that have been misallocated to your account, due to keystroke error, adjuster oversight, or confusion.

2. Ask adjusters to reducc reserves promptly when cases are resolved below amounts reserved.

3. Challenge any reserve that appears to be overstated.

4. Seek an in-person meeting with the claims representative on any reserve that appears excessive.

5. Learn the insurer's or TPA's reserving philosophy in advance to make sure it jibes with yours.

6. Seek a side agreement or understanding that sizable reserve increases must be reviewed by and discussed with you beforehand.

7. Read monthly loss runs and challenge discrepancies, erroneous data, etc. Don't let loss runs become mere doorstops.

8. Hold periodic meetings (quarterly?) with claims staff to review loss runs.

9. Make sure adjusters are prompt in dropping reserves on the basis of favorable claim developments (e.g. case dismissal, trial victory, successful summary judgment motion, claimant abandons the claim, etc.)

10. If you think a reserve is too high, seek a second opinion from legal counsel or an impartial claims or risk management consultant.

Scrub the loss data. Remember the acronym, GIGO—Garbage In/Garbage Out. Before marketing an account, spend time reviewing loss histories, coding, valuation dates, etc. Good data is important in all market conditions, and being able to demonstrate favorable loss history is a key to finding the best possible market.

Similarly, having good data can allow an underwriter to see underwriting possibilities via various loss sensitive options like deductibles or SIRs. Perhaps a product line or exposure could be carved out of the placement and treated separately by placing a higher attachment point on that particular exposure.

Invest in loss control and feature it to underwriters. Rethink and update loss control measures. Prepare a detailed report that shows why you are a much better than average risk. Organizations that disregarded loss control throughout the prolonged soft market will regret it and pay the price in this market. The key in this market for organizations with good loss history is to put the submission together in a very clear and concise format. Ensure that all pertinent underwriting information and data are included in the submission.

Justify the amount of insurance sought. Prepare detailed support for the amounts of property and business interruption insurance that you want. Why are you seeking that amount? Have you been carrying too much or too little in the past? Can you scale back limits in some areas of your operation but beef it up in others? Help the underwriter understand the thought process you put into the request for various insurance limits.

Scour the insurance market. This is particularly important for mid-size insureds. Be a savvy and persistent shopper. Whether you are in the market for a car or for property coverage, it pays to do a thorough and painstaking job. Do your homework. In past hard markets, there have always been insurance companies and/or producers that had capacity available at below market rates. This should be true again. There will still be insurers whose surplus was not heavily impacted by the September 11 terrorist attacks and who may, for example, be willing to write business at 25 to 50 percent over expiring rates rather than the 100 percent that the current carrier wants.

Differentiate yourself as a risk. Charles E. Comiskey, CPCU, CIC, CPIA, CRM, of Brady, Chapman Holland & Associates, Inc., suggests that you distinguish your organization in the eyes of insurance company underwriters. Be prepared to provide more information earlier than previously required. What makes your firm a better insurance risk than your competitors? Be prepared to make a concise but cogent case.

Make a good case about why you are in a low risk tier when it comes to potential terrorism loss. For example, The Hartford has apparently divided its book of business into three categories: highly exposed entities such as airports, bridges and tunnels; moderately exposed risks such as arenas; and low exposure risks. Doubtlessly other insurers have embraced a similar underwriting screen. Bearing this in mind, show why your business, risk, or operation falls within the low risk category.[2]

Revisit retention. Be prepared to retain more risk through higher deductibles, loss-sensitive plans, or alternative financing techniques. Now is the time to reevaluate the client's or your company's loss tolerance corridor. If the company has taken advantage of low deductibles or guaranteed cost placements, now is the time to think ahead. Accept that reality now and get ahead of the curve before everyone else comes to the same or similar conclusions.

Focus on insurer solvency. Focus on the financial soundness of your insurance company, the availability and adequacy of coverages and limits, and the required financial retention and services provided. Aggressive price negotiation may not be possible. Look more closely than ever at insurer financials Risk managers and insurance buyers should be ever more vigilant in checking out the financial health of their insurers. Chapter 8 includes a discussion of financial ratings and agencies that publish them.

Reassess your insurance budget. Are you budgeting enough? Is it a realistic amount? Revisit budgets in order to reflect the higher costs that will surely come— whether insurance premiums, increased risk reduction practices, or greater amounts of retained risk. Prepare upper management and urge them to brace for higher costs. If you took credit for lower costs in the soft market, be prepared to take some blame—perhaps undeservedly—for cost hikes in the hard market. This is a communications challenge with regard to the risk manager's dealings with upper management. Outline your action plan to mitigate, moderate, or offset the higher insurance costs.

Tread carefully on captives. Be sure, really sure that you really want to involve yourself in a captive if that is management's directive. The frictional costs are considerable, and there must be a long-term commitment to funding the entire program, such as the costs of auditing, claims administration, statutory accounting (in addition to existing GAAP expenses), etc. Consider a group purchasing program, or rent-a-captive, if traditional options within the insurance marketplace are unattractive or unavailable.

Market and package yourself to the insurer as a good risk. Some underwriters may judge a book by its cover. Nancy Germond of *The Insurance Writer*, Jefferson City, Missouri, states:

> "A strategy that has been utilized in the public sector is to put together a 'marketing package.' This is a nice looking binder that outlines all pertinent information on your entity: size, risks, loss control endeavors, loss history, demographic information, budgets, etc. It essentially tells underwriters you take the insurance process seriously and you are 'selling' the program to various markets. For public sector risk managers, PRIMA has an example of this done about four or five years ago. It won an award the year it was submitted to them.
>
> "In a hard market, underwriters will be deluged with submissions, and those submissions that do not contain 'complete' information and data needed to underwrite the risk will be tossed aside. This happened during the hard market in the mid-1980s. In a hard market, buyers or prospective buyers find underwriters tossing aside incomplete submissions; they either do not have the time or simply do not need the business, so there is no call back to the broker advising additional information is required to complete the underwriting.
>
> "In addition, it appears the long-time 'experienced' underwriters that some insurance buyers have developed relationships with over the years are a diminishing breed, and they are being replaced by these young actuaries that live in their own little numerical realm and do not have a clue about market realities. Risk managers and insurance buyers may find the insurance market over the next couple years to be extremely frustrating."

One fundamental truth in the post-September 11 hard market is that relationships with insurance underwriters are likely more important than ever before.

Build Strong Underwriter Relationships. The success of insurance professionals and risk managers depends in part on how effectively they communicate with and deal with underwriters. Adversarial relationships with insurance underwriters can hamper the effectiveness of any risk manager, producer, or insurance buyer. By contrast, building bridges with underwriters can enhance the odds of getting the best insurance coverage deal at the lowest cost.

Remember—underwriters are people too! In a perfect world, underwriting decisions have nothing to do with whom an underwriter likes or dislikes. In practice, in gray areas where the underwriter has discretion, these predilections may color the decision. Build relationships.

Respect the underwriter's time. Most underwriters have more work than they can comfortably handle. So, be selective about calling the underwriter. When phoning, identify yourself and orient the underwriter. He is dealing with dozens—maybe hundreds—of submissions and accounts. Some of these are new business. Others are renewal business. Do not expect the underwriter to have your submission or insurance application on his desk. Avoid demanding instant answers.

Know the insurance market. Is the current market hard or soft for the particular type of risk you are pitching? Who has more negotiating leverage, you or the seller? Stay abreast not only of general market conditions but trends within markets for certain lines of insurance.

Insurance professionals and risk managers talk about the insurance market as if it was a monolithic entity. In truth, it is not. There is no one Insurance Market. It is a composite of many smaller individual insurance markets. There are various submarkets that may be soft or hard, depending on just how difficult a risk is.

Make sure the insurer writes risks like the ones you're submitting. In the marketplace, certain insurers are known for liking certain types of coverage and avoiding others, and it is important to target the correct ones for each risk.

Sweat the details. Check the spelling of the underwriter's name (and his or her gender). No use getting off on the wrong foot right away. Recognize promotions and job title changes. Sloppiness in addressing the underwriter may imply—fairly or unfairly—that the risk manager is equally casual in other matters.

Be realistic on turnaround times. Understand that the leading insurance companies will be inundated with new business submissions and their response time will be slow. Build that into the workflow to avoid eleventh-hour fire drills.

Get the facts straight. Make sure the insurance application is complete and accurate. Fill in all the blanks: prior loss history, names of prior carriers, annual sales figures, payroll, etc. Give the underwriter no reason to delay a decision because of information gaps. Thoroughness and accuracy in completing the insurance application impresses underwriters, who tend by nature, training, and disposition to be detail-oriented. In addition, an error might provide an insurer with grounds to contesting coverage or allege material misrepresentation, or it may even form the basis for contract rescission.

Cooperate with requests for additional information. Underwriters loathe having to make decisions based on incomplete information, or feeling pressured to do so. Do you want a fast decision or a good one? Underwriters can often give you one or the other, but rarely both.

Two words: Be nice! Civility, courtesy, and empathy may build good karma with the underwriter. No amount of niceness, however, can salvage a terrible risk in any market—soft or hard. In the margins and gray areas, though, where it may be a close call for the underwriter on pricing or coverage decisions, courtesy may help the buyer's cause in the long run.

This also means perhaps assessing an insurance broker's personal style. Brokers should be forceful advocates of their client's cause to get the best deal for them. On the other hand, objectively appraise the broker's negotiating style to make sure it does not cross the line and become a liability.

Practice hard market disciplines in all markets. According to Greg Dodd, Risk Manager, Perot Systems Corporation, Plano, Texas:

> "A hard market comes about when insurance companies, often run by financial types and lawyers, have failed to underwrite and price insurance contracts as the risk transfers they are. A hard market is characterized by insurance companies' underwriting and pricing insurance contracts as the risk transfers they are. Businesses should function in a hard market the same as they should at all times, that is, using the proper, dynamic mix of risk management techniques, of which insurance is only one, and by negotiating, purchasing, and administering insurance purchases as the risk transfers they are."

View the hard market as an opportunity as well as a threat. Ruth S. Kamerling, President of the RKRM Risk Management Group (Pitman, New Jersey) says,

> "In a hard market, look for the ever-present but somewhat ignored risk management opportunities. They are always there, but buyers were less likely to feel the need to control loss, have written procedures, hire consultants, etc. when insurance premiums did not seem to be based on ability to secure insurance and claims. The times have changed so abruptly that in the coming months, the consumers are going to really understand the need for professional advice."

Ten Risk Management Lessons and the Terrorist Threat

- Risk managers will have to provide much more detailed information about risks and the implications of those risks for their companies.
- As insurance costs rise, corporate boards are going to hold risk managers more accountable for cost management.
- Insurers and reinsurers are going to hold risk managers more accountable for providing data on actual or prospective risks.
- Risk managers will have to raise the bar on their own creativity to find both insurance and non-insurance ways to address risks.
- Risk managers will be forced—through rising insurance costs—to take a fresh look at self-insurance, higher deductibles, and higher self-insured retentions.
- Loss control will need to receive more attention, as retentions increase along with insurance costs.
- Risk managers will need to become more familiar with captive insurance companies as one form of alternative risk transfer. They might even need to look into forming mutual insurance companies.
- Risk managers will need to expect to work more closely with their insurers to assess risks and ways to reduce them
- Risk managers who were in college (or high school) during the last hard insurance market are going to have to learn some hard market survival tips, realizing that a soft market is not perpetual
- Be prepared to include an excruciating level of detail in your application for new or renewal coverage

Protecting assets is going to continue to be more difficult as insurance companies nonrenew high-risk accounts, those with a poor claims history, or those with high but unprotected values. Risk managers and buyers should become more involved in policy development as well as program development.

Be a persistent negotiator. Any discussion on dealing with a hard insurance market should acknowledge that there might be some opportunism and price gouging. Be realistic in expectations, but stay engaged. Those who feel that genuine price gouging is occurring should consider complaining to the National Association of Insurance Commissioners (NAIC). The NAIC has established a Rate and Market Trends Oversight Working Group to monitor pricing trends and to blow the whistle on exploitative pricing trends.

Look at the surplus lines market. Firms that cannot procure terrorism cover in the conventional market may wish to see if they can find it in the surplus lines marketplace. In neither case would such coverage be inexpensive, but it might at least be available in the surplus lines marketplace. Since state insurance department rate and form restrictions do not apply to surplus lines coverage, surplus lines insurers can often tailor policies to meet a customer's specific needs.

When a standard marketplace insurer becomes insolvent, policyholders may be able to tap into the state guaranty fund to help pay claims. If a surplus lines insurer becomes insolvent, however, there is no such relief. For some insurance buyers and brokers), this is a negative. For this reason, some brokers may be leery of steering companies toward surplus lines paper. In some instances of hard-to-place coverage, state insurance commissioners have waived the usual requirements of having to get declinations from three standard market insurers before placing coverage with a surplus lines carrier.

Seven Workers Compensation Program Tips

- Start the renewal process well ahead of the renewal date.

- Take a holistic look at components of the workers compensation program to ensure they are in balance. This includes loss control, safety programs, claims management, and return-to-work efforts.

- Get creative by using deductibles, retroactive policies, self-insurance pools, or captives.

- If sizable losses propel your experience modification, learn more about programs offered by the workers compensation residual markets. These may be the sole option in a hardening market.

- Make sure your underwriting submission is thorough. It should quantify the organization's workers compensation exposures and losses and identify the systems that are in place to prevent future losses or cushion their severity.

- Get it in writing. Make sure all communications from the broker and insurer are in writing. At the least, memorialize it in writing with a confirming letter. This can be key in touchy situations when insurers have withdrawn quotes after they were presented to an insured but before coverage is bound.

- Communicate to upper management and prepare them for the expected changes.

End Notes

[1] "Some Unprepared for Terrorist Risks," *Business Insurance*, Dec. 17, 2001, p. 23.
[2] "Insurers: Returning to Underwriting," *Business Insurance*, Jan. 14, 2002, p. 18.

Chapter 10

Insurance Coverage Issues

Coverage Issues Abound

Many insurance coverage issues and disputes will flow from the September 11th attacks. They will spawn litigation between insurers and policyholders and keep platoons of lawyers busy for years to come.

Many observers predict that coverage litigation will continue for many years between policyholders and insureds over losses from terrorist attacks. In this chapter we focus on seven areas of potential coverage disputes:

- ❑ Number of occurrences
- ❑ Application of the "war exclusion"
- ❑ Terrorism exclusions—retrospective and prospective
- ❑ Debris removal coverage
- ❑ Business interruption coverage
- ❑ Workers compensation coverage
- ❑ Issues with anthrax contamination
- ❑ Pollution/nuclear exclusions

Number of Losses/Occurrences

On September 11, was there one occurrence or two (or even three or four, if one counts the Pentagon crash and the crash of Flight 93)? This question literally has multibillion-dollar implications as respects to the World Trade Center calamity. If insurers consider the planned hijackings and crashes as one occurrence, the owners of the structures are subject to a single limit of $3.5 billion on the World Trade Center towers.

If each plane crash is considered a separate occurrence, though, the limit is doubled to $7 billion.

Huge sums ride on this question. One can envision insurance carriers and reinsurers arguing that the attacks comprised one well-orchestrated and planned occurrence. Similarly, one can visualize the World Trade Center policyholders arguing that these were two separate occurrences.

The issue of "one occurrence or two" is a perplexing question, representing another facet of the terrorism event that makes it unique. There is some case law that deals with determining the number of occurrences in third-party liability situations. There also are previous cases involving whether one or two deductibles or self-insured retentions should be applied to a loss or losses arising from related causes. Insureds and insurers may stand on both sides of the fence regarding the number of occurrences, depending on the coverage profile and specific circumstances. [1]

Other Possible Situations

If a horde of vandals descended on the complex, split into two groups with one going into tower A and the other into tower B, this might also be considered one loss.

Imagine that the terrorists owned the two aircraft—rather than hijacking them. The two planes take off together, fly side-by-side to New York, and fly simultaneously into the two buildings. This might be considered one loss. But there also is an argument that this type of event would still constitute two separate occurrences.

Reading the terms of common property policies—say, a standard ISO commercial property or businessowners package policy—does not necessarily clarify the question. The property section contains the following:

> LIMIT OF INSURANCE: The most we will pay for loss or damage in any one occurrence is the applicable Limit of Insurance shown in the Declarations.

It is necessary to consider whether and how "occurrence" is defined in relation to this clause. There are different contentions about whether occurrence was defined or not on coverage written by the various primary and excess insurers on the two World Trade Center towers. However, we know that there is no definition of occurrence in the standard ISO commercial property policy; the only definition of occurrence in the

BOP policy is in the liability section. That definition is one with which we are all familiar:

> OCCURRENCE means an accident, including continuous or repeated exposure to substantially the same general harmful conditions.

If we use this definition, we need to focus on "repeated exposure to the same harmful condition," which would suggest one loss. However, various courts have shifted in their opinions on this issue, depending on the exact circumstances of the liability loss.[2]

Let us consider two more mainstream property loss scenarios in which the number of deductibles might be an issue.

First, Hurricane X damages three properties along a fifty-mile coastline owned by Company Y. The first building was damaged at 9 a.m., the second at 10 a.m., and the third at 2 p.m. Is this one loss or three?

Second, a heavy snowstorm dumps fifteen inches of snow on the roofs of various homes in a townhouse development. Freezing weather ensues, resulting in what residents in the North call ice-dam losses. That is, heat from the house melts the lowest layer of snow, but the frigid weather causes ice blockages that prevent the water from flowing into the gutters and drains. Since the water has to go somewhere, it eventually seeps back up under the roof shingles and pours into the building. Fifteen townhouses in seven different buildings (the homes are in clusters of four homes per building) suffer ice-dam damage over a three-week period. For the sake of argument, let's say that none of the fifteen losses occurred on the same day as any of the others. Is this one loss, seven losses, or fifteen losses?

In both cases, the insurance industry often might adjust these as one loss and apply a single deductible. Of course, in the case of the townhouses, if their individual owners separately insured them, each individual policy and its deductible would apply.

Factors to Consider

There seems little question that the September 11 attacks constituted a single planned event. However, there are several factors to weigh in determining whether this connection is enough to support the theory of one occurrence:

- The Pentagon was also part of this planned event. Had the Port Authority and Silverstein properties, which owned and leased the World Trade Center, owned the Pentagon, many insurance practitioners would have considered it a separate event because it involved separate locations.

- The World Trade Center complex is insured as one location, even though several structures are on site. The towers were connected by the same foundation, but they were completely different structures above ground.

- If an electrical fire broke out in One World Trade Center on the 34th floor, and twenty-five minutes later an electrical fire broke out in Two World Trade Center on the 54th floor, would an insurance company have any grounds for calling the damage one occurrence?

While everyone desires fairness and equity, in the preceding scenario the carrier would likely apply two deductibles. There were two separate planes hitting two separate structures causing two separate chains of events.

Some insurers did not wait long to seek courtroom answers, though. In late October 2001, Swiss Re filed a lawsuit in Manhattan Federal Court, claiming that the attacks were one orchestrated event carrying the maximum insurance coverage of $3.5 billion. The World Trade Center's leaseholder, Silverstein Properties, maintained that each plane crash constituted a separate event—each carrying $3.5 billion in coverage—for a total property recovery of $7 billion.

A complicating factor in the coverage dispute between Swiss Re, other insurers, and the World Trade Center owners and leaseholders was that a finalized version of the primary insurance policy did not exist as of the date of the September 11 loss. Instead, the insurer had issued a temporary document known as a slip. According to Swiss Re, the slip defined the word "occurrence" as

> ". . . all losses or damages that are attributable directly or indirectly to one cause or one series of similar causes."

Further, the giant reinsurer argued that the slip stipulated that

> ". . . all such losses will be added together and the total amount of such losses will be treated as one occurrence irrespective of the period of time or area over which such losses occur."[3]

The insured, however, argued that occurrence was not defined, at least on one or more of the policies. In addition to these questions about the meaning of the term occurrence and available limits, related insurance coverage issues arise from the September 11 attacks in New York City include:

- ❑ Whether the collapse of Building 7 hours after the twin towers fell was another, separate occurrence?
- ❑ Which collapsed tower caused physical damage to nearby properties?

Even if the insured and insurers reached agreement that the loss was one event obligating an insurer payment of $3.5 billion, other points of contention arise. A separate legal battle arose over whether the World Trade Center policyholder was entitled to collect an immediate $3.5 billion actual cash value payment, or whether it should wait to recover replacement costs as rebuilding proceeds over a period of years. The latter would obviously have cash flow benefits to an insurer; the former would have immediate cash flow benefits to the insured.[4]

War Exclusion

An immediate question arising from the WTC event was whether insurers would try to apply the war exclusion to applicable property policies. Most property forms exclude coverage for damage from "war", defining it as declared or undeclared, insurrection, etc. With President George W. Bush publicly stating that the September 11 attack was an act of war, some insurance observers wondered whether the war exclusion might apply.

It is unlikely that the war exclusion could be argued to apply since, as a matter of law, there is no war until Congress declares it. In addition, case law supports the premise that there must be war between two sovereign nations in order to trigger the war exclusion. According to the *FC&S Bulletins:*

> "While no court decisions involving the exact language of the war exclusion clause in the current commercial property forms have arisen at this time, there are decisions involving other, past war exclusion clauses that do permit some reliable conclusions to be drawn.
>
> "American courts, following British precedent, seem to adhere to a strict doctrine of what constitutes war, allowing the exclusion to be applied only in situations involving damage arising from a genuine warlike act between sovereign entities." [5]

Indeed, in the immediate aftermath of the September 11 attack, most insurance companies with substantial exposures seemed disinclined to invoke the exclusion. For example, Chubb Insurance Company was among the first carriers to expressly disclaim any intention of invoking it.

The life insurance industry generally mirrored this stance.

Many of the largest insurance companies, in fact, advanced emergency payments immediately after the disaster. State insurance departments will undoubtedly look harshly at any insurer that attempts to exclude coverage on this basis.

Some coverage attorneys who represent insurers urge caution here, however. They suggest that insurers, out of patriotic zeal, should not waive their war exclusion rights, which could be applicable to future terrorist attacks. This is needed to avoid waiver and estoppel problems. In fact, some would suggest that insurers failing to assert viable war exclusion coverage defenses might in fact open themselves up to directors and officers liability by paying such claims. As a result, coverage attorneys urge that insurers paying claims without invoking the war exclusion nevertheless do so under a full reservation of rights.

One side issue is whether insurers are acting out of a societal or patriotic obligation to treat the September 11 claims differently than they normally would. Relevant and legitimate policy exclusions that are plainly applicable should be asserted; public policy or national crisis considerations should not override the clear meaning of the policy language. Insurers owe a fiduciary responsibility to their stockholders and other policyholders to pay only covered claims. Paying noncovered losses weakens insurers, which ultimately may leave some policyholders without coverage or unpaid claims down the road.

Legal Analysis

It is worthwhile to keep in mind that, under policies that insure for risks of physical loss unless excluded, underwriters must prove that an exclusion applies to a situation in order to void coverage for the loss. And there is significant case law currently available to define war.[6]

In fact, earlier court cases have grappled with coverage issues arising from acts or war or terrorism. For example, in *Pan-American World Airways v. Aetna* 505 F.2d 999 (2d Cir. 1974), the second circut court of appeals held that terrorism is not an act of war. The court stated, "War is a course of hostility engaged in by entities that have at least significant attributes of sovereignty." The court conceded, however, that war "can exist between quasi-sovereign entities." War, the court defined, consists of "hostilities carried on by entities that constitute governments at least de facto in character."

The court in the United Kingdom's *Spinney's Ltd. v. Royal Insurance Co. Ltd.,* 1 Lloyd's Rep. 406 (Q.B. 1980), considered three elements in discussing the definition of civil war:

a) Was the conflict between two opposing sides?

b) What were the objectives of the sides, and how did they pursue them?

c) What was the scale of the conflict and its effect on public order and on the life of the inhabitants?

Other court cases have distinguished between acts of war and acts of terrorism. This is important since the war exclusion was more prevalent prior to September 11 in insurance policies than is any terrorism exclusion. For example, in the case of *Holiday Inns v. Aetna* 571 F. Supp. 1460 (S.D.N.Y. 1983), the court held that destruction of a hotel in Beirut Lebanon was not due to a "war" where the combatants did not contest the sovereignty of the state.

In *Sunny South Aircraft Service v. American Fire & Casualty*, 140 So. 2d 78 (Fla. 1962), the court rejected the "war exclusion" argument that an insurer advanced and found coverage for losses due to a hijacked airplane. Another court held comparably in *Northwest Airlines v. Globe Indemnity* 225 N.W. 2d 831 (Minn. 1975), i.e. that the war exclusion did not apply to losses from a hijacked aircraft.

Such cases further bolster the thought that the terrorist acts of 2001 did not constitute a war and, therefore, the standard acts-of-war exclusion should not apply to them.

Other Arguments

Other arguments advanced against applying the war risk exclusion to recent terrorist losses include:

No war at the time of the occurrence. A successful coverage stance must assert the war exclusion relative to the time of the event. At the time of the September attacks, the United Stated Congress had not declared war pursuant to its authority granted by Article I, Section 8, of the Constitution. Further, the President committed limited military force pursuant to the War Powers Act. This leads us to a definition of war that may vary by insurance policy form.

No recognized government or sovereign state responsible for the terrorist act. Definitions that attempt to clarify such exclusionary language generally intend to describe events that are set into motion by groups with a high level of organization and public recognition as to the entity (i.e. nations, factions, or other recognized groups). If the persons responsible are terrorists, it is doubtful that the latter would qualify.

Bad public relations for the insurance industry. The bad public relations that would follow attempts to invoke the standard war exclusion, especially in light of legal precedent, could well cause farther-reaching problems for the insurer than accepting the claims.

Although many carriers have publicly stated that they will not attempt to use this exclusion to escape payment, it is impossible to predict how all affected carriers will proceed and how the courts will react.

Future Impact

In these and possible future terrorist-caused losses, however, there are gray areas. In such situations, insurers should err on the side of coverage. In Berkshire Hathaway's third quarter 10Q form, the company's management stated that it would literally take years to resolve complicated coverage issues, as well as to develop accurate estimates of insured losses that will be ultimately incurred.

Some insurance policies contained the following pre-September 2001 language that may concern some policyholders:

> X. (1) Hostile or warlike action in time of peace or war, including action in
> hindering, combating or defending against an actual, impending or expected
> attack:
> (a)..............
> (b).............
> (c)..............
> (d) Vandalism, sabotage or malicious act, which shall be deemed also to encompass the act or acts of one (1) or more persons, whether or not agents of a sovereign power, carried out for political, terrorist or ideological purposes and whether any loss, damage or expense resulting therefrom is accidental of intentional.
> (2).............
> (3).............

Words and phrases meriting focus are:

"...in time of peace or war..." and/or

"...carried out for political, terrorist or ideological purposes..."

This again illustrates that the coverage written on a business must be reviewed specifically to determine whether exclusions will void coverage.

Terrorism Exclusions

Property policies that contain explicit terrorism exclusions, however, may be a different story. Congress may be unable to prevent insurers from enforcing a valid exclusion for acts of terrorism. According to the Contracts Clause in the U.S. Constitution, if a contract clause meets the filing requirements for the state in which the policy was issued, then it is valid. It is likely that no one can stop insurers from enforcing it.

Even prior to September 11, some policies included a terrorism exception to the war exclusion along the following lines:

> Loss or damage done by terrorists or done secretly by a foreign enemy or agent of any government (de facto or otherwise) all not in connection with operations of armed forces in or against the country where the described locations are situated is insured.

Ken Brownlee, a risk management consultant and former corporate risk manager for a large independent claim service provider, states,

> "In most cases, courts have looked very carefully at the root cause of each loss to see if the war exclusions applied. In most, they ruled against applying the war exclusion. Now, how will insurers define "terrorism" if they add a new exclusion? The problem is that you could define many terrorist acts as "extreme vandalism," hence one might be better off with a "named peril" policy than an "all risk" with lots of exclusions. Some dictionaries define "terrorism" as intimidation, coercion, etc."

We can look perhaps to some historical precedent as an analogue. Consider the insurance situation that arose following the urban riots in the 1960s. Almost every insurance company paid claims submitted by policyholders, but a few cited the insurrection exclusion as a reason for not making payment. They were unsuccessful in that contention.

By contrast, a terrorism exclusion in the policy will probably defeat coverage, but very few policies contained such an exclusion as of September 2001. Standard forms did not contain terrorism exclusions at that point in time.

ISO Filings

Since then, however, the Insurance Service Office (ISO) has introduced several terrorism exclusions that could be used with various types of policies. In those exclusions, terrorism is specifically defined.

> "Terrorism means activities against persons, organizations or property of any nature:
>
> 1. That involve the following or preparation for the following:
>   a. Use or threat of force or violence;
>   b. Commission or threat of a dangerous act; or
>   c. Commission or threat of an act that interferes with or disrupts an electronic, communication, information or mechanical system; and
> 2. When one or both of the following applies:
>   a. The effect is to intimidate or coerce a government or the civilian population or any segment thereof, or to disrupt any segment of the economy; or
>   b. It appears that the intent is to intimidate or coerce a government, or to further political, ideological, religious, social or economic objectives or to express (or express opposition to) a philosophy or ideology."

While this definition helps, it has not been tested in the court system. And it is difficult to ascertain, at the time of an incident, whether the effect is to "intimidate or coerce" a government, or whether the act is an attempt to do so. There also are thresholds of damage that must be met before the ISO exclusions are triggered. For example, there must be $25 million of damage as a result of the incident, or fifty or more people must be killed or seriously injured, depending on the line of coverage.

Some loss scenarios still may be murky, however. The Oklahoma City bombing by Timothy McVeigh had a political cause, there was substantial property damage, and fifty or more people were killed. So the ISO exclusion probably would apply to that type of loss.

But does an anti-abortion group firebombing a medical clinic that performs abortions qualify as an effort to intimidate or coerce a government? Abortion is not only a religious issue but it has become a political issue as well.

The majority of states have accepted these ISO exclusions, and they could become part of the coverage that commonly is offered to corporations. However, different definitions of terrorism could exist on company-filed forms or in coverage offered by the surplus lines marketplace. Therefore, it is important to read such exclusions carefully.

In addition, some corporations may elect to purchase terrorism coverage separately as a result of such exclusions being attached to their corporate policy. In that case, the coverage details and definitions should dovetail so that the proper coverage effect is achieved.

Risk Management Approach

One idea is for risk managers to seek advance letters or memoranda of understanding from their insurers in an effort to determine how they would treat terrorist claims, a war exclusion, and a terrorism exclusion. Others caution that the odds of getting an insurer committed to a coverage stance in advance of a loss are slim. Patricia Gerrond, CPCU, AIC, ARe, who is an insurance and coverage consultant for the Illinois Park District Risk Management Agency (Wheaton, IL) states,

> "With regard to getting a coverage interpretation about terrorism and war risks exclusion prior to the loss, I have been attempting since September 11 to get just that from our pool of domestic and international reinsurers. No luck so far. I have been trying to do the same with my question about coverage for anthrax cleanup on our property policy. My position would be that cleanup and business interruption would be covered as accidental contamination, for which

we have coverage, due to an act of malicious mischief, since we don't know that it was anything more than a domestic weirdo riding on Osama Bin Laden's coat tails. However, remember that one of the first things you learn in claims adjuster school is never answer a hypothetical question. I imagine the carriers are falling back upon that axiom to avoid being pinned down, should the investigations turn up something different."

Mike Cerf, an expert witness and consultant on insurance claims from Sherwood, Oregon, states,

> "From the perspective of a claims department, an advisory opinion or side-letter memorandum of understanding is rarely provided. Coming from a former home office claims person such as myself, such memos are simply in most cases not given. The reason being is that the carrier will not respond and provide a memo that ties them down to a given position, that is terrorist acts do not constitute "war," in advance of a claim being presented from a loss.
>
> "The reason is that, at the home office level, you try and view your overall position on given issues in a broad perspective; you try to not view single losses but exposure to a given book of business—in this case property. You do not provide coverage analysis for a loss that has not yet occurred; the reason for this is that you do not want to provide an insured the means to create by report a loss that is covered. Losses are treated on a loss-by-loss basis, not in advance of loss. That is the general mental mind-set of those that set coverage policy.
>
> "Also, you do not want something out in the market that ties you down to a given coverage analysis, keeping in mind each loss is different, the circumstances different, and the damages different. . .the view from on high (home office) is that you never provide hypothetical coverage positions."

It is also key that risk managers make sure that all excess insurance layers buy in to the primary coverage approach. The primary cover is a good start, but all layers of coverage need to be addressed and the excess layers will be especially subject to restriction.

Further, risk managers should obtain contingent coverage for business interruption that includes terrorism as a trigger.

Debris Removal Coverage Issues

Another coverage trap is debris removal coverage, since that often is limited to the actual damaged portion of the property, although undamaged property may need to be removed as well. The other catch is that removal costs may exceed the tab for reconstructing the damaged property.

In fact, the standard debris removal coverage may be insufficient for most businesses.

The problem with the standard ISO provision is that it provides 25 percent of the loss plus $10,000 for debris removal. It is important to realize that the 25 percent applies to the loss, not to the amount of insurance. It also is part of the total limit and not in addition to it. Only the $10,000 is additional insurance.

For insureds that have most of their value concentrated at one location, this is a double-edged sword. For a large total loss, they may have only the $10,000 available for debris removal; the direct loss might consume the limit. On the other hand, for a small loss, the 25 percent plus $10,000 may be inadequate because a $10,000 direct loss could generate $100,000 in debris removal; $1,000,000 in building damage might entail $500,000 in debris removal, etc. The solution lies in increasing the $10,000 limit.

For insureds with multiple properties insured on a blanket basis, the first problem is solved by the blanket limit. Insureds still face the second problem of the high expense associated with debris removal, and they should consider purchasing added coverage if it is not part of their form. Companies that do not use standard ISO coverage forms may routinely offer much more than $10,000 additional debris removal coverage, but seldom more than $1,000,000. That may not be enough.

Business Interruption Coverage Issues

Typically the coverage trigger for business interruption coverage is direct physical damage by a covered cause of loss. On and after September 11, 2001, many businesses were crippled—not due to direct physical damage—but by their dependence on the World Trade Center and surrounding firms. Others were hampered in reopening due to police cordons or debris removal operations. In many cases, civil authorities prevented businesses from accessing their premises. Business interruption policies are unlikely to respond unless an insured suffered direct physical damage or was located at a site that sustained such direct damage.

Contingent business interruption occurs when there is direct damage to suppliers or receivers of goods. It is not automatically provided in the standard ISO business income form but may be included automatically in some company-specific forms. However, contingent coverage still relies on direct physical damage. Regardless, risk managers and business owners should report any potentially covered business interruption loss as quickly as possible to their insurance companies.

One special case of business interruption coverage will emerge from the concentration of securities firms in the vicinity of the World Trade Center. Many securities firms will face thorny coverage questions that could interfere with collecting insurance proceeds. Many such firms voluntarily suspended trading. As a result, insurers could argue that the civil authority provisions in their policies do not apply to undamaged financial markets, since the firms voluntarily suspended trading.

Since the securities exchanges closed voluntarily—not due to government decree—some insurance observers believe that this was a business decision that may not trigger coverage under a business interruption policy. Policyholders and would-be claimants probably will still rely on the civil authority argument, since New York Mayor Rudy Giuliani urged residents to avoid lower Manhattan for four days after the attack. Whether this urging is tantamount to action of civil authority is open to question, debate, and interpretation. Doubtlessly courts will have ample opportunities to parse through insurance policy language and to rule on coverage disputes in this area.

All this could be moot, however, for insureds with business interruption policies that include "ingress/egress" coverage. This provision covers lost profits resulting when an insured cannot gain access to its premises due to a covered peril. It need not be due to government action or civil authority. One downside for insureds is that covered losses are typically subject to some type of sublimit or specified time period.

In order to avoid future problems in this area, risk managers and business owners should:

- ❑ Be sure to carry business interruption coverage
- ❑ Ask insurers for interpretations of the civil authority clause
- ❑ When suspending operations, be sure it is due to civil authority
- ❑ Require ingress and egress coverage
- ❑ Seek high sublimits of coverage and/or extended time periods for recovery

Review business interruption insurance coverage for elements such as

- ❑ Extra expense coverage
- ❑ Rental value interest
- ❑ Leasehold interest
- ❑ Contingent coverage for dependent properties, including contributing locations, recipient locations, manufacturing locations, and leader locations

Other related coverage questions arise from business interruption losses, including the following:

- ❑ What is a "necessary suspension" of business operations?

Many courts hold that business interruption insurance proceeds flow from financial loss due to a necessary suspension of business operations. If a business relocates its operations, this necessary suspension may be over.

❑ What insurance coverage pays for financial loss due to businesses that were not physically destroyed or damaged but were rendered inaccessible?

Some firms may be out of luck. Others may be covered if their policies contain the ingress and egress clause mentioned previously.

❑ How does the policy define the period of restoration, which specifies the amount of time in which coverage applies?

> Check to see if you have an "ingress and egress" clause in your business interruption policy. If you do not, specifically ask your insurance broker or agent to price the option for you. Unless it is prohibitively expensive, purchase it.

This period typically starts with the date of the physical damage and ends when the premises are actually rebuilt, repaired, or replaced or—if earlier—when they should, with reasonable speed and dispatch, have been rebuilt, repaired, or replaced.

Workers Compensation Coverage Issues

Under New York law, those who arrived at work before being injured in the event are entitled to workers compensation benefits. The Workers Compensation Board and courts tend to treat any injury in the workplace as arising out of and occurring during the course of employment. Coverage extends to employees injured while fleeing imminent peril in the workplace. Individuals injured before arriving at work are not entitled to benefits.

Apparently no workers compensation coverage existed for police officers and firefighters. Shortly after the loss, the New York Police Department looked for seventy-eight officers known to be at the scene, and the Fire Department of New York could not account for 200 firefighters. The New York workers compensation statute provides benefits for public employees only if they are engaged in dangerous occupations. The statute says that gardeners are engaged in a dangerous occupation, but that police officers and firefighters are not (!). The City of New York, however, traditionally has paid the medical bills of police officers and firefighters injured in the line of duty and kept them on at full pay for the duration of their disability.

Insurance Issues in Anthrax Contamination

Anthrax exposures raise a host of insurance coverage questions that transcend just anthrax, but which have ramifications for a host of biological or chemical contaminants that could find their ways into work settings. Readers can find an insightful discussion of anthrax coverage issues in "Anthrax Exposure Raises Cover Questions," Diane Richardson, *The National Underwriter*, Nov. 26, 2001.

Would insurance coverage be afforded for any cleanup of a biological agent such as anthrax if it becomes necessary? Would employees affected from on-the-job exposure be covered under workers compensation? What if the workers sought treatment, which revealed that anthrax was not in the white powder they feared?

In examining the insurance ramifications of anthrax and similar biological contamination there are some questions and possible scenarios to consider. In analyzing the risk, questions such as the following should be considered.

Scenario #1: What if either "natural" or "weapons grade" anthrax spores are discovered in an owned building and, on expert advice, the owner decides to shut down the building to decontaminate? The building may either be a target or cross-contaminated. It takes six weeks and costs $1 million. It is unsuccessful, and the owner is forced to destroy the building and everything in it. Is this a covered loss under a property policy?

Scenario #2: Three hundred employees and their families had to be tested and medicated. Three employees were infected. One dies, and two recover traumatized and are unable to work. The insured then learns that another location is contaminated. Eight customers and two supplier sites allege cross-contamination, and they and their employees sue for damages along similar lines. The insured never received another order for its product.

Assuming that this company remains solvent, what insurance coverage is likely or unlikely to respond?

- ❑ for the decontamination expense?
- ❑ property losses?
- ❑ extra expense, business interruption?
- ❑ workers compensation?
- ❑ third-party liability claims?

Regarding the second location, are there one or two occurrences? Was there one occurrence or ten for customer and supplier liability claims?

Scenario #3: Assume the same scenario as above, but in this case the government:

- ❑ quarantines the site for six months; and/or then
- ❑ condemns the site and orders demolition.

The questions are the same as above. All risk managers and business owners need to ask such questions and then review their insurance policies to determine to what extent, if any, they would respond.

In many such situations, insurance will not be the only answer. The coverage that exists may become more restrictive and harder to trigger.

Pollution/Nuclear Exclusions

Another potential source of coverage litigation may arise from insurers' pollution exclusions. Dust and debris kicked up by the attack may cause respiratory illness and other injuries, and liability claims may arise. Liability insurers may take a hard look at their pollution exclusions to see if at least some of these claims might be precluded from coverage.

In addition, it is important for insureds to understand that their exposure to terrorism goes beyond the type of terrorist risks posed by the September 11 attacks. While most of the coverage discussion has revolved around damage from them and the anthrax aftermath, other types of terrorism can also trigger coverage issues. These include loss due to cyberterrorism and nuclear exposures.

Nuclear Weapons Threat

On the latter point, for example, some discussion of future terrorist acts revolves around the use of makeshift nuclear weapons—dirty bombs or suitcase nukes. Law enforcement agents patrolled New York City's Times Square on New Year's Eve armed with radiation detectors in an effort to protect revelers from possible nuclear terrorism.

In the event that a future terrorist act takes this form, the nuclear exclusion in insurance policies could void coverage. In addition, the ISO-filed terrorism exclusions state that they apply without any regard for the amount of damage if nuclear, pathogenic, or biological weapons are employed in an act of terrorism.

Most prior consideration of nuclear exclusions were predicated on the expectation of some sort of Armageddon-type superpower exchange, where devastation is so total that insurance coverage might be the least of our worries. Now, though, the odds seem higher for some lower-level act that may be serious nonetheless.

In personal lines insurance coverage, for example, the nuclear exclusion (now part of the conditions) in the HO-3 is very comprehensive. The only loss that is covered is a fire that results from a nuclear event. But, if one's house just gets blown away or suffers radiation damage, there is no coverage.

In commercial property forms, the exclusion in the CP 10 30 reads similarly. The only loss covered is resulting fire.

ISO does offer form CP 10 37, Radioactive Contamination, which reads as follows:

The following is added to **Covered Causes Of Loss,** as indicated in the Declarations or by an "X" in the Schedule.

A. Limited Radioactive Contamination, meaning Radioactive Contamination that directly results from any other Covered Cause of Loss

1. Radioactive Contamination means direct physical loss or damage caused by sudden and accidental radioactive contamination, including resultant radiation damage to the described property.

2. We will not pay for loss or damage caused by or resulting from Radioactive Contamination if

a. The described premises contains:

(1) A nuclear reactor capable of sustaining nuclear fission in a self-supporting chain reaction; or

(2) Any new or used nuclear fuel intended for or used in such a nuclear reactor.

b. The contamination arises from radioactive material not located at the described premises.

B. Broad Radioactive Contamination, meaning direct physical loss or damage caused by sudden and accidental radioactive contamination, including resultant radiation damage to the described property.

We will not pay for loss or damage caused by or resulting from Radioactive Contamination if:

a. The described premises contains:

(1) A nuclear reactor capable of sustaining nuclear fission in a self-supporting chain reaction; or

(2) Any new or used nuclear fuel intended for or used in such a nuclear reactor.

b. The contamination arises from radioactive material not located at the described premises.

Terrorism Exclusions Take Precedence

Once again, however, the terrorism exclusions filed by ISO state that terrorism involving the use of nuclear materials is specifically excluded, regardless of the amount of damage of loss.

This chapter scratches the surface of potential coverage issues arising from the terrorist threat; it does not presume to be an exhaustive discussion of all the coverage issues that policyholders may encounter. There are gaping holes in the usual insurance coverage "safety net." Insureds and risk managers must structure their coverages very carefully to minimize these gaps and to maximize financial protection.

Realistically, though, even the most diligent efforts may fall short. Certain realities may be insurmountable. One is the price of coverage. Another is the availability—or unavailability—of insurance coverage for certain perils. The fraying patches and gaps in the insurance safety net underscore the need for careful thought on any insurance buying decision and the importance of noninsurance techniques in addressing terrorism risks.

End Notes

[1] Aylward, Michael F. *One Occurrence, Two Occurrences. . .Problem Issues in CGL*, National Underwriter Co., 2001.

[2] Ibid.

[3] "Swiss Re World Trade Center Lawsuit Analyzed," *National Underwriter*, Nov. 5, 2001, p. 46.

[4] "World Trade Center Coverage Dispute Deepens," *National Underwriter*, Nov. 26, 2001, p. 8).

[5] "War Exclusion Interpretations," General Materials, *FC&S Bulletins*, The National Underwriter Co., 2002.

[6] Ibid.

Chapter 11

Government and Insurance Industry Reinsurance Proposals

Options Introduced

Many proposals surfaced after September 11 to have the federal government provide a source of reinsurance capacity for terrorism risks.

Three major types of federal proposal emerged in the latter part of 2001:

- a federal loan program
- a federally backed reinsurance pool
- direct federal subsidy of insurers

None of the proposals was enacted, however, despite continuing signs of interest in a federal backstop and concerns about a lack of insurance to financially protect policyholders against terrorist exposures. The efforts to provide federal support for reinsurance are demonstrated by the three following initiatives.

HR 3210, Terrorism Risk Protection Act

The loan program, House bill HR 3210—also known as the Terrorism Risk Protection Act—contained the following proposed features:

- up to $100 billion in first-dollar financial assistance to the insurance industry
- the sole payment trigger would be the size of the terror-related losses
- there were no demands or standards of performance placed upon the insurance industry
- insurers would be directly responsible only for 10 percent of all covered losses up to $100 billion
- 90 percent of the first $20 billion in losses would be loaned to insurers

- insurers would reimburse such loans to the government through assessments allocated to every insurer, based on its premium revenue

Additionally, HR 3210 would suspend all state laws regulating premium rate increases, and it granted insurers unrestricted authority to raise premiums to cover 100 percent of assessments. Not even the insurance industry was unanimously in agreement over the bill. National Association of Independent Insurers (NAII) members expressed particular concern about one feature, cross-subsidization. This required that all insurance companies and their policyholders, regardless of their exposure to terrorism risks, would contribute equally to subsidize the losses. In its original form, HR 3210 would have required smaller insurance companies—insurers that provide the bulk of insurance for "Main Street businesses"—to subsidize large national commercial insurance companies.

Many of these smaller insurers had far less exposure to terrorism risks than the companies writing coverage for skyscrapers and manufacturing plants. NAII urged legislators to include language that provided the Secretary of the Treasury with authority to factor in relevant risk factors, such as territorial difference, exposure to loss, severity of loss, and premiums when determining the assessments against insurance companies and surcharges against policyholders.

The bill's final version achieved this by requiring the Treasury Secretary to consider the effect of assessments and surcharges on "urban and smaller commercial and rural areas." Assessments and surcharges would still be spread through the industry, but the payback would be a function of the exposure each company had to the risk. Increases in premium to cover assessments would not have been as severe for those who did not have the same risk as those companies insuring large national "targets."

Insurance Stabilization and Availability Act

The insurance industry advanced a different proposal to create capacity for terrorism risks, the Insurance Stabilization and Availability Act. This would have created the Homeland Security Mutual Reinsurance Company to provide reinsurance for terrorism losses in the United States to entities insured by Homeland members. Membership was to be available to any insurer that ceded all terrorism risks. The insurance program would be exempt from federal and state taxes, as well as from antitrust laws. The Act would preempt state regulation of the reinsurance company's solvency, policy terms, coverage availability, and premiums.

Under this program, which was based on the United Kingdom's Pool Re program that provides a reinsurance backstop for terrorism risks, the U.S. government would reinsure 100 percent of losses during the program's first year. Further, it would insure 100 percent of losses if Homeland's net assets dipped below 20 percent of the net assets existing at the end of the preceding year.

Homeland would function through two financially separate units—a personal lines and a commercial lines unit. It would provide reinsurance for terrorism and workers compensation war risks for losses sustained within the United States. Other salient features of the Insurance Stabilization and Availability Act included:

- ❑ The law would lapse in six years unless renewed by Congress.
- ❑ The plan would address both personal and commercial lines risks affected by terrorism.
- ❑ The Treasury Secretary would have to certify the cause of loss as being from acts of war or terrorism as a requirement to collect under the plan.
- ❑ "Terrorism" would be defined in a way to exclude any act or threat perpetrated by an official, employee, or agent of a foreign state acting for or on behalf of that state.
- ❑ Membership would be open to insurers, reinsurers, and risk retention groups in any state. Further, membership would be available to any state insurance fund, any residual market mechanism, joint underwriting association, or comparable state administered or state authorized facility.

- Member companies would retain 5 percent of each terrorism or workers compensation war risk, ceding the balance (95 percent) to Homeland, unless other terms were agreed upon.

The House Finance Committee ultimately rejected the Act, however. One reason was the public outcry over an industry bailout at taxpayer expense. Another factor was concern over excessive government entanglement in the insurance industry and increased bureaucracy.

Bush Administration Proposal

The Bush Administration advanced its own proposal. Under this, the federal government would pay 80 percent of the first $20 billion in covered loss due to terrorism and 90 percent of loss between $20 and $100 billion. In the program's second year, the federal government would pay none of the first $10 billion in loss, 50 percent of losses in the $10 to $20 billion range, and 90 percent of losses in the $20 to $100 billion range. In year three, the federal government would pay none of the first $20 billion in loss, 50 percent of losses in the $20 to $40 billion range, and 90 percent of losses in the $40 to $100 billion strata.

The Bush proposal would have divided the cost of property claims from terrorist attacks between the federal government and the insurance industry. Taxpayers would shoulder about 80 percent of the tab, and insurers would absorb the rest. One wrinkle was that the government's accommodation would have a limited shelf life of three years. Total liability for both the public and private sectors under the White House plan would have been capped at $100 billion.

The Bush administration proposal countered one championed by the insurance industry. Under the latter proposal, discussed previously, the government would create a new government-backed insurance company to manage a pool of premiums and claim payments for terrorism coverage. Once losses exceeded the sum of money in the pool, the government would shoulder the difference. This could reach sums higher than those the taxpayers might have paid under the Bush administration proposal. Both proposals addressed losses incurred under property and casualty insurance policies but not to life insurance policies.

Some observers saw certain benefits to the Bush administration proposal. These included the preservation of competitive market forces, the idea that the proposal encouraged risk management and careful insurance underwriting, and the fact that it allowed states to retain regulation of insurance companies. Still, many critics viewed it as a bailout for the insurance industry.

Reasons for Federal Proposals

Reinsurance capacity—or lack thereof—drives much of this concern. Major reinsurance companies, which backstop the policies and coverages issued by primary insurers, served notice in the fourth quarter of 2001 that they would not renew terrorism coverage after December 31, 2001. Reinsurers typically enter into treaties with insurance companies to cover certain policies or clusters of policies they issue.

Other reinsurers provide what the industry terms *facultative coverage* on individual risks. In either case, reinsurance contracts often contain exclusions similar to those included in primary insurance policies. Terrorism had not figured prominently as a reinsurance exclusion prior to September 11; it now appears with increasing frequency.

Faced with a lack of reinsurance coverage, primary insurers began to exclude terrorism in their policies in order to avoid assuming all of the risk themselves. Other carriers began to price the coverage so exorbitantly that few could afford it.

Insurance industry representatives believe that such terrorism exclusions and price-prohibitive coverage when available could be devastating to the American economy. One reason is that most lenders require insurance coverage in order to finance real estate development, plant expansions, and other financial initiatives.

There are a number of reasons for the lack of action. First, an undercurrent of concern exists that insurers are exploiting the situation to leverage a government bailout. Many are reluctant to establish a policy of rescuing certain industries, which could lead to other business sectors seeking similar assistance. Second, there is a sense that pricing risk is the very thing at which insurers have expertise, and that insurers should be left on their own to develop a market mechanism to address terrorism losses.

Third, the government does not pay for such government payments; taxpayers ultimately fund them. Hence, taxpayers in effect pay twice: once through higher insurance premiums and secondly through higher taxes. Some observers have difficulty with this model from a standpoint of equity or fairness.

Despite the fact that none of the initiatives was enacted, discussion continues about whether reinsurance support is needed and, if it is, the best way to shore up capacity.

Victim Compensation Fund

Due to concerns that the September 11 terrorist attacks could trigger hundreds of lawsuits, the September 11th Victim Compensation Fund of 2001 was established to fund payments to victims and their families without delays or litigation. The fund was included in Public Law 107-42, the Air Transportation Safety and System Stabilization Act. The fund is to provide compensation to individuals who were physically injured or to the personal representatives of those killed due to the terrorist-related airline crashes.

Web site for Victim Compensation Fund

A copy and synopsis of the Fund's workings is found at www.usdoj.gov/victimcompensation. This includes an overview of the claim process, eligibility forms and information on advance benefits.

Perhaps predictably, some personal injury attorneys struck a cautionary note about tapping into this fund, since a requirement is that recipients relinquish their right to sue.

The Justice Department announced in late 2001 that family members and survivors of those killed in the attacks would receive at least $500,000 if married and $300,000 if single. These minimums include proceeds from the special fund less any offsets from other state or federal programs, life insurance, and other compensation besides charitable gifts.

The government started to accept applications in late 2001 for victims, with the average payout estimated at more than a million dollars. Victims of the attacks can only apply for money if they forfeit their rights to sue for damages.

In an effort to encourage families to use the Fund, the trial lawyer lobby—Association of Trial Lawyers of America (ATLA)—offered free legal representation to those opting for the fund. Not all attorneys have fallen in lockstep, though. New York attorney Aaron Broder said that the victim fund "defeats the rights of the injured people and the families of the dead to achieve their day in court."[1]

Organizations Recommend Action

Various insurance and risk management organizations have advocated that their members provide input on the subject of federal backing for terrorism reinsurance. These organizations have called on risk managers and insurance professionals to help accelerate the legislative process in various ways:

- Writing to congressmen, senators, and other elected representatives, urging passage of federal legislation providing reinsurance capacity.
- Getting involved with trade groups such as NAII and RIMS (Risk and Insurance Management Society) to lend support to parallel lobbying and communication efforts.
- In isolation, the insurance and risk professional has little chance of influencing the fate of proposed legislation. Working in concert with others, though, they may be able to facilitate passage of legislation that would create a congenial (or at least tolerable) climate for underwriting terrorism risks.

End Notes

[1] "Victims: To Sue or Not to Sue," *Newsweek*, Nov. 12, 2001, p. 10.

Chapter 12

Subrogation and Recovery Options Against Terrorists

Civil Action Possible

Terrorism victims may have legal rights of civil action and money damages against the tortfeasors responsible for such acts. While many see suing as, next to baseball, the great American pastime, others might overlook the potential of financial recovery from those who commit terrorist acts or those orchestrating such events from afar. Government authorities now know that many terrorist agents or sponsors have deep pockets. No one underestimates the challenges of obtaining or collecting on such judgments, but victims of terrorism should not automatically rule out civil recovery as part of their risk management plans and options. Nor should insurers that pay losses due to terrorism overlook a viable right of subrogation.

As is typical with any massive loss—especially man-made loss as respects terrorism—the insurers paying benefits will look to culpable parties for reimbursement. This is known as *subrogation*—transferring the right of legal restitution from the insured to the insurer to pursue the wrongdoer or tortfeasor. Commercial entities, as well as their insurers, might have comparable rights of action just as would any individual. That is why subrogation and civil recovery may have a legitimate place in the panoply of risk management tools available to mitigate the financial costs of terrorism losses.

In the immediate aftermath of the September attacks, most tort reform concerns centered around the prospect of lawsuits against airlines, property owners and managers, security firms, etc. These appeared to be the most immediately targeted defendants in civil actions, defendants that were easily identified. As the proverbial and literal dust cleared, however, it also became apparent that there might be viable legal rights of tort action against the terrorists themselves.

Potential Obstacles

Despite this realization, various obstacles may hamper successful recovery and subrogation action against terrorists. These include:

- identifying the perpetrators,

- linking the perpetrators to financial resources,
- obtaining American civil court jurisdiction over terrorists,
- obtaining service of process on tortfeasors,
- maneuvering the obstacle course of laws or federal government intervention that might limit pursuit of a foreign government's assets, and
- successfully executing any judgment and collecting money to satisfy it.

Legal Precedents

Despite these seemingly daunting hurdles, precedent exists for victims of terrorism or political reprisal to seek civil redress and civil damages through the U.S. court system.

For example, hostages held in Iraq during the Gulf War sued Saddam Hussein's government. About 150 Americans held by Iraq in 1990 and used as "human shields" to deter allied air strikes in the Gulf War sued the Iraqi government in October 2001 in a U.S. court for tens of millions of dollars.

Lawyers representing the former hostages say the claims arise from injuries sustained due to an order by Iraqi President Saddam Hussein on August 2, 1990, immediately after Iraq invaded Kuwait, barring all U.S. nationals from leaving either country.

The case opened in the U.S. District Court in the District of Columbia in October 2001 before Judge Thomas Penfield Jackson. Lawyer Michael Lieder, whose firm represents the ex-hostages, said that eleven plaintiffs would take the witness stand to hold Iraq accountable for alleged human rights abuses. The plaintiffs sought tens of millions of

> Federal Laws relevant to possible subrogation and recovery options are:
>
> - *The Antiterrorism Act of 1990 (18 U.S.C. §2331, 2333).*
> - *The Antiterrorism and Effective Death Penalty Act of 1996 (Pub.L. 104-132).*
>
> For an insightful discussion of the law of suing terrorists, see "Suing Terrorists and Their Private and State Supporters," Richard K. Milin, New York Law Journal, *Oct. 29, 2001.*

dollars in damages due to the emotional and financial injuries they say they suffered in Iraq.

Winning such a case is one challenge. The other challenge is collecting on a judgment. In the claims against Iraq, attorney Lieder said that, if the judge ruled in the victims' favor, it was uncertain how possible damages would be recovered from the Iraqi government. An Iraqi government representative was not in court. Often the defendant terrorists fail to appear in court; they do not wish to be identified or do not recognize the legitimacy of foreign tribunals in judging their actions.

Foreign Sovereign Immunities Act (FSIA)

Plaintiffs can assert claims under the Foreign Sovereign Immunities Act (FSIA) 28 U.S.C. S. §1330, which allows suits against countries designated as terrorist states by the State Department when those states perpetrate acts of terrorism against U.S. citizens. The Department of State identifies the following countries as falling within the category: Iraq, Libya, Cuba, Sudan, North Korea, Iran, and Syria.

Precedent does exist for victims of terrorism or their families to win monetary redress from foreign states. Instances of individuals successfully pursuing civil tort actions against terrorist tortfeasors or their sponsors include:

- ❑ After a New Jersey girl, Alisa Flatlow, was killed by a Hamas suicide bomber in Israel in 1995 in an attack that Iran orchestrated, her father sued and received a judgment of over $20 million in seized Iranian funds.

- ❑ After Cuba shot down a "Brothers to the Rescue" plane, victims and families recovered a $300 million payment.[1]

- ❑ In October 2001, a federal judge in New York suggested that frozen assets of suspected terrorist organizations should go to families of victims of the 1998 bombings of two American embassies in Africa. These bombings killed 231 people, of whom twelve were Americans.

- ❑ Journalist Terry Anderson, held in Lebanon for six years by the Iranian-backed Hezbollah, reportedly recovered nearly $100 million in civil damages arising from his ordeal.

- In January 2002, a federal judge ordered Iran to pay $42 million to the family of a U.S. Agency for International Development (AID) officer. Hezbollah militants killed the AID officer after a 1984 hijacking of a Kuwaiti airlines flight. Hezbollah militants, underwritten by Iran, stormed the flight en route from Kuwait City to Karachi. The terrorists forced the plane to land in Teheran. Once they identified the AID official—Charles Hegna—as an American, they beat him, shot him in the abdomen, and dumped his body on the airport tarmac.

A 1996 amendment to the Foreign Sovereign Immunities Act empowered survivors of terrorist acts to sue tortfeasors in American courts.[2] While the Association of Trial Lawyers of America (ATLA) declared a moratorium on lawsuits shortly after the September 11 attacks, a few countervailing considerations arise.

First, the intent was likely to stem lawsuits against airlines and building managers. The aim was to avoid a tidal wave of knee-jerk litigation, and there is little indication that the trial attorneys had misgivings about lawsuits against the terrorist perpetrators.

Second, the moratorium may not last into perpetuity.

Third, not all attorneys agreed to comply with ATLA's suggestion. New York attorney James Kreindler, for example, already represents twenty-six families of September 11 victims. In short, victims and insurers should not necessarily view the ATLA call for a lawsuit moratorium as a huge impediment.

Sources of Funds

In the year 2000, the U.S. House of Representatives passed a bill that amended the FSIA to allow enforcement of judgments using frozen assets of countries on the terrorism list. However, the bill limited this to cases involving Iran and Cuba.

The House of Representatives in late 2001 adopted the Patriot Bill, enabling terrorism victims to enforce U.S. court judgments by collecting from the frozen assets of those states, including Iraq, said Daniel Wolf, lead counsel in the Iraqi human shield case. As Wolf said, "President Bush has stated that the United States must punish not only terrorists but also the states that sponsor terrorists. Right now, terrorism is a cheap way for those states to pursue war against Americans."

The Treasury Department has added numerous bank accounts to its list of assets it will seek to freeze and close as part of a global effort to choke off funds to the al-Qaida terrorist organization. Michelle Davis, chief spokeswoman for the Treasury Department, will not verify the number of accounts to be added to Treasury's global freeze list but confirmed several had been turned over to the National Security Council for final scrutiny before being placed on the official government list.

Other Recovery Theories

Setting aside the criminal nature of the acts of violence perpetrated by the terrorists, the latter could certainly be a defendant in civil proceedings brought by injured persons or the legal representatives of deceased individuals under an intentional tort theory. Perhaps the most likely theory would be battery, that is, an intentional, harmful, or offensive contact with the (injured or deceased) plaintiff. Civil law imposes liability for both intentional torts (civil wrongs) and negligent acts.

The individuals who actually executed the September 11 hijackings are, apparently, deceased. Moreover, they would probably lack—even if they were alive—sufficient assets to respond to damages awarded a plaintiff or plaintiffs.

Other potentially identifiable tortfeasors who are jointly liable might exist. However, even if sponsors or masterminds were identified, there would still be challenging issues pertaining to jurisdiction. If the ringleaders are abroad and not U.S. citizens, there are issues regarding the

> Optimistic Plaintiffs?
>
> David Pitchford and David Bruner of Stuart, FL sued Osama Bin Laden in October 2001 for $1.1 trillion in damages, alleging that they had suffered emotional distress due to Bin Laden's public threats against American citizens.
>
> Associated Press, October 15, 2001.

jurisdiction of American courts—jurisdiction equating to the court's power to compel a defendant to answer a tort complaint in court.

Can courts hold a foreign government liable under a negligence theory for criminal acts of its citizens? This might be speculative and unlikely. Though most agree that those responsible for injury or death from terrorism should always be held financially, as well as morally, accountable, it may be an uphill fight to carry the day and successfully recover in a civil court.

Civil Conspiracy as a Recovery Option

Civil conspiracy theory might also provide some avenue for recovery. One could sue any person involved in a civil conspiracy. It is possible that long arm jurisdiction might allow for suits against persons who themselves never entered the country if they (through their agents and coconspirators) caused harm and damage in the United States. Interestingly, if plaintiffs could find their assets, they could proceed *in rem* (see insert) against the assets without ever actually obtaining service of process on the terrorists themselves.

Some American courts have been inclined to ease the normal rules of service of process in civil cases. For example, in January 2002, a U.S. Federal judge ruled that plaintiffs suing Osama Bin Laden for the September 11 attacks could serve him formal notice of lawsuits through media outlets, including Qatar's al-Jazeera network and Afghan newspapers. Facilitating this unusual ruling was the fact that Bin Laden's location is unknown and that he has eluded American forces working to capture him. Without proper service of process, U.S. courts lack jurisdiction over Bin Laden.

> *In Rem* Jurisdiction Defined
>
> *Rem* is Latin for "thing." When a court exercises in rem jurisdiction, it exercises authority over a thing rather than a person.
>
> For example, if a divorcing couple asks a court to supervise the sale of their family home, the court exercises *in rem* jurisdiction over the house. Usually, the property must be located in the same county as the court for it to have *in rem* jurisdiction.
>
> Another example: in early marine insurance, a seaman could file a suite *in rem* directly against a ship and not the owner, seeking compensation by claiming a property interest in the ship.

The fact that civil recovery for damages against terrorists is a legal crapshoot has not deterred some victims from seeking legal redress against the wrongdoers. As of early January 2002, terrorism victims have filed two civil lawsuits against terrorists. In one, the widow of a man killed on the roof of one of the World Trade Center twin towers filed suit against Bin Laden. She identified herself as Jane Doe in order to retain anonymity and avoid reprisal from terrorists.

Another case, filed in November 2001, originates from the estate of business analyst George Smith, who also died in the World Trade Center. Additional defendants in these suits include the Islamic Emirate of Afghanistan, al Qaeda, the Taliban, and Afghanistan's former rulers.

James Beasley, a Philadelphia attorney active in both lawsuits, states that recovery is possible since the defendants' assets are frozen in the United States and overseas.

One challenge is the proof issue. If victims can connect the money to the terrorist, plaintiffs would most likely emerge after the U.S. government has already frozen the foreign government's assets. The U.S. government might allow aggrieved parties to still proceed, although that would be up to the authorities.

Osama Bin Laden, al Qaeda, the Taliban, Afghanistan, and Iraq were defendants in December 2001 in a public interest lawsuit filed in U.S. District Court. Plaintiffs seek $210 million in damages on behalf of a woman killed in the World Trade Center attack. The conservative foundation, Judicial Watch, drew upon media reports and recent books by journalists stating that terrorists met with Iraqi officials and that Bin Laden ran a training camp near Baghdad.[3] Hence, the war on terrorism is being fought not only with aircraft, ships, and ground troops, but it also may be fought in civil courts.

The larger question may not be one of getting a judgment but rather enforcing a judgment and collecting. The time and legal expense needed to successfully prosecute civil claims—or subrogation rights—against tortfeasors may lead to a hollow victory, in which the target defendants are judgment proof due to international law or collectibility issues.

U.S. Government Response

Terrorism victims can sue wrongdoers, but the legal and collection hurdles are daunting. Ironically, terrorism victims often find themselves in court at odds with attorneys from none other than the U.S. government. The government position often tends to be that the courts should dismiss judgments requiring payments from the frozen assets of hostile nations since they interfere with the U.S. execution of its foreign policy.

Some instructive examples exist in the case of American hostages held by Iran. Under formulas that courts used in one case, Iranian hostages might receive $10,000 per day for their ordeal. This would yield a figure of $4.44 million for each hostage, plus millions for family members who also joined the suit. Total compensatory damages against Iran might exceed $200 million, the largest such judgment in U.S. courts.

Hostages turned plaintiffs argue that a 1996 law helps their cause. The Antiterrorism and Effective Death Penalty Act, P.L. 104-132, allows certain terrorism victims to recover damages from nations designated by the State Department as sponsors of terrorism. The bill addresses "acts of torture, extra-judicial killing, aircraft sabotage, hostage taking or providing material support or resources . . . for such an act." Importantly, the law allows successful plaintiffs to seize commercial property owned by a foreign state in the United States.

Basic Reality

We close this discussion by acknowledging one basic reality of subrogation and recovery efforts. This is not an answer to terrorism but rather one response to terrorism. By definition, civil suit recovery is after the fact. Such a right of action exists, if at all, only *after* a loss due to terrorism has occurred. It is a reactive as opposed to a proactive strategy. It may have no deterrent effect on the tortfeasors; if they are willing to die for their cause, it is unlikely that the prospect of financial loss will cause them to pause before executing acts of terror.

On the other hand, terrorism often needs money: money to purchase airplane tickets, to train, to travel. The fewer the financial resources available to terrorists, the harder their job. Subrogation and recovery cannot be the linchpin of an effective risk management approach to terrorism. But it may have its place as part of a holistic approach to attack the terrorist threat on multiple levels. It should not be reflexively ruled out as an option, but considered carefully, weighing the costs, time involved, and odds of success.

Ten Tips on Subrogation and Recovery Options Against Terrorism

1. Explore civil liability recovery for damages due to terrorist acts.
2. Consider class action status with other plaintiffs and victims of terrorism if the facts warrant.
3. Seek legal counsel to assess rights of civil actions under federal laws such as the 1991 Antiterrorism Act and the Antiterrorism and Effective Death Penalty Act of 1996.
4. Seek legal counsel that is experienced in seeking remedies from foreign nations and governments.
5. Look beyond the perpetrators of the terrorist act to their supporters and sponsors as potential deep pockets.
6. Weigh the cost of terrorism with the transaction costs of pursuing legal action and the realistic odds of collecting on any ultimate judgment.
7. Ponder the tradeoff of pursuing civil actions in light of jeopardizing possible entitlement to damages under the September 11 victim's fund.
8. If any insurer pays damages due to terrorist acts, urge them to subrogate against the tortfeasors and provide full cooperation in building a case of culpability.
9. Do not underestimate the difficulties of enforcing a legal claim or collecting funds.
10. Realize that even successful subrogation and recovery is a reactive—not proactive—step. Enhance all other aspects of your risk management program to address the terrorist threats. Most terrorists are willing to die for their cause. The prospect of civil damages is unlikely to deter them.

End Notes

[1] "In Lawsuit Against Iran, Former Hostages Fight U.S.," *Washington Post,* Dec. 13, 2001, pp. A1, A29.

[2] "Iran Loses $42 Million Judgment in Hijack Suit," *Washington Post*, Jan. 23, 2002, p. A3.

[3] "Sept. 11 Victim Sues Bin Laden, 2 Countries," *Washington Post*, Dec. 5, 2001, p. A4.

Chapter 13

Alternative Risk Techniques

Alternative Market Interest

One result of the 2001 terrorist losses is a resurging interest in alternative market mechanisms. After a decade or more of moribund existence, alternative risk mechanisms such as captives and risk retention groups—as well as the excess and surplus lines marketplace—seem to be finding renewed interest from insurance buyers.

Alternative risk transfer means shifting the financial cost of risk through methods other than conventional insurance products. The effect of terrorism losses has reduced the availability of not only coverage for the risk of terrorism, but it also has resulted in more restrictive terms and conditions, a reduction in some lines and scope of coverage, and increased pricing in a number of types of insurance—well beyond the terrorism "peril."

Historically, alternative risk mechanisms thrive when certain lines or types of coverage are either unavailable or overpriced in the conventional insurance market. Many risk retention groups and captives were formed in the 1970s and 1980s to fill a capacity void existing among insurers. To the extent that insurance coverage is either unavailable or prohibitively expensive, buyers may seek financial protection through less conventional means.

In the past, growth in alternatives was driven by constriction in the standard market, similar to that being seen in the diminishing availability of coverage after the 2001 terrorist attacks. However, the reason to seek out alternatives is not necessarily an effort to find reduced premiums. There is no magic formula in alternative financial mechanisms for reducing rates; all must collect enough premium to pay losses and expenses. In fact, in some cases the premium rates for coverage provided through an alternative financing technique may be higher than that found in the conventional market. Alternative mechanisms must grapple with this fundamental law of solvency. Further, most alternatives will have limited capital as compared to, say, an AIG or a Berkshire Hathaway. But proper risk assessment and underwriting is paramount, which may result in some reinsurers being more receptive to specific alternative programs than traditional insured programs.

The potential for reduced reinsurance capacity looms as the greatest challenge, especially if additional terrorism losses cause further constriction in the marketplace. Therefore, captives, self-insured programs, and larger self-insured retentions will grow to create capacity, provide necessary coverage excluded by the standard market, allow more control of risk costs (whether retained or transferred), create consistency for risk management programs, and, of course, address the ultimate cost of risk for an enterprise. Alternative risk mechanisms sometimes fail to achieve, however, a reduction in the premium component of the total risk-transfer program.

According to Lyle Walker, President of Walker Risk Management Inc., Allen Park, Michigan:

> ". . . all purchasers of reinsurance, in this instance captives, will face cost increases. As to captives, they will reduce by some percentage the amount left for funding loss in the captive structure. Personally I would not expect such percentage to be a significant burden for most. After all, it is simply the end of bargain days."

Likewise, Michael R. Mead, Chairman of the Minneapolis-based Captive Insurance Companies Association noted:

> "In the past when coverage became scarce and premiums potentially unaffordable from the traditional markets, buyers have taken a new look at alternatives, seeking coverage and premium relief. Many commercial buyers have previously investigated captives, self-insurance funds or risk retention groups. Expect this pattern to recur, with some speed and urgency."[1]

Alternative Markets

Many predict a renewed interest in alternative risk financing tools such as:

- Captives
- Risk retention groups
- Risk securitization

- Self-insured retentions and programs
- Finite risk insurance and reinsurance
- Excess and surplus lines companies

These options will be explored as a means to fill holes in coverage laid bare by the fallout from the terrorist attacks.

Captives and Risk Retention Groups[2]

In the face of tighter reinsurance and more stringent underwriting guidelines, businesses may be forced to seek such alternatives as captives and risk retention arrangements. Not only may coverage for terrorism disappear; other business exposures—previously considered fairly simple to insure—may be faced with either hiking premiums or scaled-down insurance protection. The obvious question is— with these developments—where do insurance buyers go? One answer may be captive insurance companies. Both domestic and offshore jurisdictions permit captive insurance companies to operate there. Individual jurisdictions offer alternative advantages and disadvantages.

There are several ways to categorize captives: according to ownership, by type of insureds, or by the type of insurance coverage provided. Captives can be divided into five ownership categories. These are: single owner, multiple owners, pools, rent-a-captives, and profit centers.

> Captive Resource
>
> A useful resource for information about captives: www.captiveguru.com and *The Risk Funding & Self-Insurance Bulletins*, National Underwriter Company.

Although *single owner captives* are owned by one company, they may write coverage for the parent/owner, other businesses, or both. The primary reason that a single owner captive writes coverage for other risks (other than those owned by the parent or members of the financial family of the parent) is so the parent can use the premium it pays to the captive as a tax deductible expense. It also may desire to diversify the risks it is assuming. There are a number of tax implications involved with captive use, and professional advice should be solicited in that area.

Multiple owner captives can be subdivided. These include

- Risk retention captives,
- Association captives, and
- Agency captives.

Risk retention groups, which are controlled by the Risk Retention Act of 1986, write homogeneous liability risks. The insureds usually are in the same industry. In addition, the insureds under a risk retention group are required to be the owners of the captive. The coverage offerings of risk retention group captives are restricted by law and often are designed for specific needs of narrowly defined groups.

An *association or group captive* is formed by an association or homogeneous group to provide coverage needed by its members. Usually, the risks written by association captives are homogeneous. An association may form a risk retention group only if the members of the association are the insureds and the association is owned by the insured members.

Agency captives are formed by brokers and agents to provide coverage for their insureds. The captives increase the probability that the agents and brokers will have a market for their insureds. The coverage offered by agency captives may be broader than the coverage offered by association or risk retention group captives.

Rent-a-captives are a specialized form of captive operation. They are designed for firms that do not want to own a captive but want to obtain some of the advantages offered by captives. A rent-a-captive is formed by a group of investors and operated as an income-producing business. An insured pays a premium and usually pays a deposit or posts a letter of credit to back up its risks. The operators of rent-a-captives handle the operations and claims for its insureds; they also purchase the necessary reinsurance to back up the captive. At the end of the policy period (accounting period), the insured is paid a dividend that depends on its own losses, operating expenses, and the cost of reinsurance.

Profit centers are another specialized form of captive operation. They function in a manner that is similar to a rent-a-captive. A firm that wants to be part of a profit center arrangement purchases nonvoting preferred stock from the profit center. The funds paid for the stock serve as additional capital to offset the premium that is placed with the profit center. A firm desiring insurance places its coverage with the profit center through a fronting company. The fronting company reinsures the business with the profit center. The fronting company pays claims and handles most of the administrative details. It seeks reimbursement for paid claims from the profit center. At the end of the policy or accounting period, the profit center calculates the underwriting results for the firm owning the non-voting preferred stock. If the business is profitable the insured is paid a dividend on the stock; if not, no dividend is paid.

The term *profit center captive* is also used to refer to a captive that is run by a noninsurance parent. Some profit center captives write business for the parent and its financial family, while others insure customers of the parent. An example of the latter is the American Road Insurance Company, which is owned by Ford Motor Corporation and writes extended coverage warranty insurance for owners of Ford vehicles.

Insured

Captives can also be categorized by type of insured. These categories include related insureds and nonrelated insureds. Captives providing coverage for *related insureds* write coverage only for their owners and members of their economic family. This category includes single parent and multiple parent captives. Risk retention groups are a subset of related insured captives. *Nonrelated insured* captives usually write coverage for their owners, but their most important distinguishing characteristic is that they write coverage for third parties.

Insurance

Finally, captive insurance companies can be classified by the type of insurance they write. Captives usually write *property or casualty insurance.* However, they also may provide life and health insurance.

In a way, insurance buyers and risk managers may have come full circle, since captives experienced explosive growth in the 1980s when insurance companies spurned many types of insurance coverage, especially product liability.

Evaluating Captives

Chief Benefits of Captives

Many industries have pooled resources and have formed insurance programs for their particular fields. The benefits include:

- ❑ long-term sources of insurance capacity insulated from fluctuations in the conventional insurance market
- ❑ greater control over handling, adjustment, settlement, and adjudication of losses
- ❑ increased control over underwriting
- ❑ specialized loss prevention resources that many conventional insurers may lack

The transition from the conventional market to a captive is not without pitfalls, though. The mid- to late-1980s were the heyday of captive insurance organizations. Responding to tough buying conditions in the conventional insurance market, many companies and industries banded together, forming their own captive insurance companies. Often, captive members pool funds in order to pay claims and service the insurance needs of some niche market, e.g., oil companies, nuclear firms, attorneys, or psychiatrists. There may be more administrative responsibilities associated with captives than with traditional coverage, and, in some cases, the expenses may be substantial.

Factors to Weigh When Assessing Captives

Some points to consider when pondering the decision of whether or not to join a captive insurance program include:

- ❑ Does the captive offer loss prevention and control services that are not likely to be available through a business's current insurer?
- ❑ Is there a possibility of negotiating with the current insurer for better premium rates?

- ❑ How much control over settlement decisions will the business have with a captive compared with its current carrier?

Assessing Trust Factors

One of the biggest issues in joining a group or association captive is the fundamental issue of trust. Will a captive arrangement best represent the business's interests? Factors to examine when considering the trust factor include:

1. Examine the governance structure. Who makes decisions? How are those members selected?

2. Does the governing body view the captive as a revenue source or as a risk-bearing entity? This analysis will be subjective, based on interpreting the decision makers' priorities. Is its goal the generation of dividends or simply to control coverage costs? There are many excellent captive governing boards that are running group or association captives for the right reasons—to serve their industries. Those are the captives that are the most secure for the long run.

3. Where is the captive domiciled? Certain domiciles are liberal, and others are more conservative. Does the domicile's captive regulations support the business's priorities?

4. What are the departure penalties? Do you forfeit future dividends?

Using a captive for a tax advantage should be subordinate to the goal of traditional risk management (control, predictability, lower costs of risk transfer) as the insurance market hardens.

A captive may be the solution to financing exposures to terrorism, which may become uninsurable in the conventional market. Higher costs for terrorism coverage may increasingly tempt insurance buyers to look at such alternative risk transfer mechanisms as the only viable option.

Some insurance brokers might help fill the void by creating new sources of capacity themselves. For example, Aon

Corporation's special risks counter-terrorism team in late 2001 launched a new product addressing sabotage and terrorism risk. Further, insureds in this program can access risk management consulting services from an account team comprised in part of former military personnel. Aon markets the policy mostly to firms encountering sabotage and terrorism exclusions in their property and business interruption policies. Aon has structured the coverage program, but Lloyds of London is the underwriter. Similar options are also available through other channels.

Risk Securitization

Risk securitization involves issuing catastrophe bonds that bear interest. Bondholders lose their capital if specified disasters such as earthquakes, hurricanes, (or terrorist acts) occur. While the practice of risk securitization surfaced in the insurance marketplace in the latter half of the 1990s, the low cost of insurance at that time and for years thereafter made these techniques relatively unattractive. That changed starting in late 2001.

Risk securitization might be an innovative alternative risk method to handle the problem of terrorism losses. The risk of a catastrophic terrorist loss could be spread among insurers globally, and it could be a good investment. By trading the shares, an investor or insurer might be able to achieve a very good spread of risk.

Others strike cautionary notes regarding risk securitization as a way to create additional insurance coverage capacity for terrorism risks. Many observers doubt that the insurance industry will see the capital markets lining up to bet money (securitizing) that the really at-risk business(es) it is securing will not be hit by terrorist-attack loss(es). Further, the securitizations (mostly arranged by insurers and reinsurers) usually receiving publicity are fairly small (e.g., about $150 million) relative to some of today's large catastrophic exposures. One only need look, as underwriters do, at the potential affect of a few ounces of anthrax spores or a suicidal terrorist to recognize the possible catastrophic impact.

Another concern is adverse selection; it is likely that only those who could not get insurance would seek securitization.

Self-Insured Retentions

While the trend to higher self-insured retentions (SIRs) was manifest before September 11, 2001, that trend has been accentuated sharply since then. In many cases, what was once an option is no longer a choice but a necessity imposed by an increasingly tightening insurance market.

The pluses of higher SIRs include:

- lower premium costs, and
- more control over the claims-handling process.

One obvious downside is the diminished financial security if a business suffers multiple losses within that SIR. Whether an SIR is voluntary or involuntary, the risk manager subjects her corporation's earnings to more instability when she assumes more of the risk. Much hinges on the corporation's, and upper management's, risk tolerance. How much do they value peace of mind and a good night's sleep?

Christopher Mandel, Assistant Vice President of Enterprise Risk Management at the USAA Group in San Antonio, Texas, articulated one risk management lesson of the terrorist attacks:

> "Rethink your retention philosophy and get management to agree to a higher, yet financially supportable, level of risk assumption. Don't forget the emphasis you'll need to put on [loss] controls."[3]

Some organizations and industries are taking self-insured retentions—and self-insurance—to the next level by creating self-insurance pools. For example, the Delaware River Joint Toll Bridge Commission instituted a higher toll in December 2001. A portion of that will go toward creating a $300 million fund to cover terrorist losses in case of attacks on the Pennsylvania-New Jersey bridges.[4]

Finite Risk Cover

Another type of alternative risk transfer is finite risk reinsurance. There are various types of finite risk reinsurance, but they typically are arranged by insurance companies or businesses to transfer an existing portfolio of losses to another risk-bearing entity. Finite risk coverage often affects the ability of primary carriers to provide additional capacity, so it behooves risk managers to have some understanding and appreciation of finite risk reinsurance. Generally these arrangements transfer losses to a reinsurer for a specified premium, but the reinsurer's ultimate liability usually is capped. The primary insurer—or business entering into the arrangement—will be assessed if the losses develop into a larger financial burden than originally projected. Finite risk reinsurance has an ability to smooth out balance sheet gyrations that would otherwise occur in primary insurers.

Excess and Surplus Lines Carriers

One solution to the capacity crunch might be excess and surplus (E&S) lines markets.

Brokers report they are increasingly using alternative markets to cover risks that are difficult to place in the current primary market. In a special benchmarking survey released in November 2001, the Council of Insurance Agents + Brokers (CIAB) reported that 45 out of 103 respondents—44 percent of the total—said they were increasingly turning to alternative mechanisms to meet their clients' challenging needs. The report states that a significant number of those say they have turned to the excess and surplus lines market for coverage not available in the primary market.[5]

Many buyers might see coverage with a surplus lines insurer as a good news/bad news situation. The good news is that coverage might be available there when it is not available in the conventional market. The bad news is that such coverage may be on a claims-made form, which many risk managers view as inferior to occurrence form coverage. Some accounts, which would not have considered claims made coverage a year or two ago, must now give it serious thought. Another negative is that surplus lines insurers are not backed by state guaranty funds in the event of insolvency.

Very tough lines of coverage often migrating to the surplus lines market include:

- physician medical malpractice
- construction risks
- nursing homes
- property coverage in catastrophe-prone areas
- excess liability

Insurance Realities

Changes wrought by the September 11 terrorist attacks have exacerbated an already toughening insurance marketplace for risk managers and insurance buyers. This will force more risks into the alternative risk market. Risk managers who were not conversant with these techniques need to learn quickly, since the mechanisms may be one of the few options available.

Ways to Use Alternative Risk Techniques to Address Terrorism Risks

Captive Insurers

- Explore captive insurers as an option for harder to place coverages.
- Determine if industry captives exist for your segment of the industry.
- Look for captives that provide strong loss prevention/loss control services.
- Request input or veto power over settlement decisions made on claims by the captive.

- ❑ Give primary weight to the captive's size and expertise.
- ❑ Consider both agency/broker captives and those established outside of agent/broker channels.
- ❑ Find out the mechanism for leaving a captive before joining or establishing one.
- ❑ Research the captive governance structure.
- ❑ Determine whether the governing body views the captive as a revenue source or a risk-bearing entity.
- ❑ Look at the captive's domicile and the degree of oversight and regulation exercised.
- ❑ Explore insurance pools as a source of needed capacity for lines of coverage shrunken in the aftermath of terrorist acts.

Retention

- ❑ Be sure there is cash liquidity to fund losses within the self-insured retention.
- ❑ Determine upper management's risk tolerance level.
- ❑ Get upper management buy-in, in advance (preferably in writing), on retention decisions.
- ❑ Balance potential investment income with expected future losses.
- ❑ Determine the pain point at which the business can shoulder losses without exposing it to significant instability.

Excess and Surplus Lines Options

- ❑ Consider excess and surplus (E&S) lines carriers for terrorism coverage.
- ❑ Carefully check the financial statements of any E&S insurer before placing coverage with them.

- ❑ Realize that guaranty fund protection is not provided when coverage is placed with a surplus lines carrier.

Miscellaneous

- ❑ Reconsider the wisdom of having operations in certain types of building occupancies, such as high-rise structures or in high-risk zones. There is a growing perception that urban settings and business venues that are marquee-type landmarks are more likely to be terrorist targets.
- ❑ Revisit the need for private corporate jets. The cost of aviation insurance is affected by events of 2001.
- ❑ Develop and implement sound crisis response and disaster recovery plans.

End Notes

[1] "Captives Can Help Post World Trade Center Recovery," *National Underwriter*, Oct. 22, 2001, p. 23.

[2] "Self-insured Plans and Captives," *The Risk Funding & Self-Insurance Bulletins,* The National Underwriter Co., 2000.

[3] "Impact of Attacks Felt Across All Lines," *Business Insurance*, Oct. 15, 2001, p. 10.

[4] "Some Companies Insure Themselves Against Terrorism," *USA Today*, Jan. 10, 2002, p. 1B.

[5] "Special Benchmarking Survey," Council of Insurance Agents + Brokers, Nov. 20, 2001.

Internet Resources

American Psychiatric Association
http://www.psych.org/pract_of_psych/disaster_psych.cfm

Center for Defense Information http://www.cdi.org/terrorism/

Center for Internet Security http://www.cisecurity.org/

CERT® Coordination Center for Internet Security http://www.cert.org/

Chem-Bio Research Center http://www.chem-bio.com/resource/

Computer Security Institute http://www.gocsi.com/

Federal Response Plan http://www.fema.gov/r-n-r/frp/

Institute for Security Technology Studies (ISTS) (Dartmouth College) http://www.ists.dartmouth.edu/

Johns Hopkins University Center for Civilian Biodefense Studies
http://www.hopkins-biodefense.org/

Medical Information on Nuclear Biological Chemical WMD
http://www.nbc-med.org/others/Default.html

Monterey Institute Center for Nonproliferation Studies
http://cns.miis.edu/index.htm

National Council on Radiation Protection and Measurements
http://www.ncrp.com/

National Criminal Justice Reference Service
http://www.ncjrs.org/recovery/

National Domestic Preparedness Office http://www.ndpo.gov/

National Infrastructure Protection Center (NIPC) http://www.nipc.gov/

NOAA Office of Response and Restoration
http://response.restoration.noaa.gov/

Oklahoma City National Memorial Institute for the Prevention of Terrorism http://www.mipt.org/protectyourself.html

RAND Corporation http://www.rand.org/

Rapid Response Information System (FEMA) http://www.rris.fema.gov/

SANS Institute (computer security)
http://www.sans.org/newlook/home.php

U.S. Army Corps of Engineers, Building Protection Against Airborne Hazards http://buildingprotection.sbccom.army.mil/

U.S. Army Soldier and Biological Chemical Command (SBCCOM)
(http://www.apgea.army.mil)

U.S. Army Institute of Infectious Diseases, Biological Agent Information Papers http://www.nbc-med.org/SiteContent/MedRef/OnlineRef/GovDocs/BioAgents.html

U.S. Bureau of Alcohol Tobacco and Firearms, Vehicle Bomb Explosion Hazards http://www.atf.treas.gov/pub/fire-explo_pub/i54001.htm

U.S. Centers for Disease Control and Prevention Bioterrorism Page
http://www.bt.cdc.gov/

U.S. Critical Infrastructure Assurance Office http://www.ciao.gov/

U.S. Department of Agriculture http://www.usda.gov/biosecurity/home-land.html

U.S. Department of Health & Human Services Anthrax & Biological Incidents: Preparedness & Response http://www.hhs.gov/hottopics/heal-ing/biological.html

U.S. Department of State, Significant Terrorist Incidents, 1961-2001
http://www.state.gov/r/pa/ho/pubs/fs/5902.htm

U.S. Federal Computer Incident Response Center
http://www.fedcirc.gov/

U.S. Food & Drug Administration Bioterrorism
http://www.fda.gov/oc/opacom/hottopics/bioterrorism.html

U.S. Government Accounting Office, Reports on Terrorism
http://www.gao.gov/terrorism.html

White House Office of Homeland Security
http://www.whitehouse.gov/homeland/

Selected References

Commission on Engineering and Technical Systems, *Protecting Buildings from Bomb Damage: Transfer of Blast-Effects Mitigation Technologies from Military to Civilian Applications,* National Academy Press, 1995.

Committee on the Protection of Federal Facilities Against Terrorism, Building Research Board, Commission on Engineering and Technical Systems, National Research Council, *Protection of Federal Office Buildings Against Terrorism*, National Academy Press, Washington, D.C., 1988.

Critical Infrastructure Assurance Office, *Practices for Securing Critical Information Assets*, Washington, D.C., 2000.

Federal Bureau of Investigation, *Terrorism in the United States 1999.*

Federal Emergency Management Agency, "Guide for All-Hazard Emergency Operations Planning, State and Local Guide (101), Chapter 6, Attachment G – Terrorism," April 2001.

Leitenberg, Milton, Center for International and Security Studies, University of Maryland, *An Assessment of the Biological Weapons Threat to the United States, A White Paper prepared for the Conference on Emerging Threats Assessment: Biological Terrorism, at the Institute for Security Technology Studies*, Dartmouth College, July 7-9, 2000.

Massa, Ronald J. "Vulnerability of Buildings to Blast Damage and Blast-Induced Fire Damage," The World Trade Center Bombing" Report and Analysis, United States Fire Administration Technical Report Series, Report 076.

National Infrastructure Protection Center, "The Threat to the U.S. Information Infrastructure," October 2001.

National Institute for Chemical Studies, *Sheltering in Place as a Public Protective Action*, Charleston, WV, 2001.

NCRP Report No. 138, *Management of Terrorist Events Involving Radioactive Material*, National Council on Radiation Protection and Measurements, Bethesda, MD, 2001.

NFPA 101, Life Safety Code®, 2000 Edition, National Fire Protection Association, Quincy, MA, 2000

NFPA 1600, Standard on Disaster/Emergency Management and Business Continuity Programs, 2000 Edition, National Fire Protection Association, Quincy, MA, 2000.

NFPA 1620, Recommended Practice for Pre-Incident Planning, 1998 Edition, National Fire Protection Association, Quincy, MA, 1998.

Office for Victims of Crime, U.S. Department of Justice, Office of Justice Programs, "OVC Handbook for Coping After Terrorism," NCJ 190249, September 2001.

SANS Institute, "The Twenty Most Critical Internet Security Vulnerabilities (Updated) The Experts' Consensus," Version 2.501, Internet [http://www.sans.org/top20.htm], November 15, 2001.

Schmidt, Donald L., *Emergency Response Planning: A Management Guide*, M&M Protection Consultants, Chicago, IL, 1994.

Swanson, Marianne; Grance, Tim; Hash, Joan; Pope, Lucinda; Thomas, Ray; Wohl, Amy. *Contingency Planning Guide for Information Technology Systems, Recommendations of the National Institute of Standards and Technology*, NIST Special Publication 800-34, 2001.

Transport Canada, U.S. Department of Transportation, Secretariat of Transport and Communications of Mexico, *2000 Emergency Response Guidebook,* 2000.

U. S. Department of State, Office of the Coordinator for Counterterrorism, *Patterns of Global Terrorism 1999.*

U.S. Army Corps of Engineers®, *Protecting Buildings and their Occupants from Airborne Hazards*, TI 853-01, Washington, D.C., October 2001.

U.S. Bureau of Alcohol Tobacco and Firearms, *Vehicle Bomb Explosion Hazards and Evacuation Distance Tables*, http://www.atf.treas.gov/.

U.S. Department of Justice, Office of Justice Programs, Office for State and Local Domestic Preparedness Support, *Assessment and Strategy Development Tool Kit,* NCJ181200, 1999.

U.S. Department of Justice, Office of Justice Programs–Bureau of Justice Assistance, Federal Emergency Management Agency, and United States Fire Administration–National Fire Academy, *Emergency Response to Terrorism Self-Study*, FEMA/USFA/NFA-ERT:SS, June 1999.

U.S. Department of the Treasury, Bureau of Alcohol Tobacco and Firearms, *Bomb and Physical Security Planning,* ATF P 7550.2, July 1987.

United States Fire Administration, Federal Emergency Management Agency, *Emergency Procedures for Employees with Disabilities in Office Occupancies*, Emmitsburg, MD.

United States Government Interagency Domestic Terrorism Concept of Operations Plan, January 2001.

United States Postal Inspection Service, *Security Plan for Suspected Letter and Parcel Bombs,* http://www.usps.gov/websites/depart/inspect/.

Responses to Weapons of Mass Destruction Incident and the Participants Involved[1]

Events	Participants
1. Incident occurs.	
2. 911 center receives calls, elicits information, dispatches first responders, relays information to first responders prior to their arrival on scene, makes notifications, and consults existing databases of chemical hazards in the community, as required.	911 Center, first responders.
3. First responders arrive on scene and make initial assessment. Establish Incident Command. Determine potential weapon of mass destruction (WMD) incident and possible terrorist involvement; warn additional responders to scene of potential secondary hazards/devices. Perform any obvious rescues as incident permits. Establish security perimeter. Determine needs for additional assistance. Begin triage and treatment of victims. Begin hazard agent identification.	Incident Command: Fire, Law Enforcement, Emergency Medical Services (EMS), and HazMat unit(s).
4. Incident Command manages incident response; notifies medical facility, emergency management (EM), and other local organizations outlined in Emergency Operations Plan; requests notification of Federal Bureau of Investigation (FBI) Field Office.	Incident Command.
5. Special Agent in Charge (SAC) assesses information, supports local law enforcement, and determines WMD terrorist incident has occurred. Notifies Strategic Information and Operations Center (SIOC), activates Joint Operations Center (JOC), coordinates the crisis management aspects of WMD incident, and acts as the federal on-scene manager for the U.S. government while FBI is Lead Federal Agency (LFA).	FBI Field Office: SAC.
6. Local Emergency Operations Center (EOC) activated. Supports Incident Command, as required by Incident Commander (IC). Coordinates consequence management activities (e.g., mass care). Local authorities declare state of emergency. Coordinates with state EOC and state and federal agencies, as required. Requests state and federal assistance, as necessary.	Local EOC: Local agencies, as identified in basic Emergency Operations Plan (EOP).
7. Strategic local coordination of crisis management activities. Brief President, National Security Council (NSC), and Attorney General. Provide Headquarters support to JOC. Domestic Emergency Support Team (DEST) may be deployed. Notification of FEMA by FBI/SIOC triggers FEMA actions.	SIOC: FBI, Department of Justice (DOJ), Department of Energy (DOE), Federal Emergency Management Agency (FEMA), Department of Defense (DoD), Department of Health and Human Services (HHS), and Environmental Protection Agency (EPA).
8. Manage criminal investigation. Establish Joint Information Center (JIC). State and local agencies and FEMA ensure coordination of consequence management activities.	FBI; other federal, state, and local law enforcement agencies. Local Emergency Management (EM) representatives. FEMA, DoD, DOE, HHS, EPA, and other Federal Response Plan (FRP) agencies, as required.
9. State EMS support local consequence management. Brief Governor. Declare state of emergency. Develop/coordinate requests for federal assistance through FEMA Regional Operations Center (ROC). Coordinate state request for federal consequence management assistance.	State EOC: State EMS and state agencies, as outlined in EOP.
10. DEST provides assistance to FBI SAC. Merges into JOC, as appropriate.	DEST, DoD, DOJ, HHS, FEMA, EPA, and DOE.

[1] Federal Emergency Management Agency, "Guide For All-Hazard Emergency Operations Planning", *State And Local Guide (101),* Chapter 6, Attachment G – Terrorism, Table 5.

Federal Departments and Agencies: Counterterrorism-Specific Roles[1]

Federal Emergency Management Agency

FEMA is the lead agency for consequence management and acts in support of the FBI in Washington, DC, and on the scene of the crisis until the U.S. Attorney General transfers the Lead Federal Agency (LFA) role to FEMA. Though State and local officials bear primary responsibility for consequence management, FEMA coordinates the Federal aspects of consequence management in the event of a terrorist act. Under Presidential Decision Directive 39, FEMA supports the overall LFA by operating as the lead agency for consequence management until the overall LFA role is transferred to FEMA and in this capacity determines when consequences are "imminent" for purposes of the Stafford Act. (Source: Federal Response Plan Terrorism Incident Annex, April 1999.) Consequence management includes protecting the public health and safety and providing emergency relief to State governments, businesses, and individuals. Additional information on Federal response is given in the United States Government Interagency Domestic Terrorism Concept of Operations Plan (http://www.fema.gov/r-n-r/conplan/).

Office of the Director/Senior Advisor to the Director for Terrorism Preparedness

The Senior Advisor (1) keeps the FEMA Director informed of terrorism-related activities, (2) develops and implements strategies for FEMA involvement in terrorism-related activities, and (3) coordinates overall relationships with other Federal departments and agencies involved in the consequence management of terrorism-related activities.

Preparedness, Training, and Exercises Directorate (PT)

This office provides planning guidance for State and local government. It also trains emergency managers, firefighters, and elected officials in consequence management through the Emergency Management Institute (EMI), National Fire Academy (NFA), and the National Emergency Training Center (NETC) in Emmetsburg, Maryland. EMI offers courses for first responders dealing with the consequences of a terrorist incident. PT conducts exercises in WMD terrorism consequence management through the Comprehensive Exercise Program. These exercises provide the opportunity to investigate the effectiveness of the Federal Response Plan (FRP) to deal with consequence management and test the ability of different levels of response to interact. PT also manages FEMA's Terrorism Consequence Management Preparedness Assistance used by State and local governments for terrorism preparedness planning, training, and exercising.

Mitigation Directorate

This office has been assigned the responsibility of providing the verified and validated airborne and waterborne hazardous material models. The office also is responsible for developing new, technologically advanced, remote sensing capabilities needed to assess the release and dispersion of hazardous materials, both in air and water, for guiding consequence management response activities.

Response and Recovery Directorate

This office manages Federal consequence management operations in response to terrorist events. In addition, it manages the Rapid Response Information System, which inventories physical assets and equipment available to State and local officials, and provides a database of chemical and biological agents and safety precautions.

U.S. Fire Administration (USFA)

This administration provides training to firefighters and other first responders through the NFA in conjunction with the Preparedness, Training, and Exercises Directorate. The NFA offers courses pertaining to preparedness and response to terrorist events.

Department of Justice (DOJ)

Federal Bureau of Investigation

The FBI is the lead agency for crisis management and investigation of all terrorism-related matters, including incidents involving a WMD. Within FBI's role as LFA, the FBI Federal On-Scene Commander (OSC) coordinates the overall Federal response until the Attorney General transfers the LFA role to FEMA.

FBI Domestic Terrorism/Counterterrorism Planning Section (DTCTPS)

Within the FBI Counter Terrorism Division is a specialized section containing the Domestic Terrorism Operations Unit, the Weapons of Mass

Destruction Operations Unit, the Weapons of Mass Destruction Countermeasures Unit, and the Special Event Management Unit. Each of these units has specific responsibilities in investigations of crimes or allegations of crimes committed by individuals or groups in violation of the Federal terrorism and/or Weapons of Mass Destruction statutes. The DTCTPS serves as the point of contact (POC) to the FBI field offices and command structure as well as other Federal agencies in incidences of terrorism, the use or suspected use of WMD and/or the evaluation of threat credibility. If the FBI's Strategic Information and Operations Center (SIOC) is operational for exercises or actual incidents, the DTCTPS will provide staff personnel to facilitate the operation of SIOC.

During an incident, the FBI DTCTPS will coordinate the determination of the composition of the Domestic Emergency Support Teams (DEST) and/or the Foreign Emergency Support Teams (FEST). All incidents wherein a WMD is used will be coordinated by the DTCTPS WMD Operations Unit.

FBI Laboratory Division

Within the FBI's Laboratory Division reside numerous assets, which can deploy to provide assistance in a terrorism/WMD incident. The Hazardous Materials Response Unit (HMRU) personnel are highly trained and knowledgeable and are equipped to direct and assist in the collection of hazardous and/or toxic evidence in a contaminated environment. Similarly, the Evidence Response Team Unit (ERTU) is available to augment the local assets and have been trained in the collection of contaminated evidence. The Crisis Response Unit (CRU) is able to deploy to provide communications support to an incident. The Bomb Data Center (BDC) provides the baseline training to public safety bomb disposal technicians in the United States. BDC is the certification and accreditation authority for public safety agencies operating bomb squads and is in possession of equipment and staff that can be deployed to assist in the resolution of a crisis involving suspected or identified explosive devices. The Explosives Unit (EU) has experts who can assist in analyzing the construction of suspected or identified devices and recommend procedures to neutralize those items.

FBI Critical Incident Response Group (CIRG)

CIRG has developed assets that are designed to facilitate the resolution of crisis incidents of any type. Notably, the Crisis Management Unit (CMU), which conducts training and exercises for the FBI and has developed the concept of the Joint Operations Center (JOC), is available to provide on-scene assistance to the incident and integrate the concept of the JOC and the Incident Command System (ICS) to create efficient management of the situation. CIRG coordinates a highly trained group of skilled negotiators who are adroit in techniques to de-escalate volatile situations. The Hostage

Rescue Team (HRT) is a tactical asset, trained to function in contaminated or toxic hazard environments, that is available to assist in the management of the incident.

National Domestic Preparedness Office (NDPO)

NDPO is to coordinate and facilitate all Federal WMD efforts to assist State and local emergency responders with planning, training, equipment, exercise, and health and medical issues necessary to respond to a WMD event. The NDPO's program areas encompass the six broad areas of domestic preparedness requiring coordination and assistance: Planning, Training, Exercises, Equipment, Information Sharing, and Public Health and Medical Services.

Office for State and Local Domestic Preparedness Support (OSLDPS)

This office, within the Office of Justice Programs (OJP), has a State and Local Domestic Preparedness Technical Assistance Program that provides technical assistance in three areas: (1) general technical assistance; (2) State strategy technical assistance, and (3) equipment technical assistance. The purpose of this program is to provide direct assistance to State and local jurisdictions in enhancing their capacity and preparedness to respond to WMD terrorist incidents. The program goals are to:

- Enhance the ability of State and local jurisdictions to develop, plan, and implement a program for WMD preparedness; and
- Enhance the ability of State and local jurisdictions to sustain and maintain specialized equipment.

Technical assistance available from OSLDPS is provided without charge to requesting State or local jurisdiction. The following organizations are eligible for the State and Local Domestic Preparedness Technical Assistance Program:

- General technical assistance: units and agencies of State and local governments.
- State strategy technical assistance: State administrative agencies, designated by the governor, under the Fiscal Year 1999 State Domestic Preparedness Equipment Program.

- ❑ Equipment technical assistance: units and agencies of State and local governments that have received OSLDPS funding to acquire specialized equipment.

General Technical Assistance

OSLDPS provides general overall assistance to State and local jurisdictions for preparedness to respond to WMD terrorist incidents. This technical assistance includes:

- ❑ Assistance in developing and enhancing WMD response plans.
- ❑ Assistance with exercise scenario development and evaluation.
- ❑ Provision of WMD experts to facilitate jurisdictional working groups.
- ❑ Provision of specialized training.

State Strategy Technical Assistance

OSLDPS provides assistance to States in meeting the needs assessment and comprehensive planning requirements under OSLDPS' Fiscal Year 1999 State Domestic Preparedness Equipment Support Program. Specifically, OSLDPS:

- ❑ Assists States in developing their three-year statewide domestic preparedness strategy.
- ❑ Assists States in utilizing the assessment tools for completion of the required needs and threat assessments.

Equipment Technical Assistance

OSLDPS provides training by mobile training teams on the use and maintenance of specialized WMD response equipment under OSLDPS' Domestic Preparedness Equipment Support Program. This assistance will be delivered on site in eligible jurisdictions. Specifically, OSLDPS:

- ❑ Provides training on using, sustaining, and maintaining specialized equipment.
- ❑ Provides training to technicians on maintenance and calibration of test equipment.

- Provides maintenance and/or calibration of equipment.
- Assists in refurbishing used or damaged equipment.

Department of Defense (DoD)

In the event of a terrorist attack or act of nature on American soil resulting in the release of chemical, biological, radiological, nuclear material, or high-yield explosive (CBRNE) devices, the local law enforcement, fire, and emergency medical personnel who are first to respond may become quickly overwhelmed by the magnitude of the attack. The Department of Defense (DoD) has many unique warfighting support capabilities, both technical and operational, that could be used in support of State and local authorities, if requested by FEMA, as the Lead Federal Agency, to support and manage the consequences of such a domestic event.

Due to the increasing volatility of the threat and the time sensitivity associated with providing effective support to FEMA in domestic CBRNE incident, the Secretary of Defense appointed an Assistant to the Secretary of Defense for Civil Support (ATSD[CS]). The ATSD(CS) serves as the principal staff assistant and civilian advisor to the Secretary of Defense and Deputy Secretary of Defense for the oversight of policy, requirements, priorities, resources, and programs related to the DoD role in managing the consequences of a domestic incident involving the naturally occurring, accidental, or deliberate release of chemical, biological, radiological, nuclear material, or high-yield explosives.

When requested, the DoD will provide its unique and extensive resources in accordance with the following principles. First, DoD will ensure an unequivocal chain of responsibility, authority, and accountability for its actions to ensure the American people that the military will follow the basic constructs of lawful action when an emergency occurs. Second, in the event of a catastrophic CBRNE event, DoD will always play a supporting role to the LFA in accordance with all applicable law and plans. Third, DoD support will emphasize its natural role, skills, and structures to mass mobilize and provide logistical support. Fourth, DoD will purchase equipment and provide support in areas that are largely related to its warfighting mission. Fifth, reserve component forces are DoD's forward-deployed forces for domestic consequence management.

All official requests for DoD support to CBRNE consequence management (CM) incidents are made by the LFA to the Executive Secretary of the Department of Defense. While the LFA may submit the requests for DoD assistance through other DoD channels, immediately upon receipt, any request that comes to any DoD element shall be forwarded to the Executive

Secretary. In each instance the Executive Secretary will take the necessary action so that the Deputy Secretary can determine whether the incident warrants special operational management. In such instances, upon issuance of Secretary of Defense guidance to the Chairman of the Joint Chiefs of Staff (CJCS), the Joint Staff will translate the Secretary's decisions into military orders for these CBRNE-CM events, under the policy oversight of the ATSD(CS). If the Deputy Secretary of Defense determines that DoD support for a particular CBRNE-CM incident does not require special consequence management procedures, the Secretary of the Army will exercise authority as the DoD Executive Agent through normal Director of Military Support, Military Support to Civil Authorities (MSCA) procedures, with policy oversight by the ATSD(CS).

As noted above, DoD assets are tailored primarily for the larger warfighting mission overseas. But in recognition of the unique challenges of responding to a domestic CBRNE incident, the Department established a standing Joint Task Force for Civil Support (JTF-CS) headquarters at the United States Joint Forces Command, to plan for and integrate DoD's consequence management support to the LFA for events in the continental United States. The United States Pacific Command and United States Southern Command have parallel responsibilities for providing military assistance to civil authorities for States, territories, and possessions outside the continental United States. Specific units with skills applicable to a domestic consequence management role can be found in the Rapid Response Information System (RRIS) database maintained by FEMA. Capabilities include detection, decontamination, medical, and logistics.

Additionally, DoD has established ten Weapons of Mass Destruction Civil Support Teams (WMD-CST), each composed of twenty-two well-trained and equipped full-time National Guard personnel. Upon Secretary of Defense certification, one WMD-CST will be stationed in each of the ten FEMA regions around the country, ready to provide support when directed by their respective governors. Their mission is to deploy rapidly, assist local responders in determining the precise nature of an attack, provide expert technical advice, and help pave the way for the identification and arrival of follow-on military assets. By Congressional direction, DoD is in the process of establishing and training an additional seventeen WMD-CSTs to support the U.S. population. Interstate agreements provide a process for the WMD-CST and other National Guard assets to be used by neighboring states. If national security requirements dictate, these units may be transferred to Federal service.

Department of Energy (DOE)

Through its Office of Emergency Response, the DOE manages radiological emergency response assets that support both crisis and consequence management response in the event of an incident involving a WMD. The DOE is prepared to respond immediately to any type of radiological accident or incident with its radiological emergency response assets.[2] Through its Office of Nonproliferation and National Security, the DOE coordinates activities in nonproliferation, international nuclear safety, and communicated threat assessment. DOE maintains the following capabilities that support domestic terrorism preparedness and response.

Aerial Measuring System (AMS)

Radiological assistance operations may require the use of aerial monitoring to quickly determine the extent and degree of the dispersal of airborne or deposited radioactivity or the location of lost or diverted radioactive materials. The AMS is an aircraft-operated radiation detection system that uses fixed-wing aircraft and helicopters equipped with state-of-the-art technology instrumentation to track, monitor, and sample airborne radioactive plumes and/or detect and measure radioactive material deposited on the ground. The AMS capabilities reside at both Nellis Air Force Base near Las Vegas, Nevada, and Andrews Air Force Base near Washington, D.C. The fixed-wing aircraft provide a rapid assessment of the contaminated area, whereas the helicopters provide a slower, more detailed and accurate analysis of the contamination.

Atmospheric Release Advisory Capability (ARAC)

Radiological assistance operations may require the use of computer models to assist in estimating early phase radiological consequences of radioactive material accidentally released into the atmosphere. The ARAC is a computer-based atmospheric dispersion and deposition modeling capability operated by Lawrence Livermore National Laboratory (LLNL). The ARAC's role in an emergency begins when a nuclear, chemical, or other hazardous material is, or has the potential of being, released into the atmosphere. The ARAC's capability consists of meteorologists and other technical staff using three-dimensional computer models and real-time weather data to project the dispersion and deposition of radioactive material in the environment. The ARAC's computer output consists of graphical contour plots showing predicted estimates for instantaneous air and ground contamination levels, air immersion and ground-level exposure rates, and integrated effective dose equivalents for individuals or critical populations. The plots can be overlaid on local maps to assist emergency response officials in deciding what protective actions are needed to effectively protect people and the environment.

Protective actions could impact distribution of food and water sources and include sheltering and evacuating critical population groups. The ARAC's response time is typically thirty minutes to two hours after notification of an incident.

Accident Response Group (ARG)

ARG is DOE's primary emergency response capability for responding to emergencies involving United States nuclear weapons. The ARG, which is managed by the DOE Albuquerque Operations Office, is composed of a cadre of approximately 300 technical and scientific experts, including senior scientific advisors, weapons engineers and technicians, experts in nuclear safety and high-explosive safety, health physicists, radiation control technicians, industrial hygienists, physical scientists, packaging and transportation specialists, and other specialists from the DOE weapons complex. ARG members will deploy with highly specialized, state-of-the-art equipment for weapons recovery and monitoring operations. The ARG deploys on military or commercial aircraft using a time-phased approach. The ARG advance elements are ready to deploy within four hours of notification. ARG advance elements focus on initial assessment and provide preliminary advice to decision makers. When the follow-on elements arrive at the emergency scene, detailed health and safety evaluations and operations are performed and weapon recovery operations are initiated.

Federal Radiological Monitoring and Assessment Center (FRMAC)

For major radiological emergencies impacting the United States, the DOE establishes a FRMAC. The center is the control point for all Federal assets involved in the monitoring and assessment of offsite radiological conditions. The FRMAC provides support to the affected states, coordinates Federal offsite radiological environmental monitoring and assessment activities, maintains a technical liaison with Tribal nations and State and local governments, responds to the assessment needs of the LFA, and meets the statutory responsibilities of the participating Federal agency.

Nuclear Emergency Search Team (NEST)

NEST is DOE's program for dealing with the technical aspects of nuclear or radiological terrorism. A NEST consists of engineers, scientists, and other technical specialists from the DOE national laboratories and other contractors. NEST resources are configured to be quickly transported by military or commercial aircraft to worldwide locations and prepared to respond 24 hours a day using a phased and flexible approach to deploying personnel and equipment. The NEST is deployable within four hours of notification with specially trained teams and equipment to assist the FBI in handling nuclear or radiological threats. Response teams vary in size from a five person technical advisory team to a tailored deployment of dozens of searchers and scientists who can locate and then conduct or support technical operations on a suspected nuclear device. The NEST capabilities include intelligence, communications, search, assessment, access, diagnostics, render-safe operations, operations containment/damage mitigation, logistics, and health physics.

Radiological Assistance Program (RAP)

Under the RAP, the DOE provides, upon request, radiological assistance to DOE program elements, other Federal agencies, State, Tribal, and local governments, private groups, and individuals. RAP provides resources (trained personnel and equipment) to evaluate, assess, advise, and assist in the mitigation of actual or perceived radiation hazards and risks to workers, the public, and the environment. RAP is implemented on a regional basis, with regional coordination between the emergency response elements of the States, Tribes, other Federal agencies, and DOE. Each RAP Region maintains a minimum of three RAP teams, which are comprised of DOE and DOE contractor personnel, to provide radiological assistance within their region of responsibility. RAP teams consist of volunteer members who perform radiological assistance duties as part of their formal employment or as part of the terms of the contract between their employer and DOE. A fully configured team consists of seven members, to include one Team Leader, one Team Captain, four health physics survey/support personnel, and one Public Information Officer. A RAP team may deploy with two or more members depending on the potential hazards, risks, or the emergency or incident scenario. Multiple RAP teams may also be deployed to an accident if warranted by the situation.

Radiation Emergency Assistance Center/Training Site (REAC/TS)

The REAC/TS is managed by DOE's Oak Ridge Institute for Science and Education in Oak Ridge, Tennessee.. The REAC/TS maintains a twenty-four hour response center staffed with personnel and equipment to support medical aspects of radiological emergencies. The staff consists of physicians, nurses, paramedics, and health physicists who provide medical consultation and advice and/or direct medical support at the accident scene. The REAC/TS capabilities include assessment and treatment of internal and external contamination, whole-body counting, radiation dose estimation, and medical and radiological triage.

Communicated Threat Credibility Assessment

DOE is the program manager for the Nuclear Assessment Program (NAP) at LLNL. The NAP is a DOE-funded asset specifically designed to provide technical, operational, and behavioral assessments of the credibility of communicated threats directed against the U.S. Government and its interests. The assessment process includes one-hour initial and four-hour final products which, when integrated by the FBI as part of its threat assessment process, can lead to a "go/no go" decision for response to a nuclear threat.

Department of Health and Human Services (HHS)

The Department of Health and Human Services (HHS), as the lead Federal agency for Emergency Support Function (ESF) #8 (health and medical services), provides coordinated Federal assistance to supplement State and local resources in response to public health and medical care needs following a major disaster or emergency. Additionally, HHS provides support during developing or potential medical situations and has the responsibility for Federal support of food, drug, and sanitation issues. Resources are furnished when State and local resources are overwhelmed and public health and/or medical assistance is requested from the Federal government.

HHS, in its primary agency role for ESF #8, coordinates the provision of Federal health and medical assistance to fulfill the requirements identified by the affected State/local authorities having jurisdiction. Included in ESF #8 is overall public health response; triage, treatment, and transportation of victims of the disaster; and evacuation of patients out of the disaster area, as needed, into a network of Military Services, Veterans Affairs, and pre-enrolled nonfederal hospitals located in the major metropolitan areas of the United States. ESF #8 utilizes resources primarily available from (1) within HHS, (2) ESF #8 support agencies, (3) the National Disaster Medical System, and (4) specific non-Federal sources (major pharmaceutical suppliers, hospital supply vendors, international disaster response organizations, and international health organizations).

Office of Emergency Preparedness (OEP)

OEP manages and coordinates Federal health, medical, and health-related social service response and recovery to Federally declared disasters under the Federal Response Plan. The major functions of OEP include:

- ❑ Coordination and delivery of department-wide emergency preparedness activities, including continuity of government, continuity of operations, and emergency assistance during disasters and other emergencies;

- Coordination of the health and medical response of the Federal government, in support of State and local governments, in the aftermath of terrorist acts involving WMD; and

- Direction and maintenance of the medical response component of the National Disaster Medical System, including development and operational readiness capability of Disaster Medical Assistance Teams and other special teams that can be deployed as the primary medical response teams in case of disasters.

Centers for Disease Control and Prevention (CDC)

CDC is the Federal agency responsible for protecting the public health of the country through prevention and control of diseases and for response to public health emergencies. CDC works with national and international agencies to eradicate or control communicable diseases and other preventable conditions. The CDC Bioterrorism Preparedness and Response Program oversees the agency's effort to prepare State and local governments to respond to acts of bioterrorism. In addition, CDC has designated emergency response personnel throughout the agency who are responsible for responding to biological, chemical, and radiological terrorism. CDC has epidemiologists trained to investigate and control outbreaks or illnesses, as well as laboratories capable of quantifying an individual's exposure to biological or chemical agents. CDC maintains the National Pharmaceutical Stockpile to respond to terrorist incidents within the United States.

National Disaster Medical System (NDMS)

NDMS is a cooperative asset-sharing partnership between HHS, DoD, the Department of Veterans Affairs (VA), FEMA, State and local governments, and the private sector. The System has three components: direct medical care, patient evacuation, and the nonfederal hospital bed system. NDMS was created as a nationwide medical response system to supplement State and local medical resources during disasters and emergencies, provide back-up medical support to the military and VA health care systems during an overseas conventional conflict, and to promote development of community-based disaster medical service systems. This partnership includes DoD and VA Federal Coordinating Centers, which provide patient beds, as well as 1,990 civilian hospitals. NDMS is also comprised of over 7,000 private-sector medical and support personnel organized into many teams across the nation. These teams and other special medical teams are deployed to provide immediate medical attention to the sick and injured during disasters, when local emergency response systems become overloaded.

Disaster Medical Assistance Team (DMAT)

A DMAT is a group of professional and paraprofessional medical personnel (supported by a cadre of logistical and administrative staff) designed to provide emergency medical care during a disaster or other event. During a WMD incident, the DMAT provides clean area medical care in the form of medical triage and patient stabilization for transport to tertiary care.

National Medical Response Team–Weapons of Mass Destruction (NMRT-WMD)

The NMRT-WMD is a specialized response force designed to provide medical care following a nuclear, biological, and/or chemical incident. This unit is capable of providing mass casualty decontamination, medical triage, and primary and secondary medical care to stabilize victims for transportation to tertiary care facilities in a hazardous material environment. There are four such teams geographically dispersed throughout the United States.

Disaster Mortuary Operational Response Team (DMORT)

The DMORT is a mobile team of mortuary care specialists who have the capability to respond to incidents involving fatalities from transportation accidents, natural disasters, and/or terrorist events. The team provides technical assistance and supports mortuary operations as needed for mass fatality incidents.

Environmental Protection Agency (EPA)

EPA is chartered to respond to WMD releases under the National Oil and Hazardous Substances Pollution Contingency Plan (NCP) regardless of the cause of the release. EPA is authorized by the Comprehensive Environmental Response, Compensation, and Liability Act (CERCLA); the Oil Pollution Act; and the Emergency Planning and Community-Right-to Know Act to support Federal, State, and local responders in counterterrorism. EPA will provide support to the FBI during crisis management in response to a terrorist incident. In its crisis management role, the EPA On-Scene Commander (OSC) may provide the FBI Special Agent in Charge (SAC) with technical advice and recommendations, scientific and technical assessments, and assistance (as needed) to State and local responders. The EPA OSC will support FEMA during consequence management for the incident. EPA carries out its response according to the FRP, ESF #10, Hazardous Materials. The OSC may request an Environmental Response Team that is funded by EPA if the terrorist incident exceeds available local and regional resources. EPA is the chair for the National Response Team (NRT).

The following EPA reference material and planning guidance is recommended for State, Tribal, and local planners:

- ❑ Thinking About Deliberate Releases: Steps Your Community Can Take, 1995 (EPA 550-F-95-001).
- ❑ Environmental Protection Agency's Role in Counterterrorism Activities, 1998 (EPA 550-F-98-014).

U.S. Department of Agriculture

It is the policy of the U.S. Department of Agriculture (USDA) to be prepared to respond swiftly in the event of national security, natural disaster, technological, and other emergencies at the national, regional, State, and county levels to provide support and comfort to the people of the United States. USDA has a major role in ensuring the safety of food for all Americans. One concern is bioterrorism and its effect on agriculture in rural America, namely crops in the field, animals on the hoof, and food safety issues related to food in the food chain between the slaughterhouse and/or processing facilities and the consumer.

The Office of Crisis Planning and Management (OCPM)

This USDA office coordinates the emergency planning, preparedness, and crisis management functions and the suitability for employment investigations of the Department. It also maintains the USDA Continuity of Operations Plan (COOP).

USDA State Emergency Boards (SEBs)

The SEBs have responsibility for coordinating USDA emergency activities at the State level.

The Farm Service Agency

This USDA agency develops and administers emergency plans and controls covering food processing, storage, and wholesale distribution; distribution and use of seed; and manufacture, distribution, and use of livestock and poultry feed.

The Food and Nutrition Service (FNS)

This USDA agency provides food assistance in officially designated disaster areas upon request by the designated State agency. Generally, the food assistance response from FNS includes authorization of Emergency Food Stamp

Program benefits and use of USDA-donated foods for emergency mass feeding and household distribution, as necessary. FNS also maintains a current inventory of USDA-donated food held in Federal, State, and commercial warehouses and provides leadership to the FRP under ESF #11, Food.

Food Safety and Inspection Service

This USDA agency inspects meat/meat products, poultry/poultry products, and egg products in slaughtering and processing plants; assists the Food and Drug Administration in the inspection of other food products; develops plans and procedures for radiological emergency response in accordance with the Federal Radiological Emergency Response Plan (FRERP); and provides support, as required, to the FRP at the national and regional levels.

Natural Resources Conservation Service

This USDA agency provides technical assistance to individuals, communities, and governments relating to proper use of land for agricultural production; provides assistance in determining the extent of damage to agricultural land and water; and provides support to the FRP under ESF #3, Public Works and Engineering.

Agricultural Research Service (ARS)

This USDA agency develops and carries out all necessary research programs related to crop or livestock diseases; provides technical support for emergency programs and activities in the areas of planning, prevention, detection, treatment, and management of consequences; provides technical support for the development of guidance information on the effects of radiation, biological, and chemical agents on agriculture; develops and maintains a current inventory of ARS controlled laboratories that can be mobilized on short notice for emergency testing of food, feed, and water safety; and provides biological, chemical, and radiological safety support for USDA.

Economic Research Service

This USDA agency, in cooperation with other departmental agencies, analyzes the impacts of the emergency on the U.S. agricultural system, as well as on rural communities, as part of the process of developing strategies to respond to the effects of an emergency.

Rural Business-Cooperative Service

This USDA agency, in cooperation with other government agencies at all levels, promotes economic development in affected rural areas by developing strategies that respond to the conditions created by an emergency.

Animal and Plant Health Inspection Service

This USDA agency protects livestock, poultry, crops, biological resources, and products thereof, from diseases, pests, and hazardous agents (biological, chemical, and radiological); assesses the damage to agriculture of any such introduction; and coordinates the utilization and disposal of livestock and poultry exposed to hazardous agents.

Cooperative State Research, Education and Extension Service (CSREES)

This USDA agency coordinates use of land-grant and other cooperating State college, and university services and other relevant research institutions in carrying out all responsibilities for emergency programs. CSREES administers information and education services covering (a) farmers, other rural residents, and the food and agricultural industries on emergency needs and conditions; (b) vulnerability of crops and livestock to the effects of hazardous agents (biological, chemical, and radiological); and (c) technology for emergency agricultural production.

This agency maintains a close working relationship with the news media. CSREES will provide guidance on the most efficient procedures to assure continuity and restoration of an agricultural technical information system under emergency conditions.

Rural Housing Service

This USDA agency will assist the Department of Housing and Urban Development by providing living quarters in unoccupied rural housing in an emergency situation.

Rural Utilities Service

This USDA agency will provide support to the FRP under ESF #12, Energy, at the national level.

Office of Inspector General (OIG)

This USDA office is the Department's principal law enforcement component and liaison with the FBI. OIG, in concert with appropriate Federal, State, and local agencies, is prepared to investigate any terrorist attacks relating to the nation's agriculture sector, to identify subjects, interview witnesses, and secure evidence in preparation for Federal prosecution. As necessary, OIG will examine USDA programs regarding counterterrorism-related matters.

Forest Service (FS)

This USDA agency will prevent and control fires in rural areas in cooperation with State, local, and Tribal governments, and appropriate Federal departments and agencies. They will determine and report requirements for equipment, personnel, fuels, chemicals, and other materials needed for carrying out assigned duties. The FS will furnish personnel and equipment for search and rescue work and other emergency measures in national forests and on other lands where a temporary lead role will reduce suffering or loss of life. The FS will provide leadership to the FRP under ESF #4, Firefighting, and support to the Emergency Support Functions, as required, at the national and regional levels. FS will allocate and assign radio frequencies for use by agencies and staff offices of USDA. FS will also operate emergency radio communications systems in support of local, regional, and national firefighting teams. Lastly, the FS law enforcement officers can serve as support to OIG in major investigations of acts of terrorism against agricultural lands and products.

Nuclear Regulatory Commission

The Nuclear Regulatory Commission (NRC) is the Lead Federal Agency (in accordance with the Federal Radiological Emergency Response Plan) for facilities or materials regulated by the NRC or by an NRC Agreement State. The NRC's counterterrorism-specific role, at these facilities or material sites, is to exercise the Federal lead for radiological safety while supporting other Federal, State and local agencies in Crisis and Consequence Management.

Radiological Safety Assessment

The NRC will provide the facility (or for materials, the user) technical advice to ensure onsite measures are taken to mitigate offsite consequences. The NRC will serve as the primary Federal source of information regarding on-site radiological conditions and off-site radiological effects. The NRC will support the technical needs of other agencies by providing descriptions of devices or facilities containing radiological materials and assessing the safety impact of terrorist actions and of proposed tactical operations of any responders. Safety assessments will be coordinated through NRC liaison at the Domestic Emergency Support Team (DEST), Strategic Information and Operations Center (SIOC), Command Post (CP), and Joint Operations Center (JOC).

Protective Action Recommendations

The licensee and State have the primary responsibility for recommending and implementing, respectively, actions to protect the public. They will, if necessary, act, without prior consultation with Federal officials, to initiate protective actions for the public and responders. The NRC will contact State and local authorities and offer advice and assistance on the technical assessment of the radiological hazard and, if requested, provide advice on protective actions for the public. The NRC will coordinate any recommendations for protective actions through NRC liaison at the CP or JOC.

Responder Radiation Protection

The NRC will assess the potential radiological hazards to any responders and coordinate with the facility radiation protection staff to ensure that personnel responding to the scene are observing the appropriate precautions.

Information Coordination

The NRC will supply other responders and government officials with timely information concerning the radiological aspects of the event. The NRC will liaison with the Joint Information Center to coordinate information concerning the Federal response.

End Notes

[1] Federal Emergency Management Agency, "Guide For All-Hazard Emergency Operations Planning", State And Local Guide (101), Chapter 6, Attachment G — Terrorism

[2] For facilities or materials regulated by the Nuclear Regulatory Commission (NRC), or by an NRC Agreement State, the technical response is led by NRC as the LFA (in accordance with the Federal Radiological Emergency Response Plan) and supported by DOE as needed.

About the Authors

Kevin M. Quinley, CPCU, ARM, is Senior Vice President, Risk Services, for MEDMARC Insurance Company, Chantilly, VA. He has a Bachelor of Arts degree from Wake Forest University and a master's degree from the College of William & Mary. He holds the Chartered Property Casualty Underwriter (CPCU) designation plus specialty designations from the American Insurance Institute in Risk Management (ARM), Claims (AIC), Management (AIM), and Reinsurance (ARe). He has written over 500 published articles and eight books on insurance and risk management topics.

Mr. Quinley teaches classes in insurance, claims, and risk management for the Washington, D.C., CPCU Chapter, for whom he served as a past president. In 1998, he was named "The Standard Setter" by the national CPCU Society for his professional accomplishments and community involvement. A frequent speaker on topics related to insurance, claims handling, and litigation, he can be reached via the Internet at kquinley@medmarc.com (phone 703-219-2320; fax 703-385-3725). He lives in Fairfax, Virginia, with his wife and two sons.

Donald L. Schmidt, ARM, is Senior Vice President, Managing Consultant, and Operations Manager for the Risk Consulting Practice of Marsh in Boston, Massachusetts, where he provides management consulting services in the areas of risk assessment and business impact analysis, mitigation, and emergency management.

Mr. Schmidt is the author of *Emergency Response Planning: A Management Guide* published by Marsh and in part by *Facility Manager* magazine in 1994. He has been interviewed in many periodicals, including *Stores Magazine, Retail Challenge, Global Risk,* and *NFPA Journal.* He is a frequent lecturer on the subject of emergency management and has spoken to groups such as the American Society of Safety Engineers, American Society for Industrial Security, National Retail Federation, and Ford Motor Company, and at a conference sponsored by FEMA's Project Impact.

Mr. Schmidt's twenty-three-plus year career has included positions within the HPR insurance industry, heavy manufacturing, and a federal fire prevention agency. He also was a fire fighter with two fire departments.

He graduated summa cum laude from the University of Maryland with a Bachelor of Science degree in Urban Studies - Fire Science. Mr. Schmidt is a member of the Society of Fire Protection Engineers and the National Fire Protection Association, and he is a Professional Member of the American Society of Safety Engineers. He serves as a principal member on two National Fire Protection Association technical committees: *Disaster/Emergency Management and Business Continuity Programs* (NFPA 1600) and *Pre-Incident Planning* (NFPA 1620).

Don lives with his wife Sherry, son Danny, and daughter Cari in suburban Boston.

Index

Call **1-800-543-0874** to order and ask for operator BB or fax your order to **1-800-874-1916**. Ask about our complete line of products.

PAYMENT INFORMATION

Add shipping & handling charges to all orders as indicated. If your order exceeds total amount listed in chart, call 1-800-543-0874 for shipping & handling charge. Any order of 10 or more or $250.00 or over will be billed for shipping by actual weight, plus a handling fee. Unconditional 30 day guarantee.

Shipping & Handling (Additional)

Order Total	Shipping & handling
$20.00 to $39.99	$6.00
40.00 to 59.99	7.00
60.00 to 79.99	9.00
80.00 to 109.99	10.00
110.00 to 149.99	12.00
150.00 to 199.99	13.00
200.00 to 249.99	15.50

Shipping and handling rates for the continental U.S. only. Call 1-800-543-0874 for overseas shipping information.

SALES TAX (Additional)

Sales tax is required for residents of the following states: CA, DC, FL, GA, IL, KY, NJ, NY, OH, PA, WA.

The NATIONAL UNDERWRITER Company
PROFESSIONAL PUBLISHING GROUP

The National Underwriter Co.
Orders Dept #2-BB
P.O. Box 14448
Cincinnati, OH 45250-9786

2-BB

_____ Copies of *Business at Risk: How to Assess, Mitigate, and Respond to Terrorist Threats* (#8060000) $42.95

_____ Copies of *e-Coverage Guide* Book (6760000) $34.99
CD-ROM (6769000) $40.00
12-Month Internet Subscription (6769100) $40.00

❑ Check enclosed* ❑ Charge my VISA/MC/AmEx (circle one) ❑ Bill me
*Make check payable to The National Underwriter Company.
Please include the appropriate shipping & handling charges and any applicable sales tax. (see charts at left)

Signature _____
Card # _____ *Exp. Date* _____
Name _____ *Title* _____
Company _____
Street Address _____
City _____ *State* _____ *Zip+4* _____
Business Phone ( ) _____ *Business Fax* ( ) _____
Email _____ *May we email you?* _____

The NATIONAL UNDERWRITER Company
PROFESSIONAL PUBLISHING GROUP

The National Underwriter Co.
Orders Dept #2-BB
P.O. Box 14448
Cincinnati, OH 45250-9786

2-BB

_____ Copies of *Business at Risk: How to Assess, Mitigate, and Respond to Terrorist Threats* (#8060000) $42.95

_____ Copies of *e-Coverage Guide* Book (6760000) $34.99
CD-ROM (6769000) $40.00
12-Month Internet Subscription (6769100) $40.00

❑ Check enclosed* ❑ Charge my VISA/MC/AmEx (circle one) ❑ Bill me
*Make check payable to The National Underwriter Company.
Please include the appropriate shipping & handling charges and any applicable sales tax. (see charts at left)

Signature _____
Card # _____ *Exp. Date* _____
Name _____ *Title* _____
Company _____
Street Address _____
City _____ *State* _____ *Zip+4* _____
Business Phone ( ) _____ *Business Fax* ( ) _____
Email _____ *May we email you?* _____

The NATIONAL UNDERWRITER Company
PROFESSIONAL PUBLISHING GROUP

The National Underwriter Co.
Orders Dept #2-BB
P.O. Box 14448
Cincinnati, OH 45250-9786

Also available:

e-Coverage Guide

By Leo L. Clark, Esq. and Martin C. Loesch, Esq.

Learn how to help protect small and large businesses from the predominant risks of the electronic world. Chapters cover liability risks of IT consultants, intellectual property infringement, e-commerce liability risks, and more. Order with ***Business at Risk***.

NO POSTAGE
NECESSARY
IF MAILED
IN THE
UNITED STATES

BUSINESS REPLY MAIL

FIRST CLASS MAIL PERMIT NO 68 CINCINNATI, OH

POSTAGE WILL BE PAID BY ADDRESSEE

The National Underwriter Co.
Orders Department #2-BB
P.O. Box 14448
Cincinnati, OH 45250-9786

NO POSTAGE
NECESSARY
IF MAILED
IN THE
UNITED STATES

BUSINESS REPLY MAIL

FIRST CLASS MAIL PERMIT NO 68 CINCINNATI, OH

POSTAGE WILL BE PAID BY ADDRESSEE

The National Underwriter Co.
Orders Department #2-BB
P.O. Box 14448
Cincinnati, OH 45250-9786

CW01313427

To Neil

Welcome to the Shell family

Best wishes

Julian Dalzell

2019.

Talent Without Borders

Talent Without Borders

Global Talent Acquisition for Competitive Advantage

ROBERT E. PLOYHART

JEFF A. WEEKLEY

JULIAN DALZELL

OXFORD
UNIVERSITY PRESS

OXFORD
UNIVERSITY PRESS

Oxford University Press is a department of the University of Oxford. It furthers the University's objective of excellence in research, scholarship, and education by publishing worldwide. Oxford is a registered trade mark of Oxford University Press in the UK and certain other countries.

Published in the United States of America by Oxford University Press
198 Madison Avenue, New York, NY 10016, United States of America.

Library of Congress Cataloging-in-Publication Data
Names: Ployhart, Robert E., 1970– author., Weekley, Jeff A., Dalzell, Julian.
Title: Talent without borders : global talent acquisition for
competitive advantage / by Robert E. Ployhart, Jeff A. Weekley, Julian Dalzell.
Description: New York, NY, United States of America :
Oxford University Press, [2018] |
Includes bibliographical references and index.
Identifiers: LCCN 2017015632 | ISBN 9780199746897 (alk. paper)
Subjects: LCSH: Employee selection. | Employees—Recruiting.
Classification: LCC HF5549.5.S38 P59 2018 | DDC 658.3/11—dc23
LC record available at https://lccn.loc.gov/2017015632

9 8 7 6 5 4 3 2 1

Printed by Sheridan Books, Inc., United States of America

To my family: Matthew, Ryan, Jack, Ella, and Lynn, who have shown me the true meaning of talent—Rob

To Addie, Lucy, and Ben—Jeff

To my loving family: Mary Sue, Lisa and Andrew, whose willingness to share my global journey enabled me to live my dream—Julian

CONTENTS

PREFACE

Talent drives the global economy. Talent is the resource that creates economic growth through exceptional innovation, service, and performance. But talent is scarce, and finding the right talent, in the right place, and at the right time, is challenging. The types of talent a firm needs to create or implement its strategy are unlikely to be found in a single region. This means that competitive firms must be able to effectively identify, recruit, and hire the needed talent wherever it exists in the world. In this sense talent is like all scarce and valuable resources. However, talent is about people, and unlike tangible resources such as land or money, people have preferences, hopes, dreams, and relationships—and these characteristics differ across countries and cultures. Talent cannot be owned; it can only be accessed or "rented." Thus, the supreme challenge for modern firms is to develop a capability to identify, attract, and hire talent globally.

Talent Without Borders explains how to generate a competitive advantage through the effective use of global recruitment and staffing. The book offers a practical approach for how to think about acquiring talent strategically and globally. The book has several noteworthy features:

- Emphasis on linking global staffing to organizational strategy, financial performance, and competitive advantage.
- Explicit consideration of practical issues that influence the implementation of global staffing solutions.

- Practical insights based on the best science available, and the authors' decades of global human resource and staffing experience gained from different points of view—practitioner, consultant, and academic.
- Focus on analytics and data to enable evidence-based decisions.
- A global framework and model that provides a reasonable balance of generalization and localization. That is, the framework is relevant for any region of the world.
- Consideration of the entire talent life cycle, from attraction through retention.

One of the book's most unique features is its use of a lively, dynamic, first-person case that is used to convey the technical material through the eyes of "Roger." Roger is a new human resource manager in a global firm, who is promoted to a talent leader responsible for developing and implementing a strategic staffing plan in Southeast Asia. Roger must move from England to Kuala Lumpur in Malaysia. This first-person perspective brings the technical material to life in a way that is more engaging. Although Roger is a fictional character, the challenges he struggles with are very real and hence readers can learn from Roger's experience as they begin their own global staffing journey. However, for those not wanting to read the case, it can be completely ignored without any loss of the technical material.

ACKNOWLEDGMENTS

We express our sincere thanks to Abby Gross, Courtney McCarroll, and members of Oxford University Press, for their incredible guidance, assistance, support, and patience through this entire process. They have been professional and fun to work with—truly top talent in every way!

We also thank our current and former colleagues and employers for their support. The book was improved by many conversations with our colleagues—far too many to thank individually. We appreciate the assistance of the Riegel and Emory Center for Human Resources, and the Center for Executive Succession, in the Management Department at the University of South Carolina. References and content for the book were created with the assistance of many research assistants, particularly Michael C. Campion, Matt Call, Jason Kautz, Ormonde Cragun, and Youngsang (Ray) Kim. We are indebted to them for their efforts!

Finally, we thank the most important people—our families. They supported the time needed to work on this project, including weekends, late nights, and early mornings. They provided perspective and insight to the various topics and choices relating to content. And they provided incredible patience by politely enduring many conversations about the book and its topics. We couldn't have done it without them!

To all these individuals we say—thank you, gracias, merci, grazie, (domo) arigato, do jeh, danke, and terima kasih!!!

ABOUT THE AUTHORS

The authors represent a unique team of experts with extensive expertise in the research, consulting, and management of global staffing. The book offers numerous practical insights based on their collective experience.

Robert E. Ployhart is the Bank of America Professor of Business Administration and an internationally recognized expert in human resources, staffing, talent analytics, and legal issues relating to talent. He has published over 100 scholarly articles and chapters on these topics, written two books, and served as an associate editor for multiple academic journals (*Journal of Applied Psychology, Academy of Management Review, Organizational Behavior and Human Decision Processes*). He has received many scholarly awards, and is a fellow of the American Psychological Association, the Association for Psychological Science, and the Society for Industrial and Organizational Psychology. He has also engaged in an active consulting practice with domestic and global organizations. He has a PhD in industrial and organizational psychology from Michigan State University (1999), an MA in industrial and organizational psychology from Bowling Green State University (1996), and a BS in psychology from North Dakota State University (1994).

Jeff A. Weekley is a clinical professor and director of the Human Resource Management program at the University of Texas at Dallas. In this role, he is responsible for leading the education of the next generation of human resource professionals in north Texas, as well as research, professional outreach, and consulting on human resource topics. Previously, he was the global leader for talent consulting in IBM's SmarterWorkforce, leading that team in the development and implementation of systems for employee assessment, development, succession planning, and human resource analytics. Prior to beginning a 20-year career in consulting, Weekley held several senior human resource management positions with Zales Corporation, Southland Corporation, and Greyhound Lines. Dr. Weekley received his PhD in organizational behavior from the University of Texas at Dallas and his MS in industrial and organizational psychology and BS in psychology from Texas A&M University. He is a member of the American Psychological Association and the Academy of Management, and a fellow of the Society for Industrial and Organizational Psychology. He has authored numerous articles in scholarly publications such as the *Journal of Applied Psychology, Personnel Psychology, Academy of Management Journal, Human Performance, International Journal of Selection and Assessment*, and *Journal of Management* and is the coauthor of two books. His research has focused largely on talent acquisition issues, particularly in the realms of situational judgment testing, personality and values, interviewing, faking, and talent flows through organizations.

Julian Dalzell is lecturer and executive in residence at the Darla Moore School of Business, University of South Carolina, where he teaches in the MHR and PMBA programs. He has been the recipient of the Outstanding MHR Professor Award (2012), the Alfred G. Smith Award for Teaching Excellence (2014), and the Distinguished Service Award for the MHR Program (2016). He is an active speaker at national and global conferences and acts as an executive coach and talent development consultant. Julian began his Shell career in the United Kingdom in 1968 and worked in labor relations until 1979. He moved to Brunei and, after spending 10 years in Southeast Asia, focusing on recruitment, staff planning, and development, he came to the United States in 1988. He worked in compensation, as a human resource business partner, and in talent management until 1997, when he became the head of human resources for the

Gulf of Mexico Upstream operations. From 2000, Dalzell held senior human resource leadership roles as Global VPHR Shell Trading; VPHR for Shell Oil Products East, with accountability for 28 countries spanning from Dubai to China and as far South as New Zealand; Global HRVP for Supply and Distribution, based in London; and VPHR Downstream North America. In 2008 he was appointed VP human resources for Shell Oil Company. He retired from Shell at the end of 2010. Dalzell is a member of the Society for Human Resources Management and the Academy of Human Resources Development. He holds a BA in international management from Eckerd College and an MS in educational human resource development from Texas A&M University.

HOW TO USE THIS BOOK EFFECTIVELY

This book is relevant for multiple audiences: managers in human resources, general managers, and industrial/organizational psychologists; students in MBA, human resource, and industrial/organizational programs and doctoral, masters, and undergraduate courses; specialists involved with talent analytics; and most anyone responsible for, or interested in, hiring talent. The intended audience is broad and diverse, and we have prepared the book to be relevant to each audience.

- **Working managers** can go directly to the technical details in each chapter. Each chapter starts with a quick overview that identifies the key points. Figures and tables are presented in a way that enables readers to quickly digest important details. Readers can ignore the case study material without any loss of technical content. Each chapter is also intended to largely stand alone. For example, if readers do not have a need to review the statistical and measurement analytics issues in chapter 3, they can easily skip that chapter.
- **Students** will benefit from reading the case study because it introduces the reader into the "reality" of implementing a global staffing system. In contrast to generic cases, our case study with Roger allows readers to view the political and social challenges associated with conducting staffing "in the real world." Roger's

experiences are based on the authors' experiences—the challenges, tensions, errors, failures, and triumphs.

- **Analytics specialists** can focus directly on the technical material presented in chapter 3. The "Talent Analytics" chapter contains considerable depth regarding measurement, statistics, and demonstrating impact using quantitative models. For those who do not need such depth, chapter 3 can simply be ignored.

Thus, the book can be used in whatever manner best suits the audience:

- **Chapter 1:** An overview of talent and the economic and political forces that make talent a strategically valuable global resource. This chapter covers scientific theory about talent; a topic that may be more relevant for students than managers.
- **Chapter 2:** Global talent strategy, and how to connect global staffing to firm performance, strategy, and competitive advantage. Much of the focus in this chapter is making a strategic case for investing in talent acquisition.
- **Chapter 3:** Talent analytics, and how to build an evidence-based business case for the impact of staffing on operational and strategic metrics. This chapter may be less relevant for those who are not specifically required to perform talent analytics, but the material can be skimmed to help managers be conversant with analytical methods.
- **Chapter 4:** Recruiting talent globally, and how to identify and attract the best talent from anywhere in the world. This chapter explains cultural differences in applicant preferences.
- **Chapter 5:** Selecting talent globally, and how to assess and hire talent in a manner that is based on quantitative evidence and yet is fair to members from different cultures, languages, and countries.
- **Chapter 6:** Global staffing and talent management; and how to socialize, train, develop, and retain top talent in a manner that balances different cultural tensions and preferences.

CASE INTRODUCTION

An Overview of Roger Wells and Spooner Electronics

Throughout this book, we return to the case and story of Roger Wells.

Roger has worked in the United Kingdom (UK) for his firm, Spooner Electronics, for about 7 years since graduating from Christ's College, Cambridge, with a degree in History. It was quite common for companies to recruit graduates with degrees that were indicative of a quality mind and ability to think analytically but not necessarily directly related to any particular career field.

Spooner Electronics is an ambitious firm that is based in Slough, England. It now employs 32,000 employees worldwide with operations in 34 countries. The top leadership had been dominated largely by British nationals but recently they had made some progress in diversifying with the appointment of its first Asian, Chiew Hock Wah from Malaysia as VP Asia-Pacific-Middle East; Herman Van Dijk, a technically brilliant Dutchman from the renowned Delft University, who held the role of VP Technical Development; and Renee Jacobi, the only woman on the senior team, who was from the United States and was Chief Counsel. But all acknowledged that there was still work to do in this regard.

Initially its main product lines had centered on high-end small domestic appliances such as electric razors, hair driers, curling irons, and so forth. In recent years, the cost advantages of manufacturers from the Far East had led them to diversify to equipment such as blenders and coffee makers, and they even had a short dalliance with larger equipment such as washer-driers, but this was quickly aborted when they realized that they had moved outside their core area of competence.

In the early stages of moving outside the British domestic market, they had stayed close to home, mainly concentrating on the larger EC European Countries like France, Germany, Belgium, Holland, and others. The removal of trade tariffs had helped significantly, but they faced fierce competition from the established global powerhouses like Phillips, Bosch, and Keurig. An entry into Latin America had met with some success, but this was essentially done through establishing a distributor network while retaining manufacturing in the UK. Africa had not seemed a particularly promising opportunity, given the disparity of disposable income between South Africa and the rest of the continent, so the decision had been made in the early 1990s to expand into Asia. That part of the world was coming out of the deep recession caused by Black Monday in the United States—October 19, 1987—and the Spooner leadership saw an opportunity to take advantage of the signs of general market growth and the emergence of a more open China together with the potential wage arbitrage, particularly in Malaysia and the Philippines. Although Singapore and Hong Kong seemed to have some advantages as a first location (established infrastructure, transportation, etc.) the decision was made to establish a presence in Kuala Lumpur, the capital city of Malaysia. A variety of reasons lay behind that decision, but office rental rates, the availability of quality educated resources, and a national history with the electronics industry made it a sound choice.

In a manner that is not atypical for companies that move into new geographies, the philosophy adopted by Spooner was to allow significant latitude for each office to operate in a largely autonomous manner, with business and working norms framed to be consistent with local practice—albeit that they established overarching ethical norms, which did at times create some tension between HQ and the local leadership. Looking back, though, these had been manageable and the incursion into Asia had been commercially successful. Yet, the Asian crisis of 1997 had tested the autonomous model, as the inefficiencies of separate country approaches became more and more problematic. Several attempts had been made to seek synergies in a cooperative model, but the long history of independence, the different social goals and mores of countries like Singapore and Malaysia (which had led to their separation into independent countries in 1965), and the differential economic development patterns of countries in the region all made this unachievable as a ground-up initiative.

The economic meltdown in the United States in 2008 caused ripple effects around the world, and following that, Spooner decided to establish regional organizations in all of the world areas in which they had significant operations—effectively everywhere except Africa. The process had not been easy and had met with stiff resistance particularly in Europe and Asia—but the leadership had been determined to see it through and by early 2010 the various regions were established and operating, although there were still some rough edges to be smoothed out. One of the early stumbling blocks had proved to be the selection of various leadership roles. For example, a leader from the Philippines would have to overcome opposition from other Southeast Asian countries who believed their economies to be more developed. Singaporeans had to overcome a level of jealousy from their larger and more populous neighbors, Indonesia and Malaysia, with both of whom they shared common borders and who were, perhaps, jealous of that country's amazing economic success. It was against this backdrop that the decision had been made to insert a "neutral" expatriate—Roger.

Back in England, Roger had impressed his employer with his agile mind and finely honed debating skills, but also with his common sense and ability to find connections with people from all backgrounds. He had that knack of saying the right thing at the right time and these qualities were soon put to the test when he was assigned to one of Spooner's larger manufacturing facilities in Leeds in northern England, where the workforce was represented by Unite, the largest union in the UK. After initially working in more of a backroom role, he was assigned to the industrial relations team where he worked under the gentle tutelage of Gordon Baker, one of those field HR specialists who really knew their craft and had an eye for talent. They had immediately hit it off. Fueled by George Santayana's famous quote "those who cannot remember the past are condemned to repeat it," Roger soaked up as much about the history of the plant and its relationship with Unite as he could. Gordon and Roger spent hours together, looking at past agreements, disputes, and personalities, trying to think of ways to improve the plant's atmosphere. They had quite some success, and Roger's contribution to that effort was now being recognized by the chance to work overseas for the first time.

"You have worked in one section of our Company, you have only worked in the UK and only in Industrial Relations. So we are going to work on broadening your business base, broadening your functional expertise and broadening your world view all at the same time." With those words, Robert Walkely, the individual accountable for HR development in Spooner, had laid out the rationale for the assignment in Malaysia. This did not need a hard sell. Roger had made no secret of his desire to work overseas at some point, and, although he would like to have used some of his schoolboy French or Spanish, he was excited at the new adventures that awaited him in Kuala Lumpur, which he discovered through an Internet search meant "muddy confluence" (or two rivers) in the local language. He started to do some basic research on the area, local history and customs, tips on how to survive locally, and things to see and do—he even paused to buy a book on culture shock, specifically designed for Malaysia. Yet these were really first instinct action items focused on the personal living dimensions. They were also narrowly focused from a professional perspective because the role was a regional one, so in addition to the specific local information, he would need to know about several different cultures and subcultures of Asia.

SPOONER ELECTRONICS COMPANY PROFILE

HQ—Slough England

Roger's primary responsibility is to lead a team charged with recruiting and hiring talent at Spooner's quickly growing manufacturing and engineering operations in Southeast Asia. He is accountable for developing a staffing program that meets the strategic goals of the organization and the tactical goals of the regional operations—a delicate balance, to be sure.

Spooner's current operations in the S.E. Asia region cover Malaysia, Singapore, Brunei, Indonesia, Thailand, Philippines, and Taiwan. Spooner has direct sales and distribution operations in all seven countries except Brunei, where they have a distributor. Additionally, they have opened new business development offices in their targeted growth countries—China, Vietnam, South Korea, and India, but these are not part of Roger's remit.

Their manufacturing plants are located in:

- Bayan Lepas Industrial Park, Penang, Malaysia
- Cyberjaya, about 50 km south of Kuala Lumpur, Malaysia
- Kawasan Industri Pulogadung, to the East of Jakarta, Indonesia
- Gateway City Industrial Park, about an 80-minute drive southeast of Bangkok, Thailand. It has close access to Leam Chabang and the Map Ta Phut Deep Seaports.
- Laguna Technopark, Sta. Rosa, Laguna (about 38 km south of Manila), Philippines
- Wu-Ku Industrial Park, Taipei, Taiwan

They do not have a manufacturing site in Singapore because of the proximity to Malaysia and the relative cost of labor between the two countries.

Dependent on their success in breaking into the mainland China market, they intend to establish a manufacturing facility there.

They have a major engineering support center located adjacent to their Cyberjaya manufacturing site.

They have country headquarters located in:

- Jalan Dungun, Bukit Damansara, a suburb of Kuala Lumpur
- Asia Square Tower 1, Marina View, Singapore
- Jalan Jenderal Sudirman Kaveling, Jakarta, Indonesia
- Rama IV Road, Silom Bagkrak, Bangkok, Thailand
- Valero Street, Salcedo Village, Makati City, a suburb of Metro Manila, Philippines
- Taipei 101 Tower, No. 7, Xinyi Road, Taipei, Taiwan

Employment numbers by country:

Country	HQ	Manufacturing	Engineering	Sales & Marketing	TOTAL
Malaysia	250	1,850	400	120	2,620
Singapore	50	0	20	60	130
Indonesia	60	485	25	90	660
Philippines	90	1,450	25	125	1,690
Thailand	85	1,375	25	105	1,590
Taiwan	65	1,285	16	85	1,451
TOTAL	600	6,445	511	585	8,140

APME REGIONAL LEADERSHIP TEAM MEMBERSHIP

CEO, Asia-Pacific-Middle East—Chiew Hock Wah—member of the Global Leadership Team. Reports to the CEO. Based in Kuala Lumpur, Malaysia

- VP Middle East—(United Arab Emirates [UAE], Oman, Saudi Arabia, Kuwait, Bahrein, Egypt)—Ian Robertson (British, based in Dubai)
- VP Central and East Asia (India, Pakistan, Bangladesh, Sri Lanka)—Virender Tendulkar (Indian)
- VP S.E. Asia (Thailand, Malaysia, Singapore, Indonesia, Philippines, Taiwan)—Patrick Chieng (Malaysian)
- VP Oceania (Australia, New Zealand, Pacific Islands)—George Lehman (Australian)

- VP Regional Manufacturing and Technology—Bruce Little (British)
- VP Regional Sales and Marketing—Chan See Lok (Singaporean)
- VP New Business Development—Eli Esguerra (Filipino)
- VP Finance and IT—Zakiah Abdul Rahman (Malaysian)
- VP Human Resources—Alias Ismail (Malaysian)

S.E. ASIA REGIONAL HR LEADERSHIP TEAM

VPHR S.E. Asia and Human Resources and Country HR Manager, Malaysia—Ibrahim Abdul Hassan

- Country HR Manager, Singapore—Robert Wong
- Country HR Manager, Indonesia—Atmaja Baktioni
- Country HR Manager, Thailand—Suparong Kotpanow
- Country HR Manager, Philippines—Victor Rivera
- Country HR Manager, Taiwan—Chien-Fu Wang
- Regional Manager, Employee and Industrial Relations—Karim Osman (Malaysian)
- Regional Manager, Talent Acquisition—Roger Wells (British)
- Regional Manager, Policy and Benefit Coordination—Noraini Suffian (Malaysian).

S.E. ASIA REGIONAL TALENT ACQUISITION TEAM

Regional Manager, Talent Acquisition—Roger Wells (British)

- Lead, Strategy and Workforce Forecasting—Lim Beng Hian (Singapore)
- Lead Campus Relations—Mohzaini Wahab (Malaysia)
- Lead, Data Management and IT support—Jeremiah Santiago (Philippines)

- Lead, Technical Talent Sourcing—Chakthep Senniwongs (Thailand)
- Lead, Hourly Workforce Sourcing—Iwan Kasim (Indonesia)
- Lead, Commercial Sourcing—Chen Shu-Fen (Taiwan)

All of these individuals are located in the Country HR teams and have a matrixed or "dotted line" relationship with the Country HR Manager. Their direct supervision, however, comes from Roger.

Talent Without Borders

1

Overview and Introduction

THIRTY-SECOND PREVIEW

- Talent is a person's overall capacity to create impact. Talent is the sum of a person's knowledge, skills, abilities, personality traits, values, and interests that enable one to effectively perform a job and broadly contribute to the organization's effectiveness, strategy, and culture.
- Talent is an organizational resource that is intangible, malleable, and mobile—thus making it strategically valuable. Talent resources may exist individually or within collectives.
- Talent should influence the formation and implementation of strategy.
- Talent is unequally distributed around the globe.
- National culture influences what talent wants and needs, and hence influences the nature of global talent acquisition.
- The Bottom Line: Firms can achieve a global talent advantage over competitors by knowing where and how to identify, source, attract, and select international talent.

Please see **"Case Introduction" for background on Spooner Electronics, the Company's strategy, and Roger's role**.

"Hey Andrew! I have some exciting news."

With these words Roger Wells announced to his friend, Andrew, that he was soon to need a new buddy with whom to watch cricket matches.

"My employer, Spooner Electronics, has asked me to lead the team that is responsible for Talent Acquisition in Southeast Asia. I will be based in Kuala Lumpur, the capital of Malaysia, but will travel quite extensively. It is really exciting, although a bit scary. Let's go get a pint." And so they met to celebrate and talk about what this meant for Roger professionally and both of them personally.

Roger's primary responsibility would be to lead a team charged with recruiting and hiring talent at Spooner's quickly growing manufacturing and engineering operations in Southeast Asia. He would be accountable for developing a staffing program that would meet the strategic goals of the organization and the tactical goals of the regional operations.

The next morning was a Saturday, and Roger started to map out the key challenges as he saw them so as to formulate a plan to tackle the new job.

- How would he pull off the delicate task of building a team, despite his evident lack of experience in the business, the country, and the world of talent management, *and* find ways to learn from them to be better able to lead them? He knew the importance of quickly demonstrating a delicate balance of expertise and humility.
- What did talent acquisition really mean? Was it as strategic as the company executives led him to believe? How would he make a contribution that would develop a sustainable competitive advantage?
- How did the various countries and cultures view recruitment? Were there differences between countries, subcultures, age

groups, and so forth, that would require adapting image and reality to current and prospective employees?

- How do you go about developing and delivering a differentiated employee value proposition to attract talent? What does that even look like?
- What was the appropriate way to source, attract, and select talent in the region? What were the local desires and expectations about progression and mobility, and how could they be fashioned to be attractive in multiple countries?

It was a big task, and the only way to handle it was like eating an elephant—one bite at a time or as Lao Tzu, a Chinese philosopher said, "A journey of 1,000 miles starts with a single step." And so the journey begins . . .

Roger's experience is typical of many who are charged with talent acquisition in global organizations. Few are truly prepared for the challenges that sourcing, recruiting, and hiring talent will create, because few people have an understanding of talent, strategy, and global/cultural issues. This book is intended to help managers like Roger. We are going to provide managers tasked with global talent acquisition with a series of procedures, frameworks, and practices that will help them understand global talent acquisition and make sound choices about how to do it in a way that achieves sustainable competitive advantage. Let's start by thinking about what talent is and why it is so critical for competitive advantage in global firms.

TALENT IS THE DRIVER OF THE NEW GLOBAL ECONOMY

The purpose of most organizations is to create value for their stakeholders. Value is created through increasing revenues and/or decreasing costs.

Choosing the right strategy and differentiating the firm from competitors are among the most important ways to grow revenues, while improving the execution of the firm's strategy is one of the most important ways to decrease costs. Acquiring, nurturing, and retaining the right kind of talent is vitally important for differentiating the firm and improving strategy execution, especially in the modern knowledge-based economy. As economic data forecast a risky and uncertain future, the safest bet you can make is on talent. Indeed, one of the most capable HR leaders we've seen summed it up well: "Hire smart people. Smart people sometimes do stupid things. Stupid people rarely, if ever, do smart things and then only by accident."

But *where* you bet on talent requires a new way of thinking. We live and breathe in a global economy, and like other scarce but valuable resources, talent is unequally distributed around the globe. On the one hand, demographic shifts and the cost of labor are changing the division in talent between developed and developing economies. On the other hand, organizations continue to globalize to pursue strategic opportunities. Such globalization strategies require individuals with the expertise and mindset capable of adapting to different cultures and economic systems, and effectively managing resources seamlessly across borders. In both cases, managers must think about and implement talent acquisition in very different ways.

Talent is the driver of the modern global economy. That may sound strange, as the global economy remains uneven and uncertain and unemployment rates in many countries have remained frustratingly high. Yet even when financial capital is in short supply, the shortages are temporary and potentially minor relative to the shortages facing the supply of talent. Consider the following:

- Deloitte (2014) suggests there is a "talent paradox." There are challenges in finding the desired talent, and yet there is high unemployment. Causes of the paradox include high demand for specialized skills, unequal distribution of those required skills in different places of the world, and technological and demographic shifts.

- A report by Manpower (2011) predicts such critical talent challenges that it suggests "capitalism" is being replaced with what they call "talentism." They call this new future "The Human Age," with the focus shifting from financial capital to human capital.
- In a survey of over 2,000 managers in 94 countries by Deloitte (Stephan, Vahdat, Waklinshaw, & Walsh, 2014), 24% felt that talent acquisition was an urgent concern. It was the fourth most urgent issue, and it also ranked fourth in terms of showing the largest gap between readiness and urgency.
- PwC's (2014) survey of CEOs shows that the proportion who view the difficulty in finding desired talent as a threat to their organization's growth has increased steadily, rising from 46% in 2009 to 63% in 2014.
- IBM's recent survey of chief human resources officers found talent acquisition and talent retention to be two of their top five workforce-related challenges.

The "war for talent" (Michaels, Handfield-Jones, & Axelrod, 2001) was the metaphor that largely described talent acquisition during the 2000s. This metaphor conjures images of huge clashes with only two sides, and the "goals" of the war were obvious—unconditional surrender and the end of the war (think WWI and WWII). We suspect that the coming challenge will more aptly be described as ongoing battles or skirmishes for select talent. Rather than a war on a broad front, organizations will focus on winning smaller, more critical localized battles for talent (think "surgical strikes"). But these battles or skirmishes are less clear-cut than wars; there is no conclusion ("the war is over!") but only temporary victories. This more selective focus coincides with the growing recognition that not all jobs are created equal and that, in fact, only a few may be critical to a firm's success (Boudreau & Ramstad, 2007; Cappelli, 2008; Huselid, Beatty, & Becker, 2005). In the current environment, winning these select battles is a more sustainable effort than an across-the-board war. Firms can rarely compete or dominate in every job class; they must be selective and focus on those that offer the greatest impact for their businesses. Talent is the driver of the new global economy, and successful firms are those who know how to best acquire and use it.

There are several reasons that talent has become so important. First, the nature of work has changed, such that we now live in an *information age* that emphasizes knowledge, service, and data. Second, at the very time talent has become more central to firm performance, the supply has become more rare, uneven, and geographically distributed. Third, impending demographic shifts are changing the supply and flow of talent, as a large number of retiring baby boomers will exit the workforce in developed countries while a growing number of less skilled younger workers are entering the workforce in developing countries. Finally, there is incredible economic and political instability that percolates around the world.

WHAT IS TALENT?

We all think we know talent when we see it. The problem is that we are often wrong. If talent were so obvious, we would select it and nourish it and then there would be a lot less turnover and a lot higher performance. We need to calibrate our understanding of talent—we need to know *what talent is* so we can understand *what talent does* (Ployhart, Nyberg, Reilly, & Maltarich, 2014).

We define **talent** as "the sum of a person's knowledge, skills, abilities, personality traits, values, and interests that enable one to effectively perform a job and broadly contribute to the organization's effectiveness, strategy, and culture."

There are two key parts to this definition. First, talent is related to effective job performance. This means different types of talent will be required for different jobs. Usually, the elements of talent most relevant for job performance are knowledge, skill, ability, interests, and personality. Obviously, the types of talent needed for a frontline manager will differ dramatically from the types of talent needed for a senior leadership position. Or, even for the same job, the talent needed to work within one's home country must be expanded to effectively work in other countries and cultures, or in organizations with different strategies.

Second, talent must contribute to the organization's broader effectiveness, strategy, and culture. An individual with exceptional technical skills who lacks appropriate interpersonal skills is not going to effectively contribute to the firm's broader goals. Further, an employee who holds values different from the firm and its dominant stakeholders is either going to eventually quit or be destructive. Narrow technical skills are necessary but not sufficient for effective organizational performance. Usually, the elements of talent most relevant for organizational fit include personality, values, and interests and increasingly the ability to work well with others.

In some cultures and countries, the emphasis in talent acquisition is only on knowledge, skill, ability, and (frequently) personality for a specific job. We believe this is a limited and shortsighted perspective. People perform *jobs* that contain specific tasks. These job tasks are in turn embedded in a broader social context containing work roles. Jobs and roles are themselves contained in organizations that are located within *nations*. Employees may leave jobs because they can't perform the job, or don't fit the role or the organization, or some combination. Hence, fit with the job's technical requirements, work role, and the organization's strategy and culture are all necessary. Figure 1.1 provides a summary of this embedded view of employment.

We define **collective talent** as "the aggregate combination of multiple individuals' talent that broadly contributes to the organization's effectiveness, strategy, and culture."

By distinguishing between talent as an individual characteristic and talent as a collective resource, we recognize talent may exist in multiple ways. For example, star employees are those individuals who are exceptional in their careers and generate disproportionate value for firms. In contrast, a firm may not have a star employee but an exceptional team or division of employees who generate disproportionate value for firms.

Managing collective talent resources is different from managing individual talent resources. For example, when a star employee leaves a firm, he or she can take much of the value creation with him/her. In contrast,

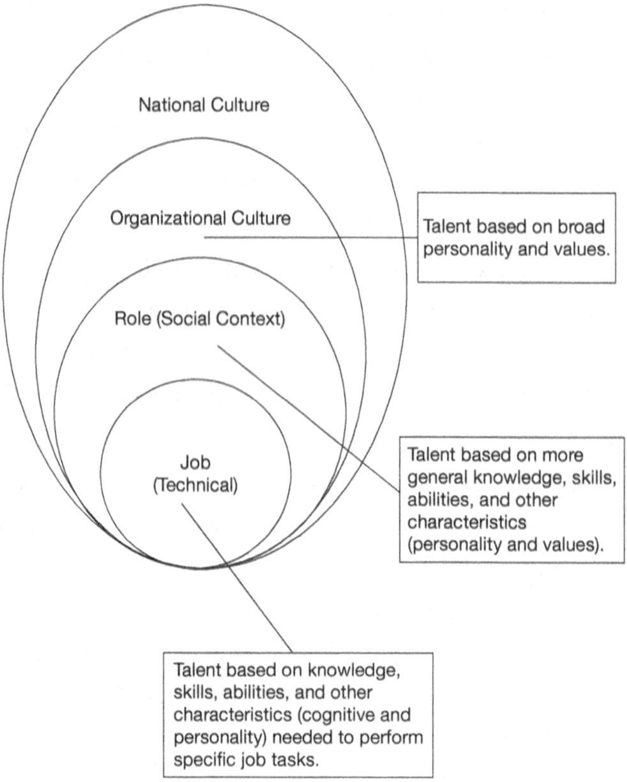
National Culture
Organizational Culture
Role (Social Context)
Job
(Technical)
Talent based on broad personality and values.
Talent based on more general knowledge, skills, abilities, and other characteristics (personality and values).
Talent based on knowledge, skills, abilities, and other characteristics (cognitive and personality) needed to perform specific job tasks.

Figure 1.1
Talent embedded within jobs, work, and context.

losing one employee from a collective may or may not matter much for the collective talent's value-creating opportunities. Indeed, in 2000, Coral Energy, the US gas and power energy trading arm of Shell, lost significant numbers of key employees. Some went to Duke and Dynegy but predominantly to Enron. Several of these took other valuable employees with them; in one case, the employee took with him all but one of his leadership team to the new company. It got to the point in early 2001 that the ability of that organization to operate effectively and in compliance with external and internal controls was in severe jeopardy. Using a variety of tools, including but not limited to financial ones, six key employees were targeted and persuaded to remain with the company for a defined period *and* to play the leading role in ensuring that their key talent did the same. It worked, and the health and financial performance of the organization recovered.

Talent, whether individual or collective, can add value to the firm. Technical, interpersonal, and managerial competence and fit with the firm are necessary for talent to add value. [Note that we recognize there will be times where the desire will be to hire someone who does *not* fit the values of the firm, for example, a leader who is expected to change the firm's culture. Yet even in this example, the goal is to hire talent needed to produce the desired change.]

WHAT MAKES TALENT UNIQUE

Managing talent and knowing how to acquire it are hard because talent itself is so difficult to understand. Compared to most other resources, like land, money, or natural resources, talent is unique because of its intangibility, mobility, and malleability.

Talent Is Intangible

Talent is intangible, and the measurement of talent is not direct or agreed on. Indeed, the very value of talent may differ by job, organization, industry, and culture. Talent is not some universal entity; talent is specific to the firm's needs today and in the future. Indeed, the nature of talent shortages differs across many industries (with the possible exception of leadership talent; Deloitte, 2014). Furthermore, some companies seek to hire based on broad skills such as creativity and problem solving (e.g., Google), whereas others seek to hire on very specific job and industry skills (e.g., oil companies looking for drilling engineers or petroleum geologists). This means that there is no universal metric that can be used to gauge talent. Even measures of intelligence (IQ) or emotional intelligence (EQ) are insufficient, because they capture only one narrow slice of a person's overall talent. The intangibility of talent often leads to confusion or disagreements about what it is and how to manage it. However, there are many ways to quantify talent and thus apply analytical techniques to understanding and leveraging it for competitive advantage. Few companies currently capitalize on these analytic tools to the extent they could. In chapter 3 we describe these analytic tools and how they can be leveraged to make talent more tangible.

Talent Is Malleable

Talent is finite when we consider the capabilities of an individual employee. But in the aggregate, such as a project team or a firm's workforce, collective talent is renewable. So long as the firm can attract and retain the desired workers, a high-quality stock of collective talent is theoretically infinite. Collective talent is thus malleable, because it can be redeveloped, reskilled, or bundled in new and different ways. For example, we can't change a person's basic intelligence or personality. But in a collective, we can change the mix of intelligence and personality, such that different types of talent can be created as the organization's needs and demands change. This requires a different way of thinking about talent. You are a talent architect; the designer of what talent should be to fit the firm's strategic goals and objectives. The talent you have is a given, the talent you want is something that can be created. Collective talent can be changed by adding new people, by removing or replacing people, by developing people, or a combination thereof. Companies known for their impressive talent (GE, PepsiCo) both attract and develop it; they create the talent they want.

Talent Is Mobile

Financial capital may be wisely or unwisely used, but for at least some time is owned (or controlled) by the firm. In contrast, talent resides in employees and is related to the firm only through formal employment contracts and informal employment relationships. Your firm's talent goes home every day, and may or may not come back tomorrow (Coff, 1997). Talent is a resource that has choice and volition. Employees may withhold knowledge or effort, they may show up late, and they may quit. The extent to which companies can engage their workforce and, as a result, motivate discretionary effort, can be an important competitive edge (Deloitte, 2014; Harter, Schmidt, Asplund, Killham, & Agrawal, 2010). As a further complication, talent mobility differs across countries, cultures, and industries. Indeed, one story suggested, "many developing countries find that their most lucrative export is people" (Gorney, 2014, p. 70). Such mobility can affect the supply, demand, and value-creating benefits of talent (e.g., Campbell, Coff, & Kryscynski, 2012).

Conclusion

Talent—the very resource that offers incredible opportunities for generating competitive advantage in the modern economy—is also the resource that is the most intangible, malleable, and mobile. Such characteristics create incredible challenges for managers because it is difficult to quantify talent, take stock of a firm's current holdings, and demonstrate how to best acquire, develop, leverage, or divest of those holdings. Further, because talent is intangible, malleable, and mobile, different countries have established different laws, regulations, and policies for how talent is to be managed and treated. These international differences in the treatment and management of talent create further complexities for firms trying to acquire or deploy talent globally.

Yet hidden within these challenges is a key for unlocking competitive advantage. Those who understand global talent acquisition can achieve an incredible advantage over competitors. This book is intended to help you understand the acquisition of talent in the global context. We hope to provide you with a new way to think about talent from an international perspective and how to better source and manage the acquisition of talent globally. But this is a book about thinking of talent strategically. As a result, we are not going to focus on specific details in some areas (such as differences in laws in each country because these laws are constantly evolving and anything said today would likely be out of date tomorrow). However, by focusing on a strategic global perspective we offer a lasting view on competing internationally through human resources (HR).

Roger had come to realize that talent was a lot more elusive than he once thought. With this in the back of his mind, he began to contemplate some very specific questions.

1. In a war for talent, especially in rapid growth environments, what options other than pay and benefits are tools that can help a company enhance its employee value proposition to attract, retain, and motivate the types of candidates it needs to deliver its business strategy?
2. If one accepts that talent is malleable and mobile, how do companies accept that employees may indeed leave them after a few years but still derive value from their contribution during their tenure at the company?
3. In multinational enterprises, there will be a tension between having a global employment brand and being attractive in multiple locations. What might be the options that would enable the best balance between these potentially conflicting forces?
4. How does one think of global talent acquisition? How does this work at the multinational level and at the local subsidiary level? Is talent the same everywhere?

WHAT IS TALENT ACQUISITION IN GLOBAL FIRMS?

Feeling overwhelmed, Roger needed a break, so he wandered the office halls asking five of his colleagues what talent acquisition meant to them. He got four answers:

"Getting the right person for the job."
"I have no idea."
"Finding someone who fits the organization."
"Finding the A-players."
"I have no idea."
"Fantastic . . . ," he thought cynically.

In many firms, there is no systematic approach to talent acquisition, much less a strategy or vision for how it will be done in a global marketplace. Talent acquisition often evolves from a firm's history and competitive environment. Some firms will use systematic processes while others will rely entirely on informal judgments; some have a plan, while others react; and some are rigid and others flexible in the extreme. Cappelli and Crocker-Hefter (1996) give numerous examples of firms in the same industry that deploy radically different HR systems. Coca-Cola, for example, uses an internal labor market, wherein employees start at the bottom and slowly work (or don't work) their way to the top. The emphasis is on internal mobility and employment stability. PepsiCo, on the other hand, relies on the external labor market as a source for experienced high performers. Its employment practices produce a highly competitive environment and ensure that only the best and brightest thrive and indeed survive. Both UPS and FedEx have similarly evolved very different practices with respect to talent acquisition and development, despite competing fiercely in the package delivery business. In a similar vein, the giant multinational oil companies Shell and Exxon/Mobil adopt different employment philosophies, for example in their willingness to hire midcareer professional staff.

The point is that there is no single best way to manage talent acquisition. As we discuss in chapter 2 (strategy), the right approach is the approach that achieves the firm's strategic goals in a sustainable manner and reinforces the desired corporate culture. Multiple strategies and their resulting practices can be quite successful if they fit with the firm's overall business strategy and with the other elements of the HR system. A brilliantly conceived talent acquisition process will be less impactful if it collides with the firm's training, performance management, or compensation programs. Targeting experienced, high performers on the external labor market, for example, may enable the firm to save money in training and development, but its compensation costs will have to increase in order to attract the talent desired. Indeed, academic research suggests that firms outperform their competitors when they employ a complementary bundle of HR practices (Jiang, Lepak, Hu, & Baer, 2012). Another survey found that companies with an integrated talent management strategy experienced 26% more revenue per employee and 41% lower turnover than those without one (see https://www.bersin.com/News/Content.aspx?id=10658). Global talent acquisition must fit within this bundle of practices to be maximally effective. A brief discussion of the nature of these practices is provided shortly.

Talent acquisition is also shaped within the context of the firm's culture, the national culture, and the political and legal environment(s) within which the firm operates. For example, when trying to create a diverse workforce, a domestic oil company like Valero (the largest refiner in the United States) only has to focus on the various elements within the United States (race, gender, veteran status, handicap status, sexual orientation, etc.). Companies like Chevron, Exxon, and Shell, which operate in multiple countries, have to adapt to laws, for example, that regulate the roles that women are permitted to perform or to advance the interests of one race or tribal group over another. The very definition of diversity and what it encompasses varies widely from one country to another. This is further complicated by the desire of governments to advance the wealth and well-being of their citizens, which often results in nationalization targets. Talent acquisition is both important and complex, as evidenced by the advantages accruing to those who get it right.

Talent acquisition means different things to different people. In some cultures managers treat talent acquisition tactically—there is a job opening and they screen candidates to ensure they can perform the technical requirements of the job. In other cultures, it is more common to think of the workgroup as a family and the emphasis will be on *both* technical competence *and* fit with the workgroup. Likewise, in some cultures it is common to screen applicants using a variety of psychometric tools (such as intelligence or personality tests), while in other cultures such tests are seldom used or possibly even inappropriate.

We find it quite striking—perhaps even scary—that talent acquisition is often treated in such a haphazard manner. First, the ability of a firm to grow, adapt, and remain competitive rests on the decisions and vision of the firm's leadership. Second, people are often a firm's largest expense, and to recover these costs one must have superior employee performance to achieve the firm's vision and mission. Third, the people within the firm will define its culture and climate, and hence determine "what it feels like to work there." Furthermore, "what it feels like to work here," in a virtuous circle, will impact who is attracted and retained by the firm. Thus, the talent of the firm will determine how it "thinks," "acts," and "feels."

Talent acquisition should be a purposeful, planned, and strategic process that helps the firm achieve competitive advantage in the global marketplace. We define **global talent acquisition** as "the strategic process of (1) *identifying* the talent competencies and values needed to effectively perform the job and contribute to the organization's effectiveness, (2) *sourcing and attracting* the people who have those competencies and values, (3) rigorously yet efficiently *evaluating* those people to *select* the best candidates, (4) providing *onboarding and socialization* experiences that shrink time to productivity, and (5) *retaining* the best people through development, engagement, and promotion."

There are several key parts of this definition (Figure 1.2 provides a visual overview of these key parts).

- **Strategic**: Global talent acquisition should be a planned, systematic, and goal-driven approach for using talent to achieve the organization's global strategy, vision, and mission. This is not to suggest that all organizations need a long-term staffing plan. Indeed, the pace of change has led many firms away from reliance on an internal labor market model, with its dependence on stability, predictability, and internal mobility, to

Figure 1.2
Elements of global talent acquisition.

---

the just-in-time acquisition of talent to meet new opportunities and unexpected contingencies. What is needed more than ever is a coherent, global process for acquiring talent. Organizations with such a process for talent acquisition find it "faster, better, and cheaper" than those without one. It is better to have a talent acquisition strategy and adapt it to changing conditions than to operate without one. As the saying goes, "If you don't know where you are going, any road will get you there." Strategic planning is discussed in chapter 2.

- **Identifying**: If talent is to add value to the firm, then one must be able to define the types of talent needed for the job and organization, both currently and in the future. While methods for identifying the knowledge, skills, abilities, and other characteristics needed for a job have long been in existence, they are unevenly applied in practice. Methods for projecting future talent requirements are less well known and, not surprisingly, practiced less often. Identifying talent is discussed in chapters 2 and 3.

- **Sourcing**: There is usually more than one source for talent, so the challenge becomes one of identifying the most cost-effective supply of talent. Some sources may be more expensive to access than others but yield a higher proportion of qualified candidates such that the unit cost of labor is actually lower. Once sources are identified, the issue becomes one of getting the desired talent to apply to and join the firm. The increased emphasis on employment branding is one response to the challenge organizations face in differentiating themselves from the competition in the attraction of talent. Sourcing is considered in chapters 2 and 4.
- **Evaluating**: One of the most difficult aspects of talent acquisition is identifying who are the *best* candidates. It is ultimately a bet, one based on limited and incomplete information. Probably no other decision point in the talent acquisition process has been more thoroughly researched. Managers are now armed with a host of methods and techniques for evaluating candidates, and the application of the right ones can greatly improve the odds of this bet being successful. Yet even advanced evaluation methods will fail if they are used in a manner inconsistent with the national culture context. Evaluation and selection methods are reviewed in chapters 3 and 5.
- **Onboarding**: Once selected and hired, the early experiences the candidates have with the firm will set the foundation for their employment relationship—for better or for worse. In some jobs, such as police officers and auditors for large accounting firms, the early experiences of a new recruit are carefully managed. Training, socialization, and enculturation follow scripted paths designed to enhance effective performance and reduce

time-to-productivity. In most cases, however, there is little thought given to systematic onboarding, resulting in great role behavior variability. As we will see, technology is increasingly being used to both systematize and personalize the onboarding experience, as noted in chapters 2, 4, and 6.

- **Retaining**: The economy has obviously changed the extent to which employees jump between jobs, and most people will hold multiple jobs in their careers—changing jobs is, to varying degrees, a global phenomenon. The investment made by some firms in its talent is so great that it is important to re-recruit these people on a regular basis. Mentoring programs, long-term incentives, engagement surveys, and many other HR programs have as an objective the reduction in undesirable turnover. Simply put, it is far cheaper to keep an effective employee than to replace him or her. Retention is reviewed in chapter 6.

Talent acquisition is a complex, interrelated set of processes all designed and implemented to help an organization execute its strategy. Now, consider the challenges of talent acquisition in the global firm. Issues such as sourcing, attraction, and retention are considerably more difficult. Imagine a global firm headquartered in France looking to expand its accounting staff. Should it try to compete in the local economy or outsource part of the accounting function to a specialized firm offering such services in India? Similarly, imagine the challenges faced by a Japanese firm trying to build and open a new plant in the United States. Where should it be located to take advantage of the best (and obtainable) talent? What role will attitudes toward unions within the local labor pool play in the selection of a site? How will HR policies and practices need to change—or should they stay the same? What mix of talent will come from the home country versus the local population? How will one ensure that important organizational cultural norms will be maintained, while accommodating the cultural differences between countries?

For example, when Western companies expand their operations into countries with different norms or cultures, one often hears, "We are an American/French/Dutch Company and we expect you to operate in American/French/Dutch ways." This seems reasonable and certainly can be advanced as a legitimate choice. For example, when a Japanese car company first set up a manufacturing plant in the southern United States, they announced that morning shifts would start with exercises in the parking lot—a practice that was common in Japan. There was an outcry: "Don't the Japanese understand that this is not consistent with the culture of the United States? That it is just unacceptable and won't work here." The obvious point is that application of standard processes and philosophies in other countries is not justified by ownership alone. Effective talent acquisition is challenging, and global talent acquisition even more so. However, with great challenge comes great opportunity for those firms who can do it effectively.

The Talent Advantage

Firms differ from each other due to the endowments of resources they have acquired or developed. Firms that have resources that are linked to their strategic goals, and are isolated or protected from competitors, may be able to achieve competitive advantage. Talent resources offer the possibility to be one such basis of competitive advantage—what is called a *talent advantage* (e.g., Boxall, 1996). It is often possible to use talent as a way to differentiate the firm from competitors, and in doing so, strengthen the business case for the importance of investing in human resources. For example, in a study of over 300 South Korean firms, we have found that firms who used more selective staffing outproduced their competitors by over $10,000 per employee and generated greater profit growth. Perhaps most importantly, selective firms recovered more quickly from the Great Recession (Kim & Ployhart, 2014). Across hundreds of business units and thousands of people, researchers have shown that units with greater talent resources show greater productivity and financial performance (Crook, Ketchen, Combs, & Todd, 2008).

Talent as Strategy

The desired types of talent should be a means to create, reinforce, and implement the organization's strategy. Organizational strategy tends to give great thought to market position, growth opportunities, and the competitive landscape, but sometimes less thought as to whether the organization has the necessary talent, with sufficient motivation and incentives, to implement the strategy. As the link between talent and organizational performance becomes clearer, the link between talent and organizational strategy (development and execution) will be become stronger. Thus, talent can be used both to inform development of the organization's strategy and to translate it into reality. This offers yet another means for the business value of human resources to be conveyed to key stakeholders.

Talent's Consequences

Thus far we have considered how talent can be positioned to show positive consequences for the organization (e.g., financial benefits or impacts on strategy). On the other hand, conveying the negative consequences of having the wrong talent, or insufficient talent, can also be an effective means to build a strong business case. Being *talent constrained*, such that a firm is unable to implement or deliver on strategic objectives, is one such negative consequence. In our own experience, we have observed many times the generally disruptive consequences of turnover (both individually and collectively). For example, the aforementioned study of South Korean firms found that collective turnover strongly weakened the otherwise positive effects of selective staffing (Kim & Ployhart, 2014; see also Hancock, Allen, Bosco, McDaniel, & Pierce, 2013). Overall, collective turnover can reduce or eliminate the benefits of talent.

Thus, building and maintaining a quality stock of talent is extremely difficult. Talent is much like the ozone layer: it takes a long time to build, but is surprisingly fragile and can erode at an alarming rate. Employee mobility creates a number of difficult challenges and dilemmas for building a long-term talent strategy (Coff, 1997, 1999). To offset these dilemmas, firms must adopt a focused emphasis on talent acquisition.

Roger sat back on the plane headed out over the Indian Ocean for the final leg of his journey to Southeast Asia. As he did so he wondered what to expect from this assignment.

- Would the people he was about to work with understand his perspectives, and how would they be different from his?
- How would he adapt to the cultural differences between the Malaysians and other Asians with whom he would interact and the ones with whom he was familiar in the United Kingdom (UK)?
- How would his ignorance of the local legal frameworks hinder his effectiveness?
- What would his new colleagues expect of him?
- What would be the challenges of working for a local manager who might not welcome an expatriate from HQ?
- How would he learn to adapt his previous parochial UK understanding of talent acquisition to one that would require him to think strategically and across national borders?

His mind was racing, but he knew one thing—others had faced these challenges before and succeeded. So there were answers and the key to his success lay in being inquisitive, being slow to form conclusions and even slower to leap to judgment as he wove his path of discovery in his new working and living environment. But what conclusions should he make about this new environment and its place in the global context?

FACTORS AFFECTING TALENT ACQUISITION IN THE GLOBAL CONTEXT

Talent needs to be considered within context—that is, the broader global trends shaping both the demands for and supplies of talent. We summarize the most important trends in this section.

Globalization

There is no question that we live within a global economy. The financial meltdown starting in December 2007–2008 in some manner touched every person on the globe. The search for capital, which includes talent, used to reside within a country but now goes beyond national borders. Talent must likewise be considered within a global context. Talent exists around the globe, but not of equal types, quality, or supply. For example, European populations are stable or in decline, while developing countries such as those on the African continent are experiencing a population boom.

Yet at the same time, many countries, particularly those in Europe and the United States, are making it harder for talent to relocate. Since the terrorist attacks on the United States on September 11, 2001, work visas and immigration into the United States have become severely restricted. As a result, workers and students from other countries are decreasingly viewing the United States as an option and are instead choosing to locate in other countries or to remain in their home country. For example, despite often high unemployment rates, the European Union is attempting to deal with a flood of immigrants from Africa seeking a better life.

Employment Trends

Employment trends vary between and within geopolitical regions, countries, and industries. However, some broad trends seem robust. A report by Manpower (2015), summarizing results across multiple countries, suggests that there will be an increasing shortage of talent in many parts of the world. In many countries, the progressive drift away from the science, technology, engineering, and math (STEM) disciplines, combined with increased demand for technical skills, is one of the reasons for the apparent shortage of technical talent. It is not necessarily a shortage of people but of people with the right skills at the right place and time. Firms who are unwilling to hire less experienced applicants and then develop them internally magnify this shortage. It is also the result of demographic and regional shifts in where talent is located. Regardless of the reasons, many employers indicate that they remain talent constrained even when unemployment rates remain rather high. A second noteworthy feature of the Manpower (2015) report is that the talent shortages are not exclusively found in knowledge work. Indeed, they found professional trade occupations (e.g., electricians, welders) to be among the most difficult to fill positions. Thus, employers should expect to face significant challenges in hiring the necessary talent in the future, regardless of economic cycle, geographic location, occupation, or industry, although technology may help lessen these strains.

Cultural Diversity

Nearly all developed countries have seen increases in their cultural diversity over the last 50 years. Developing countries are slowly seeing increasing diversity as well, as organizations invest and relocate employees to oversee international operations. For example, according to census data, whites will no longer be the major ethnic group in the United States within the next 30 or so years. Such diversity brings opportunities and challenges. The opportunities involve the potential for more ideas, innovation, and insight into different markets. The challenges involve ethnic tensions, conflict, and political instability. Employees struggle with these issues on a more personal level. For example, in many of the firms we have worked in or consulted with, an ability and openness to learn and adapt to those from other cultures was difficult to identify and instill. Cultural diversity thus requires management from the firm level all the way down to individual employees. Yet diversity takes many shapes around the world. For example, is a project team diverse if it is entirely composed of white males? What if those white males are from 11 different countries and between them speak 17 different languages?

Developing countries also make choices regarding the advancement of particular groups as part of the economic or social leveling. For example, in 1969, Malaysia experienced significant racial riots in which many people died. There have been many views about the main causes of this, but observers concluded that at least part of the issue was the unequal spread of wealth between the three predominant races—Malay, Chinese, and Indian. The New Economic Policy had as its primary goal the eradication of identity of race and social function and provided for a range of benefits targeted toward the Bumiputras (mainly Malays). This policy was similar in intent to affirmative action programs in the United States, but much more prescriptive in its demands.

Education

Shifts to service- and knowledge-based economies have stimulated the global need for a more educated workforce. Today, there are nearly 150 million students enrolled in some form of tertiary education, more than half again as many as in 2000 (Altbach, Reisberg, & Rumbley, 2009). Unfortunately, access to higher education has not been evenly distributed, and the lowest participation rates are generally found in areas where greater education is needed the most. Furthermore, English has emerged as the dominant language of scientific communication, and much of the education Intellctual Property (publishers, databases, etc.) is in the hands of the strongest universities and organizations. Most of these universities are in the West (Altbach et al., 2009), although this competitive advantage is starting to decline in the United States (http://www.infoplease.com/ipa/A0923110.html).

Education is one area that has been an early adopter of globalization as a way of life. Not only has the number of students studying abroad increased but also educational institutions have expanded rapidly by opening campuses in other countries. Although most of the current student mobility can be attributed to students from Asia matriculating to North America, western Europe, and Australia, there has also been an increase in students from North America completing at least part of their studies abroad. Universities have also adapted to take advantage of growing enrollments and globalization. Many institutions have created exchange programs with other universities to facilitate the movement of students between them. Others have opened offshore branches (e.g., Carnegie Mellon in Australia and South Korea, Texas A&M in Qatar, Georgia Tech in Singapore). The University of South Carolina has formed an innovative partnership with the Chinese University of Hong Kong (CUHK). Under the banner of the International Business and Chinese Entrepreneurship (IBCE) the program is designed to develop individuals who can operate in the Chinese business environment by taking 20 students per year who are then matched with 20 students from CUHK. The USC students receive a BS in business administration, major in a special track in the international business major, and have a second USC business major. With studies taught in both China and South Carolina, augmented by internships in both countries, the program bridges national barriers and develops transnational business and language skills that make the graduates highly competitive in the employment market. Finally, scientists and engineers are increasingly being wooed back to their home countries. The opening of economies in developing countries, combined with the economic downturn in developed countries, has created new incentives for reverse migration.

Political and Legal Factors

Country policy has long had an important impact on talent acquisition, and with expanded globalization, countries are increasingly using policy to enhance their competitive posture vis-à-vis the rest of the world. Qatar, Singapore, and the United Arab Emirates [UAE], for example, have aggressively courted foreign universities to their shores as a way of enhancing their own prestige and human capital (Altbach et al., 2009). In response to projections that it will have one-fourth of the world's total skilled workforce by 2020, India's central government has given a high priority to higher education and allocated large sums to the sector.

In addition to education, the political environment also impacts work in other even more direct ways. Tariffs, import quotas, and other restrictions are enacted by governments to protect particular industries from external competition. Some countries have created and enacted coordinated policies to support the emergence of key industries. By adopting policies and laws directed at particular industries, countries directly impact the talent acquisition issue by stimulating the demand for some types of talent over others.

Employment law, policies, and practices vary tremendously from country to country (Myors et al., 2008). Many countries within the European Union (EU), for example, are well known for providing extensive protections for their workers. Relatively rich pensions, full retirement at younger ages, and greater restrictions on firing workers all impact talent acquisition. For example, because workers are more expensive in relative terms, companies are reluctant to add them and in many cases miss growth opportunities. The use of extensive screening procedures common in parts of the EU may be a very logical response to the restrictive policies on employee terminations, which increase the costs of hiring errors. Countries that are rich in natural resources but have a relatively uneducated workforce may require that specific targets be set with respect to the proportion of local staff employed, the clear expectation being that localization of employment will increase significantly over time. One of the ways in which companies and governments can cooperate in that effort is sponsoring local staff for study overseas, particularly the United States, UK, and Australia.

Demography

The world's population has increased dramatically over the past 50 years, and that growth is projected to continue, albeit at a slower rate. The world's population in 2025 is projected to be around 8 billion, some 1 billion higher than today. The majority of this growth is projected to occur in Asia and sub-Saharan Africa; relatively little growth (e.g., 3%) is projected for Europe, North America, and Australia/New Zealand/Japan. In some countries (e.g., much of Europe and Japan), birth rates are already below the rate needed to maintain the current population. The populations of Russia, Japan, and much of eastern Europe are also expected to decline by up to 10% (2025 Trends). The combination of declining birth rates and increasing longevity, in turn, has put severe pressure on the social support systems of many countries and led to government actions to deal with them (e.g., raising of the retirement age in France). These countries, with a rapidly aging workforce and the pending retirement of many, face a looming talent shortage. For example, BASF projects that a majority of its German employees will be 50 to 65 years old by 2020. "It's become apparent that we're going to hit a wall," says CFO Kurt Bock. BASF's demographic problem is bigger than most because it mainly operates in Germany, Japan, and the United States, where older workers make up an increasingly large percentage of the population (Kimes, 2009). This same age bulge may appear in growing economies like China but on a deferred time scale due to recent economic development and social policies. Although many predict the unrelenting rise of China to become the world's #1 economy, the effects of the "One Child" policy could mitigate that as a declining population of workers supports an increasing number of those in advanced years.

One report showed that Denmark, Japan, Germany, and Italy had incoming cohorts of workers insufficient to replace anticipated retirements, and that by 2020 only Turkey and Mexico among Organisation for Economic Co-operation and Development countries will have youth cohorts sufficient to replace more than 80% of the retiring cohorts (Chaloff & Lemaitre, 2009). Conversely, youth bubbles are expected in parts of South America, Africa, the Middle East, and parts of South Asia. Countries with a more than adequate birth rate may, by default, find themselves the target of talent-hungry employers. The net effect of these trends is likely to be migration; of workers to the West, work to the East, or more likely some of both.

Technology

Probably no force has had a bigger impact on talent acquisition than advances in technology. Beginning with the personal computer, then the Internet, and now wireless smartphones and devices, technological changes over the last 20 years have fundamentally altered the way talent is sourced and selected. Since 1999, broadband costs fell steadily and dramatically (from $1,197 to $130 per 1,000 Mbps), and Internet penetration increased accordingly (from 29% to 63%; Hagel, Brown, & Davidson, 2009). This created the "perfect storm" with employers and employees increasingly taking advantage of the ease with which information (e.g., about jobs and qualifications) can be shared.

Technology has radically altered the employment contract and what it means to be employed. Many employees now work remotely and, because of the Internet, working remotely does not even require one to live in the same country. As technology fosters greater communication and connectivity, it also fosters opportunities to reduce physical barriers for those who work in knowledge or information industries.

This has an effect of bringing the world closer together and radically changing the nature of talent acquisition. For over a decade, the Internet has been the dominant form for sourcing talent, and is the primary source applicants use when researching jobs and organizations (at least in developed countries). Networking websites like Facebook and LinkedIn have become tools for recruiters and applicants. Organizations are increasingly driving applicant traffic to their corporate website, where those interested can access up-to-date information on the company and available jobs. Most large organizations, at least in the West, now use some form of applicant tracking system, or ATS, to manage the talent acquisition process. Online assessment is old hat, and it is now common for recruiting and selection activities to be performed over mobile devices. Selection assessments have become increasingly sophisticated, allowing for high-fidelity simulations and instant scoring. Together, these changes have in many ways flattened the world and allowed more organizations to reach talent globally, and vice versa. We discuss the effects of technology on talent acquisition in chapters 4 and 5.

Technology has also resulted in new forms of organization, fundamentally changing the types of skills and abilities required of employees. From the advanced problem-solving skills of manufacturing employees in a highly automated plant to the adoption of assembly-line style practices in surgery, technology has changed work and therefore the talent required to do it well. Indeed, the rapidly changing nature of work has led many to abandon traditional succession planning efforts in favor of talent pools, which offer greater flexibility and less "system maintenance."

Although we are seeing increased use of networking tools like LinkedIn and Web-based candidate search tools, they are only scratching the surface. One company has even "back-engineered" the organization chart of one of their competitors using publicly available data from sources like LinkedIn. They had done a thorough analysis of the talent of several of their key competitors to assess their capability to effectively respond to a potential new product strategy that they were considering. Forward-looking talent organizations will need to operate with that type of strategic intent to service the needs of the organizations in which they work. In short, technology has affected not only *how* talent is acquired, but *where* it is acquired. Organizations are increasingly able to migrate work (and jobs) to wherever the skills are most readily and cheaply available.

Roger had become fairly proficient in the use of technology, although perhaps not in comparison to his 12-year-old nephew. It was interacting with his nephew, combined with early discussions with colleagues, that made Roger realize that he was going to have to raise his game significantly to lead a group of younger staff effectively. That was scary on one hand but refreshing on the other, because he knew that he would have to develop his skills in delegation and trust to get the most out of his new colleagues. He also believed that the next wave of technologically and societally driven changes to how skilled workforces were attracted, mobilized, and deployed rendered obsolete some of the skills he had almost before he had mastered them. Challenging times indeed. He pondered on the words of Eric Hoffer: "In times of change, learners inherit the earth, while the learned find themselves beautifully equipped to deal with a world that no longer exists."

Roger refused to inherit a world that would soon be out of date! So he began to question himself:

1. No one really could have imagined the impact that the Internet would have on the way that business is conducted today. Few of us have the crystal balls to suggest what the next "Internet" will be, but what might be some potential game-changing events or developments? A China economic slowdown? A reemergence of a Russia with growing economic power and political will? Demographic shifts in Europe that would change the age and race profiles of that continent? Increasing instability in the Middle East and the emergence of a permanent refugee world?
2. What are the potential impacts of a growing nationalism in response to domestic stagnation and rising unemployment in countries like Greece, Italy, England, and France?
3. What do potential advances in technology offer as threats and opportunities for talent acquisition and deployment?
4. How do all these economic/societal/cultural forces interact? How can anyone realistically predict a future even 3 years away?

Roger knew that betting on talent was the key to managing an uncertain future. And he also knew that betting on talent meant he needed to know enough about talent to make a smart bet. He had made some time before flying out to Kuala Lumpur and wandered into a local university library, pulling various academic books about talent, hoping he might gain some perspective by reading the works of those who have spent a lifetime studying it.

THEORETICAL PERSPECTIVES ON TALENT

The focus of this book is on strategic, yet practical, issues underlying global talent acquisition. However, the practices we prescribe and the solutions we advocate are based on a healthy mix of scientific research and experience. The scientific theory and research that underlie our approach are summarized in this section. Those who want more detail on these theories can consult the many specific references provided.

Cultural Values Theory

Countries and regions differ from each other in a number of ways, and differences in cultural values are among the most important such differences. *Cultural values* are guiding beliefs or principles on which people within a culture are generally similar to each other, but different from people in other cultures. Cultural values closely correspond to countries, hence they are sometimes called *national cultural values*. There are many different theoretical frameworks that explain the differences in cultural values. For example, the GLOBE framework developed by Robert House and colleagues identified cultural differences across approximately 17,000 managers from nearly 1,000 organizations in over 25 countries and even more societies (House et al., 2004). For our purposes, we focus on a more concise framework created by Geert Hofstede. Hofstede began work on his cultural framework when he was working at IBM and ultimately summarized it in his book *Culture's Consequences* (Hofstede, 2001). He analyzed the cultural differences of over 100,000 employees around the world.

The framework is based on the premise that cultural values are those values that are largely shared by people within a country. Countries differ from each other on these values, which in turn means that people from different countries can be understood based on knowing what country (and thus culture) they are from. Six cultural values are the focus in his book. Table 1.1 summarizes the cultural scores for 97 countries; higher numbers indicate more of that characteristic.

- **Individualism-Collectivism (IC).** The most salient cultural dimension is individualism-collectivism. People from more individualistic cultures tend to favor personal achievement, individual accomplishment, and personal rewards. Relationships are more generic, and accountability is placed on individuals. People from more collectivistic cultures focus on group harmony, social relationships, and maintaining group cohesion.
- **Power Distance (PD).** Power distance refers to a society's willingness to tolerate power or status differentials across people. In cultures with high power distance, power is expected to be distributed unequally, there are large divisions between people based on status, and social relationships are more hierarchical. In cultures with low power distance, power is shared, relationships are more democratic, and status differences are minimized.
- **Uncertainty Avoidance (UA).** Uncertainty avoidance refers to how comfortable people are with ambiguity, risk, or the unknown. Cultures high on uncertainty avoidance tend to find ambiguous situations stressful; they tend to use rules, laws, policies, and more structured approaches to situations to minimize uncertainty. Cultures low on uncertainty avoidance are comfortable with ambiguous situations and more open to change and risk.
- **Masculinity-Femininity (MAS).** This cultural dimension accounts for the strength of gender roles in a society. Masculine cultures emphasize gender-stereotype roles, and tend to be more aggressive, dominant, and competitive. Masculine

cultures are also more focused on achievement and material success. Feminine cultures are more gender-neutral, suggesting both men and women can adopt each other's roles. Feminine cultures are more focused on empathy, work-life balance, social relationships, and humility.

- **Long-Term Orientation (LTO).** Long-term orientation refers to a culture's time perspective and sense of future. As you might expect, people from cultures with a long-term orientation tend to take a longer perspective and place greater emphasis on the future. They value planning and persistence, but also tend to view truth as something that has shades of gray and evolves over time. In contrast, people from cultures with a short-term orientation focus on more immediate goals and outcomes. They emphasize fast results and tend to view truth as a dichotomy.
- **Indulgence-Restraint (IR).** As the name implies, this dimension reflects the extent to which a society focuses on restraint versus personal gratification. Indulgent cultures are those that emphasize personal outcomes, personal freedom, and self-interest. Cultures exercising restraint are those that emphasize values, control, and strong norms.

Together, these six cultural values can be used to compare and contrast different countries. Table 1.1 provides a broad overview, but more specific country comparisons can be found at http://geert-hofstede.com/countries.html. These cultural values are important because, on average, people from a given country will manifest these values to some degree. This means that the preferences, desires, working styles, behavior, attitudes, and so on, will in part be determined by one's home country. We consider the consequences and implications of these dimensions throughout this book, as we consider their effects on global talent acquisition. At the same time, we do not mean to imply that all people from a country or culture act the same. There is frequently a great deal of variability within any country, so these dimensions should be thought of as averages rather than fact (e.g., see Table 1.1).

TABLE 1.1 HOFSTEDE CULTURAL SCORES FOR DIFFERENT COUNTRIES

Country	**Individualism**	**Power Distance**	**Uncertainty Avoidance**	**Masculinity**	**Long-Term Orientation**	**Indulgence**
Albania	20	90	70	80	61	15
Angola	18	83	60	20	15	83
Argentina	46	49	86	56	20	62
Australia	90	36	51	61	21	71
Austria	55	11	70	79	60	63
Bangladesh	20	80	60	55	47	20
Belgium	75	65	94	54	82	57
Bhutan	52	94	28	32		
Brazil	38	69	76	49	44	59
Bulgaria	30	70	85	40	69	16
Canada	80	39	48	52	36	68
Chile	23	63	86	28	31	68
China	20	80	30	66	87	24
Colombia	13	67	80	64	13	83
Costa Rica	15	35	86	21		
Croatia	33	73	80	40	58	33
Czech Rep	58	57	74	57	70	29

Denmark	74	18	23	16	35	70
Dominican Rep	30	65	45	65	13	54
Ecuador	8	78	67	63		
Egypt	25	70	80	45	7	4
El Salvador	19	66	94	40	20	89
Estonia	60	40	60	30	82	16
Ethiopia	20	70	55	65		
Fiji	14	78	48	46		
Finland	63	33	59	26	38	57
France	71	68	86	43	63	48
Germany	67	35	65	66	83	40
Ghana	15	80	65	40	4	72
Greece	35	60	100	57	45	50
Guatemala	6	95	99	37		
Honduras	20	80	50	40		
Hong Kong	25	68	29	57	61	17
Hungary	80	46	82	88	58	31

(continued)

TABLE 1.1 CONTINUED

Country	Individualism	Power Distance	Uncertainty Avoidance	Masculinity	Long-Term Orientation	Indulgence
Iceland	60	30	50	10	28	67
India	48	77	40	56	51	26
Indonesia	14	78	48	46	62	38
Iran	41	58	59	43	14	40
Iraq	30	95	85	70	25	17
Ireland	70	28	35	68	24	65
Israel	54	13	81	47	38	
Italy	76	50	75	70	61	30
Jamaica	39	45	13	68		
Japan	46	54	92	95	88	42
Jordan	30	70	65	45	16	43
Kenya	25	70	50	60		
Kuwait	25	90	80	40		
Latvia	70	44	63	9	69	13
Lebanon	40	75	50	65	14	25
Lithuania	60	42	65	19	82	16
Luxembourg	60	40	70	50	64	56

Malawi	30	70	50	40		
Malaysia	26	100	36	50	41	57
Malta	59	56	96	47	47	66
Mexico	30	81	82	69	24	97
Morocco	46	70	68	53	14	25
Mozambique	15	85	44	38	11	80
Namibia	30	65	45	40	35	
Nepal	30	65	40	40		
Netherlands	80	38	53	14	67	68
New Zealand	79	22	49	58	33	75
Nigeria	30	80	55	60	13	84
Norway	69	31	50	8	35	55
Pakistan	14	55	70	50	50	0
Panama	11	95	86	44		
Peru	16	64	87	42	25	46
Philippines	32	94	44	64	27	42
Poland	60	68	93	64	38	29
Portugal	27	63	99	31	28	33

(*continued*)

TABLE 1.1 CONTINUED

Country	Individualism	Power Distance	Uncertainty Avoidance	Masculinity	Long-Term Orientation	Indulgence
Romania	30	90	90	42	52	20
Russia	39	93	95	36	81	20
Saudi Arabia	25	95	80	60	36	52
Senegal	25	70	55	45	25	
Serbia	25	86	92	43	52	28
Singapore	20	74	8	48	72	46
Slovakia	52	100	51	100	77	28
Slovenia	27	71	88	19	49	48
South Africa	65	49	49	63	34	63
South Korea	18	60	85	39	100	29
Spain	51	57	86	42	48	44
Sri Lanka	35	80	45	10	45	
Suriname	47	85	92	37		
Sweden	71	31	29	5	53	78
Switzerland	68	34	58	70	74	66
Syria	35	80	60	52	30	

Taiwan	17	58	69	45	93	49
Tanzania	25	70	50	40	34	38
Thailand	20	64	64	34	32	45
Trinidad and Tobago	16	47	55	58	13	80
Turkey	37	66	85	45	46	49
UAE	25	90	80	50		
United States	91	40	46	62	26	68
United Kingdom	89	35	35	66	51	69
Uruguay	36	61	99	38	26	53
Venezuela	12	81	76	73	16	100
Vietnam	20	70	30	40	57	35
Zambia	35	60	50	40	30	42

Missing cells indicate a lack of available data. Higher scores indicate greater individualism, power distance, uncertainty avoidance, masculinity, longer-term orientation, and indulgence.

SOURCE: Data derived from www.geert-hofstede.com.

Individual Differences Theory

We all know that people differ from each other on all kinds of characteristics, such as height, strength, weight, intelligence, personality, values, and knowledge. In general, across all such characteristics and across all cultures and nations, individual differences tend to follow what is called a *normal distribution* (also known as a *bell curve*). On any given characteristic, most people tend to fall to the average, with people who are farther from average (higher or lower) increasingly less common.

Most of the scientific research conducted on individual differences has been performed by psychologists. The types of individual differences that are usually considered in staffing are based on job-related knowledge, skills, abilities, or other characteristics like personality (KSAOs). The KSAOs are frequently called competencies, and for our purposes we use the terms interchangeably. There are several broad classes of individual differences (Ackerman & Heggestad, 1997); these are summarized in Figure 1.3.

- **Cognitive**: Cognitive individual differences represent those characteristics describing what a person can learn, think, or remember. Cognitive individual differences explain how much one learns and how fast one learns. The main types include *intelligence* (also frequently called cognitive ability or general mental ability), verbal ability, quantitative ability, and reasoning. Cognitive individual differences also include information that is learned (*knowledge*), as well as *skill* in applying that knowledge. Cognitive individual differences can be considered indicative of what a person *can do* on a job.
- **Personality**: Personality traits represent relatively stable ways that people behave and act in different situations. There are many different traits, but modern research suggests that five are most important to a broad range of situations. The Five Factor model (FFM) of personality includes conscientiousness (reliability and persistence), emotional stability (resilience

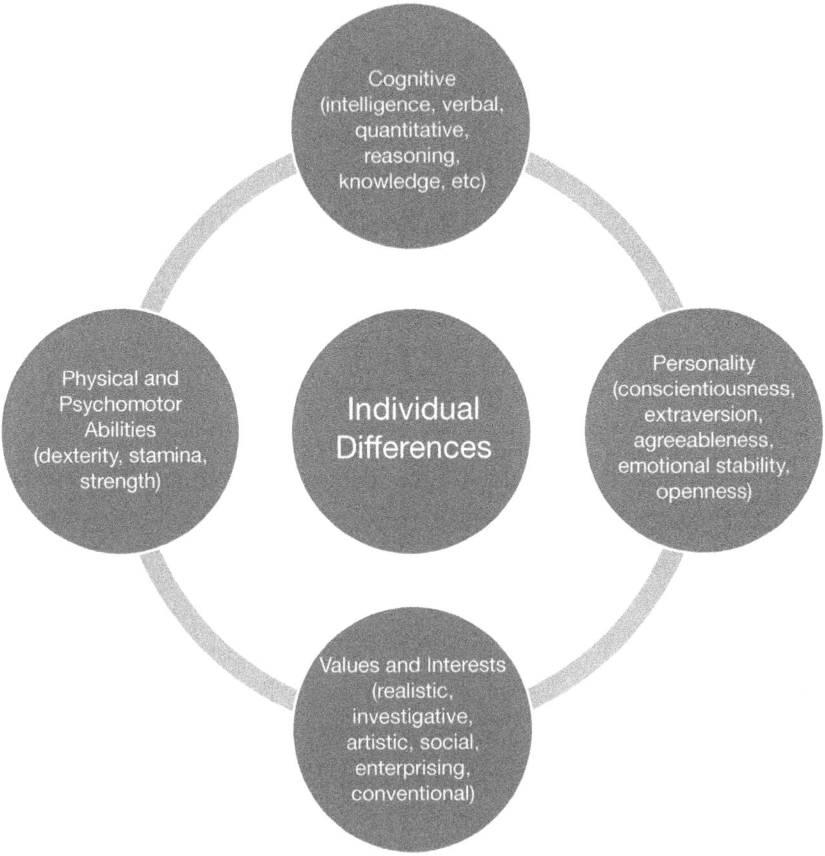

Figure 1.3
Overview of individual difference competencies.

to stressful situations), agreeableness (empathy and desire to get along with others), extraversion (preference for social activities), and openness to experience (desire to try new things and learn about new ideas). Personality-based individual differences can be considered indicative of what a person *will do* on a job.

- **Values and Interests**: Values and interests refer to the types of preferences that people have for different kinds of work. Holland (1997) has a model that has been widely used for understanding vocational interests. The idea is that people will vary in the kinds of work they prefer; some prefer work that is highly creative while others prefer work that is very structured. Holland suggests interests comprise six dimensions (RIASEC): realistic (agriculture; construction), investigative (researchers; problem solvers), artistic (creative; innovative), social (service; relationships), enterprising (leadership; management), and conventional (facts; data).

- **Physical and psychomotor skills**: Any sports fan knows that people vary greatly in their physical (e.g., strength, speed, endurance) and psychomotor skills (e.g., eye-hand or eye-foot coordination). While less and less important as work is increasingly focused on information, there are still many jobs for which physical and psychomotor skills are paramount (e.g., factory worker, agriculture, crab fisherman).

Individual differences in these characteristics are important to the extent that the characteristics are related to performance on the job and contribute to organizational effectiveness, strategy, and culture. For example, intelligence is one of the strongest predictors of job performance in nearly all jobs and countries (Schmidt & Hunter, 2004). This means that those with greater levels of intelligence tend to perform better on the job. Other individual differences, such as personality, also predict performance, organizational fit, and turnover (Barrick & Mount, 2012).

Human Capital Theory

The economist Gerald Becker (1964) won the Nobel Prize for his work on human capital theory. Unlike most economic theory at the time, which was focused on physical or financial capital, Becker was interested in understanding "human capital," or the skills and education of employees. Note that for our purposes, "human capital" and "talent" may largely be used interchangeably. Becker was interested in understanding how education related to wages and earnings. His work demonstrated that firms pay a premium for talent with advanced education, with the expectation that this "higher quality" talent will subsequently produce a greater return than those without the advanced education.

Becker's work and those of his colleagues has come to emphasize two important forms of human capital. The first type is known as *generic human capital*, which is the extent to which a firm's employees possess competencies that are relevant to a firm's current or future strategy but not specific to the organization. Generic human capital is the employees' stock of broad knowledge, experience, intelligence, personality, and related characteristics. This talent can be applied to any firm, industry, or country. A programmer's ability to write code in a common programming language would be an example of generic capital. The second type is known as *specific human capital*, or the extent to which a firm's employees possess competencies that are unique to the firm's existing strategy, jobs, and culture. Specific human capital is the employees' stock of job-, firm-, industry-, and country-specific expertise and knowledge. This type of talent is inherently tied to a specific job, company, industry, and country. A salesperson's knowledge of his or her client's preferences across the company's portfolio of products would be specific capital. If a person leaves an organization, then firm-specific knowledge is generally not very useful to competitors (organizational secrets notwithstanding!), although job and industry knowledge may be very valuable. Generic and specific talent lie on a continuum. Figure 1.4 provides an illustration of this continuum for an accountant. The figure gives the appearance that there is a causal relationship between generic and specific talent—this is intentional. Generic talent in many ways is a determinant of specific talent. For example, knowledge of basic arithmetic is needed before one can learn how to prepare a profit and loss statement. Both generic talent and specific talent have their importance, and both must work in a coordinated manner. Yet, each type of talent has some potential benefits and limitations for competitive advantage (these are summarized in Table 1.2).

A strength of specific human capital is that it cannot be easily copied by competitors. Indeed, a firm-specific talent pool is by definition nontransferrable. For example, Kia engineers will have intricate knowledge of their

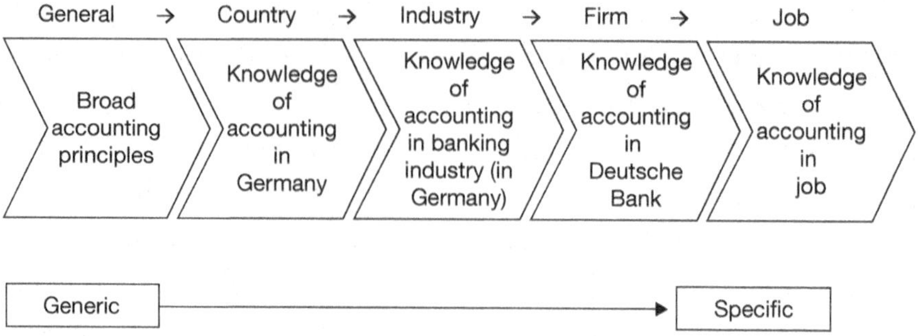

Figure 1.4
Talent lies on a continuum from generic to specific (knowledge of accounting principles is used as an example).

firm's products, services, customers, and coworkers within a given country. This represents a resource that cannot be moved to other firms (unless a competitor acquired the entire unit) or even other countries. There is a potential disadvantage of specific human capital. While specific talent is more valuable for competitive advantage in the short term, it may not be

TABLE 1.2 POTENTIAL BENEFITS AND LIMITATIONS OF GENERIC AND SPECIFIC TALENT

	Generic	Specific
Potential Benefit	▪ More adaptable to new competitive environments ▪ Contributes to performance growth during more unstable competitive environments	▪ Immobile across firms ▪ Contributes to performance growth during more stable competitive environments
Limitation	▪ Mobile to other firms	▪ Less adaptable to new competitive environments
Examples	▪ Knowledge of finance ▪ Quantitate ability ▪ Interpersonal flexibility	▪ Knowledge of regional dialect ▪ Knowledge of cultural norms in a specific firm (e.g., Unilever, Sony)

adaptable to changes in markets and the firm's strategy (a finding observed in Kim & Ployhart, 2014).

One strength of generic human capital is that it is adaptable and thus supports innovation, repurposing, flexibility, exploitation, and sustained competitive advantage (Bhattacharya, Gibson, & Doty, 2005). Also, as noted, generic talent helps determine the manifestation of specific talent. A disadvantage of generic human capital is that it is easily transferrable across firms. Thus, generic talent is more important for future growth and adaptability, and firms that know this may try to poach top talent from competitors to develop new competitive opportunities.

Generic and specific human capital exist simultaneously. Given that both types of talent offer distinct advantages, and that the disadvantages of one type are offset by the advantages of the other, firms gain even stronger competitive advantage by improving the synergy between the two types. For example, in one project with a quick-service restaurant franchise (Ployhart, Van Iddekinge, & MacKenzie, 2011), we found that the more employees were hired with high-quality generic human capital, the more they acquired specific human capital through advanced training. In turn, restaurants with higher-quality specific human capital provided better customer service and generated greater financial growth. Thus, firms that are better able to increase their generic talent, and particularly the synergy between generic and specific human capital, will differentiate themselves and create the possibility of gaining a sustained competitive advantage. It would be difficult for competitors to quickly obtain similar generic talent and impossible for them to duplicate specific talent. We revisit the relationship between generic and specific talent in chapter 2.

Resource-Based Theory

We noted earlier that firms differ in their resource endowments. Over two centuries ago, Adam Smith's central argument was that the quality of "talent" within a country would determine that country's fate and future—thus, heterogeneity in countries' talent causes differences in their success and survival. Later scholarly work by the English economist Judith Penrose (1959) emphasized that firms may grow and compete from the way they assemble and use their strategically valuable resources. More recent treatments of resource-based theory emphasize that four characteristics are necessary for resources to be a source of sustained competitive advantage, where competitive advantage may (for our purposes) simply be defined as generating above-normal returns (Barney, 1991; Wernerfelt, 1984).

To be a source of *competitive advantage*, a resource must be:

- *valuable* to the firm's vision and performance. Although it sounds obvious, we shall see that not all kinds of talent add value to the firm.
- *rare* in the factor (labor) market. The scarcity or rareness of a resource determines its price in the factor market. Rare resources are by definition hard to acquire or maintain. As an example, those with advanced degrees (e.g., MBA) are more rare and may therefore command a higher wage than those without advanced degrees.

Valuable and rare resources can lead to temporary competitive advantages. However, to be a source of *sustained* competitive advantage, a resource must also be:

- *difficult or costly to imitate.* Ensuring a resource cannot be quickly or easily copied allows firms to maintain a competitive advantage. The easiest example is one of brands. It takes years to develop the kind of brand equity that Coca-Cola, Toyota, or Siemen's have established (see www.Brandchannel.com).
- *nonsubstitutable.* If a competitor can effectively neutralize the benefits of another firm's resource, then that resource cannot be a source of sustained competitive advantage. For example, specialized retailers may compete against Walmart by delivering better, more customized service. In such a situation, the better customer service simply substitutes for Walmart's distribution advantage.

Although exceedingly rare, a sustained competitive advantage can be achieved through acquiring, developing, and retaining talent that is valuable, rare, inimitable, and nonsubstitutable. Southwest Airlines has had an enviable record in an industry fraught with bankruptcy, in large part because no one has been able to replicate its talent advantage.

It is often possible to use talent as a way to differentiate the firm from competitors, and in doing so, achieve a talent advantage. For example, note that the characteristics of talent—intangibility, malleability, and mobility—make talent valuable, rare, costly to imitate, and in many instances, nonsubstitutable. Therefore, the very characteristics that make talent so challenging from a management perspective are the same characteristics that contribute to a talent advantage.

Human Resource Systems

Human resource (HR) systems (also known in some countries as industrial relations systems or employee relations systems) will influence the nature and quality of talent attracted, selected, retained, and developed by the firm. Human resource systems have several interrelated components. Human resource *strategy* is the firm's overall approach to using the HR function to achieve the firm's strategy. Human resource *practices* are the variety of techniques and methods used to implement the HR strategy. For example, an HR strategy might involve enhancing employee competencies for teamwork and customer service. In turn, this HR function might employ practices like recruiting those with prosocial values, selecting applicants with personality and values that contribute to interpersonal skills, designing the jobs to allow discretion when interacting with customers, and training and developing customer orientation skills, and compensating employees for superior service.

The logic here is that by using a complementary set of HR practices focused on achieving some strategic goal, organizations can build the types of desired talent (Jiang et al., 2012; Lepak, Liao, Chung, & Harden, 2006). Examples of HR systems most relevant to acquisition include:

- **Recruiting**: Recruiting practices include techniques to identify and source talent, attract talent to the firm, and maintain applicant interest through the acquisition process.
- **Selection and Assessment**: Selection and assessment practices focus on techniques designed to identify who has the talent (individual differences) needed for the job and organization. Examples include interviews, intelligence tests, personality tests, and simulations.
- **Onboarding and Socialization**: "Onboarding" is an umbrella term referencing the range of activities designed to socialize employees to the firm. These may include formal classes, orientation sessions, peer or mentor programs, and self-directed learning.
- **Promotion and Internal Mobility**: Most of this book focuses on the acquisition of talent from outside of the firm—the external labor market. However, there is also an internal labor market based on existing employees who may be chosen for promotion or reassignment. This internal hiring and promotion process is equally important.
- **Succession Planning**: The volatility that exists in today's environment requires firms to be agile and adaptable. The days of developing lists of employees capable of filling each job are largely over. Instead, firms need to think about developing talent pools that can be quickly redeployed or bundled with other resources to achieve market opportunities (or respond to disruptive threats).
- **Training and Development**: Most jobs and occupations evolve over time, requiring employees to constantly learn and enhance their expertise. Talent development involves the activities required to ensure employees remain competent, or to prepare them for new assignments in the future. This may involve training, but also job rotations to learn the business from a broad perspective, and assessment activities designed to pinpoint strengths and weaknesses (e.g., 360 degree

evaluations that use evaluations from supervisors, peers, and subordinates).

- **Engagement**: Engagement refers to how satisfied, motivated, committed, and identified employees are with their firm. Engagement is based on a variety of factors, including selection, fit with the job and culture, compensation, and so on. Firms with more engaged employees can generate significantly greater financial performance than firms with less engaged employees (Harter et al., 2010).
- **Retention**: Talent is mobile, but HR practices and policies can be used to increase the chances that the desired talent remains with the firm. Such practices and policies may include pay-for-performance compensation practices, formal performance appraisal systems, retirement packages to help ensure employees remain committed to the organization, and engagement surveys to identify and rectify sources of discontent.

These HR systems are often configured to help the organization build or buy the talent needed. Internal labor markets use HR systems designed to build talent and are characterized by entry-level hiring, extensive training and development, promotion from within, job security, and compensation plans designed to enhance retention. External labor market configurations outsource development to others and are characterized by unlimited entry points, limited training and development, pay focused on performance, and minimal job security. Of course, many organizations exhibit features of both systems. For example, it is not uncommon for a firm to deploy an internal labor market with respect to its most strategically important jobs and simply hire as needed from other employers to fill less critical roles.

In the last decade, it has become increasingly recognized that HR systems should be aligned to fit with each other, and to achieve a specific organizational purpose. In this manner, high performance work systems (HPWS) are frequently used that combine aggressive recruiting, rigorous selection, and active employee development. Firms that adopt these HPWS tend to outperform those that do not (Huselid, 1995; Jiang et al.,

2012). Interestingly, the positive effects of HPWS on firm performance are consistent across all cultures and countries that have been examined to date (Rabl, Jayasinghe, Gerhart, & Kuhlmann, 2014), although the way these practices are implemented may differ. Chapter 6 discusses talent management practices, and their relationship to staffing, in more detail.

After reviewing so much academic literature before he had set out for his adventure, Roger had felt as though he knew a lot and knew nothing at the same time. "There are so many different perspectives; how can one make sense of it all?" He had left the library rather frustrated and gazed at a campus map. "I need a map for talent," he thought to himself. "I need some kind of structure or a framework to pull all this together." But what would such a map look like?

A GLOBAL FRAMEWORK

Individual differences theory and human capital theory are focused on individuals. Resource-based theory and HR practices are focused on firms. Cultural values subsume them all. In Figure 1.5, we reconcile these perspectives in a heuristic framework for thinking about talent acquisition in a global context.

First, notice there is a distinction between the multinational corporation (MNC) and subsidiaries. The MNC is where the firm's headquarters is located; it the *home* country and culture from which the firm has originated (e.g., Coca-Cola is from the United States, in the city of Atlanta). The subsidiaries are all of the business units where different operations are housed (the *host* countries), for example, manufacturing in Mexico, accountants in India, and engineering in Singapore. The subsidiaries are thus embedded within the larger MNC system, but are located in different host countries with different national cultures and political/legal systems. There are obviously many clashes and tensions that may occur when trying to balance the needs of the MNC against the subsidiaries, and this is

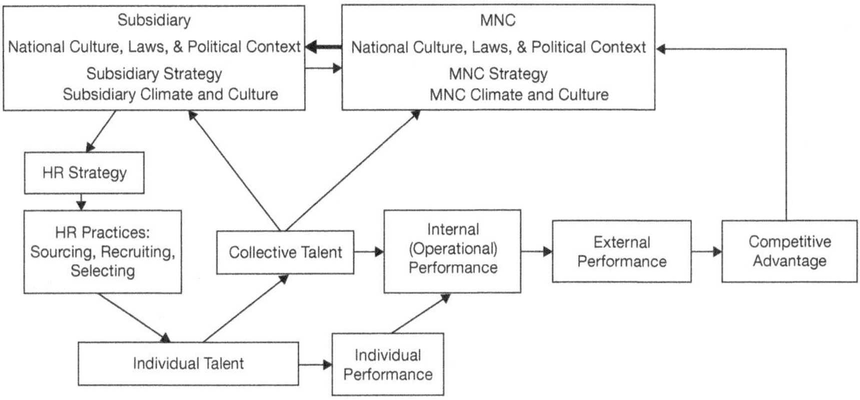

Figure 1.5
Global talent acquisition framework. For illustrative purposes, only one subsidiary is shown. The thicker arrow in the figure suggests the MNC influences the subsidiary more greatly than the subsidiary influences the MNC.

indicated by the different sized arrows in Figure 1.5. That is, the MNC has a stronger and more direct influence on the subsidiary.

Second, the figure implies that national culture, legal context, and political context influence MNC/subsidiary strategy and culture, which in turn influence the nature of HR strategy and practices used within a particular location. Importantly, the HR strategy within a given location is more directly shaped by the subsidiary than the MNC. This highlights the fact that all strategy is locally implemented.

Third, HR strategy and practices influence the way the firm will source, recruit, select, socialize, develop, and retain talent. Use of effective HR practices contributes to the development of different types of talent, first individually and then collectively. Collective talent is the combination of individual differences in cognition, personality, and values (described earlier). Most of this book is focused on explaining how to build individual and collective talent through staffing.

Fourth, collective talent is a resource that drives the firm's internal (operational) performance, external effectiveness, and competitive advantage. Thinking of talent as a resource is important because it changes the way we think about staffing. For example, in chapter 2 we introduce a talent supply chain framework to help understand how firms can best acquire and develop the talent resources needed for performance and competitive advantage in a changing, dynamic environment.

Fifth, we distinguish between different types of performance. *Operational performance* is primarily *internal* to the firm (e.g., productivity; managerial accounting metrics) and is more affected by HR activities. *External performance* is performance that is jointly affected by internal and external factors and includes market share and financial metrics (e.g., Tobin's q, stock price). *Competitive advantage* is a condition where a firm generates above-normal returns relative to competitors (e.g., sometimes also referred to as goodwill).

Hence, this framework is one that shows how to translate organizational strategy into action, and in turn, how to use staffing to transform individual differences into strategically valuable talent resources that generate competitive advantage.

STRUCTURE OF THIS BOOK

In subsequent chapters we describe four themes that provide continuity across all chapters:

Theme 1: Talent Acquisition and the Global Environment

This book is about global talent acquisition, so it will obviously have a heavy focus on issues that challenge global firms (although firms operating locally should also find value in the concepts that follow). These issues include culture and values, language, political systems, legal parameters, and demographics. However, this book does not go into great detail around specific country's political systems, laws, and so on. There are simply too many differences to cover them in any meaningful manner. Instead, we focus on how firms may think about and frame such specific issues. We offer frameworks to help managers structure and solve the types of challenges involved with global talent acquisition. These frameworks are applicable to any global organization, regardless of which country is home and which countries are hosts. We consider these topics from the perspectives of MNCs, subsidiaries, and local organizations.

Theme 2: Talent Acquisition and Business Strategy

Talent acquisition, along with other aspects of the HR strategy, must be clearly linked to the firm's strategy. Talent acquisition should form a symbiotic relationship with strategy. Global staffing should reflect strategy and it should inform strategy.

Theme 3: Talent Analytics and Metrics

We live in a world of big data, but in our opinion, many firms do not know how to handle this data. We advocate an approach to big data that is grounded in talent analytics and anchored to a bedrock of rigorous scientific research. We propose metrics that are connected to firm strategy and that are reasonable targets and deliverables for managers tasked with global talent acquisition. For example, performance, productivity, effectiveness, profit, loss, cost, diversity, speed, quality of hire, and competitive advantage, among others, are metrics that make sense for some contexts and not others. We consider these metrics and talent analytics throughout the chapters.

Theme 4: Application and a Human Touch

Years of teaching staffing to undergraduate and graduate students, working with managers (HR and non-HR), and researching staffing topics, have taught us some important lessons. One of the most critical lessons is that people remember stories better than cold facts. A good story hooks the reader and presents the material in a way that he or she can relate to and understand. Another important lesson is that describing staffing and implementing staffing are two very different things. Staffing is a technical HR function, but it is implemented by people and through people. Therefore, to make the material more interesting, memorable, and actionable, we have created a business case that, while fictional, is based heavily on reality and real-life experience. We introduce Roger Wells, a young HR manager who is tasked with moving from England to develop and implement a new regional staffing plan in Southeast Asia. Roger's story mirrors the factual topics in the book, and Roger's challenges pose many questions that readers and instructors may want to use in their class discussions. This is a book intended to teach the practice of staffing, and so walking the steps with a new manager who is struggling to implement a global staffing model should help readers better appreciate the challenges of implementing global talent acquisition systems.

2

Global Talent Strategy

THIRTY-SECOND PREVIEW

- Strategy is about making choices to differentiate a firm from its competitors.
- Talent should contribute to both the formation of strategy and the implementation of strategy.
- The key issue facing global talent strategy is the tension between local customization versus universalistic perspectives.
- There are three main types of firm performance outcomes: competitive advantage, external firm performance, and internal (operational) firm performance.
- The value of talent is determined by the nature of performance. Different types of firm performance will make different types of talent more or less valuable.

- Collective talent resources may be managed in terms of (1) quantitative and qualitative components, (2) stocks and flows, and (3) causal relationships.
- A talent alignment process enables connections from strategy to talent and identifies which types and quantities of talent are needed, now and in the future.
- A global talent supply chain framework enables leaders to manage talent alignment across borders.
- The Bottom Line: Global talent acquisition should differentiate the firm in a way where it can generate above-normal profits and sustainable growth.

Roger's first meeting with Ibrahim Abdul Hassan had not gone particularly well. Ibrahim had studied electrical engineering at MIT (Boston) and had enjoyed his university years in the United States. He had mixed well with students from all nationalities, of which there were plenty at that diverse and high-quality university. He had taken up golf and found that he had a natural talent, and spent time while on midterm breaks to travel the country. When he returned to his native Malaysia, he spent a couple of short spells working in heavy engineering contracting and also with a department of the Malaysian government that focused on developing local technical talent. However, neither of these positions really connected with his passion for advanced electronics, so when he saw an advertisement for Spooner Electronics, he was intrigued. He interviewed and was offered a role in their technical group and started to progress through the company. He had an eye for talent and had become a respected "pakcik"—literally "uncle" in the local language, but more symbolic of the type of benevolent father figure that he represented to many young local employees. The company recognized this and decided to groom him to become the VPHR for S.E. Asia and the country HR manager for Malaysia, an opportunity he seized with both hands. "A chance to make a difference," he thought.

The company sent Ibrahim to England for a year's cross-posting to expose him to the various corporate elements of human resources (HR) and build a platform of skills, but this had proved something of a turning point in his thinking about a number of things, particularly and ironically about the use of expatriates. He had also been there at a time that the company was in the initial stage of its thinking about moving toward a regionally based organization (i.e., coordinating country-level operations within a broader regional management structure). He was fiercely patriotic and proud of the quality of Malaysian talent and was largely unconvinced that any form of regional organization would actually add value. The long histories that Malaysia had with some of its neighbors, such as Indonesia and Singapore, also influenced his views about the likelihood of a regional operating model working effectively. They shared some of the traditional values that he espoused but had approached some of their internal issues related to the various ethnic groups in very different ways. He did not believe that the very real differences between the different countries in Asia were appreciated by the leaders in London, who he saw as painting a broad demographic view of the region as though it were one country. This amused him somewhat. "I wonder if I described Europe as a united place that saw the world in the same way, they would think me as ill-informed," he thought. "History shows that the French, British, and Germans have rarely agreed about anything, as multiple wars through the centuries had demonstrated."

Part of the way that Spooner decided to engage Ibrahim's allegiance to the new model was to give him a regional brief in addition to his country role. He was the HR lead for the region and in this role supervised Roger—but his real allegiance continued to be slanted to his local accountabilities.

All these thoughts clearly played into his first meeting with Roger, or at least that was how it appeared. Roger had introduced himself to Ibrahim in the manner that he had read was appropriate based on reading books and watching some videos he had found on the Internet. He had shaken Ibrahim's hand softly and then touched his own chest—a sign of my heart to yours. Ibrahim had obviously noticed that, as he had the way in which Roger had waited to be shown where to sit, and had engaged in casual "get to know you" conversation rather than dive straight into business. "So far so good," thought Roger. But then Roger broached the subject of managing staffing on a regional basis, moving high potential staff from one part of Asia to another to improve the quality of cross-border experiences without the expense of assignments to Europe. Although Ibrahim communicated in a soft-spoken, even indirect, manner, the message behind his response was very clear. He did not support the notion of cross-border talent development and clearly did not think that Roger, an expatriate with no prior experience in Asia nor in staffing, was in a position to "make improvements" to how it was being done presently. Ibrahim's perspective appeared to Roger as being clearly rooted in the notion that Malaysians knew what worked in Malaysia, Singaporeans in Singapore, Filipino's in the Philippines, and so on.

"This is going to be more difficult than I thought," Roger said to himself as he left Ibrahim's office. "Ibrahim and I have very different opinions on this issue. If I am going to get my vision implemented, I need to make a strong business case for how my plan helps advance the company's strategy, while staying true to local and regional goals and values."

A firm's strategy should set the tone and direction for all subsequent plans, goals, and activities within the organization. Strategy is the force that connects and unites diverse organizational functions and people together to achieve common goals. Strategy defines how a firm is competitive and makes money (or achieves alternative outcomes, such as with not-for-profit organizations). Strategy defines what the firm values and what it does not value, what it is going to do and what it is not going to do. Strategy defines the nature of the relationships between employees, the organization, and other stakeholders. These points are great in theory, but as the saying goes, the devil is in the details. Conceiving strategy and strategic goals is relatively straightforward (we didn't say easy!). Implementing strategy is monumentally more difficult. A company like Google may have as its mission, "to organize the world's information and make it universally accessible and useful," but *how* it does this requires innumerable decisions and judgment calls. Organizational strategy informs those judgment calls.

For example, a leader within Shell had been recruited from a fast-moving consumer goods industry. After one year, he was asked what impressed him negatively or positively about Shell compared to his past employers. He said that one thing stood out and that was the approach to problem-solving and decision-making (the following figures are illustrative):

> When making a decision to construct a new bottling plant, which may cost $50 million, you need to do your due diligence and assess the likely Return On Investment (ROI) of that, where to locate it, et cetera. But if you get it wrong, the impact taken in context of the overall financials of a large company is significant but not company breaking. Speed was more important, because if someone did not drink a bottle of soda one day, they were not going to drink two the next day to make up for it! Our decision-making processes reflected that, and not just in business-centered issues. In HR we looked to be decisive and use judgment in developing the best appropriate solution but not languish forever getting each decision perfect. In Shell, if you make a decision to build a liquefield natural gas plant, enter a project like the Sakhalin project in eastern Russia, build a major offshore platform in the Gulf of Mexico, et cetera, these are multiyear multibillion-dollar commitments, and getting them wrong can have major impact on long-term profitability. So you make sure that you have examined every angle, all the data that you can possibly gather, test potential decisions against multiple possible scenarios, et cetera, before finally bringing the project to a final investment decision (FID). My issue is that the same rigor is imposed on decisions in the HR field, where the consequences of

getting it wrong may not be desirable, but they are neither financially disastrous nor life threatening.

What is appropriate rigor for one set of decisions does not mean it should permeate every decision.

Professor Patrick Wright, our colleague and a leading scholar of strategic human resources, sums it up nicely: "HR should be driven *for* the business, *not by* the business!" Human resources and the staffing function are tasked with implementing a firm's strategy through its people. The HR functions strategy, policies, and practices should be directly aligned to the firm's strategic goals. It should be clear how staffing contributes to the firm being successful and differentiated from competitors. Who, where, when, and how to hire should be informed by a firm's strategy. Yet we see few organizations with such clear alignment. In fact, in many of our experiences, HR is often embattled with other functions. A report by McKinsey (Guthridge, Lawson, & Komm, 2008) found that line managers often believe HR is misaligned from the firm strategic goals, or at least not held accountable for them. In contrast, HR managers tend to believe their efforts are closely aligned with business goals. Hence, there are large differences in the perception and reality of strategy implementation.

These issues are exponentially more complicated when thinking about strategy and staffing in global firms. Multinational corporations (MNCs) will have a corporate headquarters, usually in the *home* or parent country. However, they will have multiple subsidiaries located in *host* countries. The management of these subsidiaries will to varying degrees be affected by differences in cultural values, governance structures, and legal considerations. Further, the staffing of MNC and subsidiary operations may comprise a mix of parent country nationals (PCNs; employees from the home country), host country nationals (HCNs; employees in the host country), and third country nationals (TCNs) who are from neither the parent country nor the host country. The multiple perceptions around this may be rooted in a number of things, but at their heart is a fundamental divide. The MNC headquarters will be seeking to optimize the use of talent across the enterprise. This often results in an efficiency-driven approach, where consistency in policy and practice is encouraged to control costs and complexity. This will contradict a belief held in the host countries that each country is different and to be locally relevant and effective requires a country-by-country approach to staffing. The trap here is that both perspectives have some validity. The most successful companies find ways to optimize their talent acquisition processes across the enterprise *and* be relevant and effective locally. In this chapter we consider these various issues and present a variety of frameworks and approaches for structuring the challenges in a manner that can lead to solutions.

In chapter 1 we noted how talent is a valuable resource. As a resource, talent is a capacity for action (Ployhart, Nyberg, Reilly, & Maltarich, 2014). Regardless of whether they are individual employees or collectives, talent resources are potentially valuable, rare, and costly to imitate and substitute. These characteristics are what give talent the ability to generate sustainable competitive advantage. Therefore, it is important to move away from thinking of talent as a characteristic of a person and instead think of talent as a resource that can be deployed for organizationally relevant purposes. This elevates talent to a degree of importance similar to financial capital, natural resources, and organizational brand. If managers think about talent only in terms of characteristics of individuals, then talent's value leaves when the employees leave. But if talent is thought of as a collective resource to be managed, then talent exists beyond the tenure or employment of any one person or group of people. This dramatically changes the way we think about global talent strategy, because it broadens the perspective from a focus on a small number of individual expatriates to the management of a renewable collective resource.

Figure 2.1 provides an overview for thinking about talent in terms of strategic resources. First, a firm's strategy, goals, and policies determine the relevant performance metrics. Second, the relevant performance metrics determine which kinds of talent resources will influence those metrics (i.e., are valuable). Third, knowing the needed strategic talent resources enables one to source the talent wherever it is situated. Finally, understanding cultural differences allows a firm to strike a balance between universal MNC staffing practices and subsidiary/culture-specific staffing practices.

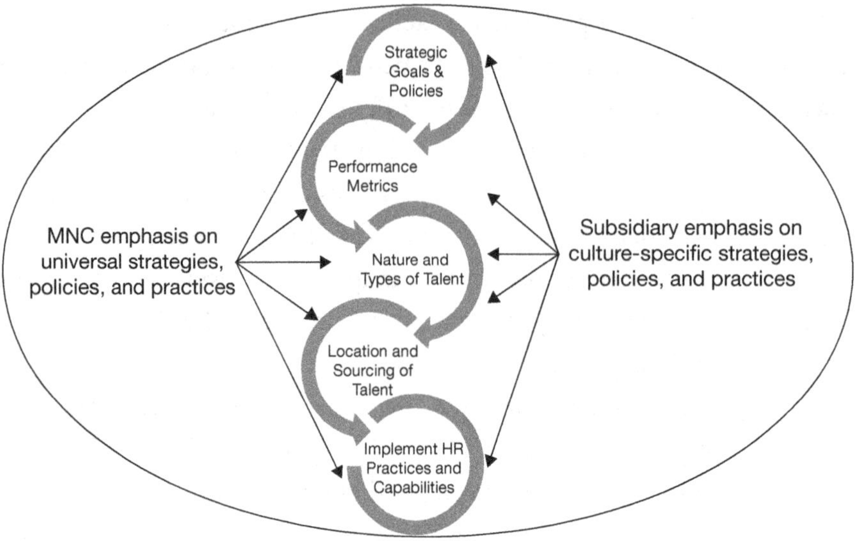

Figure 2.1
A strategic resource-based framework for talent, nested within the tension of MNC-focus versus Subsidiary focus.

Thus, in contrast to many approaches that start with HR practices and move toward performance, our approach starts with firm strategy and performance and then moves toward practices—all while balancing MNC and subsidiary goals.

Roger had a bit of time to attend to personal matters. He spent a day looking for apartments somewhere near the offices of Spooner Electronics, which were in Damansara Heights. Roger was single, and most of the people who lived in the area close to the offices were married with children so he had managed to find a nice two-bedroom apartment that was a short commute and would still give him reasonable access to the livelier parts of the center of Kuala Lumpur. He would move in a week or so, but for now was sitting by the pool at the Shangri-La Hotel, which was located in the "Golden Triangle" of the city, close to nightlife, great restaurants, and one of the major sights of the city, the Petronas Twin Towers, one of the world's tallest buildings, which dominated the skyline.

Sipping on one of the local beers, Roger typed on his laptop some key questions that were on his mind after this meeting with Ibrahim.

1. What did "regional" talent look like? Was talent broadly available in the regional market? Were other companies struggling with the same dilemma? Advice from Peter George, a former colleague in the UK, came to him—"when faced with a problem, start with looking to see if others had solved the same problem before." Is there guidance on this issue? How can I leverage local networking?
2. Assuming that the talent did indeed exist, at least in nascent form, what was the value proposition that would differentiate Spooner Electronics from other companies who were making their way in this somewhat unfamiliar new world? Was there even agreement internally about what that looked like? Different phrases and options came to his mind. "Job for life." "Playing with the coolest technology." "The coolest job you ever held." "Work hard, have fun, make a difference—at Spooner you really can have it all." But these felt more like buzzwords or a cheap TV advertisement. What *was* the Spooner difference? A compelling way to differentiate Spooner was key, and this would require serious attention from regional thought leaders.
3. How would his regional HR leadership team blend the cultures of the various countries in some way? He had talked in London about how other regions had tackled the issue, but they all felt very Eurocentric and he was unconvinced that the "command-and-control" approach would work in this part of the world, where consensus was such an important element. For sure, the Hofstede work describing the high power distance ratios of many of the countries would come into play, but that was more likely to produce compliance and he really wanted—in fact needed—commitment.

4. He was to be in charge of staffing in Asia, but it was readily apparent from initial briefings that there were two major dimensions to that challenge—what he would call differentiating talent and high-volume staffing. The focus of differentiating talent would need to be on quality, with the cost of each hire being subordinate to the quality of the outcome of the hiring process. Differentiating talent would have to focus on talent that would truly move the business and make Spooner move toward a top-tier market player. But what were these strategic jobs? While the quality of high-value staffing could not be ignored, the primary goal would be to leverage scale across the region to drive down the overall cost to the company.
5. He needed to understand how Spooner made money in the region, and this required him to recognize the key performance metrics used to operationalize Spooner's global and local strategy. Roger was familiar with the corporate performance metrics and who was accountable for them, but he needed to understand how the regional metrics tied back into the global metrics.
6. The final issue was perhaps the most difficult one he had to wrestle with. He had to find a way to convince his new colleagues of the benefits of moving more rapidly away from the country focus and embracing the new regional model. He knew that there was the natural resistance to initiatives that emanate from HQ and the questioning of their applicability in the local context. It had been plain to him in London that this challenge was a key test of his ability and was a critical step in more substantial moves that were afoot to leverage scale and the manner in which the business was run.

So there was much to ponder, and Roger knew he would need a high-quality team to tackle it. And there had to be some magic "sauce" that would serve multiple issues. His team had to be constructed with an eye to ability as the first parameter, but it also needed to have chemistry, have the experience necessary to understand the region's complexities, and be built in a way that worked to build bridges across boundaries, not walls that defended them.

Roger closed his eyes and let the rays of the early evening sun massage his mind as he prepared for the journey that lay ahead.

STRATEGY IS ABOUT DIFFERENTIATION

Figure 2.1 suggests that a firm's strategy and its strategic goals set the direction and tone for all subsequent talent activities. Firms can of course adopt a variety of different strategies, and these are usually based on their existing resources relative to perceived market opportunities and competitive pressures. That is, strategy may be set based on internal (e.g., resources) and external factors. However, most strategies are ultimately seeking to create value for stakeholders. Value creation will in turn most frequently be created through a process of differentiation. Differentiation is what enables a firm to generate above-normal returns and create more value. For example, a firm may charge price-premiums because of perceived product quality or superior service. Therefore, strategic actions are, at their core, actions designed to differentiate the firm in its relevant market.

Differentiation may be generated through the creation of new products (e.g., introduction of the iPhone), improving existing products (e.g., electric cars), superior customer service (e.g., Singapore Airlines), or price (e.g., Walmart). The firm's strategy will be the "vision" through which differentiation is expected to occur. Resources that a firm controls or has access to provide the means through which differentiation occurs (Barney, 1991). A key task for firms is thus to protect those resources from imitation or substitution. Talent fits into this formula because many strategies in the modern economy are heavily dependent on talent, and talent is itself a resource that is difficult to imitate and for which few substitutes exist. Thus, one must create a talent strategy that connects talent resources to performance.

WHAT PERFORMANCE OUTCOMES OR METRICS ARE RELEVANT?

Thinking of talent as a resource means talent can be used to conceive of, implement, or change strategy (Barney, 1991). The question is, *how* does this occur? How can hiring employees, staffing teams, or choosing key managers contribute to successful strategy implementation? The answer is found in a simple principle: *The value of talent is dependent on the types of performance outcomes to be influenced because these outcomes operationalize the firm's strategic and tactical goals.*

Aligning talent resources to the performance metrics used to operationalize and monitor strategy execution is challenging. A change in a firm's strategy will result in different performance metrics becoming more (or less) important. This, in turn, could mean that what was once a critical type of talent is supplanted by a different type of talent. For example, in 2002, Shell Oil acquired Pennzoil Quaker-State for a considerable amount of money. A financial analyst even went so far to say that acquiring Pennzoil moved Shell into a new and highly competitive environment because they were in different businesses (i.e., consumer products and oil, respectively). This was a fair commentary (see Banerjee, 2002). The dynamics of a large multinational integrated oil company are very different from those required to compete in the field of fast-moving consumer goods. And that realization prompted a serious examination of the existing lubricants talent and a conclusion that some significant external hiring of different talent was required to be able to compete successfully. Targeted hiring from companies like Coca-Cola and Proctor and Gamble became a significant focus.

The value of talent is thus determined by the nature of performance. This means talent can't be approached with a "get it and forget it" mindset. It is critical to know what makes the business money, how it makes money, what metrics are valid leading indicators of successful performance, and thus what kinds of talent are needed to influence those indicators. Easier said than done, but we've developed a performance framework to make this process more tangible. Figure 2.2 provides an overview of this performance framework. There are several important distinctions.

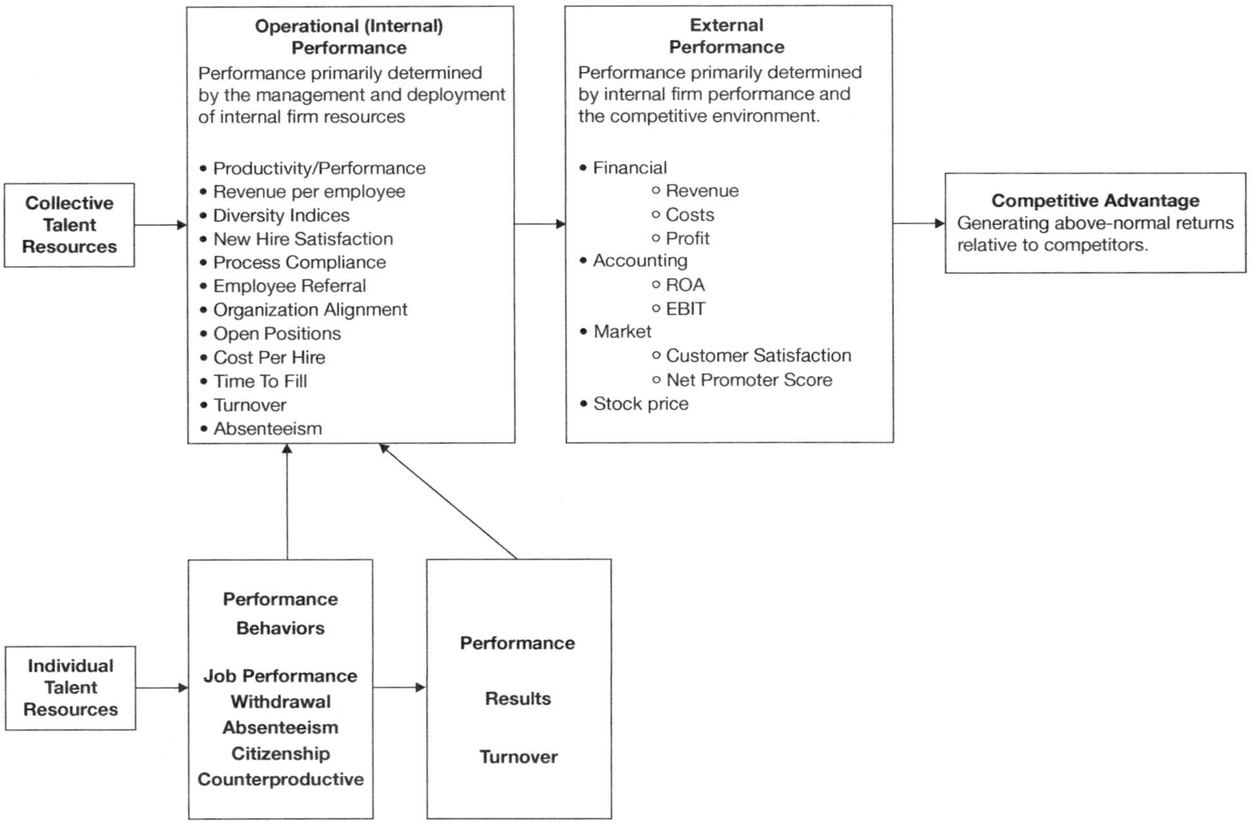

Figure 2.2

Performance management framework. Note that items in the boxes are only examples.

Performance Exists at Multiple Organizational Levels

The framework recognizes important differences between individual performance outcomes and collective performance outcomes. Individual-level outcomes are those focused purely on a person's job performance (and also includes absenteeism, accidents, turnover, etc.), whereas collective performance outcomes are those focused on group, strategic business unit, subsidiary, or firm-level performance. Individual and collective performance outcomes are related but are not identical or interchangeable. Individual performance outcomes are usually easier to change, but they are imperfect indicators of collective performance outcomes, and usually underestimates of it as well. In other words, the measure of collective performance is not the same as the sum of individual performances. For example, the costs of collective turnover can be much larger than the costs of individual turnover when employees need to work interdependently. The reason is because when people work collaboratively, one person's turnover creates disruptions that affect coworkers (e.g., reduces cohesion; creates extra work for those who remain). Subsequent integration of new members can also slow down work production and change group dynamics—not always for the better. Thus, one cannot calculate the firm-level costs of individual turnover based on the simple aggregation of individual turnover.

It is important to manage both individual and collective performance because different talent resources may be required for each type of outcome. For example, intelligence and technical knowledge may be critical for individual performance, but personality and shared knowledge may also be critical for collective performance. Furthermore, simply enhancing individual performance may not translate into greater external collective performance (e.g., profit) or competitive advantage. Individual and collective performance can have different talent resource determinants, and it is necessary to manage both types of performance in a coordinated manner.

There Are Different Types of Collective Performance

It is helpful to distinguish between internal (operational) performance, external performance, and competitive advantage. *Internal (operational) performance* refers to a firm's operational efficiency, productivity, and related outcomes that are influenced primarily by how well a firm manages and deploys its talent resources. Included within internal operational performance are managerial accounting metrics, as well as HR metrics such cost-per-hire, sales-per-employee, or hours of internal and external training per employee (this last is a measure of activity, not impact). *External performance* refers to a firm's standing on financial, accounting, or market-based metrics. External performance is influenced by political, economic (e.g., recessions, unemployment) and competitive factors external to the firm, as well as performance internal to the firm. *Competitive advantage* is a special type of external performance and occurs when a firm is able to achieve supranormal or above-normal returns, relative to competitors (Peteraf & Barney, 2003). Obtaining a competitive advantage requires differentiation from competitors.

There Are Different Types of Individual Performance

Performance behavior is what employees do, and results are the consequences of those behaviors. There are many different types of performance behaviors. As described in Ployhart, Schneider, and Schmitt (2006), common examples include task performance (e.g., technical performance, teamwork), citizenship performance (e.g., voluntarily helping others and supporting the organization with discretionary behaviors), adaptive performance (e.g., flexibility), and counterproductive performance (e.g., theft, sabotage). Results are the consequences or outcomes of behavior (e.g., providing better customer service increases customer purchases).

The Effects of Talent Are Both Direct and Indirect

Internal performance is a strong determinant of external performance and—indirectly—competitive advantage. However, recognize that internal performance is not the only determinant of external performance. External factors, such as the actions of competitors, will also influence external performance. There are two implications: talent is likely to be more strongly related to internal than external performance, and talent may influence external performance *indirectly* through its effects on internal performance. This may seem obvious, but it is important to remember and reinforce when we are thinking about talent. Too many HR departments focus on internal HR-focused metrics. These may be useful for gauging the efficiency and effectiveness of the HR function, but they are often not very useful for demonstrating the strategic value of talent. Take cost-per-hire as an example. Cost-per-hire implies that talent is a cost that is to be minimized. While minimizing talent acquisition costs is always important, it is more important to spend as much as justified to obtain a return on talent. A sole focus on reducing cost-per-hire may result in not obtaining the best talent—only the cheapest. Imagine if professional sports teams based their staffing decisions on cost-per-hire!

National Culture Can Influence Which Types of Performance Matter Most

Understanding that talent relates to different performance metrics in varying degrees has important implications for understanding cultural differences in the management and strategic value of talent. For example, firms from cultures that take a short-term orientation are likely to favor metrics that show fast returns on investments (e.g., short-term profits; change in stock prices). Those from more individualistic cultures are more likely to evaluate and reward individual rather than team performance. As culture is likely to influence firm strategy in subtle (and sometimes not so subtle) ways, one should be cognizant that firm strategy is affected by culture and thus the performance metrics likely to be favored. We return to this topic in greater detail toward the end of this chapter.

Conclusion

The value of talent is dependent on the performance metrics that operationalize the firm's strategy. Therefore, every attempt should be made to link talent to the strategic outcomes that operationalize the firm's strategy. One should also realize that strategy permeates the way talent is valued and managed. The direct effects of talent will be, all else equal, strongest for internal performance, moderate for external performance, and possibly small for competitive advantage. Yet, the indirect effects of talent are important because these effects operate through other variables in a causal chain such as that shown in Figure 2.2. Realizing that talent's consequences flow through this performance chain has three implications. First, both the direct and indirect effects of talent are important and interconnected. Second, talent will relate to some performance outcomes more strongly than others. The performance outcomes more closely connected with talent will be more strongly related than those outcomes farther away. Finally, it is essential to link talent to those metrics that are most important. Because a strong, direct relationship between talent resource and competitive advantage is unlikely, understanding the *causal process* through which resources have their effects becomes paramount.

A TALENT RESOURCE FRAMEWORK

We now turn to developing a framework for talent resources. Chapter 1 discussed different ways that talent has been studied. These included individual difference approaches from psychology, resource approaches from strategic management, and generic and specific human capital approaches from economics. We noted how there are different types of talent that reflect different psychological characteristics (e.g., cognitive, personality).

In this section we blend these perspectives to develop a talent framework that allows us to align talent to performance. Table 2.1 provides an overview of this talent framework.

Talent Quality

Talent quality refers to whether there is a sufficient capacity of the necessary type of talent needed to achieve a particular desired performance outcome. In general, the greater the quality, the greater the likelihood of enhancing performance. Talent quality may exist within an individual or within a collective. For example, individual engineers may differ with respect to their knowledge of calculus, and those with greater knowledge will generally be more effective in solving differential equations. Similarly,

TABLE 2.1 OVERVIEW OF THE FIRM-LEVEL TALENT RESOURCE FRAMEWORK

Quantity	Quality: Low	Quality: High (Quality Threshold)	
High	Weak effect on performance	Moderate effect on performance	
			Upper Quantity Threshold
Quantity	Moderate effect on performance	Strong effect on performance	
			Lower Quantity Threshold
Low	No effect on performance	Moderate effect on performance	
	Low	High	
	Quality		

a team of engineers may form a collective talent resource, such that a team with complementary skills and a willingness to work together will outperform teams that lack either skills or cohesion. However, there is a *talent quality threshold* that exists. This threshold functions as a tipping point. Below this threshold the quality is insufficient for performing the task; above the threshold quality and performance should be more linearly related.

Talent Quantity

Talent quantity refers to whether there are enough people needed to accomplish a performance outcome. Obviously, talent quantity becomes more relevant for collective talent resources. Tasks are performed by individuals, but increasingly tasks and work are structured in collaborative team-based structures. Sometimes quantity is referred to as a "critical mass" or "staffing level," but the implications are the same whatever the nomenclature. Too few employees means that performance on some collective performance outcome is likely to suffer—even if quality is sufficient. Too many employees means that talent is being wastefully applied, and in some cases too many employees will actually harm collective performance (e.g., increasing time to completion; infighting or indecision; diffusion of responsibility). Thus, there is a "sweet spot" of talent quantity—an *upper talent quantity threshold* and a *lower talent quantity threshold*—that must be present for talent to influence performance outcomes. This optimal zone for talent is a threshold that needs to be considered so that the desired quantity levels are specified. For example, accidents and mortality rates increase when hospital units are understaffed.

Talent Stocks and Flows

Recall in chapter 1 that talent can be mobile and malleable. This is important, because as a firm shifts to anticipate or respond to environmental and competitive pressures, the types of performance outcomes vital to strategy and success will likely change. In turn, the types of talent that were valuable one day can become irrelevant the next. For example, there is not much demand today for workers with deep knowledge of Fortran or COBOL, even though at one time having proficiency in these programming languages led to good job opportunities. The quality of talent may change over time, the quantity of talent may change over time, and they both may change over time. Talent is neither constant nor assured. Hence, it is helpful to distinguish between talent stocks and flows. *Talent stock* is the quality and quantity of talent at any given point in time. *Talent flow* is the change in the quality and quantity of talent over time. Both are important, but in different ways.

Talent stocks are helpful to determine where an organization is today. Talent flows are helpful for understanding where the organization may be tomorrow and how talent drives changes (e.g., growth, decline) in internal and external performance. For example, many firms in developed countries face an impending "brain drain," when scores of baby boomers who have decades of deep knowledge and experience begin retiring in large numbers. This exodus of experience is creating a strain on firms' talent management systems, as the flow of talent leaving key positions may be greater than the flow of talent replacing them. Nyberg and Ployhart (2013) refer to this as an *erosion* of talent. In contrast, ensuring that replacements or new employees are of higher quality than existing or former employees is known as an *expansion* of talent. One can review these inflows and outflows to predict when talent constraints or opportunities may occur, in much the same way that operations managers monitor supply chains.

Talent Interrelationships

Chapter 1 noted how talent resources can reflect different types that range from generic to specific. Generic talent resources are applicable to many different types of jobs, contexts, organizations, and cultures. Examples of generic resources include basic cognitive abilities (e.g., verbal, quantitative) and personality. Specific talent resources are those only applicable to a particular job, context, organization, or culture. Examples of specific resources include knowledge (e.g., employment laws in India), skills (e.g., speaking fluent Mandarin), or social relationships (e.g., knowledge of one's immediate coworkers). Research on strategic human capital resources (Ployhart & Moliterno, 2011) and individual differences (Hunter, 1983) suggests that talent resources follow a causal chain where generic talent resources determine specific talent resources (e.g., high levels of cognitive ability enable the acquisition of knowledge on employment laws in India).

However, if we consider that resources are a capacity for action, then it becomes apparent that the capacity or quality of generic talent resources will set the upper limit on the capacity (or quality) of specific talent resources. Thus, the causal relationships between different types of talent are important, because the joint relationship offers new insights into how talent will influence performance outcomes. The higher-quality generic talent resources a firm can acquire, the more likely it will be able to develop specific talent resources. Further, thinking of these causal relationships in terms of applicant flows helps one realize that it will take time to build a generic talent resource capacity. It will take additional time for this capacity to enhance specific talent resource capacity. Finally, combining the causal relationship between generic and specific talent, with the relationships among internal (operational) performance, external performance, and competitive advantage, leads to a model shown in Figure 2.3. This model illustrates the causal interrelationships among talent resources, showing how the capacity of generic resources may constrain the capacity of specific resources and thus performance.

To illustrate, Example 1 in the figure shows how reducing the capacity of generic talent can subsequently lower the capacity of specific talent and ultimately lower performance. Assume the capacity of generic talent is at 80% and specific talent 50%; and the effectiveness of internal (operational) performance 30%, external performance 10%, and competitive advantage 1%. We can use these estimates to examine the amount of slippage that occurs in other parts of the causal sequence, and hence how much value could be gained by enhancing talent resources. With generic

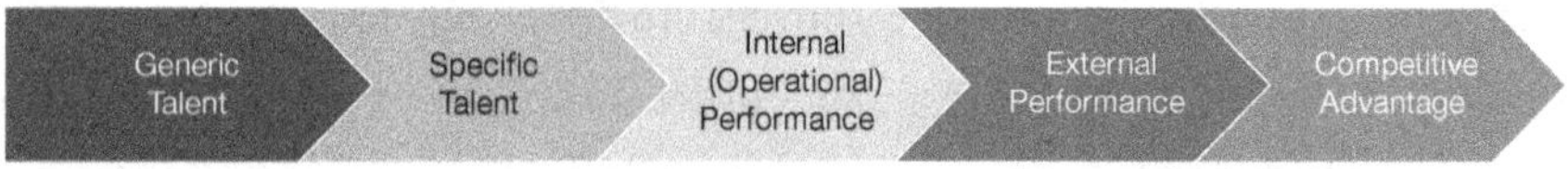

Example 1: How diminished capacity effects are cumulative. The solid line represents the baseline capacity of generic talent and hence the maximum possible capacity for all subsequent variables. Dashed lines represent the actual capacity for each variable.

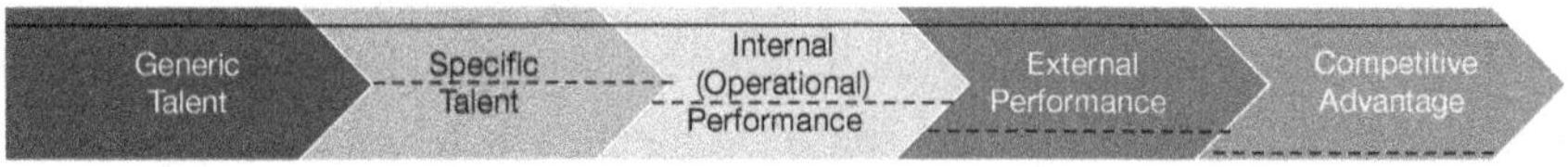

80%	50%	30% Percent Current Capacity	10%	1%	Capacity
80%	30%	20% Percent Under Capacity (relative to prior resource)	20%	9%	Direct Capacity Reduction
80%	30%	50% Percent Under Capacity (relative to generic talent)	70%	79%	Baseline Capacity Reduction

Figure 2.3
Causal interrelationships among talent and performance resources.

human capital at 80% capacity, the highest capacity possible for any subsequent resource or performance effectiveness metric is 80%. However, this is unlikely to occur due to slippage and inefficiencies in other parts of the causal chain. One way to consider this slippage is in terms of the *direct capacity reduction*. Under this model, we compare the capacity of a resource to the resource directly prior to it. For example, the direct capacity reduction for external performance is: 30% internal performance – 10% external performance = 20% less effective. Another way to consider slippage is in terms of the *baseline capacity reduction*, which compares any resource back to the baseline generic resource capacity. For example, the baseline capacity reduction for external performance is: 80% generic talent – 10% external performance = 70% under capacity. Thus, reduced generic capacity reduces all subsequent capacities to some degree, and this insight can be used to determine where to invest in increasing capacity (in this example, by reducing the slippage between generic and specific talent resources). Getting accurate estimates of these capacities can be difficult, but we have had good success using a combination of interviews with experts, surveys, and reviews of workforce planning data.

Conclusion

Conceptualizing talent as a resource helps managers understand how to acquire, deploy, and bundle talent for strategic purposes. As a capacity for action, talent can be considered in terms of quality and quantity. Too little quality or quantity lowers or even constrains a firm's performance. Yet it is important to think of talent quality and quantity with a temporal lens. Talent resource flows will generally be the strategic focus because they determine whether the necessary talent is being acquired in sufficient quality and quantity to enable superior performance. With the possible exception of mergers or purchasing another firm or business unit, changing talent resources will require a longer-term emphasis on enhancing the flow of talent (quality and quantity) into the organization. We return to the importance of talent capacities in later chapters, where it will be shown that firms with greater talent capacities will generate above-normal profits by building higher-capacity generic and specific talent resources, reduce costs by more effectively implementing strategy, and be more adaptive and flexible to changing competitive conditions.

"Talent as a capacity for action," thought Roger. This opened some new ways of thinking about talent within the regional focus he was struggling to understand. For one thing, it moved the conversation past simple headcount numbers, to focus on what talent really is—a strategic resource. As a thought experiment, Roger estimated talent capacities for some high-volume jobs in different countries, as seen in Table 2.2:

TABLE 2.2 QUALITY OF HIGH-VOLUME JOBS IN PACIFIC COUNTRIES

	Generic		Specific	
	Quantity	**Quality**	**Quantity**	**Quality**
Malaysia	Low	High	Low	High
Indonesia	High	High	High	Moderate
Philippines	Moderate	Low	Low	Low

Even though this was admittedly a quick and rough estimate, it made Roger appreciate just how strategic it was to choose between investing in generic or specific talent. However, it also made him realize that different countries had different talent capacities. In this manner, he could think about bundling or redeploying talent to different regions as a means to better diversify the talent portfolio. For example, he could try to encourage some employees from Indonesia to relocate to the Philippines, and then perhaps Malaysia as a means to build the quantity and quality of their specific talent resources. He found this interesting and decided to take it a step further by connecting it to the broader strategy.

ALIGNING TALENT TO STRATEGY

With a firm's strategy as the guide and performance metrics as the targets, it becomes possible to determine the degree of alignment between strategy and talent. Assessing alignment does two things. First, it enables one to examine existing talent needs and strengths (stocks). Second, it allows one to forecast talent needs in the future (flows). In our experience, three to five years is about the maximum one can reasonably forecast talent needs.

Our approach to talent alignment builds heavily on balanced scorecard approaches, most notably Kaplan and Norton (2004) and Becker, Huselid, and Ulrich (2001), but the underlying logic actually goes back much further to personnel psychology (e.g., Guion, 1961). The focus is to work backward from performance to talent, or "outside in" in the language of Ulrich et al. (2012). The idea is that strategy has identified the few, key performance metrics that will be used as targets for building a talent strategy. The task is to work from the performance metrics back to the kinds of talent resources that can most impact those metrics. Figure 2.4 provides an overview of this process.

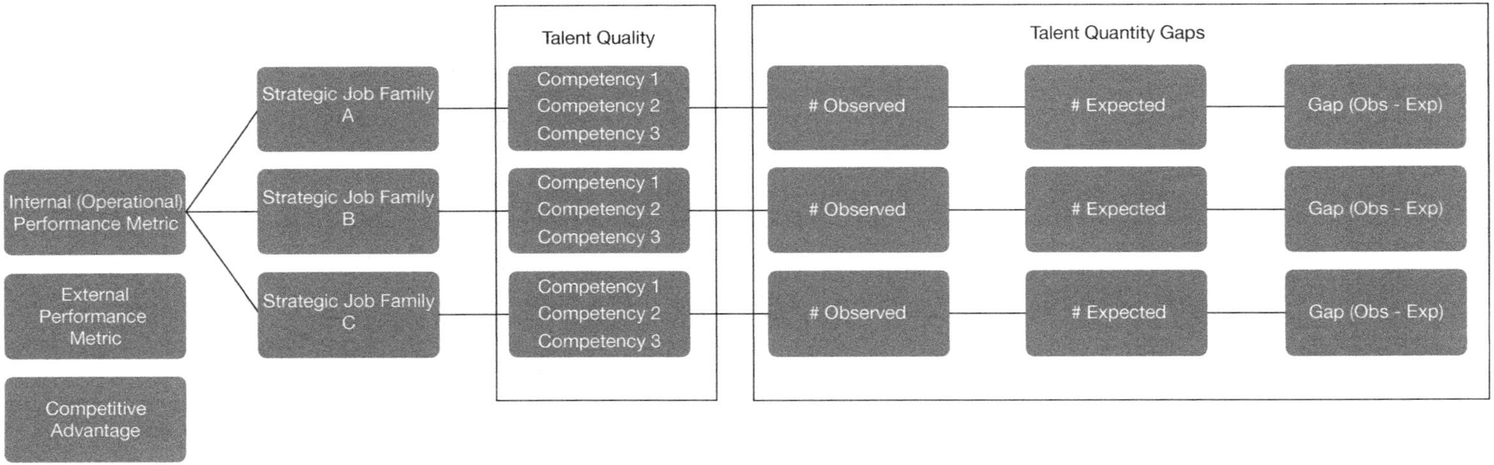

Figure 2.4
The talent alignment process. Note that only one internal (operational) performance metric is illustrated.

One first starts with thinking about how the firm creates a competitive advantage, then identifies the types of external performance metrics that contribute to competitive advantage, and ultimately the kinds of internal (operational) performance metrics that impact external performance. It is easy to let the list of key performance metrics grow long, so one approach we have found helpful is to focus attention on no more than three performance metrics at any one time. For example, in one firm the emphasis of management was on sales growth. The leading indicators of sales growth were (1) the sales funnel (the number, value, and maturity of proposals outstanding), (2) the renewal rate (on existing licenses), and (3) the R&D pipeline (where new products could boost sales). These three metrics could provide a reasonably accurate projection of *future* sales.

The second step is to identify the strategic job families that most directly impact the performance metrics. All jobs in a firm are important, but not all jobs are necessarily *critical* in terms of impacting the key internal (operational) performance metrics. For example, one might think that pilots are the most critical job family for an airline. While there is no denying that pilots are critical, performance in these roles does not vary much, so they are not necessarily the most critical if the key performance metric is customer satisfaction. In contrast, cabin crew jobs and ticketing counter jobs are most important for influencing the customer experience. Thus, for each performance metric, identify only those jobs where performance variation *most strongly* relates to performance change. Continuing the previous example, the strategically important jobs were in sales (driving the funnel) and customer service (impacting renewals), plus and a handful of jobs in product development. When making these judgments it is helpful to think of the *value* each job family brings to strategy, and the *uniqueness* by which the talent housed in that job is specific to the firm (Lepak & Snell, 1999). By juxtaposing value and uniqueness, one can identify which jobs are truly strategic (those high on value and uniqueness), versus those that are moderate (high value–low uniqueness or low value–high uniqueness), versus those that are not (low on value and uniqueness). It is helpful to start with no more than three strategic jobs, per performance metric, as well.

The third step is to identify the critical types of talent necessary within each of the strategic job families. This question relates back to talent quality discussed earlier. These types of talent may be specific to a job, or may cut across jobs. There are multiple ways to acquire this information. Often this information will be based on existing job analyses or competency models. Other times, job or competency information may be incomplete (this will nearly always be true when planning into the future). In this situation focus should be on *projected* talent competencies needed, given different economic scenarios. We describe these approaches in the next section of this chapter. The outcome of this step is the identification of a talent quality threshold within each strategic job family.

The fourth step is to estimate the number of people needed to have the minimum talent quality levels identified in the prior step. Hence, this fourth step focuses on the quantitative component of talent resources for each strategic job family. This estimate obviously requires some judgment about staffing levels, both in the present and in the near future. Once these target expectations are set, one needs to identify the number of people who actually have the necessary competencies within each strategic job family. Again, if there is an existing talent management process or competency model in place, then it may be possible to leverage this existing information to determine whether there is a sufficient quantity of talent. For example, it is not uncommon for professional services firms to know the average amount of revenue that can be supported by a single lawyer or accountant. This, in turn, can be used to project the number of lawyers or accountants needed to meet expected demand. If such systems are not already in place, then managers and line directors will need to estimate the necessary numbers of people with the needed talent within each function. The outcome of this step is the estimation of talent quantity thresholds for each strategic job family.

The final step involves identifying talent gaps within each strategic job family. The task is one of comparing the number of people who meet threshold on the critical talent competencies to the number of people who are needed. If this number is positive then there is a surplus; if this number is negative then there is a deficiency. Plans can then be developed for reducing the surplus or deficiency. However, these gaps are obviously estimates of talent stocks at a given point in time. Such "stock statistics" are valuable, but more insight is likely to be gained by understanding talent flows. Therefore, considering historical data or creating future scenarios to model talent flows and determine if a surplus or deficiency in talent is growing over time is advisable.

This basic model must be adapted to think of talent alignment in a global firm. One approach is to simply duplicate this process at the subsidiary level, and hence determine talent alignment against each subsidiary's performance metrics. We call this *emic alignment*, because it ensures alignment only within the local subsidiary and does not necessarily consider talent alignment to the broader MNC (see Figure 2.5a). In contrast, a second approach is to modify the procedure just described into what we call *etic alignment*. In this approach, strategic job families are replaced or supplemented with subsidiary talent (see Figure 2.5b). Some MNCs may have similar strategic job families in multiple locations, while other MNCs may have a different strategic job family in each subsidiary location. The etic alignment approach is, in effect, transforming the "two-dimensional" structure proposed earlier into a "three-dimensional" structure, because location and subsidiary operations are overlaid across the broader talent alignment process. Figure 2.5 provides a contrast between the emic and etic talent alignment processes. Notice that in Figure 2.5b, the only strategic job families that influence the MNC performance are Jobs A in the first subsidiary and Jobs A and C in the second subsidiary. This illustration highlights the fact that only a few job families within any given subsidiary are strategic at the MNC level. The most important implication is that strategic jobs are likely to differ between the MNC and subsidiary levels.

The etic or emic talent alignment processes enable organizational leaders to understand how the quality and quantity of talent they can access contributes to the firm's performance metrics and differentiation strategy. They identify areas of surplus and deficits, both as a stock and a flow. Modeling different scenarios allows further insight into the stock and flow of talent.

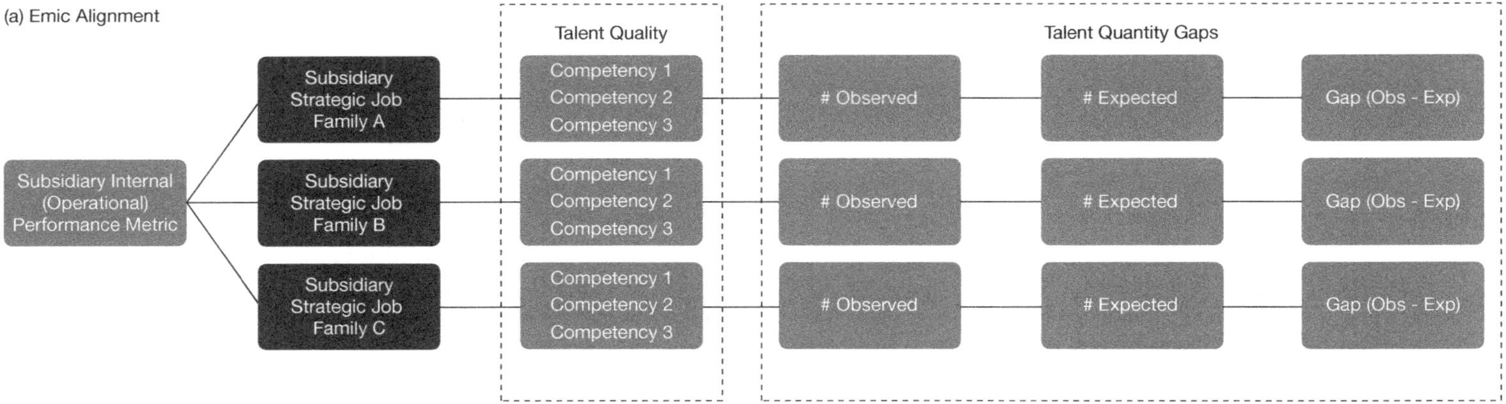

Figure 2.5
Emic (a) and etic (b) talent alignment processes. Note that only one internal (operational) performance metric is illustrated.

(b) Etic Alignment. The darker lines indicate those strategic jobs (in black) aligned to the MNC internal (operational) performance metric.

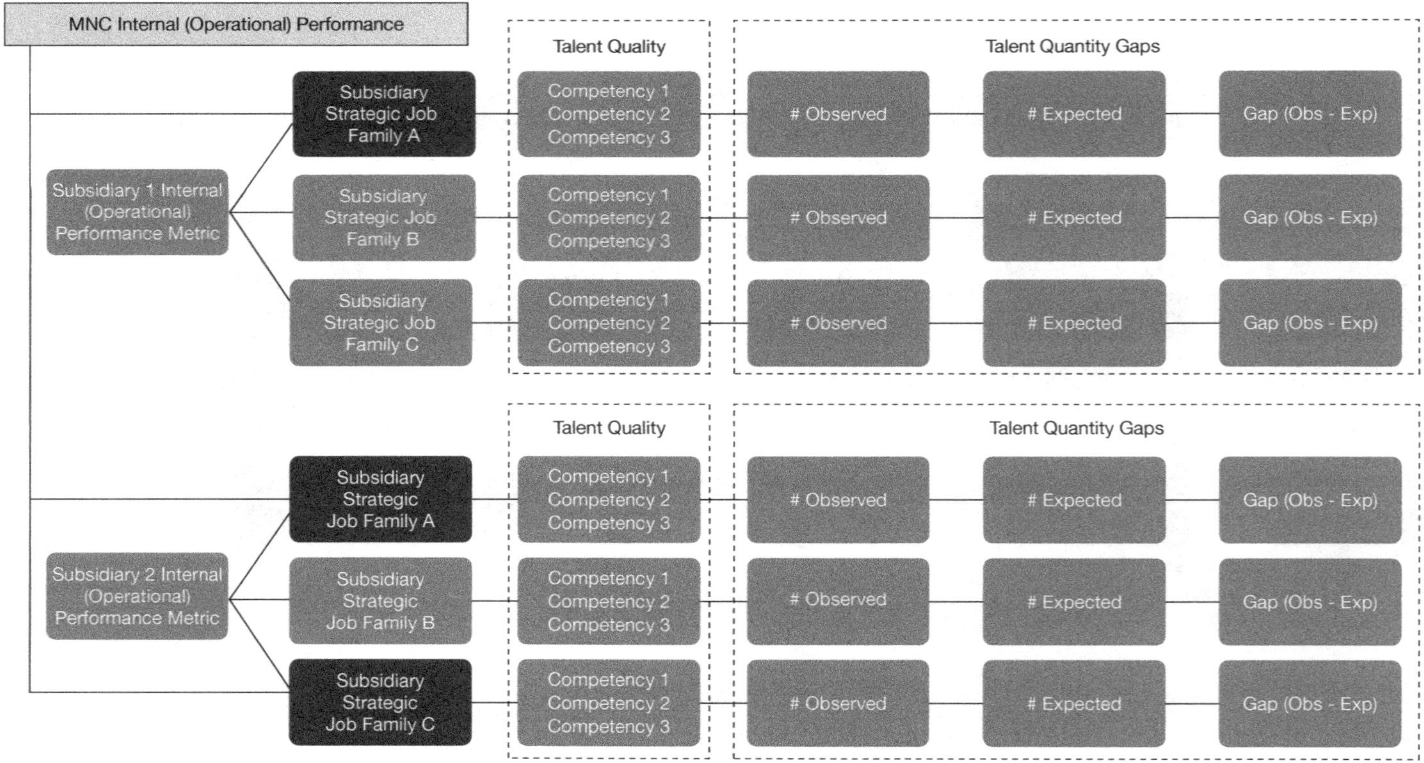

Figure 2.5
Continued

DETERMINING THE NATURE AND STRATEGIC IMPORTANCE OF JOBS AND TALENT

A critical step in the talent alignment process requires the identification of strategic job families and the strategic competencies within those families. This obviously requires an understanding of the nature of jobs and how these jobs contribute to strategic goals and performance metrics. Such an understanding is critical not only for determining whom to recruit and hire but also for determining training and development needs, succession planning, setting compensation levels, and talent management. This process requires a lot of judgment, but there are ways to structure and even quantify the process to increase objectivity.

Job analysis is the process used to develop a clear understanding of the important tasks and talent requirements for different jobs. Nearly every job analysis will start by reviewing existing content and documentation about the job. If such information is lacking, an excellent place to get baseline information about the occupation is the US Government's Occupational Information Network (O*NET; www.onetonline.org). This information is freely available and contains task and talent specifications for nearly every occupation in the US economy (although developed within the United States, the system is applicable for jobs in many different countries and cultures). The next step requires understanding the specifics of a given job. There is no one best way to do a job analysis, and there are many variations (Ployhart et al., 2006). For example, some approaches will survey employees about their job requirements, other approaches will observe employees to identify the "critical incidents" about the job, and most will involve meetings with job experts. Table 2.3 summarizes the major types of job analysis approaches and who participates in these approaches. In reality, most job analyses use a blend of these approaches to obtain the most comprehensive information about the job as possible.

Regardless of which approach is followed, a good job analysis will comprehensively identify the critical tasks needed to perform the job, then link these to the critical types of talent needed to perform the critical tasks. Table 2.4 lists the major steps in a job analysis. Notice that the

TABLE 2.3 THE “HOW” (METHODOLOGY) AND “WHO” (SAMPLING) OF MAJOR JOB ANALYSIS APPROACHES

Job Analysis Approach	Strengths	Limitations	Comments
How (methodology)			
Review of Existing Job Documentation and Content	■ Provides baseline information of job. ■ Economical use of time and resources.	■ Usually only a broad summary; not comprehensive. ■ Assumes everyone performs the job similarly. ■ Describes how job should be done, not how it is actually done.	■ Provides a helpful starting point but insufficient without supplementing with additional information. ■ O*NET website is almost always a good place to start.
Job Observation	■ Economical use of employee time and resources. ■ Can see how work is performed, with what tools, and in what context.	■ Observing behavior may change behavior. ■ Observing behavior in some jobs is difficult or dangerous. ■ Some knowledge work is not really observable.	■ Helpful to at least observe some samples of actual work behavior, if for no other reason than to provide context.

Critical Incidents	■ Structured technique that identifies (a) antecedent of behavior, (b) behavior, and (c) consequence of behavior. ■ Can be collected by interviewing employees or observing behavior. ■ Provides detailed information about tasks.	■ Same potential limitations as job observation. ■ Can be difficult to provide detailed information about specific tasks. ■ Can be expensive and time consuming to implement.	■ Very useful technique for developing simulations because the work context is a part of the observation. ■ Often helpful to at least observe or interview some employees using this technique.
Focus Group Meetings	■ Fast and efficient way to collect information. ■ Allows different perspectives to be considered and discussed. ■ Nuances between how people perform job more easily identified.	■ If participants don't feel they can trust the interviewer, then they will not provide accurate information. ■ Can be difficult to reach consensus about tasks or types of talent. ■ Sometimes difficult to stimulate discussion.	■ Most important issues are to (a) ensure a representative sample of employees and (b) ensure they are open and honest in their discussion. ■ Usually conduct sessions in groups of 5–8. ■ Conduct meeting with peers only; no supervisors present.

(continued)

TABLE 2.3 CONTINUED

Job Analysis Approach	Strengths	Limitations	Comments
Surveys	▪ Allows broad access and input to job analysis. ▪ Can conduct quantitative analysis and make precise specifications. ▪ Can be cost-effective if administered over the Internet. ▪ Increases access and participation. ▪ Large amounts of data collected quickly.	▪ Sometimes generates low response rates. ▪ Participants sometimes do not complete survey honestly or accurately. ▪ Requires knowledge of survey design and analysis. ▪ Requires large samples.	▪ Need to ensure the sample is representative. ▪ Evaluate validity of scores to identify response bias, faking, etc. ▪ Try to use surveys whenever possible because they provide the most comprehensive and inclusive view of the job.
Who (Sample)			
Job Incumbents	▪ You must obtain information from incumbents. Because they are the ones actually doing the work, they are the only ones who can say how it is performed. ▪ May identify important differences between employees.	▪ It can be expensive and time consuming to pull employees off their jobs. ▪ Employees can be reluctant to provide information about how they perform their jobs.	▪ Only incumbents can describe how the job is actually done. ▪ Let incumbents describe how they do the job honestly; don't lead them to describe how it *should* be done.

			■ You must have incumbents provide job analysis information about the job tasks and behaviors. ■ Different incumbents may perform job differently; these differences must be recognized as real and potentially important.
Supervisors	■ Often provide a unique perspective about the nature of work. ■ Help identify contextual or coordination challenges about the job.	■ It is expensive and time consuming to pull supervisors off their jobs. ■ May be hesitant to describe job as it exists, rather than what they want it to be.	■ Best to describe how job should be done. ■ Offer a useful perspective but not critical for a job analysis.
Direct Reports	■ Sometimes helpful for understanding a "bottom-up" view of the job. ■ Provide a different perspective, particularly if the job has extensive management or leadership elements.	■ It can be expensive and time consuming to pull employees off their jobs. ■ May be hesitant to describe job as it exists.	■ Offer a useful perspective but not critical for a job analysis.

(continued)

TABLE 2.3 CONTINUED

Job Analysis Approach	Strengths	Limitations	Comments
Customers	▪ Can be helpful for understanding an "outside" view of the job. ▪ Provide a very different perspective, and one unaffected by company policies or politics.	▪ Response rates will likely be low. ▪ Quality of information may be low. ▪ Collecting information may be difficult.	▪ Rarely used, but may offer additional insights. ▪ For jobs that have a substantial customer component, this perspective may be extremely valuable.

process starts by identifying the important job tasks, or those activities that are critical for performing the job. *Tasks* are the discrete actions one actually performs on the job. One will usually develop a comprehensive list of tasks, and then seek to identify which of those tasks are truly important. Determining task importance is facilitated by having subject matter experts rate the tasks in terms of criticality, importance, frequency, and/or time-spent, as shown in Table 2.4. From the important tasks, one then identifies the important types of talent, including ability, personality, knowledge, and skills. Notice something very important here—*the critical types of talent are determined by the nature of the job tasks.* This removes much subjectivity from the staffing process, because it ensures the process is job related. From the list of talent, we identify those types that are critical for performing the important tasks and needed at the start of the job (because our focus is on selection, not development).

The conclusion of a job analysis will be clear linkages between critical tasks and the critical types of talent needed to perform those tasks. This is known as a task-talent matrix, an example of which is illustrated in Table 2.5. This matrix becomes the foundation for any HR activity that will be based on talent because it defines the critical types of talent needed for the job. For example, consider developing a staffing system for personal financial advisors. The task-talent matrix will determine which types of talent should be used for recruitment and hiring. Imagine some critical types of talent are knowledge of financial accounting, quantitative ability, extraversion, and persuasion and influence. The job analysis can be used to determine the relative importance of these characteristics, perhaps finding something like the following: 43% knowledge of financial accounting, 14% quantitative ability, 14% extraversion, and 29% persuasion and influence. This information can then be used to determine how much to emphasize each characteristic in the selection process, thus ensuring that the selection process is entirely job related. For example, using only a multiple choice test with 100 questions, 43 questions would reflect financial knowledge, 14 questions would reflect quantitative ability, 14 questions would reflect extraversion, and 29 questions would reflect persuasion and influence.

TABLE 2.4 MAJOR JOB ANALYSIS STEPS

Step	Purpose	Examples (personal financial advisor job)
1. Identify Tasks	▪ Create comprehensive list of tasks using the methods described in Table 2.3.	▪ Use financial information to provide consultation to client. ▪ Prospect new customers. ▪ Cross-sell products and services to existing customers.
2. Focus on Critical Tasks	▪ Have incumbents rate the criticality, importance, frequency, time-spent, and/or consequences of mistakes, etc., on each task. ▪ Identify the smaller subset of truly critical tasks. ▪ Group critical tasks into approximately 5–15 task clusters.	▪ Criticality: How critical is this task (1 = not critical, 5 = extremely critical). ▪ Frequency: How frequently do you perform this task (0 = never, 5 = hourly). ▪ Time-Spent: How must time do you spend performing this task (1 = very infrequent, 5 = very frequent). ▪ Importance: How important is this task (1 = not important, 5 = very important.
3. Use Critical Tasks to Identify Talent	▪ Create comprehensive list of talent linked to tasks: abilities, personality, knowledge, skills, etc. ▪ Each type of talent is linked to a critical task.	▪ Knowledge of financial accounting. ▪ Quantitative ability. ▪ Extraversion. ▪ Persuasion and influence.
4. Focus on Critical Talent	▪ Have incumbents rate each type of talent in terms of importance for performing the job and whether it is needed at the start of the job. ▪ Reduce the total list of talent into a smaller set containing only critical types of talent.	▪ Importance: How important is this type of talent for performing the job (1 = completely unimportant, 5 = extremely important). ▪ Needed at entry: Is this type of talent needed the first day on the job (0 = no, 1 = Yes).

TABLE 2.4 CONTINUED

Step	Purpose	Examples (personal financial advisor job)
5. Develop Task × Talent Matrix	■ Show the linkages between each task and each type of talent. ■ This step is usually conducted by HR personnel and reviewed by job experts (incumbents and supervisors).	■ See Table 2.5.

The job of a personal financial advisor is used as an example.

TABLE 2.5 A SAMPLE TASK-TALENT MATRIX

Critical Tasks	Critical Types of Talent			
	Knowledge of financial accounting	Quantitative ability	Extraversion	Persuasion and influence
Use financial information to provide consultation to client.	X	X		
Prospect new customers.	X		X	X
Cross-sell products and services to existing customers.	X			X
Relative importance:	(3/7) = 43%	(1/7) = 14%	(1/7) = 14%	(2/7) = 29%

The critical tasks and talent types are based on a personal financial advisor job, as an illustration. The "X" in each cell represents a substantive relationship between the task and talent. Relative importance is determined dividing the number of "X's" in each column, by the number of total cells with an "X" in them (seven in this example).

NOTE: The last row must sum to 100%.

By definition, job analysis provides a very deep examination of job and talent requirements for a specific job. This information is important for understanding the job but can hinder a broader understanding of the talent demands in an organization—cutting across jobs and even hierarchical levels. Competencies were developed in response to the need to have a broader and more strategic view of organizational talent. *Competencies* are critical types of talent that span multiple jobs and are linked to longer-term strategic goals. Thus, when organizations think of the talent that differentiates their firms from competitors, they are actually thinking in terms of strategic competencies. Indeed, the talent alignment process described earlier focuses exclusively on competencies. The focus of competencies should be on identifying the kinds of talent needed to implement and achieve the firm's strategy. Box 2.1 lists the types of questions we have found helpful in identifying these strategic competencies.

Box 2.1 QUESTIONS FOR DETERMINING STRATEGICALLY AND FUTURE-ORIENTED COMPETENCIES

1. What competencies are currently vital for effective performance in the job/role?
2. Do these competencies differ depending on which culture or country the job/role is performed in?
3. How do these competencies align with the firm's strategy?
4. Can you map each competency onto specific strategic goals?
5. What are the main competency gaps present within new members for this job/role?
6. What competencies are likely to be important for this job/role in the next 5 years?
7. What competency gaps do you anticipate for this job/role in the next 5 years?
8. What competencies drive job/role performance?
9. How do these future competencies or competency gaps differ across countries and cultures?
10. Which competencies differentiate your firm from competitors?

Job analysis and competency models should work in tandem, as both are helpful for comparing and contrasting the talent demands for different jobs, determining whether jobs with different titles share the same underlying talent demands and competencies, and determining whether culture influences the nature of the job and thus the talent needed to perform the job. For example, we worked with a client organization that administered the same entry-level test for engineers, but used different cut scores for each of the different engineering job families. Their thinking was that the jobs differed in their complexity. To illustrate, they assumed production engineers had higher talent demands than engineers working in the customer service and troubleshooting functions. Further, they felt the talent demands in some geographic locations were greater than others, due to differences in the operational focus of different facilities. We conducted a job analysis using a stratified, representative sample of job incumbents from each location. The conclusion of this job analysis was that while the tasks differed slightly, they were functionally similar enough that a single selection system and cut score would be appropriate. This saved money by reducing redundancy and simplifying the hiring process. We also found that the talent demands were similar across each location, adding further savings. The fact that this information was obtained from their own job incumbents helped managers and recruiters accept the system, even though many of these individuals had been strong vocal supporters of the old "tiered" system. Similar insight can be found for comparing the task and talent demands across similar jobs in different countries, cultures, and subsidiaries. For example, one might expect cultures higher in collectivism to perform work in a more collaborative fashion than those in individualistic cultures, and those in cultures with greater power distance to have more formal work roles (and thus greater similarity across incumbents) than those in cultures with low power distance.

Roger began to develop a plan with respect to talent alignment. He knew Spooner's global strategy very well because he had received extensive briefings on that as part of his preparation for this new assignment. He also had a reasonably good sense about Spooner's key performance metrics and how his regional operations contributed to the overall metrics. Spooner differentiated itself on product innovation, product reliability, and cost. Within this broader umbrella, Roger saw the Southeast Asian operation contributing manufacturing operational efficiency and technical innovations. Hence, Roger adopted an alignment strategy that began with those two internal (operational) performance targets. He vetted these metrics against the HQ and the subsidiary leadership team, who were comfortable with a focus on these two as primary operational metrics.

From there, Roger held numerous focus group meetings with general managers, assistant and frontline managers, and employee groups. These focus group meetings were held with a cross-section of functional backgrounds, ages, experiences, and demographic and cultural diversity. This was done to increase the transparency of the process, gain input from experts, and build support for the system (since employees had a say in the design of it). The focus group meetings identified the strategic job families that most impact each performance metric.

- Manufacturing operational efficiency
  - Plant assistant general managers. These managers are charged with overseeing the day-to-day operations of the plant, and were universally viewed as critical to the achievement of the plant's production and efficiency goals.
  - Assembly team associates. These are the employees who work in groups of 10–15 members assembling the electrical components for production.
- Technical innovation
  - Engineering R&D team members. These are members of cross-functional teams. The teams are staffed by small groups (about 10–15) of engineers with highly specialized training. Each member can serve on multiple teams, and management is based on a matrixed arrangement.

- R&D team leaders. The team leader for each R&D team plays a critical role. Given that that team members are highly specialized, they often have difficulty communicating with each other. They also sometimes fail to see "the forest from the trees" because they immediately gravitate to highly technical issues. The team leader needs to keep the team on track and focused on the broader R&D objectives.

The assembly line associates constituted a high-volume hiring position, while the other three positions were more likely keys to differentiating the organization.

Roger also held focus groups with members from the strategic job families, following similar protocols for ensuring that the membership was stratified with respect to diversity, experience, and culture. From these meetings Roger identified the *key* types of talent required for each job (it is assumed that job-related knowledge is required for all jobs):

- Plant assistant general managers
  - Communication and influence
  - Performance management
  - Teamwork
- Assembly team associates
  - Conscientiousness
  - Attention to detail
  - Teamwork
- Engineering R&D teams
  - Technical competence
  - Teamwork
  - Adaptability
- R&D team leaders
  - Cultural agility
  - Teamwork
  - Performance management

Note that this list of talent competencies is for each strategic job. Teamwork is a competency shared by all strategic job families and hence is a strategic competency. Performance management is shared among the managerial jobs. The remaining competencies are job family–specific. These competencies define talent "quality" in each respective job family. Interestingly, Roger learned that the assembly team associates had quantitative scores for each competency. This is because these were high-volume hires, and an outside vendor had assisted in developing a selection process and had performed an empirical evaluation of talent competencies. The other jobs were lower-volume, and in the case of managerial positions, were usually hired from within. Hence, expert judgment was required to identify the talent quality thresholds for the other three positions that differentiated the firm.

Finally, Roger surveyed the appropriate managers to identify the expected and observed talent quantity numbers. They did this based on numbers needed today, but also flow numbers projected in three years assuming (1) no growth, (2) expected growth, and (3) above average growth. Roger had learned from his mentor the importance of providing equal consideration to "above forecast" and "below forecast" predictions. Even though many managers tend to ignore above forecast predictions, it can create problems as severe as below forecast predictions. The following data in Table 2.6 were obtained and organized according to what some refer to as a "gumball" chart (diagonal lines = acceptable, vertical lines = risky, horizontal lines = deficit).

Roger was (pleasantly) surprised to see that, based on this more objective analysis, he was in good shape with respect to plant assistant general managers. This was a relief because everyone he interviewed had a gut feeling that there was an insufficient pipeline to offset demographic shifts on the horizon. The strong pipeline was probably a result of management giving so much careful attention to internal development and succession planning for this role. On the other hand, Roger was very concerned about the numbers for assembly team associates

TABLE 2.6 GUMBALL ORGANIZATIONAL CHART

	# Observed	# Expected	Gap (Stock)	Gap (Flow-no growth)	Gap (Flow-expected growth)	Gap (Flow-above avg growth)
Plant assistant general managers	38	38	0			
Assembly team associates	4,918	5,500	–582			
Engineering R&D team members	45	47	–2			
R&D team leaders	7	12	–5			

and R&D team leaders. The strong economy in Southeast Asia had made competition for general skilled labor positions difficult. Further, the reason his group had a reasonable number of R&D team members was mainly because they were expatriates from other areas. They could continue this plan unless there was above average growth. However, this masked a larger problem of having enough R&D team leaders. These leaders had to have sufficient cultural agility to work with members from diverse expertise, functions, and cultures. There were few who could balance the technical and with interpersonal, and it was not always possible to promote from within.

These estimates gave Roger cause for concern. "More risk than I'd like" he thought. Now, what to do about it?

MANAGING THE TALENT SUPPLY CHAIN

We are now to a point where it is possible to consider the management of talent resources. We start by developing a talent supply chain that is shown in Figure 2.6. The supply chain begins with the sourcing of talent. *Sourcing* involves defining the desired kinds of talent for the job and company,

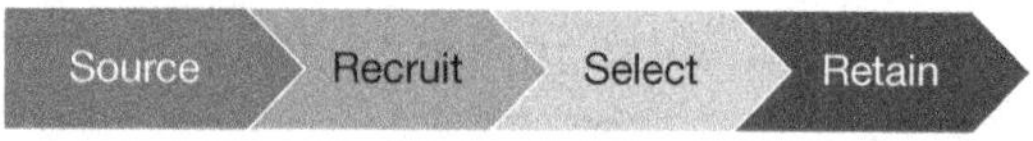

Figure 2.6
The talent supply chain.

identifying where this talent is located, and determining what the potential candidates want and value. *Recruitment* involves trying to attract the desired candidates to the firm by developing the recruiting message and tailoring it to the desired talent. Recruitment also includes the kinds of practices and procedures used to convey the recruitment message to candidates. *Selection* involves assessing candidates on job-related characteristics in culturally appropriate ways. The selection process concludes with a hiring decision and employment offers for some candidates. *Retention* involves keeping the hired talent engaged and satisfied, and developing them for future opportunities within the firm.

Specific practices and steps for managing the talent supply chain are presented in chapters 4 (sourcing and recruitment), 5 (selection), and 6 (talent management). The chapters provide frameworks and general guidance for conceptualizing and implementing best practices. Although many firms treat each of the elements in the supply chain as a separate function, for present purposes it is important to think about the supply chain more holistically and strategically. The term "silos" is often used to reference functions that do not integrate with other functions. For example, it is not uncommon in large firms to have different managers responsible for recruitment and selection. While such specialization may offer some efficiencies and cost savings, a potential risk is that these functions develop practices that work against each other. In one project, we found that the selection managers and recruiting managers were pursuing very different objectives. The recruiting managers were trying to fill available jobs at the lowest possible cost per hire. The selection managers were focused on hiring the highest-quality candidates without concern for costs. The predictable conflict that ensued is indicative of the many and varied criteria that can be used to judge the effectiveness of talent acquisition systems.

It is critical that the elements in the talent supply chain be integrated and work in a synergistic manner. Doing so offers additional opportunities for enhancing performance and competitive advantage. First, a more lean supply chain reduces waste and increases efficiency, which saves costs. Second, few organizations have the capability of managing the entire supply chain, and competitors would need to copy this capability or build it themselves (which takes a lot of time and learning). A competitor would have to copy or substitute all of the elements in the supply chain and their interrelationships (Dierickx & Cool, 1989). Hence, managerial and HR capabilities to manage the talent supply chain enable "hidden" forms of competitive advantage.

Considering the nature of the supply chain with the resource-based talent framework provides even more novel insights. For example, should a firm source generic or specific talent resources? On the one hand, if the firm wants to source specific talent resources, then it may take longer to attract and select such candidates because there are fewer of them in the labor market (and they are also likely more costly). Yet the organizational benefits on performance and competitive advantage should occur more quickly because specific talent is the immediate determinant of performance. On the other hand, sourcing high-quality generic talent resources could be done more quickly and more cheaply because there are more such people in the labor market. The disadvantage is that it will take more time and resources to convert that generic talent into specific talent. Of course, if the environment changes, then there are many benefits to having generic talent because it can more easily be rebundled and redeployed toward new strategic goals (Kim & Ployhart, 2014).

There is probably no issue more fundamental to HR strategy than the decision as to whether to "build" the required talent internally or buy it when needed on the external market. Many companies have adopted an internal labor market approach, wherein recruitment and selection are done at entry levels, generic capital is acquired, promotion is purely from within, and compensation systems reward stability, development, and tenure. These organizations have adopted a model that enables them to build firm-specific talent over time. Internal labor markets tend to be found in industries with high barriers to entry (e.g., extensive capital requirements that prevent competitors from popping in and out of the market), such as utilities, oil and gas, auto manufacturing, and government.

Others have effectively outsourced talent development to the external labor market. These organizations buy talent as needed, resulting in entry at all levels of the organization, little emphasis on training or development, and compensation systems that reward current performance. These organizations emphasize speed and flexibility in their talent acquisition processes. External labor markets tend to be found in smaller companies and in industries with modest barriers to entry or those undergoing rapid change.

These two extremes, though, ignore the majority of companies, which practice elements of both internal and external labor markets. A common approach is to adopt an internal labor market for strategically important jobs, thereby exerting tighter control over a mission-critical resource, while filling less impactful jobs internally or externally, depending on availability. Even this common-sense approach is being challenged, as companies have begun increasingly to use alternative employment arrangements (even with strategically important jobs).

A second important decision embedded in an HR strategy is the choice to "secure" the talent resource or "rent" it. For jobs that are not critical to the organization's success, outsourcing is a common solution to ensuring support (indeed, recruitment process outsourcing is considered later, in chapter 4). Routine, nonessential activities such as back-office administration, call centers, and many IT services have frequently been outsourced to an independent organization, often in another country. The global market for outsourced services exceeded $100 billion in 2014, roughly 50% growth from a decade previously (Statista, 2014). Reasons for outsourcing usually involve costs (e.g., specialists can provide the service cheaper than the organization can do it) or flexibility (i.e., the ability to add or shed staff quickly as market conditions change), but one survey reported that 59% of organizations outsourcing work cited "access to talent" as one expected benefit (Brown & Fersht, 2014). And despite the objections of politicians in many developed countries, the growth of outsourcing suggests that it will continue to be a viable option for completing some types of work.

Another way to acquire talent is through a joint venture (JV), essentially an agreement between two organizations to pool resources for a period of time to accomplish some specific objective. Although use of JVs declined somewhat following the recession, they remain a popular way for organizations to rapidly acquire talent in geographies in which they do not currently operate (KPMG, 2012). In emerging markets, a joint venture partner not only provides a ready-made work force but also more importantly can help mitigate local country risk through knowledge of customers, culture, laws, and key stakeholders. Although outsourcing and joint ventures are well beyond the scope of this book, it is important to remember they represent a very viable means of "acquiring" talent to fill a specific need (e.g., administrative processing or entrance into an emerging market). Technology has introduced a host of additional options for securing various services.

Cappelli and Keller (2013) identified a variety of work arrangements, other than outsourcing, that entail less-than-direct employment, ranging from independent contractors to leased employees to vendor on premises employees. Companies such as Guru, Elance-oDesk, and Upwork provide an exchange for freelancers and employers to find one another. One survey (Freelancers Union & Elance-oDesk, 2014) estimated that roughly 14% of the US workforce were independent workers, roughly double the number from the previous decade. Although numbers are difficult to come by, the phenomenon seems to be on the rise in Europe as well (Horowitz, 2013). Despite working without the safety net typically provided by regular employment, one study found independent workers to be more engaged than regular employees (Rasch, 2014). Further, although on-demand employment has primarily targeted the consumer market (e.g., Uber offering transportation services, Taskrabbit help with chores around the house, and Handy with cleaning services), it appears headed to the business market as well—for example, some are experimenting with an on-demand sales force to economically get at hard to reach customers (Wall Street Journal, 2015). We expect the growth in independent workers to continue to grow and become an increasingly important source for talent.

These examples illustrate how knowledge of talent resources, the talent supply chain, and the options for acquiring talent directly or indirectly provide a unique perspective on deploying talent for performance and competitive advantage. However, as difficult as this may be, it becomes even more challenging within a global context.

MANAGING THE GLOBAL TALENT SUPPLY CHAIN

The ideas underlying the talent supply chain can be expanded to consider a global context. The basic elements including sourcing, recruitment, selection, and retention, will remain the same but become broader in scope. For example, the sourcing of talent will obviously be geographically distributed across cultural and national borders in global staffing. Likewise, the kinds of recruitment and selection practices an organization follows may need to be adapted to different cultural and legal contexts. This means the basic supply chain does not significantly differ in a global context, but rather the number and complexity of relationships within and across the supply chain does.

The *management* of a global supply chain is dramatically more difficult than a domestic one. The core issue to be balanced is whether the sourcing, recruitment, selection, and retention practices that are used to manage the supply chain will be applied universally across cultures or localized to specific cultures (or regions). Every global firm is confronted with this issue, and there is no single best way address it. Further, some practices may be applied universally while others are customized locally. The greatest challenge with the global supply chain is essentially a strategic one.

For example, the trendy clothing retailer H&M has a distribution channel of over 3,000 stores around the globe and employs over 100,000 employees. The firm, which started in Sweden, has maintained a strong culture based on Swedish values and leadership. As this firm has expanded into new markets, particularly those outside of Europe, it has confronted a tension between allowing more "localized" management styles versus those more "centralized" from its Swedish origins. How does a firm such as H&M balance these tensions? The answer starts by recognizing how strategy is translated into action in global firms.

Specifically, when describing the different modes of managing the global talent supply chain, it is helpful to understand how strategy, policies, and practices are related. Figure 2.7 illustrates the interrelationships among these components. *Strategy* defines where a firm chooses to compete and how it chooses to compete. Strategy is usually operationalized in terms of goals and objectives. *Policies* refer to the broad guidelines that translate strategy into action. Each function can have its own set of policies. The *HR policies* are those guiding principles that explain how HR contributes to strategy execution. For example, the HR function may emphasize high performance, employee loyalty, or customer service (Lepak, Liao, Chung, & Harden, 2006). In turn, these policies help determine which specific practices will be used. *Practices* are those specific actions that influence employee attitudes, perceptions, and behavior. The HR practices are those discussed in chapter 1 ("High Performance Work Systems" section) and include recruitment and selection. An HR policy emphasizing high performance might use selection practices focused on skills, job-training, and merit-based pay; a policy emphasizing loyalty might use selection practices focused on fit, long-term development, and seniority-based

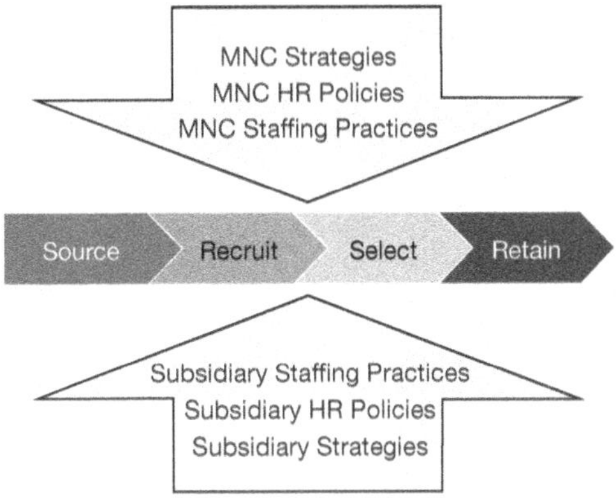

Figure 2.7
The global talent supply chain.

compensation; a policy emphasizing customer service might use selection practices focused on service orientation, interpersonal training, and customer satisfaction–based compensation.

In MNCs, strategy, policies, and practices may be further distinguished between those held by the home country headquarters/corporate and those held by subsidiaries. Although it may be natural to think of these relationships as hierarchical, such that the parent organization dictates the nature of the policies and practices in the subsidiaries, this does not have to be the case. Perlmutter and colleagues (1969; Heenan & Perlmutter, 1979) have observed four modes for MNCs to organize themselves (Bartlett & Ghoshal, 1989; Caligiuri & Paul, 2010; Colakoglu, Tarique, & Caligiuri, 2009; Collings & Scullion, 2006; Ployhart & Weekley, in press; Prahalad & Doz, 1987). These modes are shown in Figure 2.8.

On the one extreme, the *ethnocentric* mode occurs when the parent company's strategy, policies, and practices are applied universally across subsidiaries. Policies and practices will be implemented the same way regardless of country or culture. The various functions will be highly centralized within the home country headquarters. In the extreme it means that key employees of subsidiaries will be PCNs from the home country. On the other extreme, the *polycentric* mode occurs when the parent

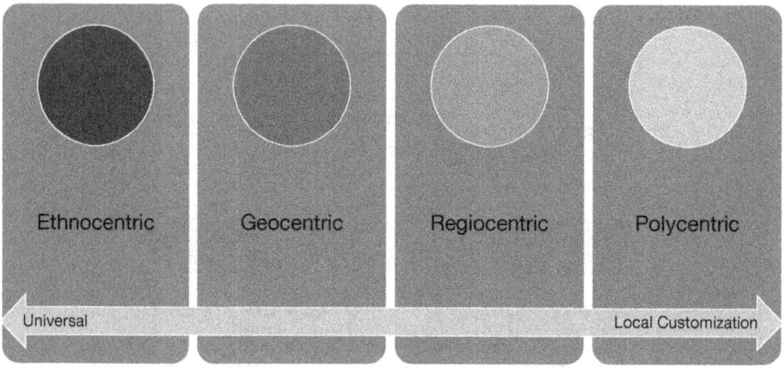

Figure 2.8
Modes of organizing MNCs. The shading of the circles reflects the amount of local customization. Darker colors imply more universal approaches (e.g., the entire circle is the same color), lighter colors imply more local customization (i.e., more variation).

company's strategy, policies, and practices are entirely customized to local cultures, laws, and political systems. Subsidiaries will be staffed with HCNs, even in key positions. Each subsidiary will have its own policies and practices tailored to its unique environment. Blending these two approaches are *geocentric* and *regiocentric* modes. Both modes will have some universal elements, but with some degree of local adaptations and customization. The geocentric mode will tend to favor more centralization whereas the regiocentric mode will tend to favor more localization. Both geocentric and regiocentric modes will employ PCN and HCN, but there will be more HCN in the regiocentric mode.

At H&M, the management style is very similar to the geocentric mode. H&M tries to maintain its strategy, values, and culture around the globe. For example, the company's strategy and policies emphasize high ethical standards and human rights for all employees, regardless of whether these standards are legally required within a given country. The firm maintains a relatively flat hierarchy with a strong Swedish influence, yet it recruits retail associates locally and within compliance of local laws. The tension thus comes mainly from the ranks of middle management located in the subsidiaries; where choices must be made between promoting internally from the local labor force versus importing talent from the home country. This tension will always be present, and it is only a question of where, within the organizational hierarchy, it is situated.

Conclusion

The global talent supply chain provides an actionable framework for managing the global talent acquisition process. It provides a way to think about opportunities for leveraging connections between the elements of the supply chain, managing the supply chain globally, and anticipating tensions between the MNC and subsidiaries. The supply chain prompts a series of questions that are intended to think about the management of each element within the chain, as well as the chain holistically. Figure 2.9 provides an overview of these questions.

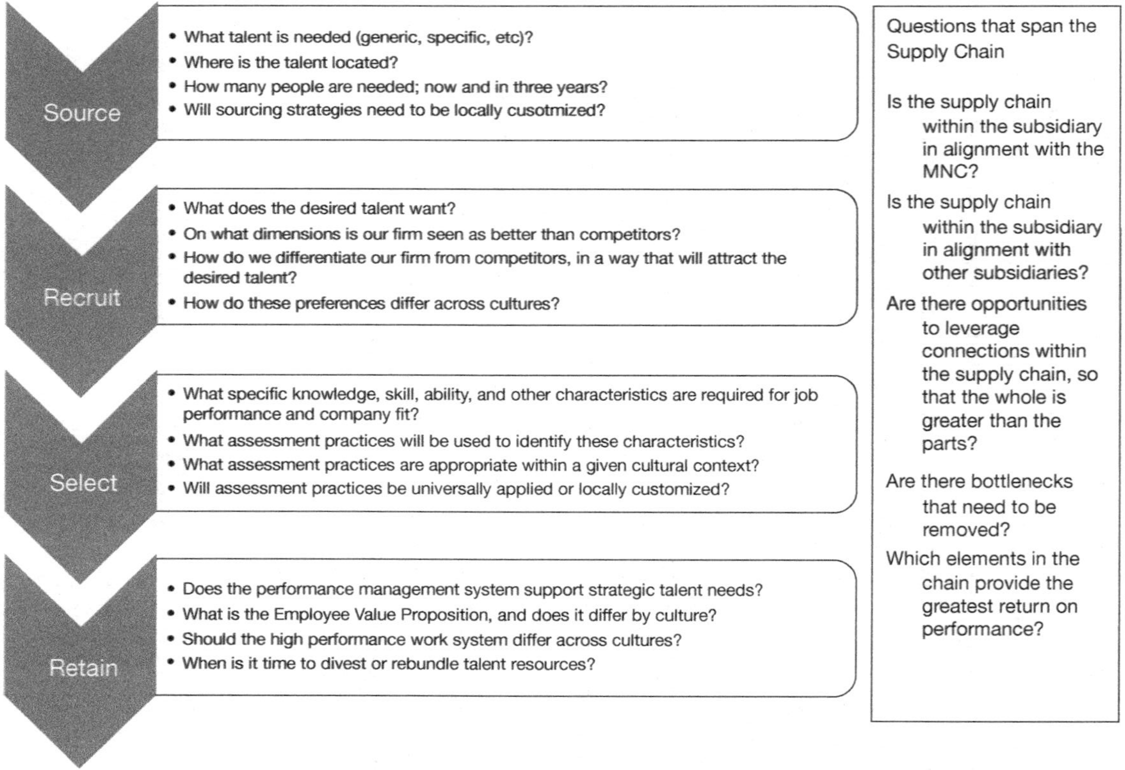
Source
What talent is needed (generic, specific, etc)?
Where is the talent located?
How many people are needed; now and in three years?
Will sourcing strategies need to be locally cusotmized?
Recruit
What does the desired talent want?
On what dimensions is our firm seen as better than competitors?
How do we differentiate our firm from competitors, in a way that will attract the desired talent?
How do these preferences differ across cultures?
Select
What specific knowledge, skill, ability, and other characteristics are required for job performance and company fit?
What assessment practices will be used to identify these characteristics?
What assessment practices are appropriate within a given cultural context?
Will assessment practices be universally applied or locally customized?
Retain
Does the performance management system support strategic talent needs?
What is the Employee Value Proposition, and does it differ by culture?
Should the high performance work system differ across cultures?
When is it time to divest or rebundle talent resources?
Questions that span the Supply Chain
Is the supply chain within the subsidiary in alignment with the MNC?
Is the supply chain within the subsidiary in alignment with other subsidiaries?
Are there opportunities to leverage connections within the supply chain, so that the whole is greater than the parts?
Are there bottlenecks that need to be removed?
Which elements in the chain provide the greatest return on performance?

Figure 2.9

Summary of key global talent supply chain management questions.

Roger had not yet been exposed to the various techniques used to measure talent, but he had looked at some external articles and had managed to obtain some internal documents that described the outlines of the company methodologies in talent management. So Roger was determined that this new chapter of life needed to start with an approach that started with his desired outcomes. But what were those desired outcomes? And how would he explain the effects of talent on those outcomes to any cynical and angry line managers in terms that THEY would understand? Using the basic ideas underlying the global talent supply chain, he jotted down the areas on which he would start to look for data:

1. Supply—Map the potential available talent—Upstream.
   a. Identify the supply lines of relevant talent by country
   b. Segregate the data by technical and business discipline
   c. Apply qualitative assessments of the various learning institutions and assess talent quality
2. Supply—Map in-house talent—Downstream
   a. Demographic analysis by discipline and country
   b. Talent analysis by country and discipline
   c. Adjust for known short-term plans like plant closures or business/country exits
   d. Using historical data, establish attrition patterns to predict probable future patterns (regression analyses?)
   e. Do in-depth demographic analyses of existing workforce and project known data (retirement rates in countries where the age was set by policy and/or legislation)
3. Demand—Understand present and future demands
   a. Using business plans for the next 3 years, augmented by known long-lead-time projects, do high and low case estimates of staffing requirement by level and discipline
   b. Test outside probability scenarios that could prove to be a major impact (e.g., new build plants, changes in technology, outsource possibilities, etc.)

4. Balancing
   a. What were the critical shortfalls by
      i. Discipline
      ii. Country
   b. What could be predictable influences, for example
      i. Government interventions such as increased pressure on immigration and work permits
      ii. Government entries into the market through nongovernmental organizations (NGOs)
   c. What assumptions could be made in respect of
      i. Graduate and other recruitment successes
      ii. Reductions in attrition rates

CONCLUDING THOUGHTS ON THE MNC-SUBSIDIARY TENSION

While the Internet, access to education outside home countries, improved ease and cost of travel, and so on, have dramatically accelerated the spread of knowledge across boundaries, the notion that "one size fits all" is surely fallacious. There are more similarities than differences, in our view, across cultures, but some of these differences can make the application of common processes, systems, and policies difficult—even illegal. For example, one such critical difference is how cultures resist change. A conversation with a senior manager from a large global IT company led to a discussion of whether resistance to change was expressed consistently from one country or region to another. "Oh heavens no," he said. "In Asia there is a tremendous appetite to please, but the issue can be that there will be a tendency to implement change even if many will know that there is a major failure coming. In Europe, they are typically happy with whatever they have—so anything that changes is, by definition, a bad idea. In the US, everyone is busy convincing you that they have a better idea." So the manner in which change is managed in a global context is even difficult because different cultures react to change differently.

Another important difference is one of national aspirations. In the same way that national governments look to control and optimize their natural assets (minerals, forests, land, etc.), they also have a desire to protect and nurture their human resources. Such desires are evidenced by the control of the right to work, demands for specific nationalization targets, and so on. Governments look to increase the wealth of their citizens or even just to provide work. For example, when commenting on the excessive number of government employees working in a rarely traveled border crossing, a colleague from that country noted that employment opportunities for people with relatively little formal education were few. The issue was not about efficiency, which a Western mind considered important, but about the provision of meaningful employment for as many citizens as possible.

Yet despite some important cultural differences, effective use of talent, and the costs of acquiring, retaining, and developing it, demand a broader approach in many cases. It is *not* a given that large conglomerates need to be managed as one, especially if there are little natural linkages between them or interdependencies. But the issue of universal versus local customization presupposes that some level of decision has already been made to operate on a global basis. In our experience, large multinational companies opt for HR systems that work, to greater or lesser degrees, across national boundaries. And that decision brings with it natural tensions between the MNC headquarters and the subsidiaries, many of whom may have operated with a good deal of independence over years or even decades. If you then inject the added spice of different cultures, national laws, and aspirations, the recipe becomes a complex one.

Roger collected his thoughts and prior notes and started to develop a more formal, regional, and strategic talent plan—which he knew was going to be a tough sell to Ibrahim. It needed to be made using logic and data, but would also need to take into account the issues of national pride, the "not invented here" syndrome, the history of the region, and the fact the he was an outsider. He recalled a cartoon depicting a dog on a high wire with the caption—"as he approached the half-way point, two thoughts occurred to Oscar. He was an old dog and this was a new trick." In Roger's case he was a young pup trying to convince older dogs that he knew better tricks.

So he outlined some key early conclusions based on his initial conversations, his preassignment briefing in England, his library background research on the region, and knowledge of the company as a whole.

1. First and foremost, any strategy that he was going to formulate needed to be rooted in the overall regional HR strategy, which, in turn, needed to be rooted in Spooner Electronics' business strategy. This alignment would be ensured by linking to the key performance metrics. Based on his conversations thus far, the forward strategy was to use Malaysia and the Philippines as the two areas where the majority of the high-volume jobs would continue to be located. This was based on two main factors:
   a. The availability of skilled and semiskilled labor. The Klang valley in Malaysia and the areas surrounding metro-Manila in the Philippines had multiple technical schools that produced a steady stream of technically focused certificate or associate degree equivalents.
   b. Cost. The prevailing wage rates in those two countries were materially less than Singapore.
   c. He needed to examine "talent alignment" across all the countries in the region to assess the "as is" position. The sources for that would be any existing data and records that existed currently on a country-by-country basis. While that data, and even assessments, may require significant later modification once a broader view was obtained, he had to start somewhere.
2. The second key part of the business strategy was to make significant incursions into markets that showed massive growth potential but were only nascent in terms of Spooner's presence—China, South Korea, and Vietnam. Each posed different, but significant,

challenges in terms of legal frameworks and governmental involvement, culture, local competition, and language capability.

3. His view was the technical competitive advantage of Spooner Electronics was best served by two linked approaches:
   a. Attract generic talent, for example engineers with disciplines that were connected but not necessarily immediately applicable.
   b. Sustained investment in training and development to convert that generic talent to the specific talent that was required to design and produce high-quality products.

   This approach only made sense if there was long tenure to reap the benefit of the investments made. So he needed to ensure that the various related HR systems supported that model (service-related benefits like pensions, promotion systems, etc.). He also would need some data on items such as turnover rates of high talent to assess whether these systems were working as they were intended, variations by country, and so forth.

4. Based on the outcome of the talent review, he needed to think of how to ensure that his "A players" were in "A jobs"—in other words, were his high-performing, high-potential resources deployed in a way that they added the greatest value to the company? This obviously necessitated a more serious talent alignment effort. He realized from his early engagements with Ibrahim that not all of the GMs and country-based leaders would necessarily welcome the notion that the outcome might not be that local talent stayed local. He also realized that ensuring optimal enterprise-based deployment *and* individual development required a discussion about moving individuals operating satisfactorily but who were at their ultimate capacity, perhaps even letting them go, to allow their roles to be used as a key development for those with greater upward mobility. He was certain that these would be very difficult conversations—and that he would need significant backing from the most senior regional leadership. He was also uncertain whether such an approach

would work in the cultural and legal frameworks in which he was to work—but it had to be explored.

So the challenges loomed, and he needed to start building the solutions in more depth. He also needed a team to make that happen. So how could he do this and provide some building blocks of regional cohesion? How might he develop bridges between his staffing group, embedded in HR, and the technical functions that he needed to support? A spark came to him. Clearly his team needed diversity in the classic sense—balance of HR and technical, balance of nationality, gender, and so forth. But could he leverage balance of location in a way that would engender mutual support and "skin in the game" between different parts of the region? If he located each member of his senior team in four or five countries—say one in Thailand, one in Singapore, one in the Philippines, and one in one of the key emerging countries—each with regional accountabilities, could that build some cohesion and mutual trust? He decided that he would at least start with that as his prime option. He knew it was risky, but with the availability of communications technology, why not locate his talent in different key locations so that they could be advocates of the regional model, develop relationships with key local leaders, and show how an organizational model of this type could indeed be made to work?

His final thought was that he needed reinforcements to help him in communicating with his line leadership on THEIR terms. So he wanted someone with a solid financial background to develop sustainable systems to track cost performance and other relevant financial data. He also wanted a data analyst with high-level IT skills who could interface with the various country systems to load into an integrated system and enable his team to understand the present state and provide the basis of a vision going forward and, to repeat that phrase, "let the data do the talking."

"Much to do," he thought, but by now he was beginning to see a plan emerging from the mist.

3

Talent Analytics

THIRTY-SECOND PREVIEW

- Talent analytics refer to the application of statistical or quantitative models to provide evidence-based insights about talent.
- Scores are the "currency" of talent analytics. The quality of talent analytics is dependent on the quality of scores provided by measures.
- Scores differ with respect to the information they can provide. The more information available in the score, the more sophisticated modeling approaches can be applied.
- Scores must be reliable and valid if they are to be useful for talent analytics, decision-making, or forecasting. Reliability is whether scores provide consistent information under similar

circumstances. Validity is whether the inferences we draw from scores are accurate.

- Different talent analytic models are useful for answering different types of questions. Using the appropriate type can be used to build a business case for talent and make more accurate predictions.
- Talent analytics can help identify and understand subtle cultural differences in talent quality and quantity.
- The Bottom Line: Talent analytics can help identify hidden sources of competitive advantage, but only if the scores used in the analytics are reliable, valid, and modeled appropriately across cultures.

Roger's ears were ringing! He had just put the phone down on a general manager (GM) who was, to put it mildly, irate. Roger had only been in Kuala Lumpur for a few months and already he was being pulled into a power struggle between HR and a GM from one of the facilities Roger's team served. In this particular case, the GM was upset because a member of his team, Zainoridah Hassan, had agreed to serve on Roger's staffing team. Roger had done his due diligence and Zainoridah had been highly recommended because she was a respected engineer but had a knack for identifying and developing talent. Zainoridah had enthusiastically said yes to Roger's request, but the problem was that Roger had failed to follow local norms and customs. Roger should have asked the GM first, and vetted the decision with several of the GM's team. This would have taken into account the respect for both seniority and consensus that is so typical of Malaysian culture. He needed to assemble a cross-functional team to lead and implement his vision for staffing and recruiting dominance in the region, so in some ways he was willing to live with the GM's frustrations . . . for now. The result of not following these protocols was that the GM adamantly refused to let Zainoridah join Roger's team. It was an early and salutary lesson about operating successful in his new environment.

However, the phone call about Zainoridah was just the opening to a salvo being fired by several plant GMs toward HR. The GMs were under increasing pressure to cut costs and raise productivity, which was nothing new, except that talent in the region was getting really hard to find and wages were exploding upward. So the GMs turned to HR and expected them to fix the problem. Since the staffing costs within each plant came out of the GMs' budgets, they were very critical of HR expenditures and demanded a reasonable return on investment. Of course, typically being trained in engineering, they also generally felt HR was a soft science based more on feelings than facts. The GMs collectively wanted HR activities linked to tangible business outcomes. This was something the local HR group was unaccustomed to.

Even though his old boss, Gordon Baker, used to tell him to "let the data do the talking," Roger had to wrestle with presenting a reasoned and data-based argument to support his recommendations. The problem was that Roger was fairly unfamiliar with HR analytics. He had participated in leading some employee engagement surveys back in England, but he was really just a custodian and the surveys had been implemented for years without too much concern (or consequence). In Roger's new role, he understood the need for pushing for a set of talent analytics that would help demonstrate HR's effectiveness. He had a vision, and with his team he was formulating a plan for recruitment and selection, but he knew that plans and strategies are fictional until you put some quantitative discipline around them. He also felt that a quantitative focus would help him better negotiate with the home country nationals employed in the firm. "There may be cultural differences," he thought, "but numbers are numbers, and we are a business run by numbers." If he could link talent to key performance metrics, then surely numbers would be the great equalizer, and showing some early successes would be important for his plan and reputation.

Roger knew his limitations. He was no quant guy; he was more a guy who had an appreciation of statistics. Several of the high-volume hiring contracts were up for renewal or renegotiation with new vendors. How could he distinguish the vendors with solid products from those with empty promises? Roger, candidly, had not developed the first idea about how to work through the various analytical issues in a methodical manner, much less how to determine which ones were really selling programs that might look good on the surface but had no real validity in practice. "Validity" he said to himself. "I know the word but how do I probe to understand what that means in the real world?" How could he demonstrate return on investment (ROI) in talent in a manner the GMs and the finance-types would find credible? He had some vague recollection of statistics that he had taken at high school as part of his math class, but it was indeed vague. Words and phrases like "standard deviations" and "statistically significant" were floating around in the dark recesses of his mind, but for the life of him he could not recall their specific meaning nor in sufficient detail to be useful.

As Roger cogitated on these issues, it made him more conscious that the general approach to issues in staffing should be more data based, that he would use that data in his discussions with the GMs, line leaders, and his HR colleagues to challenge their current approaches with, at least potentially, less friction. He needed to develop a shared understanding with his client groups about what "success" would constitute and how it would be measured. What were the key criteria? Several came to mind, but were they the right ones? And did he have any baseline data to determine from where progress would be measured? Cost per hire? Time from vacancy creation to fill? Percentage use of external versus internal sourcing? Retention after hire? That might be some initial relevant data, he thought, but the three keys to a productive conversation with his numerate managers would be:

1. Were the data relevant to his managers (and did the same data apply to most jobs)?
2. Could the existing data be used to form a baseline comparison and track progress?

3. Above all—what was the action plan to change the "as is" to a "desired state?"

Coming in as the new gun was going to be tough, and he knew the cultural divides were going to require careful attention. But perhaps a bridge could be found in seeking common data—even if the interpretation of the data and conclusions drawn may vary. Roger thought to himself, "Time to dust off the old stats textbook!"

Never in history have employees or organizations produced so much data or had so much access to data about each other. Many of us go about our daily lives having innumerable interactions with the digital world in the form of e-mails, Internet browsing, credit card purchases, phone calls, and TV viewing—all of which leave digital footprints. Like archeologists who study thousand year-old footprints in remote jungles and deserts, today's digital archeologists are quantitative investigators trying to make sense of the colossal mountains of data each of us generates daily. Today's buzzword is "Big Data" and the belief is that Big Data leads to big profits. It is an appealing story, but one that's too good to be true. Simply having access to massive amounts of data is not necessarily meaningful for business purposes. Most of what is in Big Data is just noise. However, for those who know what kinds of data to look for and how to leverage these data to assist in decision-making and forecasting, there is a possibility of finding hidden insights. These insights will not be discovered by Big Data itself but rather by a surgical approach to data called talent analytics.

Talent analytics refers to the application of quantitative tools and decision-support methods to provide evidence-based insights for managing talent and demonstrating how talent relates to strategic goals. Too many talent decisions are made based on subjective judgment, and HR has often held a reputation of being a "soft" science relative to finance, accounting, or operations. This is slowly changing, and advanced quantitative methods are being used to understand how talent affects, and is affected by, organizational and environmental factors. Using talent analytics is not simply a quantitative exercise; rather it requires a deep understanding of how to measure talent, the types of business performance and strategic outcomes that may relate to talent, and how to tie the pieces of data together in a way that tell a story.

Roger was determined *not* to be one of those managers who claimed that "talent is our most important resource" if he could not find valuation approaches that he himself would find credible and believable. He knew this standard would require him to find approaches to valuing talent as an asset, but he was only familiar with talent as a cost on a balance sheet. As Roger started to think about tying analytics and metrics to his talent acquisition strategy, he began to further refine the relevant questions that needed to be addressed.

- Assuming he could measure talent, how could he show the recruitment and selection plans would have impact on increasing its value? He knew very well the concept of ROI, but he didn't have the slightest idea how to integrate that concept when it came to talent.
- If he could convey impact, then what types of analytic models could be leveraged to demonstrate such impact? Could he quantify a linkage between talent and operational/strategic business outcomes GMs and line managers would find credible and convincing?

MEASURING TALENT AND PERFORMANCE

Galton is claimed to have said, "Whenever you can, count." Quantifying concepts such as talent, customer satisfaction, and performance provides opportunities for leveraging analytical tools to make better decisions.

Accurate measurement of talent and performance offers a necessary foundation for talent analytics because measures provide the scores used as inputs for data analysis (see Figure 3.1). Scores are simply information we get from measures, and different measures will provide different scores. That is, measures are used to provide scores that provide information for making decisions. For example, there are many different kinds of scores used in professional sports, and each type of score provides different information relevant for different purposes. In American football, the number of first downs and time of possession indicates the level of control that one team has over another. In baseball, statistics such as Earned Run Average measure a pitcher's performance, and Batting Average measures a batter's performance. In soccer, shots on goal and percentages of passes successfully completed are tracked. Each of these statistics provides insight to a particular aspect of an individual or team performance, but the ultimate focus is the win/loss column. There are some parallels in business. For example, the HR profession has been accused

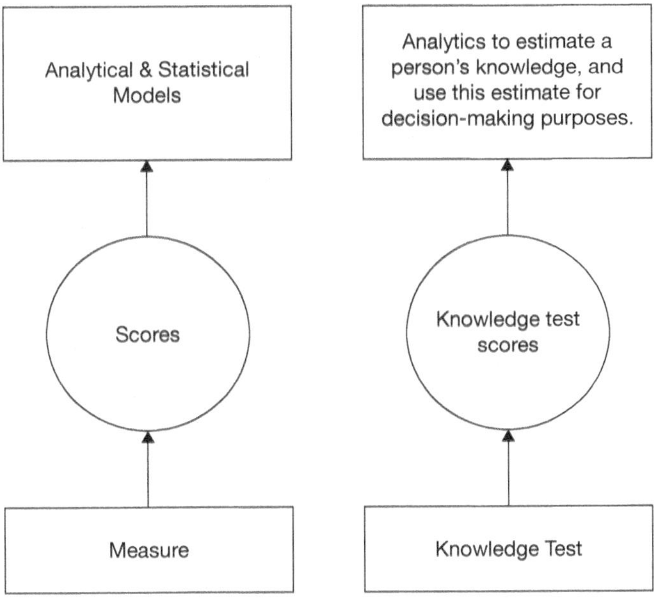

Figure 3.1
How talent analytics are based on scores provided by measures.

of measuring activity instead of impact. When asked about leadership training, for example, HR leaders might reply with the number of training days delivered, the cost per person trained, or the percentage of staff trained. None of these answers the question regarding how effective that training has been. Has it changed the performance at either the individual or company level?

Despite a long history demonstrating the importance of measurement in psychology and (to a lesser extent) human resources, there is little attention given to measurement in talent analytics. Big Data is a case in point; the amount of attention devoted to analytical methods is substantially greater than the amount of attention paid to measurement. The problem is, without accurate measurement, the analyses are potentially meaningless or wrong. As the saying goes, "Garbage in, garbage out."

Suppose we want to study bank branch talent defined in terms of customer and market knowledge. We further want to link this knowledge to bank branch sales. Clearly we need a measure of collective knowledge. Some common examples include years of experience, scores on a knowledge test or assessment, interview scores, or scores based on performance in a simulation exercise. Each of these approaches has advantages and disadvantages. Experience is easy to measure and obtain, but is also deficient because different employees may learn more on the job than others, given the same tenure (not to mention some employees may retain more knowledge than others). Test scores may be more comparable across people, but what if the test developer fails to include business-to-business transactions as part of the customer knowledge construct? Interview scores could have the same concerns but also be influenced by impression management and faking (particularly if they are unstructured). Simulation scores should eliminate impression management, but we don't know whether a person's behavior in a relatively brief simulation will be the same as his or her daily behavior on the job. Finally, none of these measurement issues considers the fact that we want to link *collective* knowledge to branch level performance—not individual knowledge. Similar concerns also affect performance measures. Branch sales is an objective measure but may also have sources of inaccuracy. For example, a branch's sales may be impacted by the number of nearby competitors or the average household income in surrounding neighborhoods.

The key point is that there is no perfect measure. Instead, we must think about whether measures provide scores that are useful for a given purpose—a concept known as *validity*. Measures do not provide information, but the scores that are produced from these measures provide information. For example, when employees complete an engagement survey, we analyze the scores that represent the employees' responses to the survey items. Stated differently, it is the scores produced from measures that are the data points for talent analytics. Thus, we really need to think about scores and the various factors that influence scores. We hope that the scores obtained from a measure are true reflections of the attribute—that is, they provide information about that attribute relevant to our specific purpose (e.g., scores on the survey reflect true differences in engagement). However, scores on all measures are also affected by contamination and deficiency. *Contamination* is the degree to which scores from a measure are influenced by factors we do not care about or want to study (e.g., engagement scores are affected by not having enough time to complete the measure). *Deficiency* is the degree to which scores from a measure neglect factors we do care about (e.g., engagement survey items don't measure job satisfaction). Contamination and deficiency are prevalent in scores from all measures, but not to equal degrees. Figure 3.2 shows an example of true scores, contamination, and deficiency.

Because there is no such thing as a perfect measure, managers must identify measures or scores that most accurately serve their needs. This means managers should not accept measures as a given, but search for measures (or develop their own) that will serve as useful indicators. There are several characteristics of measures that are important to consider, including measurement scales, reliability, validity, objectivity, and cultural equivalence.

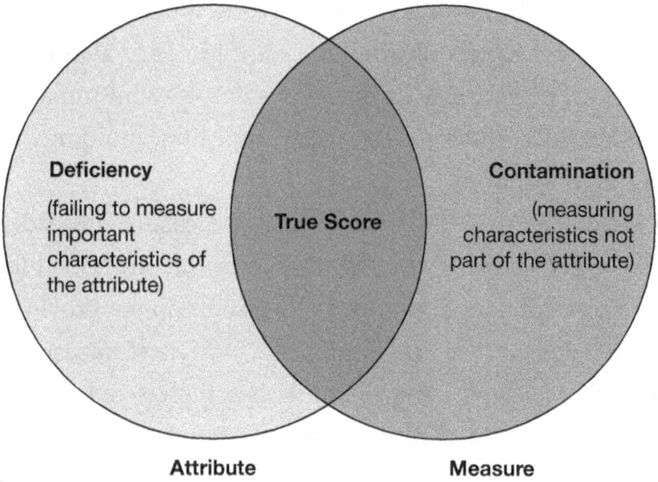

Figure 3.2
Contamination, deficiency, and true score variance.

Scales of Measurement

Measurement means assigning numbers to represent properties of some phenomena. This is a complicated way to say "we want to assign a metric to phenomena so we can better understand those phenomena and make better decisions and predictions as a result." Take the weather for example. We can walk outside and say it is cold. The problem is it is difficult to convey with any precision what cold "is," and what is cold to someone in the Amazon is likely different to someone in Siberia. Further, knowing it is cold today does not allow us to very accurately predict what the weather will be like in the future. However, if we assign scores to the weather in the form of temperature degrees, then we can be much more precise—"It is -9 degrees Celsius" provides specific information that is consistently understood by all who know the metric system.

Measures produce scores that are the inputs for data analysis. *Scales of measurement* refer to the "richness" of information scores will provide. There are four scales of measurement, and these are illustrated in Figure 3.3. *Nominal* scales exist when the scores represent simple categories but provide no additional information. Examples include gender, race, or

Scales of Measurement	Information	Examples
Nominal	Less	Pick one: ____Male ____Female
Ordinal	↓	Rank from highest to lowest: ____ Rick ____Jamal____Kim
Interval		Strongly Disagree 1 2 3 4 5 Strongly Agree
Ratio	More	Financial currency like Euros, Dollars, Yen, etc.

Figure 3.3
Scales of measurement. More powerful statistics—and hence more information—are possible as one moves from nominal to ratio scales of measurement.

country. With nominal scales, the scores only tell us which category something is from, but they do not tell us anything about rank order, which is better/worse, or more/less. *Ordinal* scales provide slightly more information because they tell us about category membership and rank order. Examples are forced-ranking performance assessment programs (rank order employees based on their performance), school grades (A, B, C), or even "top ten" lists (e.g., the top 10 most livable cities). However, ordinal scales do not tell us whether the differences between each rank are the same. For example, what if your workgroup has two star employees and five average employees? A forced-ranking system will require you to rank these seven employees top to bottom, even though the top two employees may be four times more productive than the remaining five employees. Therefore, an interval scale is needed when you need to account for both rank order and ensure the differences between each rank order are the same. The most common example of an interval scale is the familiar 5-point or 7-point scales used in surveys and performance management systems. These numbering systems provide the difference between a one and a two, to be the same as the difference between a two and a three, or a four and a five, and so on. However, ordinal systems do not have a true zero point. The final measurement scale, ratio, has all the characteristics of the ordinal scale, but also includes a true zero point. Examples include money (euros, dollars, yens), temperature, or weight. The scales of measurement are important, because the more information provided by a measure, the more insight we have into the concept, the more fine distinctions we can make, and the more sophisticated analytical models can be used to make predictions.

Reliability

Imagine working with a colleague who is highly inconsistent. You can't believe the colleague about deadlines, deliverables, or quality of work because they differ every time. Sometimes quality is high; sometimes it is low. Sometimes deadlines are met; sometimes deadlines are missed. It doesn't matter whether this person is lying or telling the truth; the bottom line is that this is not someone you can rely on. There is a similar concept in measurement known as reliability. *Reliability* refers to the consistency of scores. For example, if you administer the same measure under the same conditions, will you get the same scores? High reliability is needed to enable meaningful inferences in talent analytics. Reliability is expressed quantitatively, such that reliability = true score/(true score + error). This puts reliability on a scale from 0.00 to 1.00, where 0.00 indicates no reliability and 1.00 indicates perfect reliability. Contamination or deficiency can influence reliability; frequently lowering it but sometimes raising it. There are different ways of estimating reliability that make different assumptions about what is "true score" and what is "error." These various methods are summarized in Table 3.1.

Different types of scores will have different levels of reliability. For accounting or financial data, reliability is frequently close to 1.00 (perfect reliability). Perfect reliability does not mean the financial numbers are constant; rather it means that we would obtain those same numbers under *identical circumstances*. For intangible phenomena such as talent, reliability is never close to 1.00 but instead usually in the 0.50 to 0.95 range. We want reliability to be as high as possible. To the extent scores are unreliable, it weakens our ability to make fine distinctions and hurts our ability to make future predictions. All else equal, we can make more accurate predictions when reliability is .80 than when it is .75. The level of reliability minimally needed is dependent on the consequences of the decision. If we were trying to decide whom to hire based on psychometric assessments, we would want reliability to be very high (e.g., 0.85 or higher). If we are simply doing some scenario planning and rough forecasting, we can tolerate lower levels of reliability (say, 0.70).

TABLE 3.1 SUMMARY OF DIFFERENT TYPES OF RELIABILITY. RELIABILITY RANGES FROM 0.00 TO 1.00, WITH HIGHER NUMBERS INDICATING GREATER RELIABILITY. THE MINIMUM ACCEPTABLE ESTIMATES ARE ONLY ROUGH GUIDELINES AND EXCEPTIONS MAY APPLY

Types of Reliability	Minimum Acceptable Estimates	Design	True Score	Error
Test-retest	.50 and higher	Administer same measure to same people at two different points in time	Shared variance between the administrations	Unshared variance between the administrations
Internal consistency (coefficient alpha)	.70 and higher	Administer one measure, with multiple items or indicators, at a single point (e.g., 20-item knowledge test)	Shared variance among the items	Unshared variance among the items
Interrater	.50 and higher	Two or more raters (e.g., interviewers) provide scores using the same measure (e.g., interview rating form)	Shared variance among the multiple raters' ratings	Unshared variance among the multiple raters' ratings
Intrarater	.40 and higher	The same rater (e.g., interviewer) observes the person two or more different times, but uses the same measure (e.g., interview rating form)	Shared variance among the one rater's ratings	Unshared variance among the one rater's ratings

Validity

Validity is a very complex scientific concept, but in simple terms, *validity* is the extent to which scores obtained from a measure are consistent with the intended purpose of those scores. For example, given a test of mechanical reasoning, validity will be the extent to which scores on the test actually tell us something about an applicant's true level of mechanical reasoning. There are different types of validity, and these are summarized in Table 3.2. Keep in mind that validity is a question relevant for scores obtained from any and all measures, not just those related to talent

TABLE 3.2 SUMMARY OF DIFFERENT TYPES OF VALIDITY

Type of Validity	**Example**
Criterion-related: Empirical relationship between predictor and outcome (criterion) variables. Usually estimated using correlation or regression.	Knowledge test scores are correlated with job performance scores. If the correlation is statistically and/or practically significant, it provides criterion-related validity evidence.
▪ **Concurrent**: Empirical relationship between predictor and outcome variables, when both types of data are collected on job incumbents.	A knowledge test is administered to 500 current employees. Job performance evaluations are collected on the same group of people. If the correlation between the knowledge test and job performance is statistically and/or practically significant, it provides criterion-related validity evidence.
▪ **Predictive**: Empirical relationship between predictor and outcome variables, but predictor data is collected on job applicants and outcome data is collected after some of these applicants are hired and become job incumbents.	A knowledge test is administered to 500 job applicants. 300 of these applicants are later hired. After a sufficient amount of time has passed, job performance evaluations are collected on the 300 individuals hired. If the correlation between the knowledge test and job performance is statistically and/or practically significant, it provides criterion-related validity evidence.

TABLE 3.2 CONTINUED

Type of Validity	Example
Content: Expert judgment about the conceptual overlap between predictors and outcomes.	A group of staffing experts works with job incumbents to determine whether the content of a job knowledge test is important and needed for performance on the job. If the experts believe the items on the knowledge test measure important aspects of knowledge needed for the job, there is content validity evidence.
Construct: Whether scores on some assessment represent the underlying attribute they are intended to represent.	A researcher examines whether a knowledge test is related to similar types of knowledge, and unrelated to personality traits that should theoretically have no relationship with knowledge.
▪ **Convergent**: Scores on an assessment are positively and strongly related to scores representing similar attributes.	Knowledge test scores are positively related to scores based on similar kinds of knowledge (i.e., large positive correlations).
▪ **Discriminant**: Scores on an assessment are negatively related or weakly related to scores representing different attributes.	Knowledge test scores are unrelated to personality scores.
Face: Whether an assessment appears to measure the intended attribute (is not a "true" kind of validity; entirely based on perception).	Based on their perceptions of the selection test content, applicants believe their knowledge was measured by the test.

acquisition. When you are trying to decide whether the current interview process provides information about an applicant's likelihood of success, you are asking questions of validity. When you are trying to decide whether return on assets is a useful metric for evaluating a firm's strategy, you are asking questions of validity. Thus, validity is the litmus test that should precede data analysis—it asks the question of whether the data are substantively meaningful. Contamination and deficiency both damage validity. Unreliability also reduces validity because predictions based on scores cannot be more accurate than the reliability on which those scores are based. Reliability sets the upper limit on validity, and scores from an assessment cannot be more valid than they are reliable. For example, a test of mechanical reasoning cannot predict performance more accurately than it predicts itself. Thus, scores need to be both reliable and valid.

Objectivity

Measures may be objective or subjective. *Objective measures* are those that require little to no human judgment in creating the scores, and hence can be verified. Examples of objective measures include financial data (e.g., revenue, cost), company size, time of day, and so on. Clearly there can be human judgment in how we interpret these numbers, but the creation or generation of the numbers is pretty much outside of human interpretation. On the other hand, *subjective measures* are those that are highly dependent on human judgment and difficult to verify. Common examples include employee attitude measures, psychological assessments (e.g., intelligence, knowledge, personality), supervisor ratings of employee performance, and interviewer ratings. We may be able to measure these characteristics with reasonably high reliability and validity, but it is difficult to truly verify their accuracy. Generally speaking, objective measures are more tangible and reliable, while subjective measures are usually more intangible and less reliable.

There is a clear preference for objective measures in business, and because HR data are frequently subjective, HR is often seen as a "soft science" relative to finance and accounting and thus of lesser standing and rigor. We have three responses to this criticism. First, HR data can be highly reliable and valid, and hence very useful, even when they are based on subjective data. Second, even though scores may be based on objective or subjective measures, the inferences one makes from those scores (and the decisions one makes based on those scores) still require human judgment. There are innumerable examples of poor human judgment surrounding financial and accounting data, which is why they are so closely regulated. Finally, it becomes possible to link these subjective data with objective data to build sophisticated connections between people and business outcomes. An illustration that stems from the Big Data movement is collecting and analyzing the digital byproducts people leave as they interact with digital technology: cell phone transactions, Internet activity, and instant messaging are common examples. Firms are now starting to mine these massive data to learn whether there are patterns and connections that offer new ways of reaching customers or understanding employees. These data are objective—they are collected as a byproduct of human choice and behavior—but they are still based on people.

Cultural Measurement Equivalence

One final point about validity needs to be made within the context of global talent measurement. Given differences in language and cultural differences in the meaning attached to similar concepts, it is vital that scores be comparable across cultures, countries, and languages. This is an issue of *measurement equivalence*. For example, suppose we want to use a measure of situational judgment to select entry-level managers. Situational judgment tests (discussed in chapter 5) present respondents with short workplace situations and then ask how they would respond in each of those situations (usually, the responses are different behaviors that could be effective or ineffective in that situation). In this example, the firm spends considerable money developing and validating the situational judgment measure in North America. They also believe that company culture should value good judgment identically across cultures. Consequently, they decide to roll out the North American situational judgment measure in their international operations in Taiwan, China, India, and Russia. They then go on to find that scores differ across cultures and countries. Are these differences real? The answer requires three conditions to be met.

First, any measure developed in one language needs to follow the appropriate translation protocol. It is standard practice to use a process of translation into the new language, and then take that measure and "back translate" it into the original language (Brislin, 1970). Different translators should be used for the translation and back-translation steps. This process of translation–back translation ensures that the language, intent, and meaning of the measure are preserved. In our example, the firm must hire expert translators and go through the translation process for *every relevant language*. This can be quite costly, and we have seen considerable variability in the quality of translations, so it is always helpful to pilot test the translated measures before using them operationally.

Second, even after following appropriate translation procedures, it is critical to ensure the item is interpreted the same across different countries, languages, and cultures. In our example, are situational judgment scores in China equivalent to those in the United States? Assume we find large differences on the following question, such that Chinese respondents are most likely to choose option A, while those in the United States are most likely to choose option D:

> You are leading a project team that has performed extremely well. Tasks were completed ahead of time, and the project was finished under-budget. Your supervisor sends an e-mail to your peers congratulating you on such outstanding performance. However, she fails to mention the other team members who also worked very hard on this project. What do you do?
>
> a. Ask your supervisor to send a follow-up e-mail acknowledging the accomplishments of the team.
> b. Tell your team you are sorry they were left off the e-mail, and ask them what they would like to do about it.

c. Look for other opportunities to publicize the accomplishments of the team.
d. Accept the credit because you are the team leader.

These differences are likely caused by cultural differences in individualism-collectivism. In China, option A is likely to be most appropriate but in the United States the prevalent answer would more probably be D. Thus, cultural values will influence the meaning ascribed to different behaviors and actions.

Finally, assuming the first two conditions are met, it is then necessary to establish the measurement equivalence of the scores from different countries, languages, and cultures (Little, 1997). For example, is a score of "3" in China equal to a score of "3" in the United States? There is absolutely no point in comparing any differences in scores based on tests, assessments, surveys, or performance ratings, until measurement equivalence is demonstrated. Assessing measurement equivalence requires sophisticated analytical models using item response theory and/or confirmatory factor analyses. The details of these models are beyond the scope of this book, so ensuring that experts trained in measurement and cultural equivalence do such analyses is important.

Conclusion

Talent analytics is based on scores obtained from measures. If the analytics are going to offer useful insights, then measures must be used that provide scores that are reliable and valid. Regardless of which type of performance or talent we care about, we need some way to measure it. Different measures will provide different scores, some of which are relevant for our specific purposes. Thus, one should first understand the reliability, validity, and appropriateness of the measures before building sophisticated analytical models. This is a vital point usually lost in the discussion of Big Data. No amount of analysis can make up for bad data: Garbage in, garbage out. Table 3.3 provides a checklist of key issues to consider with talent measurement.

TABLE 3.3 A CHECKLIST OF KEY ISSUES AND QUESTIONS TO CONSIDER WITH GLOBAL TALENT MEASUREMENT

Measurement Issue	**Questions to consider**
Measurement Scale	_____ Have you used the most informative scale possible (preferably interval or ratio)?
	_____ Does the scale make sense for the concept you are trying to measure?
	_____ Does the scale make sense in each language and culture?
	_____ Will respondents be likely to use the entire scale? If not, then consider using a simpler scale to save time and energy.
Reliability	_____ What is the appropriate form(s) of reliability for assessing the scores (i.e., test-retest, internal consistency, interrater, intrarater)?
	_____ For each form of reliability, do the scores reach an acceptable minimum threshold?
	_____ What are the likely causes of contamination and deficiency? Are there any ways to reduce contamination and deficiency?
	_____ Are reliability estimates similar across cultures? If not, why?
Validity	_____ What form(s) of validity evidence are available for the scores (i.e., criterion-related, content, construct)?
	_____ Do the forms of validity reach minimum standards? If not, why?
	_____ Is the validity evidence credible; should more evidence be provided?
	_____ Does the validity evidence support using the measure as intended?
	_____ Are there language or cultural differences in validity evidence? If so, why?

TABLE 3.3 CONTINUED

Measurement Issue	Questions to consider
Objectivity	_____ Given goals and constraints, is it appropriate to use an objective measure instead of a subjective measure?
	_____ If both objective and subjective measures are available, do they have sufficient reliability and validity evidence?
	_____ What are the potential sources of contamination and deficiency for objective and subjective measures; and can using both offset the weaknesses of the other?
Cultural Equivalence	_____ Has the measure been translated and back-translated by experts? Has the measure been pilot tested within each culture/country/language? Have any errors been fixed?
	_____ Has the meaning of the items been considered within each culture/country/language?
	_____ Has evidence of measurement equivalence been established? It is impossible to put any faith in the scores until such evidence is demonstrated.

THE VALUATION OF TALENT

It is common for organizational leaders to say that their people are the firm's most important resource. The problem is that it's hard to say what talent resources are actually worth, because there is no commonly accepted way to demonstrate the financial value of talent. Fulmer and Ployhart (2014) reviewed a variety of approaches that have tried to do so over the last 50 years. Table 3.4 provides a brief summary of these approaches, and they span areas as diverse as economics, accounting, human resources, human resources accounting, and organizational psychology. While each approach is reasonable given its assumptions, the problem is that none of

TABLE 3.4 TRADITIONAL APPROACHES TO THE FINANCIAL VALUATION OF TALENT

Approach	Key Points	Implications
Financial Accounting	Financial accounting, based on generally accepted accounting principles (GAAP), only allows talent to be recognized as a cost.	Financial accounting frames all talent investments as costs. Hence, talent is not operationalized as an asset and the strategic value of talent is not seen as an asset but rather as a cost.
Managerial Accounting	There is more discretion in treating talent as an asset in managerial accounting, but any such reporting is more idiosyncratic to each firm because it will focus on topics/questions relevant to that firm.	Managerial accounting can operationalize talent as an asset. Hence, talent can be treated as an asset for use in internal (operational) metrics.
Financial Market	Stock prices, market-based measures, unrecognized good will, residual income valuation, and related approaches may estimate the collective valuation of talent because investors may consider information beyond accounting metrics.	These estimates may reflect the unrecognized benefits of talent but are confounded with other unrecognized assets (e.g., company brand). However, they are always based on investor expectations at a given point in time.
Human Resource Accounting	Attempts to put a direct value estimate on talent, primarily at the collective level and primarily based on the changes or outcomes talent generates.	These approaches are interesting but are not recognized in accounting standards. The estimation approaches can also be difficult to operationalize and require more subjective judgment than alternative approaches.

TABLE 3.4 CONTINUED

Approach	Key Points	Implications
ROI-Utility Analysis	There are a variety of approaches that attempt to demonstrate the financial ROIs in talent. For example, one can estimate the firm-level financial benefits of making more effective hires. New approaches exist (see Boudreau & Ramstad, 2007), but have not seen widespread implementation.	These approaches can be useful for comparing and contrasting different HR practices such as recruitment and selection programs. Selection ratio, quality of hires, variability in job performance, and cost all factor into these kinds of estimates. One potential limitation is that utility models assume one can sum individual performance to collective performance. Such an assumption may over- or underestimate the actual firm-level ROI. Attempts to integrate these models into accounting frameworks are often met with skepticism. It is better to use these models for comparing and contrasting different HR practices, for internal decision-making purposes.

SOURCE: Adapted from Fulmer and Ployhart (2014).

these approaches really captures the nature and breadth of talent and talent resources. For example, generally accepted accounting principles (GAAP) prohibit the reporting of talent information as anything other than costs because talent is intangible and difficult to verify. Firms can of course adopt their own supplemental valuation metrics, but it is not clear whether there are any organizational benefits for doing so. There have also been attempts to implement standards for such voluntary reporting, but to date these efforts have not been successful. Similarly, thinking of talent in terms of financial "goodwill" has intuitive appeal (e.g., the difference between purchase price and perceived value in an acquisition cost), but again this is not universally recognized or even potentially comparable across countries or industries. Finally, none of the frameworks are able to model and place valuation estimates that capture the interdependent nature of work.

Thus, lacking clear approaches to putting a value on talent that are commonly understood and accepted, managers must use other metrics and analytical techniques to make a business case for talent. We introduce several analytical techniques that can be useful in the next sections.

Realizing that finding a commonly accepted valuation approach to talent was unlikely, Roger began to consider how he could build a business case using metrics important to the firm's business leaders. Firstly, his intuition was that a country-by-country approach to recruitment and talent development was suboptimal, and that a regional approach—which was bound to be regarded with suspicion and probable opposition—offered major opportunities for improved efficiency and effectiveness. Besides, it was with that mandate that Roger had been sent from London, albeit somewhat covertly. Roger's initial skirmish with the GM convinced him that data was the ONLY way to advance that argument. So Roger outlined an approach to start the journey.

1. Decide what the final preferred data outcome would be. Start with perfection in a perfect world and work down from there based on reality of the present. Reality, in this case, meant he did not have all the types of data he wanted, but could access plant and operational data (e.g., productivity), marketing data (e.g., customer satisfaction), and sales data by location. He also believed it would be fairly easy to get raw data on turnover and

number of hires as well as cost performance in Malaysia. Getting the same from other countries could prove more difficult. So some of the data were there, even if not in ideal formats. The question now was whether he could convince the gatekeepers of this data to share it. How would he convince them it was worth their time and effort?

2. A second step would be to determine how easily the gaps in data could be filled. Did all the regional countries under his responsibility have the data in the same format? What was their capability to produce missing data?
3. Next, he would need to figure out what capability existed to distill and massage the data into a format that would inform dialogue on talent management across the organization and region. He needed to do a talent assessment of his own team to test that capability. The data analyst on his team would have that as a prime early deliverable.
4. Finally, how would he choose among the multiple staffing vendors fighting aggressively for his business? Many of their assessment systems appeared sophisticated, but also very similar. He wanted quality and had a budget to pay for quality, but how would he determine whether the money was being spent wisely?

He had lots to do, and he would need to guide his team carefully as they mined the data and worked out which tools would be useful to understand, interpret, and present that data. What Roger really needed was a way to demonstrate talent's impact.

CONVEYING TALENT'S IMPACT

In this section we consider the topic of how one can convey impact (the types of talent analytic models used to demonstrate impact are discussed in the next section). Sophisticated analytical tools are only helpful to the degree they provide credible information to make informed decisions. Conveying impact is based on (1) estimates of *effects* and (2) estimates of *significance* (see Figure 3.4). Further, estimates of significance are both statistical and practical (i.e., is there a meaningful effect). Ideally, all three pieces of information will be available to provide evidence of impact. Losing any one of the outer boxes will weaken evidence of impact.

Effects

The most important piece of information resulting from talent analytics is something known as an "effect." An effect is a quantitative estimate of whether an intervention, event, or phenomenon has a consequence on

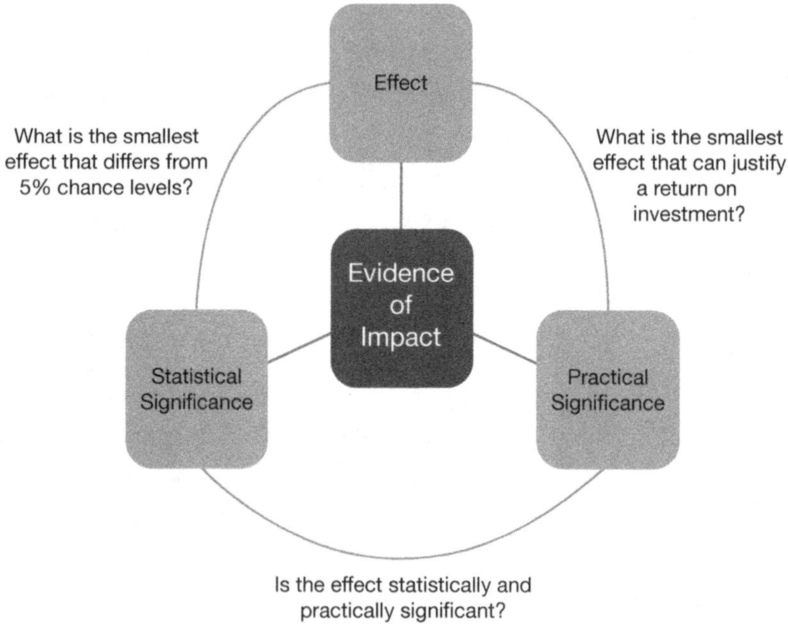

Figure 3.4
Illustration of interrelationships among effects, statistical significance, and practical significance.

some other event or phenomenon. For example, if a new recruitment program has an observable impact on applicant quality, the new recruitment program has "an effect" (assuming other conditions, such as an increase in unemployment, have not caused the change in applicant quality). For present purposes, effects are always estimated quantitatively and they are on a scale of small, medium, and large. The information we care the most about is estimating the effect, because the effect is our numerical estimate of whether talent matters.

There are no universal conventions on what are considered irrelevant, small, medium, or large effects. What is considered small or large must be judged within the context of project goals and purpose. For example, change models will often produce what seem to be small effects (e.g., 1% of variance explained in profitability over a quarter). However, a 1% increase in profitability over one quarter is a 4% annual increase in profitability and may be quite important from a business perspective (particularly if the benefits are compounded). On the other hand, explaining 20% of the variance in an employee attitude survey may not be seen as important if job satisfaction is only weakly related to profitability. Thus, help is needed in gauging the significance of effects.

Significance

The term "significance" means different things to different people. We should all understand that statistical significance is not equivalent to practical significance. *Statistical significance* means that some effect (e.g., implementing a staffing system and finding better performance) is different enough from zero (i.e., no effect) by greater-than-chance levels. Most of the time researchers will call an effect statistically significant if the odds of it occurring due to chance are less than 5% of the time (denoted as $p < .05$, where "p" refers to probability). Statistical significance is mainly used as a guide for decision-making in scientific research—it provides a common standard to help researchers determine which effects are large enough to pay attention to and which are not. However, statistical significance is also a function of sample size, and when analyzing data based on large numbers of people (e.g., thousands), even trivial effects are statistically significant. The more observations used in an analysis, the greater the likelihood that an effect will be statistically significant even if the effect is very small.

Consequently, practical significance is equally important. *Practical significance* is whether an effect is large enough to consider it useful for a specific applied purpose. For example, assume a new psychometric assessment has a correlation of .01 with job performance (as discussed shortly, 0.00 = no relationship and 1.00 = perfect relationship, so .01 is a very weak relationship). This effect would not be statistically significant if it were based on 25 applicants, but it would be statistically significant if it were based on 15,000 applicants. The question becomes, is .01 a large enough effect to be practically meaningful or important? (In this example the answer is "No.") Managers thus need to focus mainly on whether the effect is practically significant. Practical significance is a judgment call about whether the effect is large enough to be meaningful—this is frequently a question about return on investment (ROI). Effects, statistical significance, and practical significance are relevant to all statistical models.

Return on Investment

Effects and significance enable managers to understand impact, but they (by themselves) do not allow inferences of whether such impact generates benefits that will offset the costs associated with producing those benefits. *Return on investment* is a broad concept and focuses on whether the benefits of talent investments are offset by their costs. For example, many firms use third-party vendors or recruitment process outsourcing as a means to reduce costs yet maintain reasonably strong effects for talent acquisition. It may be possible to achieve stronger effects for a fully customized talent acquisition system that is developed in-house, but the added cost in doing so is likely to be extreme and negate the potential benefits.

Calculating the ROI of talent is challenging because, as noted earlier, it is difficult to put a credible estimate on the valuation of talent. Most

ROI efforts are focused on establishing whether the costs of a talent investment produce gains in job performance sufficient to justify the costs. This is known as *utility analysis*, and there are a variety of formulas and approaches that can be used to make such estimates (see Cascio & Boudreau, 2008). However, most utility analysis models are intended to compare the benefits of an HR practice (e.g., selection system) relative to other HR practices. Utility analysis is well suited for estimating ROI across different staffing systems. Utility analysis has some intuitive appeal and is generally accepted among staffing experts and HR managers, but the estimates are often not seen as credible by those trained in finance and accounting. One problem is that many utility analysis approaches are not strictly based on accounting principles. Another problem is that utility analyses often require estimating a dollar value for differences in individual job performance (e.g., how much more is a top performer worth than an average one?), and then aggregating this performance to the firm level. Not only are such estimates often questioned but also such an approach neglects the types of synergies or process losses that may occur in collectives as well as the value or costs associated at the collective level (e.g., economic market factors affecting profitability). Because these models have most frequently been applied to evaluate specific selection practices, utility analysis models are reviewed briefly in chapter 5 (see Cascio & Boudreau, 2008, for more details).

TALENT ANALYTIC MODELS

There are a variety of analytical tools and methods that can be used to quantify the effect of talent—that is, its business value and strategic impact. There is no single best tool; rather the "best" tool is the one that will provide relevant and credible information to the stakeholders who are affected by the information. However, remember that all analytical tools are approximations used to summarize and convert large amounts of data into information that can improve prediction, forecasting, and decision-making. We introduce the simpler quantitative methods first, followed by those that are more sophisticated.

Descriptive Models

Many data analysts march straight toward sophisticated analytical models, but a simple summary of the data in terms of a distribution, average (mean), and variability (standard deviation) still contains a great deal of insight. A distribution is a summary of how scores "stack up" across observations. Many types of data follow what is called a normal distribution (also known as a bell curve). In a normal distribution, scores stack up in the middle and become fewer as one moves to either tail of the distribution. Talent is usually normally distributed, except in those instances where we may "choose" talent based on specific criteria such as only hiring those in the top 5% of a distribution. A normal distribution has a number of very useful properties. First, it allows us to talk in terms of averages (also known as the *mean*). Most scores will cluster toward the middle of a distribution; the mean is the most common score within a normal distribution. The mean is important because, if we don't know anything more about a sample, the most accurate prediction we can make is to bet the average. Stated differently, if you are meeting someone from a new culture and don't know a single thing about that that person other than where she or he is from, then the most accurate prediction you can make about that person is what you know about her/his home country's average values (see Table 1.1 in chapter 1). Second, knowing the mean is helpful, but the spread of the distribution is also important for making sense of the mean. The spread of a distribution is represented by the standard deviation, which is essentially how much the average score deviates from the mean. When there is a large spread to the distribution, the mean becomes less helpful in describing any one score (although it is still the best summary of any given score, lacking outside information). Together, the mean and standard deviation provide a very quick and powerful window into the nature of a distribution—and hence the underlying phenomena of interest. Figure 3.5 shows an example of a normal distribution.

Of course, there are many instances where a distribution will naturally be skewed toward one tail or the other. Such skewed distributions

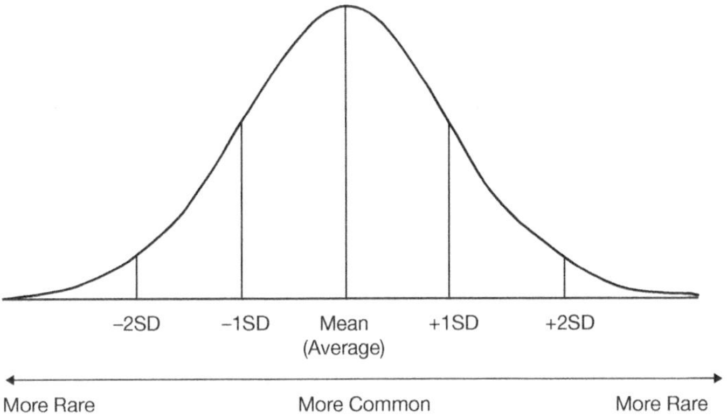

Figure 3.5
The normal (bell curve) distribution. SD = standard deviation. The farther away a score is from the mean, the more unusual or rare it is. Stated differently, it is less likely for a score to be at one of the extremes than it is to be in the middle of the distribution.

are often found with financial data. For example, income and salary are usually skewed, such that few people make a large amount of money, and many people make a small to moderate amount of money.

It is important to know the shape, mean, and standard deviation of a distribution, because it allows us to know where any given score is in relation to other scores. For example:

- What was the average tenure of employees hired last year?
- What is the average time-to-hire for each recruiter?
- What is the cost-per-qualified-candidate for those sourced via LinkedIn?
- What was the average performance of those hired at a specific location?

These examples highlight that knowing the distribution, mean, and standard deviation of key internal and external metrics enables firms to understand their relative positions and benchmark/differentiate appropriately.

Difference Models

Difference models are those that seek to demonstrate whether the mean scores in two or more groups differ from each other in meaningful ways. Sometimes we want to examine whether the means of groups are different from each other. For example, employee survey scores in Asian countries are often more favorable and give a more optimistic outlook than the somewhat lugubrious view of life in Western Europe. In other instances we want to examine whether mean scores differ on the same group measured at two or more points in time. For example, we may want to examine whether mean performance changes after implementing a new staffing system. Most people think in terms of mean differences because they want to know if one condition is better or worse than another. This situation is most often encountered when presenting the results of a new staffing system; managers want to know if the hires that have gone through the new system are of higher quality, are performing better, or are staying longer than those from the old system. The preferred analytical tools for difference models are the *t*-test and analysis of variance (ANOVA), and these tools can be used to address several questions:

- How does the average tenure of employees hired last year compare with that of the year before?
- Are there any differences in the average time-to-hire across recruiters?
- How does the cost-per-qualified-candidate compare by recruitment source?
- Is the average performance of new hires at each location different?

For example, several years ago we worked with a call center that was changing from paper assessments to Internet testing (Ployhart, Weekley, Holtz, & Kemp, 2003). The call center hired applicants based on a battery of personality, experience, and judgment tests. They had historically administered these tests using paper surveys, but the Internet allowed them to administer the tests online, thus taking advantage of greater convenience, speed, and cost-efficiencies. It was important to see how test-takers might be affected by Internet testing, which was pretty novel at that time and unfamiliar to many people. We compared the test scores across three groups of people: 425 job incumbents who completed the paper assessments, 2,544 applicants who completed the paper assessments, and 2,356 applicants who completed the Internet assessments. We expected mean differences between job incumbents and applicants on the personality tests because applicants tend to respond more desirably than incumbents (who already have a job and hence have no need to embellish their responses). However, we really wanted to know whether the scores between the two applicant groups were comparable.

To illustrate, we report the findings just on the personality trait conscientiousness (this was measured on a 5-point interval scale, where higher scores indicate greater conscientiousness). We found scores were highest when applicants took the test using a paper assessment (mean = 4.01; SD = .39), followed by when applicants took the assessment on the Internet (mean = 3.85; SD = .49), and lastly as an incumbent taking the test on paper (mean = 3.53; SD = .41). These mean differences were each statistically significant from each other ($p < .05$). Furthermore, the differences were practically significant because the difference in SD units was 1.12 between applicants and incumbents when taking the assessment on paper, 0.65 between applicant Internet–incumbent paper, and 0.36 between applicant paper–applicant Internet. Thus, use of the Internet actually reduced differences that might exist due to applicant embellishment and resulted in applicants responding more like incumbents (i.e., more honestly). Table 3.5 provides a summary of these findings.

Relationship Models

There are many types of relationship models, but the most commonly used are based on the correlation coefficient. The correlation coefficient is a single quantitative number that estimates the strength of a linear relationship

TABLE 3.5 DIFFERENCE MODEL EXAMPLE

			Correlations				
	Mean	SD	1	2	3	4	5
Incumbent Paper-and-Pencil							
1. Conscientiousness	3.53	.41	(.63)				
2. Agreeableness	3.37	.42	.50	(.67)			
3. Emotion Stability	3.53	.48	.54	.59	(.73)		
4. Biodata	66.20	10.22	.25	.25	.32	(.75)	
5. SJT	7.32	3.84	.43	.24	.30	.19	(.46)
Applicant Paper-and-Pencil							
1. Conscientiousness	4.01	.39	(.64)				
2. Agreeableness	3.60	.35	.45	(.68)			
3. Emotion Stability	4.21	.40	.63	.53	(.72)		
4. Biodata	68.54	9.56	.31	.28	.35	(.78)	
5. SJT	10.44	3.28	.28	.25	.26	.22	(.43)
Applicant Web							
1. Conscientiousness	3.85	.49	(.75)				
2. Agreeableness	3.47	.36	.57	(.74)			
3. Emotion Stability	4.00	.51	.76	.60	(.80)		
4. Biodata	66.24	10.78	.28	.31	.39	(.79)	
5. SJT	9.79	4.16	.39	.31	.38	.21	(.62)

Values in parentheses are internal consistency reliabilities.

SOURCE: Adapted from Ployhart et al. (2003).

between two variables. Correlations range from -1.00 (perfectly inverse relationship) to +1.00 (perfect relationship). Zero indicates there is no (or little) linear relationship. Thus, the larger the number in absolute terms, the stronger the relationship between two variables. This is another way of saying that two variables or phenomena covary. For example, there is a positive correlation between cognitive talent and job performance, such that people with more cognitive ability perform better on their jobs. There is a negative correlation between the personality trait conscientiousness and absenteeism, such that those who are more conscientious tend to miss work less often. However, the correlation will be reduced if the scores on the two variables are not normally distributed, are unreliable, or have a small spread (lack of variance) to their distributions.

Many advanced analytical models are based on the correlation. By far the most common and most useful is the regression model. The regression model is powerful because it is capable of combining multiple variables in an optimally weighted manner. The regression model allows estimates of how much variance in the outcome is explained by a set of predictor variables (this estimate is the squared correlation, or *r*-squared). The regression model also allows us to make predictions and forecasts about future events, but always with the caveat that we should not forecast too far in the future from the data we have available today. That is, the further the prediction in the future, the less accurate it is likely to be, all else equal. There have been numerous studies comparing the predictions based on regression models to those based on human experts (e.g., regression predictions versus predictions of medical experts in diagnoses). The results are quite consistent—so long as the same information is used, the regression model will make more accurate predictions. Stated another way, the most accurate human predictions are those that the regression model will provide. One study summarized the results of 17 empirical research papers and found that the regression model prediction is at least 25% more accurate than human expert prediction (Kuncel, Klieger, Connelly, & Ones, 2013). However, we should never blindly accept the results of any analytical model, so the best approach is to use the regression model to combine information in the statistically optimal manner, and then have human experts examine the evidence and predictions to make a decision consistent with the business goals. Regression models could be used to answer questions such as:

- How will average time-to-hire change if the worst recruiter left?
- What is the cost-per-qualified-candidate if only the two most efficient recruitment sources are used?
- What is the average performance of new hires if hiring is done at only three locations?

Regardless of whether correlations or regression models are used, top-down hiring is the optimal way to make hiring decisions based on relationship models. For example, if there is a positive relationship between intelligence and job performance, and both are normally distributed, then the best way to hire people is to rank order them based on intelligence scores and then hire top-down until all openings are filled (e.g., hire the highest-scoring person first, then the second-highest-scoring person, and so on until there are no more vacant positions). Top-down hiring is the optimal way to make hiring decisions because it will result in the greatest expected performance. Any deviations from top-down hiring (e.g., hiring the seventh-highest-scoring person first because he is the son of your friend) will result in lower expected performance.

To illustrate the application of relationship-based statistical models, suppose we examine the predictors of employee customer service performance. The sample consists of candidates who applied for jobs in one of three different retail firms. All employees completed a selection test inventory measuring judgment, conscientiousness, agreeableness, and extraversion. They were also evaluated by their immediate supervisors in terms of job performance. The correlations of these scores and job performance were as follows: situational judgment ($r = .18, p < .05$), conscientiousness ($r = .12, p < .05$), agreeableness ($r = .16, p < .05$), and extraversion ($r = .15, p < .05$). These effect sizes were large enough to be practically significant and meaningful. This might lead one to believe that all four of the predictors should be used as part of a hiring process.

However, in a regression analysis that considered all of these effects simultaneously, only the relationship between situational judgment and performance was statistically significant (regression weight = .14, $p < .05$). The other three predictors (conscientiousness, agreeableness, and extraversion) were unable to improve prediction beyond situational judgment alone. Thus, if these firms wanted to maximize the efficiency *and* predictiveness of their selection system, they would only use the situational judgment measure, rank order candidates from highest to lowest, and then fill available jobs by hiring those with the highest judgment scores first. The results of such an analysis on five fictional candidates are shown in Table 3.6.

Causal Chain Models

The regression model is very useful for combining information in a statistically optimal manner, but it does not specify a causal sequence among the variables. However, many times we want to understand the causal chain between variables. For example, in talent supply chain models, sourcing talent influences attracting talent, which in turn influences selecting talent and then onboarding talent. The higher the quality that exists early in this process, the higher will be the quality of the final slate of applicants who will be hired. For causal chain models, there are a variety of analytical tools available, but one of the most useful is called path analysis. In path analysis, a set of variables is specified in a casual network. This approach is powerful because it allows one to understand how small changes early in the chain can produce large effects later in the chain.

We provide an example of a causal chain using a study we conducted on 238 quick service restaurants (Ployhart, Van Iddekinge, & MacKenzie, 2011). We reasoned that restaurants that were more selective with their hiring would see greater returns on training (operationalized as the number of employees within each restaurant who completed advanced training). In turn, we expected restaurants with more trained staff to manifest higher customer satisfaction, which should in turn produce greater growth

TABLE 3.6 RELATIONSHIP MODEL EXAMPLE

Variable	Correlation with Performance	Regression Weight
Situational Judgment	.16*	.14*
Conscientiousness	.09	.04
Agreeableness	.10*	.07
Extraversion	13*	.09

Hiring decisions are based on talent needed for customer service performance.

* $P < .05$

SAMPLE HIRING DECISION FOR FIVE FICTIONAL CANDIDATES

Candidate	Situational Judgment	Conscientiousness	Agreeableness	Extraversion	Regression Score[1]	Hiring Rank Order (regression score)
#1	75	23	85	42	21.15	3
#2	23	65	65	69	16.58	5
#3	50	22	33	91	18.38	4
#4	78	15	58	65	21.43	2
#5	85	22	71	49	22.16	1

For convenience, scores for each assessment range from 0 to 100.

[1] regression score = (situational judgment * .14) + (conscientiousness * .04) + (agreeableness * .07) + (extraversion * .09)

in sales-per-labor-hour over time. Thus, we proposed a talent profit chain where selection→training→customer satisfaction→sales-per-labor-hour growth. Each of these variables was measured over time, and the variables where lagged to strengthen causal inferences. Figure 3.6 shows the results of this analysis, where the numbers indicate the strength of the relationship between each pair of variables (higher numbers indicate a stronger effect). The numbers thus signify the direct effects. However, selection has an indirect effect on sales *through* training and customer satisfaction. Improving staffing by 1% would produce a 2% increase in sales-per-labor-hour each quarter; a sizable annualized return. Further, ensuring all employees within a restaurant met minimum hiring standards would reduce labor costs by 18%. Thus, even though selection did not have a direct effect on sales-per-labor-hour, it produced an important, cumulative indirect effect.

Change Models

The models discussed so far focus primarily on predicting a performance outcome at a single point of time (although the predictors of such outcomes may be measured earlier in time). Change models are different in that they emphasize the prediction of *change over time* in some performance outcome. Common examples include growth in market share, increases or decreases in profitability, changes in customer satisfaction or employee productivity, and so on. In our opinion, understanding how

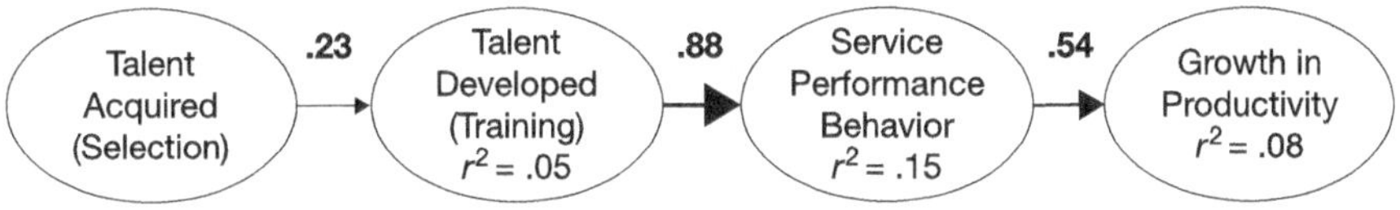

Figure 3.6
Causal chain model example (adapted from Ployhart et al., 2011). The size of the arrow indicates the magnitude of the effect. Variance explained (r-squared) is provided in each outcome variable. The numbers above each arrow indicate the size of the effect.

staffing and talent contribute to growth or change is one of the most important goals of talent analytics. Predicting growth is perhaps the most powerful means of analytical modeling, because the prediction of trends gives insight into where a firm can generate the strongest returns on its investment.

We illustrate the importance of looking at change models within the context of store-level talent. In a study conducted within a large retail establishment (Ployhart, Weekley, & Ramsey, 2009), we examined the relationship between average talent within a store and a variety of store financial and operational metrics (e.g., controllable profit; same-store-sales) over three quarters. Hence, we sought to examine whether improvements in talent over time created improvements in operational and external performance over time. Using a sample of 114,198 employees nested within 1,255 stores, we modeled the relationship between changes in talent with changes in store performance. The results found that improving talent related to performance growth positively, but with diminishing returns. In general though, stores with one standard deviation above average talent produced approximately $4,000 more sales-per-employee than did stores one standard deviation below average talent. Interestingly, these effects were only observed when examined over time; simple correlations suggested there was no significant relationship and hence obscured understanding of the dynamic relationship between talent and store performance.

Another example is provided by Kim and Ployhart (2014). Here, we modeled whether firms that were more selective in their hiring and used more internal training outperformed competitors before, during, and after the Great Recession. Using a sample of 359 South Korean firms, we found that firms who were more selective in their staffing and used more internal training generated greater profit than their rivals before and after the Great Recession, even though these same firms showed a bigger drop when the recession hit. Stated differently, firms with more selective hiring and internal training will generate more *profit growth* than competitors before a recession, and will recover more quickly after the recession starts. Figure 3.7 shows these findings.

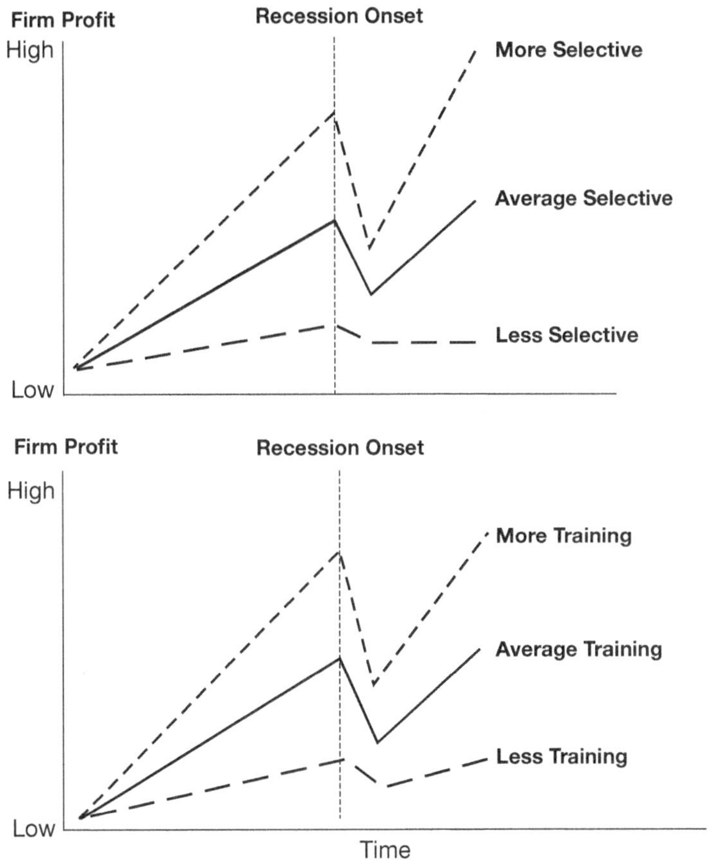

Figure 3.7
Change model example (adapted from Kim & Ployhart, 2014). The trend lines shown in the figures are approximate.

Multilevel Models

Multilevel models are intended to statistically aggregate employee-level data to strategic business unit (e.g., stores, departments, divisions) or firm levels. Multilevel models can be applied to any of the models used above. For example, we may have 10,000 employees nested within 500 stores. We can use multilevel models to determine whether the individuals within those stores are sufficiently similar that we can then take the mean level of talent *within* each store to create a new store-level aggregate talent score.

In this example, we might average the talent scores within each of the 500 stores, and then run all subsequent models at the store level (based on a sample size of 500 instead of 10,000). The benefits of this approach enable one to link talent to strategic outcomes and metrics that usually reside at levels higher than individual employees. For example, in one study we looked at differences in personality traits across 12 retail organizations and over 80 jobs within those organizations (Ployhart, Weekley, & Baughman, 2006). We found personality was fairly homogeneous within each of the firms and jobs, to the extent that we could distinguish firms and jobs based on the average personalities of those within them. Further, these firm- and job-level personality traits manifested differences in job performance and satisfaction beyond those effects found for individuals. This means that employee job performance and satisfaction were partly determined not only by their own personalities but also by the average personalities of their coworkers and the organization more broadly! For more information on multilevel modeling, see Bliese (2000) and LeBreton and Senter (2008).

Big Data Models

The newest analytical approaches are those being used within the broad umbrella known as "Big Data models." As noted earlier, the pervasive use of personal digital devices (e.g., smartphones) has created volumes of data on individual behavior and actions, and much of this data may be used to make smarter business decisions. Big Data generally refers to data that is high in volume (amount), velocity (speed at which the data is generated), and variety (types, including numerical and text). Now that such data is available, Big Data analytics can be used to identify patterns in the data. Many of the analytic models that can be used are based on "machine-learning" principles, which means the models search through data according to algorithms and look to identify meaningful patterns. Thus, a common feature of many Big Data models is the use of computer-based iterative methods for finding patterns without as much human direction. Examples of Big Data models include decision trees (to model a chain of decisions or choices), network analysis (to model social networks), text mining, Bayesian or resampling approaches, ensemble models (combinations of analytic models), machine learning (models that learn autonomously after being programmed with basic rules), random forest/classification models (to identify subgroups or patterns), and spatial association models. Big Data models are powerful approaches that offer incredible potential, but they also raise a number of serious concerns when applied to staffing. For example, does the application of Big Data models on employee data violate privacy laws? Should an employee consent to having his or her data used? How might Big Data operate in a global context, where countries differ in their laws and policies? Guzzo, Fink, King, Tonidandel, and Landis (in press) raise a number of practical concerns with Big Data that need to be considered. Currently, there is almost no guidance on how to handle Big Data in an ethical and responsible manner. We believe this topic will dominate the HR profession over the next decade (see Ployhart, 2015).

Conclusion

Depending on the manner in which data is collected, different analytical models may be used to generate insights from those data. Note that all of the models discussed in this chapter assume at least an interval scale of measurement (alternative models exist for nominal and ordinal scale data). Simple models, such as distributions or differences, are easy to explain but offer less information. More powerful models, such as causal chain or change models, offer insights of more interest to business partners because they speak to how much talent impact may exist and which levers to pull to generate the greatest impact. The appropriate analytical model is dependent on the quality of the data, the types of data, and importance of the question to be addressed.

All of the analytical models discussed in this chapter provide an estimate of effect, but they express these effects in different ways (see Table 3.7).

TABLE 3.7 TYPES AND DEGREES OF EVIDENCE FROM DIFFERENT ANALYTICAL MODELS

Model	Effect (and symbol)	Estimates	Size
Distribution	Mean (M); Standard Deviation (SD)	z-scores (or standardized) scores convert all scores into SD units: z-score = (score – M) / SD	Z-score will have a mean of zero and an SD = 1. More positive (above the mean) or negative (below the mean) scores are farther away from the mean and hence more extreme.
Difference (*t*-test, analysis of variance or ANOVA)	Standardized Mean Difference (*d*)	*d*-value: $(M_{group1} - M_{group2})$ / (sqrt $[(n_1 - 1) * s_1^2 + (n_2 - 1) * s_2^2]$ / $(n_1 + n_2 - 2)$])	*d*-values .30 small .50 medium .80 large
Relationship			
Correlation	Correlation (*r*)	(Covariance XY) / $(SD_x * SD_y)$	.10 small .20 medium .30 large

Regression	Regression weight (*b*); *r*-squared (r^2)	Regression weight: Relationship between variable and outcome, conditioned on all other variables in the regression equation *r*-squared: explained variance / total variance	Regression weight: How much of a change in outcome given a one unit change in predictor. Depends on other variables in the model. *r*-squared: .01 small .04 medium .09 large
Causal Chain	Path weights (*b* or β or γ)	Path weight: Relationship between variable and outcome, conditioned on all other variables in the path equation	How much of a change in outcome given a one unit change in predictor. Depends on other variables in the model. Can estimate direct and indirect effects.
Change	Growth or decline over a period of time	Slope, or rate of change over a period of time	How much of a change in outcome given a one unit change in predictor, per time period. Depends on other variables in the model and length of time (months, quarter, years, etc.).

(*continued*)

TABLE 3.7 CONTINUED

Model	Effect (and symbol)	Estimates	Size
Multilevel	Aggregation of individual employees to business unit or firm levels.	Intraclass correlations (ICC): Variance at level / total variance	How much variability in talent is explained by between-business unit differences. ICC: .01 small .10 medium .25 large

NOTE: n = sample size.

Mean difference models express effects in terms of a standardized difference between two groups. This is known as a "*d*" value and converts any difference into standard deviation units, where larger numbers indicate a larger effect (i.e., larger mean differences). Correlational models express effects in terms of strength of relationships (correlations) or variance explained (also known as "*r*-squared"). Causal chain models provide two types of effects. There can be both direct effects (X leads directly to Y) and indirect effects (X leads indirectly to Y through M) between variables. In this manner one can evaluate the total effect of a variable far upstream in the causal chain on the outcome. Finally, change models enable an estimate of growth or decline over some period of time. Big Data models blend elements of all these approaches.

DEGREES OF EVIDENCE

All analytical models will offer estimates of effects, statistical significance, and practical significance. But not all models will offer the same degree of evidence. Models based on experimental designs manipulate or use random assignment to control for confounding effects. For example, a firm may implement a new staffing system in one plant while using the old staffing system in a different, but closely comparable, plant. Differences in applicant quality between the two plants should be attributable to the difference in the staffing system. In contrast, correlational designs are those that do not involve manipulation or random assignment. For example, a firm may implement a new staffing system in both plants and observe differences that occur after the system is implemented. This is valuable, but it is not often certain whether the changes are due to the staffing system or other external changes that co-occur (e.g., change in unemployment rate, other regional differences such as education levels, etc.).

Generally speaking, models based on experimental designs offer more evidence than correlational designs, because they rule out more extraneous effects. Yet, experimental designs are more costly and resource-intensive to conduct (plus, if there is a positive effect for the staffing system, it is not fully realized until adopted by all groups). This leads to our final point: the type of analytical approach you should adopt is one based on the amount of evidence needed to support a decision, but in practice is often driven primarily by what data are available. Table 3.8 provides an overview of the benefits and costs associated with different degrees of evidence. We summarize Table 3.8 with one simple fact: it costs more resources (money, time, people) to have more evidence and thus be more confident in hiring decisions. As a manager, you need to carefully consider how much evidence is truly necessary, versus how much evidence is desired. If you want

TABLE 3.8 DEGREES OF EVIDENCE

		Benefits	**Disadvantages**
Strong Evidence	▪ Experimental designs ▪ Change models ▪ Multilevel models ▪ Outcomes at the firm level ▪ Outcomes based on external performance metrics	▪ Stronger inferences of causality ▪ Outcomes of compelling strategic importance ▪ Easy to convey business impact	▪ More resources (expensive, time consuming, large samples) ▪ Harder to find direct relationships with objective external performance metrics ▪ Effect sizes tend to be smaller
Moderate Evidence	▪ Short-term change designs ▪ Relationship models (regression) ▪ Causal chain models ▪ Outcomes at the business unit level	▪ Moderate inferences of causality ▪ Outcomes of reasonable strategic importance ▪ Some assumptions required to show business impact	▪ Resource efficient (moderate cost, time efficient, moderate sample requirements)

TABLE 3.8 CONTINUED

		Benefits	Disadvantages
	▪ Outcomes based on internal (operational) performance metrics		▪ Moderate difficulty in finding direct relationships with performance metrics ▪ Effect sizes tend to be smaller for external than internal performance metrics
Weak Evidence	▪ Relationship models (correlation) ▪ Difference models ▪ Distribution models ▪ Outcomes at the individual level (job performance)	▪ Weak inferences of causality ▪ Outcomes of tactical, but not strategic, importance ▪ Many assumptions required to show business impact	▪ Relatively fast and inexpensive ▪ Straightforward to find relationships between talent and individual subjective job performance metrics ▪ Effect sizes tend to be smaller for objective performance than internal performance

more certainty about a hiring decision, it will require more resources and it will cost more money.

We believe in modeling the firm-level or strategic business unit–level consequences of talent whenever possible. Modeling the data at this level usually allows a linkage to key strategic outcomes and metrics that are indicators of a firm's strategic goals. Further, relevant alternative variables (e.g., size, market saturation) that may be favored by others trained in operations or marketing can be included in the analysis. Firms also usually track these business-level metrics internally and externally, hence the data are available. It is also advisable to conduct more than one type of analysis—different models may yield different insights. We reiterate points raised in the chapter on strategy: start with the metrics and performance outcomes most directly indicative of strategic goals, and then link talent to those outcomes.

Roger was glad he fought hard to get an analytics person onto this team to provide insight and focus on the data-gathering mission the team was charged with. Now, the data that was available had been collected and it was time to put together an evidence-based argument. Even if leaders had different views on the data and on potential actions to be taken in response to it, the data had to be the basis to start the conversation. So he determined to follow a path that would tell a compelling story and create a shared commitment to the vision.

- The talent analytic plan would start by setting a baseline using existing and historical data.
    - First, identify the key performance outcome metrics on which to evaluate the efficacy of the talent acquisition plan. These would take two forms.
        - The first were performance metrics of key interest to GMs and leaders within the various lines of business. These external metrics were the ones linked to strategic objectives, which included growth in sales, customer satisfaction, and market share.
        - The second were performance metrics of key interest to HR and GMs. These internal (operational) metrics included productivity, scrap rates, turnover, and accidents.

  - Second, in the high-recruitment-volume jobs, he would use talent scores based on the past or current vendors' assessment systems. These were usually some combination of basic knowledge and experience assessments, sometimes supplemented with personality and ability testing.
  - Third, in the key managerial, or low-volume jobs, he would use what existing information was available on skills and abilities (e.g., scores in training programs).

However, before worrying about any of this, the reliability and validity of the scores used to estimate the quality of talent would need to be assessed.

- The nature of the performance metrics would drive the talent linkage approach. Because all the primary performance outcomes were at the strategic business unit or regional levels, the talent data would have to be aggregated to those levels using multilevel models.
- The nature of the jobs would dictate the amount of evidence required, in order to balance cost versus ROI. The key managerial jobs might seem the natural ones for more evidence, but the reality was that the high-volume jobs actually could generate greater returns because turnover was so high and they had less information on the incumbents of those roles. In contrast, the key managers had already been successful in their prior operations and so they had more information available about them—they were a safer bet.
- Linking the collective talent resources to the business unit outcomes would be based on analytic models most appropriate to the data. Change models would be used to link collective talent resources in the high-volume jobs to growth in sales, customer satisfaction, market share, and productivity. Regression models and correlations would be used when data were not available over time. Mean differences would be reported between plants and regions to gain perspective on the variability in talent.

- The effect sizes observed by the baseline results would provide the comparison for choosing among different vendors, recruitment plans, and selection practices. At a minimum, the effect sizes had to be better than what was currently in place, or at least no worse.

Overall, it was felt that this data-driven plan would set a solid baseline of evidence to pinpoint which specific regions/plants/operations should be the focus of the first rollout of Roger's plan. The baseline results would allow the team a means to compare and contrast different recruitment and selection policies and practices. The data would be used to argue for a regional plan. For example, mean scores and variability within and across countries would be used to demonstrate that there was more similarity than differences across the similar jobs in different countries. Certainly cultural differences are important, but may not always be important when looking at the nature of work and the performance of specific jobs. Such analytics help move the conversation from one of opinion to evidence. Let the data do the talking.

4

Recruiting Talent Globally

THIRTY-SECOND PREVIEW

- Recruitment refers to the strategies and practices used by firms to (1) identify and source desired talent, (2) attract the desired talent, and (3) ultimately ensure they accept job offers.
- The goal of global recruiting is to generate a sufficient pool of qualified talent who *fit* with the job, organization, and national culture.
- Recruitment can be a strategically valuable way to differentiate the firm from competitors by attracting top talent.
- Recruitment is a dynamic process that occurs across distinct stages. Applicants are looking for different types of information at each stage.
- Talent is unequally distributed around the globe, and applicant preferences differ across cultures. Companies need to make

strategic choices about whether practices should be tailored to each culture or be universally applied.

- Applicants have needs and wants that vary across cultures. However, most job choice decisions are based on wants; but needs are used to eliminate firms.
- Global recruiting practices should be monitored using metrics that are linked to the firm's strategy.
- The Bottom Line: Talent is a rare and valuable resource that varies in supply around the world, so attracting it effectively, efficiently, and consistently can be a source of global competitive advantage.

Roger sat back on the plane and looked idly at the clouds passing by him. He was making his first visit to Singapore to meet members of the local senior management and members of his team there. He had heard much about Singapore, a vibrant city-state that was a true economic wonder and was sometimes described as "Asia for babies" because of its greater Western orientation. It was only a 40-minute flight from Kuala Lumpur, but several colleagues had given him a heads up that there was a significant difference between the two countries. Malaysia had a more ethnically balanced population where the ruling party was a coalition made up of Malay, Chinese, and Indian subparties, but real political power predominantly lay with the ethnic Malays, who constituted about half of the population. In Singapore, on the other hand, the Chinese made up nearly 75% of the population with the Malays and Indians constituting 13% and 9%, respectively.

After even his first week in Kuala Lumpur he realized how much he had to learn about the skills needed in his new role, quite apart from the cultural assimilation that had only just begun. His experience in the work of recruitment was "limited." So what experience could he lean on that would at least start the voyage of discovery into this new professional world? Suddenly it occurred to him that he had at least one relevant experience—his experience of being recruited from

Christ's College some 10 years previously. He had wanted the chance to make a difference to those with whom he worked; he wanted to join a company that developed its employees over long careers; he wanted to travel internationally; he wanted to work with bright people who shared his values of honesty, integrity, and respect for people; he wanted to work for a company that left the countries in a better place than before their entry there. So he started to wonder whether those were unique perspectives or ones that were more broadly held and, if so, how could he frame a compelling proposition to the potential talent that would be looking at his company in Asia?

Finally, he was struck by the thought that while he had made a concentrated effort to overcome his lack of knowledge of Malaysia and the region through extensive reading, they mainly focused on general tendencies which, while thoroughly researched, could overlook the obvious fact that individuals were just that and could vary significantly one to another. How would he use these questions to frame up possible ways to attract the very best talent to his company?

For sure, it was a lot to ponder, and Roger knew he had much to learn—and the key to success was to start with an open mind and a willingness to be grounded in his own view of the world BUT dismiss the possibility that it was the ONLY view. He would need to adapt but remain grounded; he was to be like a chameleon with a solid spine and that was not going to be easy.

Recruitment refers to the strategies and practices used by firms to (1) identify and source desired talent, (2) attract desired talent, and (3) ultimately get them to accept job offers. Recruitment has historically been "organization-centric" such that the focus was on efficiently processing candidates who came to you. This model is largely one-directional and linear, and assumes that the candidates are actively looking for a new job. Although this organization-centric model is still in use, recruitment is becoming much more "applicant-centric," such that the focus is on finding ways to attract and entice specific talent to the firm. This means going to where the talent is and offering them a value proposition that your competitors cannot match. Many of these are passive candidates, meaning they are not actively looking for a job but would consider a new job if the right opportunity came along. Recruitment is like a dance between the firm and candidates, with each party leading in some instances and following in others.

If modern recruitment means going to where the talent is, then organizations need to think globally, because talent is unequally distributed around the globe. As a consequence of the factors noted in chapter 1, there are large national and regional differences in the availability, surpluses/shortages, and types of talent. Given that talent must be available in the right quantity, right quality, and at the right place and time for it to enhance firm effectiveness (see chapter 2), it becomes vital for firms to know where and how to acquire talent globally. The challenge is that differences in national culture, political systems, laws (including but not limited to those related to immigration and work permits), language facility, and labor markets can affect firms' attempts to source talent. For example, differences in cultural values across countries may require different recruiting practices and messages to be used from the same multinational firm. It is perhaps for these reasons that many firms now use recruitment process outsourcing (RPO), which means relying on an outside vendor to handle the recruitment process.

Recruitment is big business, because acquiring the right kind of talent enables firms to better compete and perform. One report suggests that the global market for recruiting exceeds $450 billion (Koncept Analytics, 2015). Chapter 2 suggested that recruitment will determine the quality of a firm's human capital potential and ultimately its ability to adapt and compete. Recruiting will set the upper limit on what a firm may achieve through its people. Yet, if recruiting fails to attract qualified candidates, then future hires will not have the capacity to perform well and fulfill the organization's goals. If recruiting is successful in converting applicants to employees, but fails to identify people who fit the company's culture, then high turnover costs will offset recruiting investments. Thus, effective recruiting requires ensuring a good fit of applicants to jobs *and* the organization. Effective *global* recruiting additionally requires ensuring a good fit of applicants to the national culture, company culture, and work environment. Adding the often-discussed issue of changing attitudes and expectations of different generations, this becomes a complex and multidimensional problem—like three-dimensional chess.

Recruiting is undergoing a period of revolution and renaissance enabled by technological developments. At no time in history has it been as fast, easy, and inexpensive for organizations or individuals to learn about each other. Technological advancements make it possible for organizations to source and reach qualified applicants anywhere in the world—and vice versa. Furthermore, the public nature of sites like LinkedIn even allows companies to make strategic assessments of the quality of key talent in the competitors. Wherever there is access to the Internet, people can learn about job openings or apply for a job.

As the plane began to descend, there were lots of questions spinning in Roger's mind:

- Where does a top firm find top talent? What practical constraints are there? How can firms source talent globally or regionally?
- How does one recruit talent to a firm when the local people and the global firm can have different cultural values? How can a firm do this better than competitors?
- What is the appropriate resourcing mix between internal and external candidates? What are the messages, both conscious and unconscious, that companies send to existing staff by making choices about recruits, leaders, and so forth?
- Who should do the recruiting? Line leaders? HR staff? Search firms? Or was there a time and a place for all of these to be key agents in the process?
- How will I know if my recruiting system is successful? How do I balance cost-effectiveness versus outcome effectiveness?

Different criteria are likely to lead to different answers relative to the same question.

Roger was somewhat familiar with these issues, but really didn't know how to start thinking about recruitment in a regional context. He had mainly been an observer to recruitment activities—quietly present in meetings with other HR leaders when the topic had been discussed, but never formally responsible for developing and implementing a recruitment plan. And even these conversations were limited to the single country in which he grew up. So to get past the self-doubt, he thought about the various issues and jotted down a list of "to dos":

1. Recruiting is strategic but what does that mean? He went back to his thoughts about the criticality of linkage. The recruiting strategy must be consistent with the Spooner business strategy and the overall talent strategy.
2. I have to figure out how to get everyone on board. In a consensus culture, which is very much the norm in the region, how will I convince my team and the leadership of the need to approach it that way? Am I assuming that the current system is broken? How will I secure resources, financial and people, to execute an overarching strategy? Who would be allies in this effort, and how do I identify them? Who could be resistant, and how do I need to manage/deal with that?
3. I know, or at least believe, that national cultures will vary. How then can I implement a regional recruitment plan? Should we adapt to the different cultures, or do I start with mapping our company culture and look to present that in a way that appeals to the different geographies? What would that look like? Why would the top talent "want" to work for us? How are our competitors doing that, and how can we do it better?
4. Any process we use for recruitment will need to be both efficient AND effective. I must establish which is the earlier priority and how I will measure the success of what I design? NOTE TO

SELF: Must remember what a cultural expert wrote. In collective cultures, like Malaysia, it is about the "we not me." Hmm.

Roger had reviewed the staff numbers by region before his flight (see Introduction to Case in the Frontmatter; Employment Numbers by Country; XXIV). In Asia, despite the traditional reputation for affiliation and loyalty, turnover was frequent but unevenly distributed across the countries. This turnover was largely fueled by the combination of growing economies, many new entrants to the region, and a relative shortage of the types of skills that companies like Spooner coveted. Based on prior scenario planning, he quickly realized that Spooner would need to be an aggressive recruiter in the region.

But the real elephant in the room was the expansion agenda around the strategic job families (discussed in chapter 2 and reproduced in this chapter). Roger would need to develop a special recruiting plan to source and attract candidates in these key jobs, in a region where they were already in high demand. He would have to identify what made Spooner distinct and attractive relative to competitors, but in each of these job families (Table 4.1).

TABLE 4.1 "GUMBALL CHART" OF EXPECTED AND OBSERVED TALENT QUALITY FLOW NUMBERS

	# Observed	# Expected	Gap (Stock)	Gap (Flow-no growth)	Gap (Flow-expected growth)	Gap (Flow-above avg growth)
Plant assistant general managers	38	38	0			
Assembly team associates	4,918	5,500	-582			
Engineering R&D team members	45	47	-2			
R&D team leaders	7	12	-5			

diagonal lines = acceptable, vertical lines = risk, horizontal lines = deficit.

Roger had never had any formal budgetary accountability, so he was somewhat overwhelmed with the financial scale of the department—and how he could demonstrate the value add to his line colleagues, who often only saw the tip of the iceberg in terms of activity but also saw the cost impact reflected in their own budgets. He had looked over the figures and even to him they seemed pretty high.

Staff costs, with burden such as pensions and other benefits, totaled nearly $1.5 million. His own costs were nearly a third of that, given the various taxation, leave travel, housing, and other expenses that combined to triple the effective cost of his salary of $150,000. Travel was the next largest element, nearly $100,000, with advertising set at about the same. Finally, there was an allocated budget (which had had apparently come under significant challenge at the annual plan review), of nearly $250,000. The approval had only been given with the condition that for every dollar spent on vendor outsourcing, a savings of $1.25 had to be shown in internal cost reductions. All in all, the budget was over $1.9 million, and in an industry where every cent counted, that was a LOT. The pressure to create a plan that was effective would be great—Roger knew he would need a strategy.

RECRUITMENT IS STRATEGICALLY VALUABLE

The primary means through which firms compete is by increasing profits and/or decreasing costs in a manner that differentiates them from competitors. A firm increases productivity or decreases costs by improving operating efficiency and strategy execution. Good talent increases the odds of both, so effective recruiting can itself be a strategically valuable capability. Firms that are able to attract and retain rare but important talent also derive an indirect benefit, that of keeping talent away from their competitors. Whether it is through enhanced execution of one's own strategy, or denying the competition the resources needed to execute their strategy, recruiting can be strategically valuable.

- **Recruitment Shapes the Talent Supply Chain.** The talent supply chain presented in chapter 2 starts with sourcing and recruitment. Small changes in recruitment strategy, practice, or implementation can cascade into large and irreversible consequences at later stages in the supply chain. We are familiar with many firms that have cut or drastically reduced graduate recruitment because of short-term business conditions. In 10 years, when perhaps what is most needed for success of the enterprise is staff with 10 years of relevant experience—the heartbeat of many organizations—there will not be enough. The talent supply chain is path-dependent. As the function that determines the flow of the talent supply, recruiting's consequences are disproportionately large relative to recruiting's inputs. In turn, this means firms that get recruitment right will be creating talent resources that will be difficult for competitors to imitate. Even if they can, it will take a long time to recoup the benefits.
- **Recruitment Determines a Firm's Talent Potential.** *Talent potential* is the extent to which a firm's employees possess the competencies needed to execute the firm's strategy, develop new products and services, and exploit opportunities now and in the future. Firms with greater talent potential are better able to create, identify, and capitalize on market opportunities to achieve a competitive advantage. Firms with insufficient talent potential become *talent constrained*, meaning they are unable to implement strategy or pursue new market opportunities. When that occurs, buying talent from the external labor market is often the only solution, albeit an expensive one. For example, Dana Petroleum, a medium-sized oil and gas company based in Aberdeen, has interests spanning Europe and Africa. Owned by the Korean National Oil Company and in need of internationally experienced staff, they recruited heavily from competitors. They achieved a massive infusion of talent with international perspectives, and this approach was seen as a key element of their strategy. They also sought local staff who understood the

specific cultures of the focus countries. As one can imagine, the longer-term task was to meld all of these with a culture that was both unique and unifying.

- **Recruitment Provides Access to Rare Talent Resources.** By definition, talent is rare because talent means having the right competencies needed for the job that contribute to the firm's broader effectiveness. But shortages of talent exist in many occupations, making talent even more rare. The ups and downs of performance in different industries affect short-term academic and career choices, which in turn impact long-term supply-demand patterns. Consider the postrecession construction downturn, which led to highly skilled plumbers and electricians being open to different employment options. One large home improvement company saw this a strategic business opportunity to employ some of these skills in their stores, so that customers could have inquiries answered by staff that had practical and deep expertise in the products being sold. Although there is debate over whether talent shortages are real or created by overly selective firms, the consequences are the same—talent shortages limit a firm's talent potential and ultimately its ability to compete and adapt.
- **Global Recruiting Practices are Difficult to Imitate.** Recruitment is focused on acquiring valuable and rare talent resources, which are unequally distributed around the globe, and are time consuming to imitate because of the talent supply chain. Recruitment is even more difficult to imitate in a global context because effective recruitment is based on a unique mix of the firm's culture and operations with the cultural values of employees in subsidiary countries. These cultures and operational processes must be complementary.

Conclusion

Recruiting is of profound importance for competitive advantage, and yet few firms think about it strategically. Because recruitment affects the quality and quantity of talent that enters the firm, effective recruitment increases a firm's talent potential and hence reduces costs on all subsequent HR activities. Yet recruitment can't work well if there are insufficient talent pools. The nature of economic, demographic, and societal shifts highlighted in chapter 1 all suggest talent can be a scarce resource that is geographically distributed. Therefore, talent pools are found in different places around the world and the effective acquisition of talent in the modern economy must adopt a global perspective. Effective global recruitment is necessarily a firm-specific blending of multinational (MNC) culture and values with the cultural values of subsidiary employees and candidates. It is highly unlikely there is a "simple formula" for effective global recruitment because recruitment will be tailored to the firm, subsidiary, and cultures. This in turn requires each firm to figure out its own unique strategy and, for the firms that can do it right, ensure their approach is reasonably protected from competitors. Understanding this perspective begins with an understanding of fit within a global context.

RECRUITMENT IS ABOUT FIT

Recruitment is about the fit between a person and the broader environment. *Fit* is a broad term that refers to the match between a person's talent competencies, interests, values, needs, and wants, with task demands, workgroup values and demands, organizational values and practices, and cultural values. Good things happen when there is a match between the person and the broader environment—the person is more productive, satisfied and engaged, and likely to stay. Many bad things can happen when there is a mismatch or a lack of fit—performance declines, satisfaction and engagement suffer, and turnover increases.

Fit occurs in different ways. *Supplementary fit* occurs when the individual's characteristics are similar to those of the broader organizational context. For example, if a project team needs everyone to speak Portuguese, then a person's knowledge of Portuguese supplements the group's knowledge. *Complementary fit* occurs when the individual's characteristics address important gaps or holes in the broader environment. For example, if the project team members all speak Portuguese, but they need someone who can also speak Vietnamese, then an individual's Vietnamese language knowledge will complement the skills of the group. Thus, fit means having the same characteristics (supplementary) or having different but interrelated characteristics like those in a jigsaw puzzle (complementary) (Muchinsky & Monahan, 1987).

Another way to think about fit is in terms of task and social distinctions (Kristof, 1996). *Demands-abilities* fit has a task or "can do" focus and refers to the match between the knowledge, skills, and abilities of a person, and the task-related demands of the job. For example, if the job requires technical knowledge of chemical engineering, then a person who has more of that knowledge is a better fit for the job. *Needs-supplies* fit has a social or "will do" focus and refers to the match between the person's preferences and desires and those that the organization needs. For example, if the organization's culture emphasizes teamwork, then only those with a social, teamwork orientation are likely to fit within the firm. Again accepting that national tendencies are just that, the more collectively minded countries, such as much of Asia, can find it difficult to adapt to a company where the western European or North American individualist mindset prevails. Thus, fit occurs for both task and social characteristics, and both types of fit need to be considered in recruitment.

Companies are increasingly recruiting and screening candidates based on fit. IKEA, a global household furnishings retailer, has offered visitors to its company website job portal the chance to take a quiz and determine whether they fit the firm's unique Swedish culture. As questions are answered correctly, the respondent is rewarded by pieces of furniture populating a room. Those ending with a sparsely furnished room are discouraged, but not prevented, from applying (candor being one of the company's defining cultural values). As another example, one airline that emphasizes exceptional cabin service and consistent friendliness has a motto: "If it moves, greet it." They instill this service first by very careful recruitment and selection. Thus, one may think of overall fit as a match between a person and the totality of his or her environment, with this overall match being affected by more specific elements of fit.

Research in social psychology provides overwhelming evidence that similarity breeds liking and attraction (McPherson et al., 2001). People tend to get along better and want to spend time together when they have similar interests and values. This becomes a virtuous cycle, leading to even greater similarity. Organizational research suggests that work units of all sizes, from small groups to organizations, will evolve toward homogeneity (similarity) over time (Schneider, 1987). This is because people will be attracted to organizations that are similar to them (needs-supplies fit), applicants with the necessary task and social characteristics will be hired by firms (demands-abilities fit), and applicants will choose to remain at firms that match their interests. This attraction-selection-attrition (ASA) model is important to understand because it provides insight into how fit may be used to transform individual talent into collective talent resources. Figure 4.1 illustrates how this occurs. The processes of ASA create homogeneous talent resources because they build a critical mass of employees with similar levels of talent. It moves the focus from a collection of talented individuals to a collective talent resource that can be deployed to achieve competitive advantage.

All recruiting strategies and activities should be targeted on improving fit because doing so will increase the likelihood of the right types of applicants (1) having a favorable image of, (2) applying to, and (3) accepting job offers from, the organization. Of course, sometimes it is necessary to hire talent that is different from the firm. We consider this topic in a later section. But fit is affected by a number of factors, national culture, and the stage at which recruitment occurs.

RECRUITMENT OCCURS IN STAGES

Earlier we noted that the modern approach of recruitment is one of enticing a partner to dance: you have to approach the person, put forward your best impression so that s/he will want to dance with you, and then move in sync with the person so that you don't make a fool of yourself. And of course, assuming you both like each other, the dance leads to spending more time together!

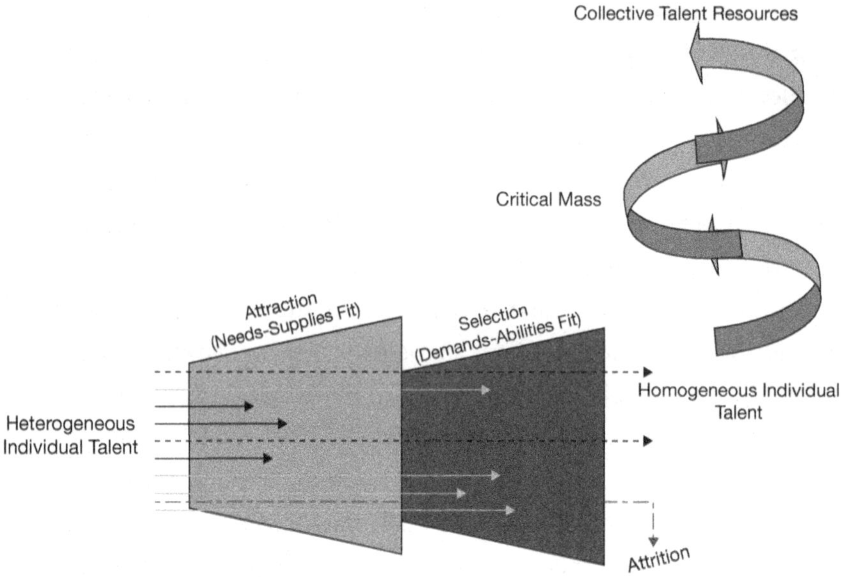

Figure 4.1
How attraction-selection-attrition create collective talent resources.

Like dancing, recruitment evolves over time as each learns more about the other, each tries to manage the impressions they leave on the other, and each tries to stay in sync. Yet there is an information asymmetry present because applicants and organizations clearly know more about themselves and are presenting their best impressions to the other. Thus, the types of information observed and scrutinized will change over time. There are three distinct stages of recruitment, and organizations should realize that different types of information are more or less important to applicants at different stages (Barber, 1998). Table 4.2 shows an overview of these stages and the types of fit dominant at each stage.

Stage 1: Source and Convince to Apply

The first stage involves the organization finding and then convincing qualified candidates to consider their firm for employment. For low-skilled

TABLE 4.2 STAGES OF EXTERNAL RECRUITMENT AND FIT, AND THE DIFFERENT GOALS AND CHOICES MADE AT EACH STAGE

	Stage 1: Source & Convince to Apply	**Stage 2: Persuade to Remain**	**Stage 3: Influence Job Choice**
Organization	Identify talent pools Focus on attracting a large number of qualified applicants Emphasize organizational image/brand/reputation Emphasize needs-supplies fit	Test and evaluate applicants based on job-related characteristics Emphasize job-relatedness, speed, fairness of selection process Emphasize demands-abilities fit	Provide competitive offer to applicants who pass assessments. Provide reasonable explanation for those who do not pass assessments Emphasize instrumental/job/compensation information along with company information Emphasize needs-supplies and demands-abilities fit
Applicant	Screen/narrow set of potential organizations based on needs	Focus on performing well in selection assessments Decide whether to stay active in the selection process	Decide whether to accept offer; negotiate offer by focusing mainly on wants

jobs in high-unemployment areas, this may require nothing more than announcing an opening. In tight labor markets with a limited supply of qualified candidates, a far more active approach is required. In either case the candidate is trying to decide whether she or he should apply to a given firm. The candidate is attempting to deduce, from a potentially large number of organizations, which should be pursued. In the case of a passive applicant, the candidate is trying to determine whether it is worth it to take the risk of joining a new employer and leave the old one behind. Top talent has a lot of choices and they tend to be more critical of even slightly negative information. The organization, on the other hand, is trying to persuade qualified talent to apply, deter unqualified talent from applying, and to do the latter in a way that does not tarnish the firm's reputation. The organization therefore emphasizes factors that increase perceptions of meeting the applicants' needs and attracting applicants to the firm, while signaling to unqualified candidates the unlikeliness of receiving an offer.

Stage 2: Persuade to Remain

Even a fast selection process will still take a few days or (more commonly) a few weeks to complete. During this interim period between application and offer, the firm must maintain the candidates' interest in the organization while it completes its due diligence on each candidate's qualifications and fit relative to the other candidates and the job requirements. In some ways the tables have turned and, after being recruited, the candidate is now subjected to a variety of screening tools (e.g., interviews, assessments, background checks; discussed in chapter 5). Each of these activities sends a signal about the firm (e.g., "I will be joining a selective group of high achievers." or "This company is impersonal and just looking for ways to disqualify me."). The candidate may not like this and question the value of pursuing employment with the firm. The organization must therefore emphasize the job relatedness and relevance of the selection process, and its value to both the organization and the applicant (e.g., "to ensure the people we hire find the work satisfying and meet our high expectations"). In doing so, it increases demands-abilities fit.

Stage 3: Influence Choice

The final stage consists of the organization extending job offers to select candidates. The candidates with offers must now decide whether to accept an offer with a particular firm relative to an existing job or potential offers from other firms. The organization is at the point where it is emphasizing very specific information about the job, employment offer, and related factors. The applicant is carefully screening all of this information, comparing it to other employment options and personal desires, and potentially negotiating changes to the terms and conditions of the offer. Both the applicant and the organization now focus on very specific and detailed information—the candidate's fit with the job and organization.

Thus, across the stages of the recruitment process different decision processes are taking place for applicants and organizations. Starting at the early stages, applicants are focusing mostly on broad company-wide information (e.g., reputation, type of industry, location, culture), and then proceed to more specific details of the job and organization (fit with workgroup, supervisor). The applicant is trying to narrow choices at the first stage and focus on the available job openings for which he or she may be qualified. In contrast, organizations are trying to cast a large but targeted net and widen choices at the first stage, so that they can then efficiently winnow the candidate pool to the most qualified group. Recognizing that candidates are engaging in very different decision processes at each stage allows an organization to tailor its recruiting information to the type of most value to candidates in that stage. Customizing information to candidate needs at a given stage will be more effective at influencing candidates' attraction and choice. Customizing the message to the stage can differentiate the firm's recruiting message from competitors who ignore the stages or fail to align the information to the stage. To further complicate matters, different cultures will have different expectations for how fast these stages will occur. For example, those from cultures with a long-term orientation will be more patient, perhaps even more expecting, of a more lengthy recruitment process. We discuss these cultural influences next.

A MODEL OF GLOBAL RECRUITMENT

Fit is at the core of recruitment, but a number of factors influence fit and hence become the levers that managers can try to pull to enhance fit and thus recruiting effectiveness. Figure 4.2 shows the main factors that influence fit, attraction to organizations, and ultimately job choice in a global context. Further, notice that these relationships are strongly affected by national culture, overall culture of the MNC, and local employing (subsidiary) company culture.

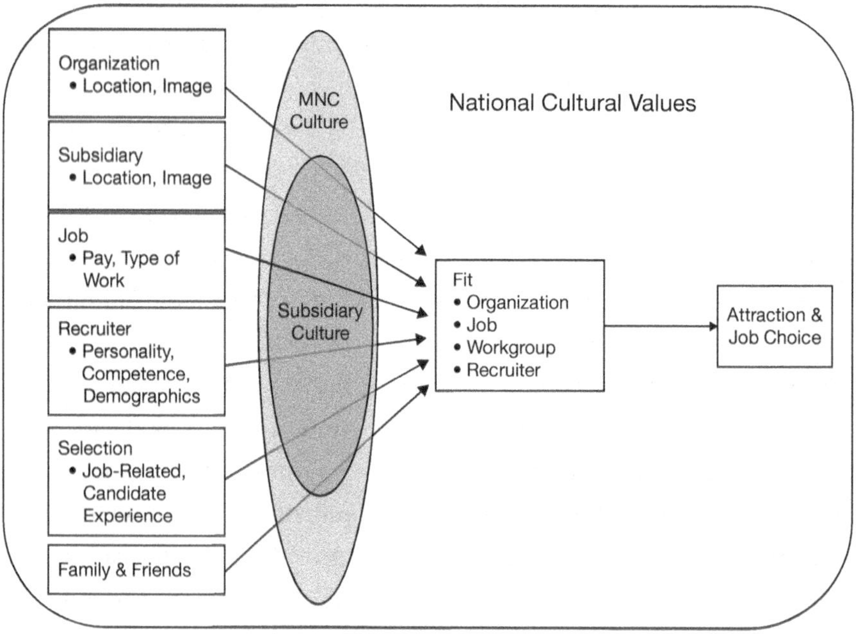

Figure 4.2
Global recruitment model.

Figure 4.2 shows there are many factors that contribute to perceptions of fit. Fit is itself a multidimensional phenomenon, representing many different types (e.g., supervisor, workgroup, organization). As noted earlier, attraction may be formed before a formal working relationship exists (e.g., as an applicant), and will be tweaked or changed over time as the person gains more direct experience with the organization (e.g., as a new hire). Thus, the determinants of fit shown in Figure 4.2, and fit itself, may fluctuate and change over time in response to new information and experiences. That said, the main parts of the global recruitment model are explained below.

National Cultural Values

National cultural values—individualism/collectivism (IC), power distance (PD), uncertainty avoidance (UA), masculinity/femininity (MF), long-term orientation (LTO), and restraint/indulgence (RI)—influence all parts of the global recruitment model. First, cultural values will influence the importance of fit for organizational attractiveness and job choice, as well as which types of fit matter most. For example, organizations in cultures high in collectivism will likely attend more to workgroup fit than those from more individualistic cultures. Second, cultural values will influence the extent to which the recruitment factors (e.g., community, type of work) influence fit perceptions. Continuing the previous example, those in collectivistic cultures such as South Korea will likely pay more attention to the opinions of family and friends than will those in more individualistic cultures like New Zealand. Finally, cultural values will affect the extent to which organizational (MNC) culture and subsidiary culture clash. Indeed, differences between the cultural values of the MNC and subsidiaries usually require the recruiting strategy (and certainly the tactics) to be adapted from the MNC to local cultures. This means that MNCs will have to adapt the information presented in recruiting messages to speak to the desired talent within a different cultural context. Recruiting messages aimed toward those in South Korea may need to differ from recruiting messages aimed at those in New Zealand. Thus, in a global context there is the potential for misfit between one's own cultural values and those of the MNC and the culture where the new job resides. The greater the "distance" between the two cultures, the more they are dissimilar, the greater the misfit, and the less attractive the job or organization.

MNC Culture

The culture of the MNC—the overall organizational culture—will influence fit. Like national culture, organizational culture represents the values of the firm that are largely shared, are assumed, and implicitly guide employee thought and action. As it relates to recruitment, different organizational cultures will attract different types of people. The challenge for global recruitment is thus to understand the MNC's culture and leverage those aspects of culture that the desired talent wants from an employer.

In its simplest form, culture is the invisible force that guides employee behavior when no one is looking. There are a few different ways to conceptualize MNC culture. One approach is a framework offered by Quinn and Rohrbaugh (1983). In this framework there are four cultural types that characterize most organizations. *Clan* cultures are internally focused and emphasize collaboration, *adhocracy* cultures are externally focused and emphasize creativity, *hierarchical* cultures are internally focused and emphasize control, and *market* cultures are externally focused and emphasize goal setting and competitiveness (Hartnell, Ou, & Kinicki, 2011). Clearly, these MNC cultural varieties will influence the types of people who may be attracted to a firm, such as those from cultures high in power distance preferring organizational cultures that are more hierarchical than collaborative.

Alternatively, one might conceptualize a MNC's culture in terms of its core values. The organizational culture profile (O'Reilly, Chatman, & Caldwell, 1991) approach conceptualizes culture in terms of values that can be summarized in the following dimensions: Innovation, Stability, Respect for People, Outcome Orientation, Attention to Detail, Team

Orientation, and Aggressiveness. Firms differ on these seven values, and hence these values can be used to determine which are most important to top candidates (we give an example of how to do this shortly).

Regardless of the type of cultural framework employed, it is clear that companies differ in their culture and that these differences have important implications for who will be attracted to the firm.

Subsidiary Culture

Subsidiaries will likewise have cultures, and these may or may not be consistent with the broader MNC culture. In the case of acquisitions, the acquiring firm's culture often slowly takes over the acquiree's culture. For example, the Brazilian steel manufacturer Gerdau acquired the US firm Ameristeel and in doing so, began to impart its own family-run culture onto the US firm. In other instances an MNC will be opening a new facility (sometimes called a "greenfield" site), but even here there can be cultural differences. For example, the Japanese manufacturing firm Denso operates a facility in Battle Creek, Michigan, United States. In this facility, much of the dominant Japanese national cultural values are imparted within the Michigan facility, such as collective approaches to HR management and use of employee uniforms to minimize individual differences within workgroups. For local employees who will never leave their home country, the subsidiary culture is largely equated with the MNC culture. For managers, who must straddle the MNC and subsidiary cultures, a lack of fit is highly apparent.

The strength of MNC culture and subsidiary culture will influence the extent to which there are culture clashes. Employees may differ in their underlying cultural orientations and values, but if the MNC has a strong culture, it may override national cultural differences. For example, Sonoco is a world leader in packaging with 300 operations in 35 countries. Their strong company culture, emphasizing collaboration and teamwork, has been instrumental in helping it accommodate cultural differences around the globe yet maintain the vision of the firm. Applicants know what to expect with Sonoco, and hence those who aspire to be part of a successful global operation realize that fit with the company's culture provides a strong foundation for accommodating national cultural differences.

Local Community

The characteristics of the local community where one lives can influence fit perceptions. It is obviously easier to recruit talent to desirable locations than undesirable locations. The challenge comes from the fact that not everyone has the same opinion about what makes a location desirable. And it is not just the physical location that matters. Technology has made it easier for even remote workers to maintain relationships with family and friends, so in some ways geographic barriers are being leveled. For example, IBM has a very large proportion of employees that work remotely. Many of these employees share administrative help that is not colocated with any of them. Similarly, to the extent there is a critical mass of people who share similar values and interests in the local community, then even those from other cultures may find the local community to provide good fit. "College towns," which are small towns in rural or remote locations that primarily exist because of the presence of a university (e.g., Oxford, UK; Cornell University in Ithaca, NY), are a good example. College towns are often considered attractive places to live because they are safe, have a low cost of living, and offer many cultural and intellectual amenities. Because of the dominance of people who work at the university and the focus on academic excellence, they often have good school and daycare systems, sometimes associated with the university itself.

"Talent hot spots" or "hubs" exist where a geographic location is recognized as having a mass of select talent. Silicon Valley in the San Francisco Bay area (United States) is a well-known example of a high-tech talent hub. This area houses the headquarters for such leading hi-tech global companies as Apple, Cisco, Google, Hewlett-Packard, Intel, and Yahoo, to name just a few. Silicon Valley is a hub for tech talent and draws people from around the world. For firms located in Silicon Valley, recruiting talent to the area is relatively easy, because top talent will be attracted to the opportunities that exist and the similarity they will have to others in the region. Of course, the downside is that a firm who recruits talent to them will also have to compete with many other firms in the same geographic area to retain them. Conversely, organizations located in small towns (Walmart in Bentonville, Arkansas, for example) may find the pool of candidates willing to relocate there smaller than if it were in a larger metropolitan city. This raises such questions as: "Which is more costly: Recruiting talent into a less desirable location or retaining it when recruited into a talent hotspot?"

The more positive connections an employee has in the community, the more likely that a person will tolerate lower levels of fit with the organization, supervisor, or coworkers (Mitchell, Holtom, Lee, Sablynski, & Erez, 2001). For example, if one's spouse has an excellent job, and the kids are all happy in school and have good friends, and family live nearby, then it may be difficult for an employee to seriously consider leaving the area (and he or she may thus try to repair a deteriorating workplace relationship). Positive community features, such as good schools and hospitals, safe and environmentally friendly communities, strong local economies, and access to desired recreational activities, can be powerful recruiting tools and are difficult for competitors to copy.

Organizational Characteristics

These include the "symbolic" characteristics of the firm that influence preferences (e.g., brand, reputation, image) and the "instrumental" characteristics of it that impact basic needs (e.g., size, locations). Making one of the various "Best Places to Work" lists is highly desirable for many organizations, as it enhances their ability to attract talent. Similarly, the flight of businesses from downtown metropolitan areas to the suburbs is in part motivated by the increased ability to attract talent. These factors will obviously be related to the organization's values and culture, but they are not the same. For example, many people graduating from today's MBA programs pursue work in the technology or financial sectors because they are seen as being more interesting and personally profitable. Those companies in the manufacturing sector are often viewed as dinosaurs and hence face challenges recruiting MBA talent as compared to those in other industries. We return to a discussion of symbolic and instrumental factors in a later section.

Job Characteristics

Features of the job include such instrumental factors as pay, type of work, task responsibilities, working conditions, hours, and so on. People vary in their preferences for different kinds of work, and a job that is one person's dream may be another's nightmare. For example, someone from a high uncertainty avoidance culture will be more comfortable with structured problems that can be solved by looking back at a successful outcome and applying the same solution again. They will look for the *right* solution. People from low uncertainty avoidance cultures will look for a range of possible solutions and choose what they see as the *best* solution. In an environment where there are already solutions, they will be bored. Yet both are motivated to succeed. Individual differences in personality are related to work preferences and these preferences determine the kinds of jobs one will pursue. For external applicants, the job and organization are often one and the same, as the differences between the two are difficult to determine from the outside. It is for this reason that organizational image is so critical to applicant attraction. For internal applicants (i.e., current employees) this is not the case, so stronger effects might be found for job characteristics in promotional situations (i.e., internal candidates may focus more on job fit than organizational fit as they contemplate a new job). Other factors such as the new boss come into play. Are they high potential so they represent a possible sponsor? Have they a good reputation for developing staff? How do they treat staff generally?

There are a variety of useful frameworks for thinking about person-job (or person-occupation) fit. A very popular framework is Holland's

(1997) RIASEC model (see chapter 1). In this model occupations differ with respect to six dimensions (realistic, investigative, artistic, social, enterprising, and conventional), and people have preferences for occupations that favor one of these dimensions. For example, those who are more extraverted tend to prefer social or enterprising occupations, while those that are more conscientious tend to prefer conventional occupations (Barrick, Mount, & Gupta, 2003). Other models include the US Occupational Information Network (O*NET; www.onetonline.org). This system is online, freely available, and offers a very broad perspective that is applicable to many countries and cultures (in fact, the model offers a self-assessment of the RIASEC model that anyone can take for free). The O*NET system links the preferences and competencies of people to the task and social demands of work, as well as broader environmental and contextual information. Job characteristics are critical for gauging fit, particularly misfit, because those who can't do the work or can't stand the work will leave very quickly.

Recruiter or Hiring Manager

In large organizations, it is common for full-time recruiters or hiring managers to be the main point of contact with the firm. In smaller firms, it may be a line manager or a representative from the HR department who is the point of contact. In either instance, these individuals can have a strong influence on how applicants perceive their fit with the organization. Interestingly, research suggests "surface" features like racial similarity do not have much of an effect on attraction. Rather, "deep" features such as values and personality are much stronger determinants (Chapman et al., 2005). Once the company's unique organizational image has been identified, recruiters and hiring managers should be chosen based on their ability to exemplify those characteristics, including cultural values.

For example, the "Big Four" accounting firms in the United States typically use alumnae to help recruit new talent from their alma maters. Usually employed as auditors, tax specialists, or providers of services in other lines of business, these alums are used as "part-time" recruiters. They are familiar with the school and its traditions, the program and professors, and the best places to eat. They also know the firm and the job, as most joined the firm in that job straight off the same campus at which they are now recruiting. That said, it is a safe bet that not every graduate of a particular college or university is asked to play the role of part-time recruiter. Those with the best social skills and "presence" are tapped most frequently.

Selection Process

The nature of the selection process (discussed in chapter 5) may also influence fit and attraction, because how candidates are evaluated and treated can be a signal for how the firm treats its people. For example, highly organized and rigid talent acquisition procedures invoke images of a bureaucratic organization, while minimal and excessively quick selection procedures can invoke images of desperation. Conversely, a well-coordinated, rigorous, and even lengthy process can signal exclusivity and competence to those chosen. Consider Cisco as an illustration. As a technology leader, their talent acquisition process makes heavy use of social media to source candidates, using such tools as LinkedIn, Facebook, Twitter, and Four Square. Their selection process is likewise technology heavy, where interviews are conducted using their Telepresence system (rather than fly in candidates for face-to-face interviews). Now, imagine if Cisco used a low-tech process involving paper and pencil testing. It is quite likely that candidates would take the selection process as a signal that the Cisco organization is not a very innovative firm, despite its product line, or one that does not invest much in its people.

Of all the pieces of a selection system that can influence attraction and the candidate experience, two appear to be most important. First, candidates must feel they are treated with honesty and respect. Second, the process must be appropriately timed. By appropriate, we mean that the length and intensiveness of the process should correspond to the importance of the job. Selection of top executives is expected to be a more lengthy and intensive process than selection of entry-level retail positions. The best way to accomplish both goals is to set realistic expectations for candidates by providing them with the general nature and timing of the process. Explain, within limits, the steps in the process, what will be expected of them at each step, and when they are likely to be informed of a decision. Setting realistic expectations costs almost nothing and can do wonders to enhance the acceptability of a selection process. Setting expectations is particularly important for global applicants because they may be coming from cultures with very different talent acquisition practices. Providing a positive applicant experience, regardless of the outcome for the applicant, is important because the Internet makes such information and experiences public knowledge. For example, Glassdoor provides information on the company, both positive and negative, based on others' experiences with the hiring processes.

Family and Friends

It is surprising that many organizations still focus nearly all of their recruiting energy on the candidate, to the neglect of also recruiting the candidate's family. Yet family and friends have powerful influences on applicant choice and behavior, and these influences are even greater in more collectivistic cultures. For example, India is a very collectivistic culture, and family and respected community members (*jan pehchan*) play a critical role in a candidate's job choice decisions (Ramesh & Gelfand, 2010). Further, given its collectivistic nature, Indian employees are more likely to show loyalty and stability with a single employer. For these reasons, a firm wishing to recruit Indian workers must establish itself as an employer who can offer stability, trust, and status. Further, it must convince the candidate's family and social network about the value of the new position. Thus, in India and other collectivistic cultures, recruiting must be focused in a much broader manner (to target the candidate's family) than is necessary in individualistic cultures (which usually only targets the candidate).

The effects of family continue. Most people will marry spouses of similar interests and values (the social psychology of attraction suggests similarity breeds liking), and many will work in similar occupations (e.g., professional, managerial, etc). A person with high career aspirations will likely choose a spouse with similar aspirations and thus create what is known as a "dual career" couple. Recruiting a dual career couple means one spouse is usually the "trailing" spouse. Organizations may enhance their recruiting effectiveness by making attempts to assist the trailing spouse, sometimes even recruiting the trailing spouse to win over the candidate. Thus, positive impacts on organizational attractiveness can be made by directly influencing the opinions of family and friends.

Conclusion

There are many ways an organization can influence fit and ultimately attraction to their firm and job choice. The key is to focus on those elements that can be controlled and that offer the potential for favorably differentiating the firm from competitors. For example, an organization's size or location cannot be changed, but the manner in which it presents its image can. Likewise, recruiting directly not only to the desired talent but also to spouses and family members can do wonders to increase attraction. Remember that employees work in organizations, but they live in communities and most of them have families. Emphasizing only instrumental job information, as has been the traditional approach, is but a small (albeit important) factor influencing fit. Smart companies identify other drivers of fit over which they have influence and then leverage these characteristics throughout the recruiting process.

Figure 4.2 suggests that national cultural values permeate and influence all aspects of the global recruiting process. To illustrate and summarize these potential effects, Table 4.3 lists a variety of ways that cultural values (individualism, power distance, uncertainty avoidance, masculinity, long-term orientation, indulgence) may influence the recruitment model. For example, the more individualistic the culture, the more candidates are

TABLE 4.3 OVERVIEW OF HOW CULTURAL VALUES SHAPE RECRUITMENT PRACTICES

	National Cultural Values					
	Individualism	**Power Distance**	**Uncertainty Avoidance**	**Masculinity**	**Long-Term Orientation**	**Indulgence**
MNC Characteristics ■ Culture ■ Brand	Individualistic attracted to: ■ firms rewarding personal achievement; promotion; individual pay and compensation [e.g., market cultures; outcome orientation] Collectivistic attracted to: ■ firms with well-known brands; organizations rewarding group settings, harmony; collective outcomes; job security [e.g., clan cultures; respect for people, team orientation, stability]	Low PD attracted to: ■ firms rewarding close contact with supervision; fast promotion opportunities; flatter structures [e.g., nonhierarchical cultures; innovation, respect for people] High PD attracted to: ■ firms with prestigious brands; bureaucratic and formal organizational structures; status symbols [e.g., hierarchical cultures; stability, attention to detail]	Low UA attracted to: ■ looser and more nimble structures; less formal firm cultures [e.g., adhocracy cultures; innovation] High UA attracted to: ■ firms with more formally established lines of authority; stability of employment; standard or centralized HR practices; stability and security of firm [e.g., cultures with low adhocracy; stability, outcome orientation, attention to detail]	Masculinity attracted to: ■ firms emphasizing achievement; competition; toughness, material outcomes; distinct and stereotypical gender roles [e.g., market cultures; outcome orientation, aggressiveness] Femininity attracted to: ■ firms emphasizing blended gender roles; balance personal and work life; humility; relationships [e.g., clan culture; respect for people, team orientation]	Short-term orientation attracted to: ■ firms emphasizing short-term goals; quick promotion and salary growth opportunities [e.g., market culture; innovation, outcome orientation] Long-term orientation attracted to: ■ firms emphasizing employee development; commitment to employees; loyalty [e.g., clan culture; stability]	Restraint orientation attracted to: ■ firms with transparent control; strong cultures with clear norms; brands representing strong values [e.g., hierarchical cultures; stability, attention to detail] Indulgence orientation attracted to: ■ cultures that enable more personal freedom and discretion; brands emphasizing tolerance and self-interest [e.g., adhocracy cultures; innovation]

(continued)

TABLE 4.3 CONTINUED

	National Cultural Values					
	Individualism	**Power Distance**	**Uncertainty Avoidance**	**Masculinity**	**Long-Term Orientation**	**Indulgence**
Subsidiary Characteristics ■ Culture ■ Brand ■ Cultural difference from MNC	Individualistic: ■ not too concerned about cultural difference from MNC Collectivistic attracted to: ■ local firm with well-known brand and reputation; local "in group"; job security	Low PD attracted to: ■ subsidiaries with more local control and discretion High PD attracted to: ■ subsidiaries with clear connection to MNC; prestigious brand reputation	Low UA attracted to: ■ subsidiaries with more local control and discretion High UA attracted to: ■ subsidiaries with policies and practices consistent and standardized with MNC; stability and security	Masculinity: ■ not too concerned about cultural difference from MNC Femininity attracted to: ■ subsidiaries with harmonious relationship with MNC; supportive of local community	Short-term orientation attracted to: ■ subsidiary with aggressive goals Long-term orientation attracted to: ■ stable and planned relationship with MNC; supportive of community	Restraint orientation attracted to: ■ subsidiary with high ethical standards; strong local norms Indulgence orientation attracted to: ■ freedom to deviate from MNC; local discretion and interests
Community Characteristics ■ Location ■ Quality of life	Individualistic: ■ more focused on personal recreational opportunities Collectivistic: ■ more concerned about location or community; attracted to values and relational opportunities within a community	Low PD attracted to: ■ easy access to community leaders; ability to easily become community leader High PD attracted to: ■ formal authority channels and functions; community roles seen as high status	Low UA: ■ relatively unconcerned about stability of community High UA attracted to: ■ communities with more stable or standardized systems (e.g., local government, schools)	Masculinity attracted to: ■ communities with achievement opportunities Femininity attracted to: ■ quality of life; balance	Short-term orientation attracted to: ■ dynamic; rapid growth and change Long-term orientation attracted to: ■ systematic planning; planned communities; stability	Restraint orientation attracted to: ■ communities with strong moral norms Indulgence orientation attracted to: ■ communities with lax moral norms; personal freedom

Job Characteristics ■ Task demands ■ Working conditions	Individualistic attracted to: ■ personal financial awards, achievement, and advancement opportunities Collectivistic attracted to: ■ teamwork; communal relationships as part of the job; collective rewards and identification; job security and development	Low PD attracted to: ■ opportunity to shape job; collaborative work with supervisor; opportunities for input; access to senior leadership High PD attracted to: ■ formal and well-defined work roles and job tasks; directive supervision; clear hierarchy; status of job	Low UA attracted to: ■ achievement opportunities in job; challenge and risk; variability High UA attracted to: ■ security and stability of job	Masculinity attracted to: ■ competitive and challenging jobs; jobs requiring aggressive advancement Femininity attracted to: ■ harmonious working relationships; security, stability	Short-term orientation attracted to: ■ fast-paced jobs and careers; fast advancement and achievement opportunities Long-term orientation attracted to: ■ investment in skills and development; stability and commitment to employees; loyalty	Restraint orientation attracted to: ■ jobs with clear and strong standards Indulgence orientation attracted to: ■ loose standards acceptance of variability across people
Recruiter Characteristics ■ Competence ■ Personality ■ Cultural background	Individualistic: ■ focus more on recruiter competence than social relationships Collectivistic attracted to: ■ recruiters perceived to be part of similar "in group"	Low PD attracted to: ■ recruiter who is more informal, interactive, "give and take" High PD attracted to: ■ recruiter highly directive and controlling of process; recruiter seen as high status	Low UA attracted to: ■ informal interviews or engagements with recruiters High UA attracted to: ■ formal recruitment interviews and engagements; recruiters with experience to illustrate stability	Masculinity attracted to: ■ recruiters emphasizing ambition; aggressive mannerisms; Femininity attracted to: ■ nonaggressive, warm, and friendly recruiters	Short-term orientation attracted to: ■ quick response times; recruiter emphasizes speed and advancement Long-term orientation attracted to: ■ more stable recruiter; focuses on developing skills	Restraint orientation attracted to: ■ recruiters with clear moral and ethical standards Indulgence orientation attracted to: ■ recruiters willing to fulfill applicant's personal preferences

(continued)

TABLE 4.3 CONTINUED

	National Cultural Values					
	Individualism	**Power Distance**	**Uncertainty Avoidance**	**Masculinity**	**Long-Term Orientation**	**Indulgence**
Selection Process ■ Job-Related ■ Candidate Experience ■ Fair ■ Timing	Individualistic: ■ expect staffing process to be based more on competencies; see processes based on "who you know" as unfair; process favors own outcomes	Low PD attracted to: ■ more interactive and flexible recruiting practices High PD attracted to: ■ highly structured staffing practices; practices seen as highly selective	Low UA attracted to: ■ more unstructured and customized staffing procedures High UA attracted to: ■ more standardized, structured, and formal staffing procedures; preference for faster process; share more information about staffing process	Masculinity attracted to: ■ aggressive recruiting tactics and messages; emphasize competition and achievement; use active recruitment	Short-term orientation attracted to: ■ short, fast, efficient; process focuses on skills needed immediately	Restraint orientation attracted to: ■ staffing processes based on fairness; clear ethical standards
	Collectivistic: ■ more likely to expect staff process to be based on personal contacts; personal contacts or advocates seen as fair and appropriate; process fair to all			Femininity attracted to: ■ nonaggressive messages and tactics; relational recruitment practices; use passive recruitment	Long-term orientation attracted to: ■ process of appropriate length; process focuses on potential for development	Indulgence orientation attracted to: ■ staffing processes that benefit the individual

Family & Friends ■ Spouse ■ Family ■ Friends & important relationships	Individualistic: ■ relatively uninfluenced by opinions of others except perhaps spouse; family opinions less important Collectivistic: ■ more influenced and affected by relationships and opinions of others; focus on communal harmony; family opinions very important	Low PD: ■ influence not dependent on person's status; more collaborative feedback discussions High PD: ■ influence more based on person's status; more directive feedback and opinions	Low UA: ■ less concerned about differences of opinion from important others High UA: ■ more consensual opinions of important others	Masculinity: ■ more likely to have primary earner make decisions; less influence by others (including spouse) Femininity: ■ more likely to make decisions based on consensus; strong influence by family and friends	Short-term orientation: ■ focus on advice about immediate job options Long-term orientation: ■ advice about long-term career; development and growth	Restraint orientation: ■ focus on opinions that emphasize "doing the right thing" Indulgence orientation: ■ focus on opinions that emphasize only "what's best for the candidate"
Recruitment practices and message should focus on . . .	Individualists: focus on personal rewards (usually financial); nature of job Collectivists: focus on relational elements of work; social networks, "in groups," identification with firm; firm brand and reputation; appeal to social network	Low PD: focus on accessibility to leadership; flat hierarchical structure High PD: focus on brand prestige; distinctions based on status; clear hierarchy	Low UA: focus on achievement, variety, challenge High UA: focus on security and job stability	Masculinity: focus on achievement, earnings, competition, aggressive messages; nature of job; active and passive recruitment Femininity: focus on relations, harmony, work-life balance, security, group identification; passive recruitment	STO: focus on immediate success, fast outcomes, job success LTO: focus on growth, development, career success	Restraint orientation: focus on moral and ethical foundations; norms Indulgence orientation: focus on individual outcomes; personal choice; personal gratification

NOTES: Entries in this table are drawn heavily from Guo and Miller (2010); Ma and Allen (2009); Stone, Stone-Romero, and Lukaszewski (2007); Caligiuari and Paul. (2010); Tarique, Schuler, and Gong (2006); Collings and Scullion (2012). Examples in brackets "[]" illustrate connections to cultural, occupational profile, or job characteristic frameworks.

FINAL CAVEAT: remember that there is as much variability within cultures as between cultures, so do not take these differences to be universal!

likely to focus on job-specific information. In collectivistic cultures, candidates will more strongly attend to family, community, and organizational information. Thus, the ways a firm can increase attraction will differ by culture and should be tailored to the dominant cultural framework. However, companies frequently give too much emphasis to them in recruiting and fail to also emphasize the importance of the organization's culture. Job characteristics are clearly important, but so is the broader workgroup and organizational context that the person will be joining. The job, workgroup, and organization thus need to be in alignment. These examples are not exhaustive and only serve to highlight an important fact—candidate fit is going to be affected by national cultural values, and so firms must consider and accommodate such values in the design of their recruitment program.

Roger's arrival in Singapore had so far been going smoothly. He was struck by the fact that most signs were in both English and one or more of the local languages, including Malay, Mandarin, and Tamil.

Even after his first few meetings, it was obvious that the regional recruiting plan would have to balance the tension between customization to specific countries and cultures, yet maintain some consistency across cultures. Roger came to understand that his focus would first be on recruitment in the strategic job families:

- Plant assistant general manager recruitment would favor internal candidates because they needed the expertise of the plant and region to be successful. This meant that he was already reasonably assured that internal candidates would fit with the country and company cultures, so he could focus on job fit.
- Assembly team associates required the largest recruiting efforts because of the volume. These would primarily be local hires, but external candidates. Hence, he could focus less on country culture fit, and focus more on organizational and job fit. He would need to emphasize what is distinct about Spooner in each geographic location.

- The R&D team members and leaders would primarily be external hires, and they would be sourced internationally. This meant he would have to heavily emphasize fit, first, with country cultural values, and second, with the organization and job.

Now, how to do it?

STEPS FOR EFFECTIVE GLOBAL RECRUITMENT

There is an old saying. “All politics are local.” The same logic applies to recruitment—all recruitment is local, and what works in one culture won’t necessarily work in a different culture. For an MNC the key is being able to manage different approaches to meet a global need. Thus, recruitment in a global context is exponentially more difficult than domestic recruitment because of national differences in politics, laws, and cultural values. These cultural differences must be recognized within the recruitment strategy and accommodated within recruitment system design. The degree of centralization versus customization is dependent on whether the firm is following an ethnocentric, geocentric, regiocentric, or polycentric approach (discussed in chapter 2). In this section we discuss the seven steps that must be followed to effectively recruit in a global context, regardless of whether the firm emphasizes centralization versus customization. Figure 4.3 summarizes these steps, and they are discussed in some detail below.

Step 1: Define the Types of Talent Needed

Firms must first identify what the desired talent is. That is, the needed competencies, both generic and specific, need to be defined in advance of any recruiting efforts. This is obviously a strategic exercise, because a firm should be thinking about what skills and competencies it needs now and what it will need in the short- and long-term. Specific steps for defining the types of talent needed were discussed in chapter 2. Surprisingly, a large

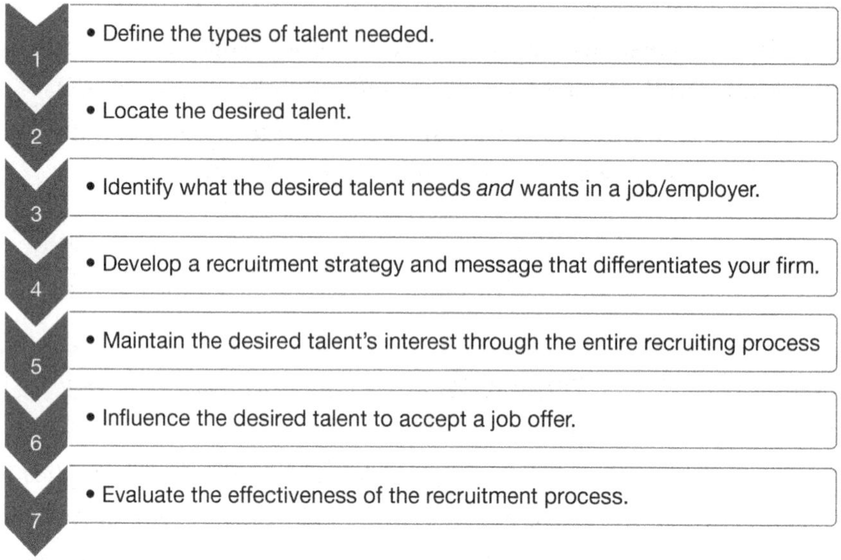
1
• Define the types of talent needed.
2
• Locate the desired talent.
3
• Identify what the desired talent needs *and* wants in a job/employer.
4
• Develop a recruitment strategy and message that differentiates your firm.
5
• Maintain the desired talent's interest through the entire recruiting process
6
• Influence the desired talent to accept a job offer.
7
• Evaluate the effectiveness of the recruitment process.

Figure 4.3
Steps for global recruitment.

number of firms don't know what kinds of talent they need now or in the future. A study by McKinsey suggested that less than 20% of top managers could define a high performer within their *own* organization (Chambers et al., 2007). If firms can't even define talent internally, how will they know what to look for externally? And even more difficult, how might this talent look when it comes from different cultures?

In our own experience with many global organizations, we have found it helpful to consider these issues from a variety of strategic perspectives to forecast talent needs. For example, firms have partnered with us to identify talent needs for general managers currently and over the next 5 or so years. We interviewed a number of their executive leaders who represented a broad cross-section of functions, levels, countries, and cultures. We also examined industry forecasts on talent needs, and considered these within the context of their strategy and competitive environment. The result was the identification of several core competencies that would define the general manager role for the next 5 years. Many of these competencies overlapped with their existing competency model, but there were also some important differences. One critical difference was a growing need for cultural agility within general managers; another was an ability for building effective yet diverse teams. These were competencies that were generally known as being important, but became recognized as strategically vital to their future growth.

The focus on short and long term is vital for understanding future talent needs. For example, Shell used an HR functional excellence tool that provided competences of different roles at different levels (specialist, generalist, etc.). A key exercise in its development was to conduct focus groups with a range of HR leaders and ask questions about the experiences that had taught them the most. The end product was a series of experiences that enabled the learning of broad competencies.

The key point is that what defines critical talent today is not necessarily the critical talent for the future. Recruitment must start strategically by defining talent needs within the context of the firm's strategy. The focus should be on identifying the kinds of talent needed to implement and achieve the firm's strategy. This means recruitment does not start with job requirements, as is usually the case, but instead starts with firm and HR strategy (see chapter 2).

Step 2: Locate the Desired Talent

Locating talent pools is challenging (assuming any such pools exist in the first place). There are three major questions to be considered in this step, and the answer to each of these questions has important strategic ramifications.

First, do the talent pools exist internally or externally? *Internal recruiting* is done through promotions, job transfers, and job rotations, and focuses on those already in the firm. Internal recruiting is often tied to employee training, development, and performance management (chapter 6). Most recruiting that takes place actually occurs internally (i.e., the promotion or transfer of employees within the firm). Estimates vary widely (in accordance with the nature of the job and the state of the economy), but range from 30% to 60% of all jobs being filled internally. Internal recruiting is particularly effective because those hired internally have better fit and job satisfaction, perform better, achieve high performance more quickly, and stay with the firm longer, than external hires (e.g., Bidwell, 2011). The reason, of course, is that there is more information on internal hires (so hiring decisions may be more accurate), and those already employed by the firm fit the organizational culture and strategic direction. Internal recruiting also benefits employee engagement by signaling the existence of career opportunities within the organization.

In contrast, *external recruiting* is done from the general labor market, focusing on talent outside the firm. Sometimes there is a shortage of qualified internal talent; other times there is a desire to bring in talent with fresh perspectives. It may also be an attempt to address newly required competence, as with Shell's acquisition of Pennzoil-Quaker State and the need for skills in consumer goods. Or, it could address demographic issues such as age bulges or a lack of diversity driven either by legislation or a firm's own desires. Finally, it may just be the firm's HR strategy to outsource development to competitors and hire externally as needed. Regardless, it is usually more difficult, time consuming, and expensive to hire externally because less information is known about external candidates and candidates are unfamiliar with the firm's culture.

Typically, there needs to be a mix of internal and external recruitment to maintain a healthy balance. Overuse of internal recruiting can create a "sameness" to the organization. There are times when a change in perspective is needed, particularly when change or innovation and creative solutions are required. For example, in many sports teams with failing records, it is the head coach that is the first to go because the team needs a fresh perspective and new thinking. In reality, both internal and external recruitment should be used, and knowing when and how to use both in a complementary manner can itself be strategically valuable. Box 4.1 provides some suggestions for considering when to use internal versus external recruitment.

Second, where are the talent pools geographically located? We've noted that pools or hotspots exist in different parts of the world and for different

Box 4.1 ADVANTAGES AND DISADVANTAGES OF INTERNAL AND EXTERNAL RECRUITING

- Use internal recruiting when:
  - the organization has high need for specific knowledge and has created an internal labor market (i.e., entry at the bottom and promotion from within)
  - filling leadership roles where consistency and constancy is desired
  - existing skill sets are adequate or there is sufficient time to retool
  - there is a succession plan identifying the desired slate for a particularly developmental role
  - there are strong time and/or cost pressures reducing the ability to search broadly
  - there is a lack of external talent
  - existing competencies can be developed for future plans
  - there is a strong support network to help facilitate interactions with the local community
- Use external recruiting when:
  - a change in vision or culture is desired
  - there are strong supportive elements to integrate the person into the organization
  - variability in performance can be tolerated
  - new markets/products or technologies demand skills unavailable on the internal labor market
  - the jobs are entry-level (and must be filled externally)
  - existing competencies are insufficient for future plans
  - there is no time to develop talent internally
  - trying to deny key talent from competitors

types of talent. Want to recruit talent from the high-tech sector—go to Silicon Valley. Want to recruit talent from the financial sector—go to London or New York—because financial and accounting talent is gravitating to "superhubs" that are transforming the talent landscape in those industries (Kops & Lyon, 2013). There are many excellent sources for identifying talent pools. At a country level, we are particularly impressed with the Global Talent Competitiveness Index Report published by INSEAD (Evans & Lanvin, 2013). In this comprehensive report, the top three talent-competitive countries were Switzerland, Singapore, and Denmark. At the city level, a report by the Economist Intelligence Unit identifies the cities most likely to be talent hotspots by 2025 (Economist Intelligence Unit, 2013). The top three cities include New York, London, and Singapore. Table 4.4 lists the top 10 countries and cities identified in these reports. Of course, one must consider the specific types of talent rather than relying

TABLE 4.4 TOP TALENT POOLS: COUNTRY AND CITY

Country Global Talent Competitiveness Index (INSEAD)	City Talent Hotspots (Citigroup)
Switzerland	New York
Singapore	London
Denmark	Singapore
Sweden	Hong Kong
Luxembourg	Tokyo
Netherlands	Sydney
United Kingdom	Paris
Finland	Stockholm
United States	Chicago
Iceland	Toronto

SOURCE: Adapted and data obtained from INSEAD (http://knowledge.insead.edu/talent-managment/the-worlds-most-talent-competitive-countries-3006) and Citigroup (http://www.citigroup.com/citi/citiforcities/pdfs/hotspots2025.pdf).

on broad indices such as these, but they provide a glimpse into the global pattern of talent and where it is located.

Finally, firms need to consider whether they will focus on passive candidates, active candidates, or some mix of the two. An important change that has occurred in the last 10 years, stimulated primarily by technologies such as social networking, is the ability to source passive candidates. Active candidates are those formally applying for jobs; passive candidates are those who may be looking but are not committed to formally applying. One recent global survey by LinkedIn (https://business.linkedin.com/talent-solutions/blog/2014/03/active-vs-passive-candidates-the-latest-global-breakdown-revealed) found that 25% of candidates were reasonably active, while 60% were open to considering another job (the other 15% were completely satisfied and uninterested in changing jobs). This should come as no surprise, given that the employment relationship has eroded so much over the last 20 years and online resumes (e.g., LinkedIn) enable the efficient sourcing of passive candidates.

Historically, the recruiting model was one where an organization would post a job, applicants would learn of the job and apply, and then a filtering process would occur. The old model was very much a one-way street (see the left side of Figure 4.4). Technology has transformed recruitment such that the new model is more open-source and open-access (as in a cloud). Firms may now recruit both active and passive candidates by pushing recruiting messages through different channels (see the right side of Figure 4.4). In this new model, talent and mobility is much more like a two-way street, with active candidates in one lane and passive candidates in the other.

The questions of drawing applicants from internal/external talent pools, from different locations, and passive/active candidates are interrelated. Internal or external recruiting may be done domestically or internationally, with passive or active candidates. For example, it is pretty typical for MNCs to rotate top talent through different geographic and cultural assignments as a means of exposing them to the broader organization and fostering a global mindset. Some also adopt a philosophy that promotes movement across separate business units so that high potentials obtain

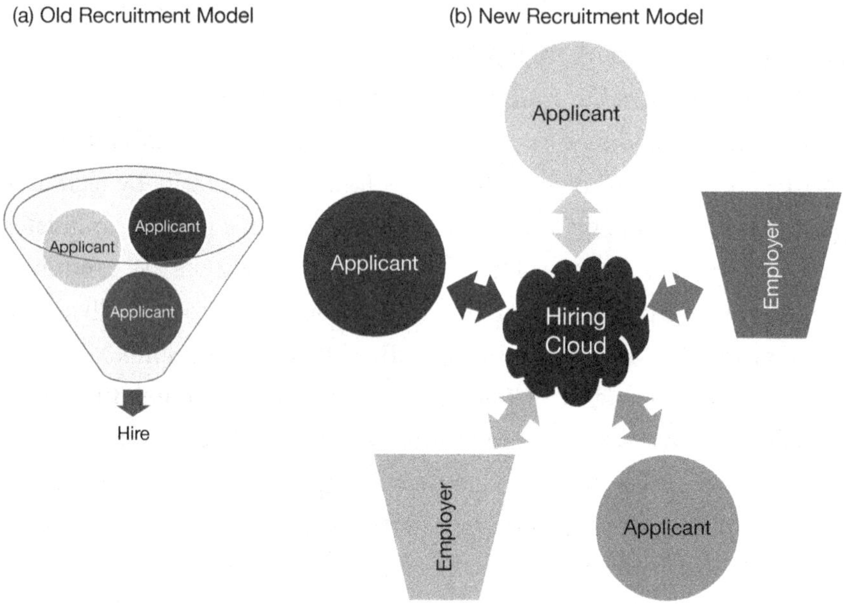

Figure 4.4
The old (on the left) and new (on the right) recruitment models.

a grasp of the different business units they will be required to manage as they reach more senior roles. In this sense, certain lines of business or operations in a single country may become an internal talent pool. Alternatively, technological advancements in communication (e.g., social networking) have made it easier and more cost-effective to reach external talent around the globe and source it from different locations. We return to the opportunities and threats provided by technology in a later section.

Step 3: Identify What the Desired Talent Needs and Wants in a Job/Employer

All applicants have needs and wants, but you can't be concerned with satisfying them all. You must focus on what are your *desired* talent needs and wants, because such understanding is vital for effective recruiting. For example, high-achieving people generally will prefer challenge and opportunities for advancement (Cable & Judge, 1996). However, such preferences are further affected by cultural differences in values. For example, the types of advancement offered in more collectivistic cultures will differ from those in more individualistic cultures (Guo & Miller, 2010). Those in more individualistic cultures will emphasize personal gain and reward, while those in more collectivistic cultures will emphasize group achievement and harmony.

A distinction may be drawn between talent needs and wants. Needs are what a candidate must have to consider a job; wants are what a candidate prefers in a job. Both are important, but after needs are met, attention turns to preferences (wants). In addressing candidate needs, firms should focus on those *instrumental* characteristics that a job and firm must offer for the candidate to seriously consider it. Examples of instrumental characteristics include pay, benefits, nature of the job tasks, where the job is located, organizational size, advancement opportunities, and related factors. In addressing candidate wants, firms should focus on those *symbolic* characteristics that a job and firm must offer to address candidate preferences. Examples of symbolic characteristics include company culture, brand, reputation, image, and related factors.

Instrumental factors are obviously important in recruiting, as one is unlikely to accept a job that doesn't pay enough to support one's lifestyle, or is located in a place one is unwilling to live. As a result, many firms adopt recruiting strategies that emphasize instrumental factors like pay, and they compete with other firms primarily in terms of offering higher pay (if possible) and/or signing bonuses. Such an approach is short-sighted as increasingly higher pay creates a growing drain on monetary resources, yet does little to enhance commitment to the organization. Indeed, if an employee is attracted to a firm based solely on pay, it will only be a matter of time before she or he is lured away to a competing firm offering more money. Even using retention bonuses, incentivizing a candidate by providing a bonus payable at the end of a period of time (like 2 or 3 years), can essentially be bought out through an appropriately pitched sign-on bonus. If is just about the money, then the danger is that the entire employment "contract" is just that—it is about the money and money is easy to replicate. Competing for talent using instrumental factors alone is not a sustainable way to achieve a sustainable competitive advantage.

In our research, we have found that most applicants quickly screen out offers that do not meet their basic needs. For example, a job may be located in a region that the candidate will not consider or the pay range may be unacceptable. For those opportunities that remain, candidates give much more deliberate attention to both needs and wants. Yet surprisingly, candidates are often poorly able to articulate what actually influences their decisions. They will often overemphasize the importance of instrumental characteristics and underemphasize the importance of symbolic characteristics in their decision-making. Using a methodology called "policy capturing" in our research, we have varied instrumental (e.g., salary) and symbolic characteristics (e.g., culture) to find complex decision strategies where both instrumental and symbolic factors are important, and in ways the candidates themselves could not articulate (Harold & Ployhart, 2008).

Within a global context, understanding what the desired talent needs and wants is even more critical because needs and wants will be highly affected by cultural values. First, cultural values may influence the importance of instrumental characteristics. For example, the importance of pay relative to personal time varies greatly across countries, with many European cultures valuing them more equally than the United States. Second, cultural values may influence the importance of symbolic characteristics. It is here that culture is likely to have its strongest influence. For example, those from cultures with a long-term orientation are going to place a greater premium on symbolic characteristics that value commitment, loyalty, and development (see Table 4.3).

Step 4: Develop a Recruitment Strategy and Message That Differentiates Your Firm

You want your recruiting efforts to differentiate your firm in a way that favors your strengths and cannot be easily imitated by competitors. This means *you should focus on your firm's symbolic characteristics and, more specifically, those characteristics that advantage your firm above your competitors in the minds of your desired talent.* This is important for several reasons. First, it focuses on symbolic characteristics and hence is not easy for competitors to imitate (e.g., Google's culture, 3M's innovation). Second, it focuses purely on the desired talent (which should contain a more homogeneous set of preferences). Third, it focuses on only those characteristics that your desired talent actually wants (rather than trying to satisfy all candidate preferences). This is a much more precise, surgical approach to recruitment.

The real power of differentiation using symbolic characteristics can be understood when considering how people choose between jobs and organizations that are highly similar. Consider an applicant who has offers from two different bank branches representing two different organizations. The branches are located across the street from each other; and they offer nearly identical pay, benefits, and working hours. How does the applicant choose? It turns out that "symbolic" characteristics become a powerful means of influencing applicant choice. In a study involving Belgian students and bank employees (Lievens & Highhouse, 2003), the researchers found that people make "trait inferences" about different banks. Just as we use terms like "strong," "weak," "energetic," "intelligent," and "industrious" to describe others, we use similar terms to describe organizations—essentially *personifying* organizations. In this study, organizational differences in trait inferences did a better job predicting potential job choice than information about the job (i.e., instrumental characteristics). Trait inferences thus serve as symbols.

If people personify organizations and hence differentiate them in ways similar to how we differentiate people, then symbolic characteristics can be constructed to influence applicant perceptions, just as individuals present images to convey how they want others to perceive them. Smart organizations can leverage this knowledge to achieve competitive advantage. There are four main steps to this process. Table 4.5 describes the process for how this can be done, and uses Spooner Electronics as an example.

After identifying the symbolic (and possibly instrumental) characteristics to emphasize, the task becomes one of conveying the information

TABLE 4.5 HOW TO IDENTIFY AND LEVERAGE ORGANIZATIONAL IMAGE FOR COMPETITIVE ADVANTAGE

Step	**Example (using Spooner Electronics)**
Step 1: *Identify* the top talent.	
Identify the top talent that your firm seeks to acquire. Consider new hires, top applicants, or even customers.	Within each country, Spooner HR managers, marketing managers, and line leaders, identified key stakeholders representing top talent. The initial focus was on the four strategic job families; other job families would be examined gradually over time. These included high performing employees who were recently hired, key suppliers and customers, and longer-term employees. This was a project done jointly between HR and marketing. Many of the sampling approaches discussed in chapter 2 (Table 2.3) for job analysis are applicable here.
Step 2: *Survey* the top talent.	
A. Among the group of top talent, interview or survey them in terms of how they perceive your organization, and then how they perceive competing organizations. Ask them to first describe your firm using adjectives and attributes (in their own words). Then, have them use adjectives and attributes to describe competing firms. After they exhaust the potential attributes, provide them with additional attributes based on your own benchmarking. Summarize the adjective list, focusing only on those attributes that the top talent believes is important for their job choice decisions.	Focus groups can be run with participants. Many of the sampling approaches discussed in chapter 2 for job analysis are applicable here. Based on the focus groups, Spooner Electronics was generally seen as being better on these characteristics, relative to competitors: stability, successful, analytical, conscientious, open, caring, and focused. Competitors were seen as being more: innovative, faster, trendy, visible, and extraverted. Note that these perceptions may differ for each job family, but care must be taken to ensure the company messaging is consistent.

B. Prepare a survey that lists the main attributes. Have a different group of top talent complete the survey.

Step 3: *Identify* your organizational image advantages.

For each attribute, look for major areas where the firms are different from each other. Focus on attributes where your firm has a clear advantage over competitors.

Many of the survey approaches discussed in chapter 2 for job analysis are applicable here. Spooner used the following (these are just three example items from one job family):

Please provide your opinion Spooner Electronics using the following attributes:

Stable	1	2	3	4	5	Unstable
Successful	1	2	3	4	5	Unsuccessful
Trendy	1	2	3	4	5	Dated

Spooner Electronics put together the following benchmarking data (higher numbers are better):

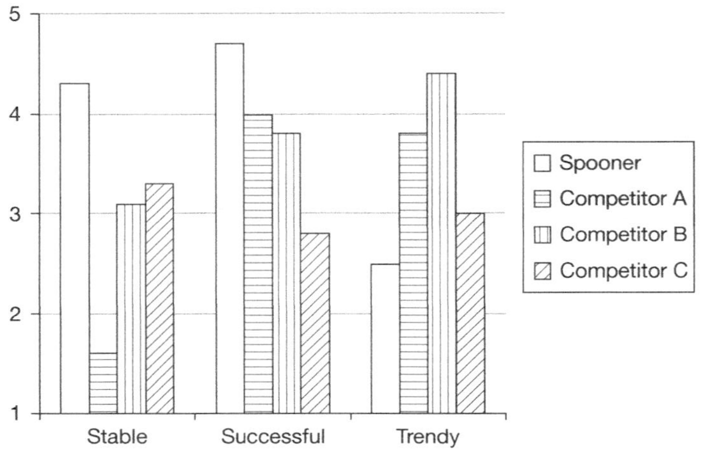

TABLE 4.5 CONTINUED

Step	**Example (using Spooner Electronics)**
Step 4: *Leverage* your message.	
For those attributes where your firm is better than competitors, leverage this advantage by emphasizing it in your recruiting messages.	Notice that Spooner is favorably differentiated in terms of stability and success. These are themes where Spooner should focus its recruiting messaging. Its marketing images and language should emphasize stability and success, and these themes should be infused through all recruiting efforts. This might include taglines such as "Stability. Success. Spooner." and reinforce such themes using video testimonials where employees constantly discuss the company's stability and success.

in an appropriate manner. There are lots of ways this can be done, but most of the scientific research on recruitment has focused on two particularly important ways to do this: realistic job previews and employee value proposition.

Realistic Job Previews

The first approach is to provide an accurate portrayal of the job and the company, including the positive and negative features of each. You want to develop a recruiting message that is realistic because an overly favorable (unrealistic) recruiting message will lead to turnover and frustration from both parties. Use of realistic job previews (RJPs) presents a balance of positive and negative information about the job and organization, so that candidates can make an informed decision about the firm and thus ensure those who pursue the firm will truly fit the job and culture. The RJPs are not job descriptions, but convey the realities of work within a firm, "the good, the bad, and the ugly." Such realities may include working hours, travel, accountability, unusual demands, chaotic schedules, organizational culture, and career and developmental opportunities.

When RJPs are done effectively, they encourage the desired applicants to apply and discourage unqualified or "unfit" applicants from applying. Because they discourage applicants who will not find the job or organization to fit with their needs and values, they also tend to enhance satisfaction and commitment for those who join the firm. The challenge with using RJPs is to provide accurate information that doesn't discourage the *desired* talent. The most qualified candidates also have the most offers and so they are looking for reasons to screen out organizations. These folks tend to overweigh negative information. This means that a RJP with too much negative information (e.g., long hours) could actually turn off some otherwise desired applicants.

Firms must carefully balance the benefits and drawbacks of a job and company culture. However, some negative aspects of jobs are, for certain people, desirable. For example, a high degree of challenge and risk will be appreciated by those with an achievement orientation but will be discouraging to those with a high need for security and stability. Positively framing some job aspects, such as high accountability and challenge, helps ensure they speak to the kinds of people who would like such work. Candor about aspects of the job that are negative also helps enhance perceptions of the organization's honesty. We emphasize that use of positive or negative framing should not be equated with being misleading or lying to candidates. Table 4.3 illustrates how what is considered desirable or undesirable in a job or company can be affected by cultural values. Thus, one must carefully evaluate how the RJP message is being perceived by candidates.

The RJPs can be delivered in a variety of formats and media: question and answer, purely descriptive, job observation, written text, video, and so forth. Generally, the richer the communication medium the better, meaning that video-based depictions of the key features will be more impactful than a written description of them. Some of the better RJPs we have seen have involved video interviews with current incumbents at the work site, who describe in their own words both the positive and negative aspects of their jobs.

An RJP should also focus on those elements of the job and culture that top talent desires, but where the firm is better than competitors. The point is to differentiate the firm in a positive way, in the minds of the desired talent. This becomes much harder as talent becomes more geographically distributed. For example, talent sourced in Japan will likely prefer Japanese firms to foreign firms, even though both jobs may be physically located in Japan (Guo & Miller, 2010).

Employee Value Proposition

The second approach, the employee value proposition (EVP), emphasizes what the employer offers in return for the employee's productivity. The EVP is the sum of the employee's experience and includes pay/benefits, career growth and training, company culture and values, leadership's vision and competitive strategy, the working conditions and physical environment, and of course the job itself (Yates & Sejen, 2011). The EVP should be aligned to the organization's brand, reputation, and image. There are differences between brand, reputation, and image, but for our purposes we refer to them cumulatively as organizational brand. *Organizational brand* represents the way the firm is perceived by stakeholder groups (including, for example, applicants, management, employees, customers, suppliers, policymakers, and the general public). Marketing research shows that organizational brand is a powerful means of differentiating the firm from competitors, with corresponding effects on profitability and customer loyalty. Employees will often identify with their employer, and such identification is stronger when there is a clear brand presence. For example, top brands such as Coca-Cola or Unilever will have an immediate global recruiting advantage because the brands are so highly recognizable. But it should not be assumed that even very well-known brands have that reputation in all countries.

One survey found that roughly a third of organizations globally have a formal EVP (Yates & Sajen, 2011). Among high-performing organizations, this rose to 42% (as compared to 32% of average performing and 28% of below average performing organizations). Another study found a 21% increase in compensation needed to attract candidates to organizations with an unattractive EVP, versus an 11% premium required to entice candidates to an organization with an attractive EVP (CEB, 2015). A global survey of professionals found that over half cited the company's reputation as the most important factor when considering a new job (Srinivasan, Gager, & Ignatova, 2014). While it is impossible to say that an EVP causes greater performance, having one does clearly articulate to prospective employees "the give and the get." A formal EVP, consistent with the organization's corporate brand, helps ensure that new hires join the organization with reasonably accurate perceptions of the organization's purpose, its strategy for achieving that purpose, the employees' role in it, and knowledge of "what's in it for me?"

Organizational brand and EVP have a powerful influence on applicants' choices and behaviors. Brand perceptions may be formed quickly and reinforced through repeated interactions with the company's personnel, products, and services. They are also influenced by the organization's advertising and marketing. Similarly, the EVP should be reflected in recruitment messaging and interactions with recruiters, hiring managers, and other organizational members. Yet both organizational brand and the EVP are ultimately perceptual, and can be constructed, reconstructed, or adapted to fit different cultural contexts. Such cultural customization is important because culture-specific or culture-sensitive advertising is more effective than generic advertising. Relatedly, organizational brand perceptions are particularly powerful determinants of applicant choice and behavior when applicants don't have specific details about a firm. Once applicants make contact with the organization, the EVP takes over for the organizational brand, and guides subsequent interest and behavior.

Overall, many firms know what their competitors pay because it is fairly objective and easy to verify, but many firms do not know if they lose talent to others that have more attractive cultures. Using symbolic characteristics to develop a recruiting strategy and program, within the context of cultural values and preferences (Table 4.3), helps position the firm in a strategically valuable manner because it (1) differentiates the firm by playing to its strengths, (2) makes it difficult for competitors to copy the strategy, and (3), most importantly, sends a message that top talent wants to hear. Indeed, because the desired talent helped identify the attributes that differentiate the firm, they will be implicitly drawn to them and attracted to them. This means the organization can be very subtle in its recruiting message, yet be strongly attracting the desired talent while simultaneously discouraging the talent unlikely to fit the organization.

Step 5: Maintain the Desired Talent's Interest Through the Entire Recruiting Process

We have noted how recruitment occurs in stages. Keeping the staffing process a reasonable and appropriate length helps tremendously in keeping talent interested in the firm. This is especially true when the timeline for the recruitment process is explained in advance. Unexpected delays in recruitment usually translate into negative perceptions by candidates—particularly the top candidates. Maintaining candidate interest thus requires explaining the nature and length of the recruiting process, and ensuring an appropriate amount and type of contact between the firm and candidates. Technological advancements such as social media have dramatically changed the ways in which such "high-touch" recruiting can occur.

For example, recruiting speed can be a differentiator in a very hot market for talent. In the late 1990s working in New Orleans, Shell was losing potential recruits to competitors that were operating more nimbly because Shell required campus recruits to be vetted by a technical staff planner. This was despite the fact that the campus recruiters were selected as being among the best of the young talent. The solution was to give campus recruiters a certain number of wild cards that they could use to make offers to talent they deemed as exceptional. The plan worked. In one instance, a particularly strong candidate was given a formal letter after an interview. She accepted the next day with a comment that she wanted to work in an organization that could make quick decisions when needed.

It is important to understand that the images, activities, and experiences candidates are exposed to during the recruitment process send a signal about what it will be like to work in the organization. Psychological research suggests bad experiences are more influential in our thinking (Baumeister, Bratslavsky, Finkenauer, & Vohs, 2001). People expect the treatment they receive to carry forward after they join the firm. As a result, negative treatment, even if relatively minor, is overweighed. This means that even one small "mistake" in a recruitment process can have large negative ramifications on recruiting the desired talent. In general, it takes many exemplary positive experiences to make up for one negative experience . . . but in recruiting, there may not be an opportunity to make up, because the candidate will drop the firm from consideration. The recruiting process must be seamless in its execution and delivery—just as is the provision of high-quality customer service. This is even more true when the candidate has no existing knowledge of the company or experience with its products and services. For example, if you apply for a job with Unilever, you already have an impression about the organization. If you apply for a job with a firm you've never heard of, your impression is based entirely on the firm's recruiting practices, public media, and information about it on the Internet. Firms should strive to ensure their recruiting messages (e.g., RJPs) are consistent with their organization's message and employee value proposition, even if there may need to be some country-shaped focus. Large firms with well-established brands and images should leverage that public awareness in the recruiting process. Small firms or firms without brand recognition can use recruiting to construct the image they wish to present.

Step 6: Influence the Desired Talent to Accept a Job Offer

Provided steps 1–5 have been implemented appropriately, the final slate of candidates should be higher quality and fit well with the job, organizational culture, and national culture. Now the challenge becomes one of providing an offer that the desired talent finds compelling. The nature of such offers is highly dependent on the supply of talent, nature of occupation, organizational policy, and cultural values.

For example, the offers that attracted traders to Enron in the late 1990s and early 2000s were focused on the financial rewards and the highly competitive, innovative practices of the company. This manifested itself in significant earning opportunities, immediate sign-on bonuses, and a "What have you done for me lately?" mentality that rewarded individual contributions. The longer defined contribution plans, long service benefits, career development plans, and so on, offered by some of their competitors struggled to compete.

As another example, the Chinese market at the moment is something of a paradox, perhaps symbolic of the generational shift. The recruitment market is overheated, and younger Chinese employees are looking for rapid and visible promotions—even if they are more in form than substance (e.g., being promoted from general manager to senior general manager, even if the actual job duties have barely changed). Those companies that are fixed on a belief that the culture is so collectivistic that "face" does not matter, lose high talent to companies who have more finely tuned their approaches to both the attraction and development of talent.

Step 7: Evaluate the Effectiveness of the Recruitment Process

Companies realize the importance of acquiring top talent and they are willing to pay for it, but this doesn't mean that money is always well spent. Many firms believe recruiting must be slick and sophisticated; they also tend to believe that the way to attract the best talent is by paying more than competitors. Such thinking is a race to the bottom. All levels of talent will have a threshold on salary that must be met (i.e., needs). However, for firms that can pass that threshold, it becomes a level playing field and the focus shifts to candidate "wants." The way to tip the talent in your favor is to ensure your firm offers more of the qualities the desired talent wants than competitors can offer (see step 4). These qualities may not be costly—they might even be free—but it is important to convey these qualities to applicants in a way that differentiates the firm. Thus, attracting high-quality talent in a manner that is sustainable is vital for achieving competitive advantage; otherwise the acquisition costs will offset (or even overwhelm) the talent benefits.

Therefore, as discussed in chapters 2 and 3, it is critical for firms to demonstrate the return on their recruiting investments. The challenge is that "return" can be gauged according to a variety of different recruiting metrics. Demonstrating recruitment effectiveness is not a simple task. There are many different ways of benchmarking and measuring recruitment efforts and each serves a different purpose (see chapter 3). We highlight the most common metrics, making a distinction between metrics that provide information about efficiency (e.g., cost) and those that provide information about effectiveness (e.g., how it differentiates the firm). However, keep in mind that the "best" metric is the one most directly tied to a given firm's strategic goals and objectives (see chapter 2). A summary of these metrics is provided in Table 4.6.

Efficiency Metrics

Efficiency metrics emphasize the firm's economical and productive use of resources. These are primarily internal performance metrics. Improving

TABLE 4.6 COMMON RECRUITING METRICS

Type of Metric	Examples
Efficiency Metrics	
Cost	Overall recruitment function cost, cost per applicant, cost per hire, cost per recruiter.
Time-to-hire	Time between initial job positing and hiring decision; time between offer and decision.
Conversion Rates	Ratio of the number of potential applicants to those who apply; ratio of those who apply to those who receive offers; ratio of those who receive offers to those who accept.
Effectiveness Metrics	
Manager Reactions	Perceptions of whether the candidates being sourced are of appropriate quality; manager support for the recruiting system.
Applicant Experience	Perceptions of the system's job relatedness, fairness, invasiveness, timeliness, and so on.
Job Performance	Job performance metrics such as supervisor and/or peer evaluations, sales, productivity, accidents, and so on.
Unit Performance	Evaluates recruiting practice effects on branch, division, subsidiary, and other metrics (e.g., controllable profit). Compare within and across units.
Diversity	Representation in potential applicants, applicants, and hires in terms of racial, gender, cultural, functional, and related forms of diversity.
Individual Turnover	Individual level includes length of employment and reasons for turnover. Estimates for voluntary, involuntary, and overall turnover. Consequences/costs of individual turnover.
Unit Turnover	Collective turnover rates within a unit or organization, along with consequences/costs of turnover rates on unit metrics. Estimates for voluntary, involuntary, and overall turnover rates.
Quality of Hire	Subjective or objective assessments of the quality of the general and/or specific talent. Often based on test scores, education, specialized skills, or training/certifications.

efficiency metrics enhances external performance and may even contribute to competitive advantage in the short term. Common examples include:

- Cost. There are many ways to estimate recruiting costs. The simplest is the overall cost of the recruiting system, which should include all direct (e.g., advertising) and indirect (e.g., time off work for employee website testimonials) expenses. Other cost estimates attempt to identify cost at different stages of the recruiting process or by recruiter. For example, cost per applicant can be used to estimate the costs associated with sourcing candidates (i.e., converting them from potential applicants to actual applicants). Cost-per-hire can be used to estimate the costs associated with filling an open job and to evaluate the performance of the recruiter responsible for filling the job.
- Time-to-Hire. Good recruiting processes are timely, and maintaining estimates on speed of placement is helpful for ensuring the system is being run efficiently. There are multiple ways of estimating speed, but a common metric is the overall time lag between the date a job requisition was approved and the date the successful candidate begins work. Shrinking time-to-hire is akin to just-in-time manufacturing, wherein resources are made available as near to the time needed as possible. Another important metric is the time between when an offer was made to when it is accepted (a metric gauging desirability or quality of offer).
- Conversion Rates. Conversion rates represent the "hit rates" of the recruiting process. There are several variations that provide different windows into recruiting accuracy at different stages: the ratio of the number of potential applicants to those who apply (estimating sourcing effectiveness), the ratio of those who apply to those who receive offers (estimating recruiting quality), or the ratio of those who receive offers to those who accept (estimating

closure effectiveness). A focus on conversion rates shifts the emphasis from generating a large pool to one of generating a qualified pool.

Effectiveness Metrics

Effectiveness metrics emphasize the firm's ability to increase performance and differentiate itself from competitors. Improving effectiveness metrics may enhance competitive advantage over a longer term. Common examples include:

- Manager Reactions. The opinions of managers who must "live" with the new hire are an important gauge for evaluating the system. Clearly, managers who feel new hires are incompetent are not going to support future hiring recommendations and can undermine the credibility of HR more generally. Similarly, hiring managers may be the first to report inefficiencies in the process, as they will often be new to it and view it with an unbiased perspective. The best way to ensure their commitment and support is to involve the relevant managers in the design of the system.
- Applicant Experience. Just as the opinions of hiring managers are important, so too are those of the applicants to the organization. After all, if talent is in high demand and the recruitment process sends a signal about "life in the firm," then applicant experiences matter a lot! Efforts to enhance employment brand will do little if the applicant experience is largely negative. Today, prospective applicants can learn much about an organization, not only via social media sites such as LinkedIn and Twitter, but through websites like Jobvent and Glassdoor, which enable employees past and present to provide an insider's take on the organization's employment practices. They also provide space for unsuccessful applicants to discuss their experience with the process.
- Job Performance. Job performance can be measured in a variety of ways, including performance reviews by supervisors,

productivity, service or sales, days tardy or absent, accidents, upward mobility, and so on (see chapter 2; also discussed in chapter 5). The emphasis is on individual job performance, using the metrics the firm already has in place for performance management purposes.

- Unit Performance. Indices of unit performance represent a different way of evaluating recruiting effectiveness (see chapter 2). For larger firms that may have multiple units (e.g., divisions, branches, stores) or MNCs with subsidiaries, the recruiting budgets can be compared relative to other metrics like conversion rates, diversity, and so on. Thus, the focus is on using a common metric to compare recruiting practices across units. Unit performance is a particularly valuable metric for demonstrating the strategic importance of recruitment.
- Diversity. Diversity can refer to many different things, such as diversity in race, gender, culture, values, personality, functional expertise, and so on. It is also defined differently across the world. Much of the focus is on physical aspects of diversity, which is relatively easy to measure. The more subtle aspects of diversity, such as thinking patterns and personality types, are much trickier. Many organizations struggle to achieve a diverse workforce for strategic or compliance goals, and consequently, diversity becomes an important index of the effectiveness of recruiting. An example is Google, who voluntarily made public the challenges they have in creating a diverse workforce. One can examine diversity with respect to applicants, hires, and the variety of time-to-hire and conversion rates mentioned above.
- Turnover. Effective recruiting means developing a good fit between the person and the firm, and as fit improves, turnover decreases. This is true for both involuntary turnover (due to poor demands-abilities fit) and voluntary turnover (due to poor needs-supplies fit). Individual-level turnover can be monitored as an index of how well the recruiting process has developed a

good match. Metrics may include length of employment and both voluntary and involuntary turnover rates.

- Unit-level turnover. This is the collective voluntary, involuntary, or total turnover rate within a strategic business unit, subsidiary, or organization. It may also be monitored to determine whether there is something systematic within a unit or location that is creating a mismatch; units with higher turnover rates may help identify differences in applicant pools that require greater segmentation. Unit-level turnover is helpful for identifying regional and cultural differences in turnover rates, which may help inform a mismatch of recruiting practices to the cultural values within a given location.
- Quality of Hire. Organizations have begun to focus on the rather vague concept of quality of hire. The simplest, but most subjective, way to assess quality is to ask hiring managers about candidate quality. More objective and quantifiable approaches emphasize use of scores on assessments such as tests, interviews, certification exams, or training criteria (chapters 5 and 6). Sometimes job performance is used as a quality metric (as noted above) and the most thorough approaches combine measures of performance, potential, and engagement. Metrics to measure quality are many and varied, and the attractiveness of the concept will ensure that organizations continue to pursue this line of evaluation. Quality of hire is thus likely to be defined in a very firm-specific manner, and should be linked to specific operational and strategic goals.

Thus, there are many different types of metrics available. Generally, the effectiveness metrics are better at conveying the business value of recruiting to non-HR managers. The efficiency metrics are most useful for evaluating and monitoring the performance of the recruiting system. Both efficiency and effectiveness metrics are needed to provide cost-benefit analyses, or comparing the return on different recruiting practices and demonstrating impact. It is important for firms to keep track of as many metrics as appropriate and to consider those metrics carefully. For example, a firm finding that source-applicant conversion rates are starting to slip while applicant-hire conversion rates remain constant could signal that the sourcing strategy is losing its effectiveness. Further, a low conversion rate (hires/applicants) could mean that the offer is unattractive or that too many candidates are being sourced. Similarly, a global firm can compare the effectiveness of recruiting methods across different countries. Costs of recruiting by region could be calculated and combined with conversion rates to show the relative cost and effectiveness of recruiting within each region (of course, data on pay rates, staff search costs, and so on, would need to be included). Systematically monitoring recruiting effectiveness ensures the talent acquisition strategy remains supportive of the organization's goals. Just remember that metrics are not the end goal, but should be used as indicators to inform manages how well recruitment is performing relative to operational and strategic goals (chapter 2).

To illustrate, when Shell opened a Shared Service Center in Manila, we looked at the predicted turnover rates for centers like ours, in general and specifically in Metro Manila. For the sake of illustration, say that 25% was the competitive number. One of the assumptions we made was that we could beat that target, and we projected part of our cost targets on the assumption that the targeted level of turnover would mean only so many staff would move up to levels higher the pay scale. As it turned out, we had *significantly* lower turnover and the cost impacts on salaries was an issue with which to wrestle. We concluded it was a great problem to have, but it shows how measures need to be tested against multiple different dimensions.

Conclusion

Thinking about recruitment strategically is becoming more common but is still not something many firms do well. They think they are doing well when they are spending more money on recruitment, but in our observations many firms are more wasteful of dollars in recruitment than in most other HR functions (and certainly, other business functions!) because it is so difficult to quantify quality of hire. This is a really poor way to manage a resource as valuable as talent, and the same discipline that is paid to other resources such as financial capital needs to be paid to talent resources.

One of the most difficult questions an HR manager has to answer from other leaders is whether the recruitment system has attracted the best staff. Even allowing for lack of definition on what "best" constituted, the answer is always a struggle. One could measure the success and failure rates of the employees recruited through the various processes, or one could provide data on attrition. One could even link recruitment metrics to other strategic business metrics. But how does one get data on how successful candidates would have been who were *not* offered a job or who turned the firm down?

Roger settled in his room for the evening, exhausted from the day's conversations and meetings with the regional team. However, he did not want to lose the momentum of what had been a productive day with lots of good questions raised. Although the choice had already been made to focus on the strategic jobs, he still struggled with some of the more basic questions:

- How would recruitment be a linchpin to enable the overall business strategy to be achieved?
- What were the aspects of recruitment that really needed to be worked along global lines, which ones should be focused on regional consistency, and which one totally shaped the local environment?
- What was the brand of the company? If he were asked to capture the essence of the corporate culture to someone who had no or minimal contact with them before—how would he do so in 60 seconds or less—buzzwords?
- How would he answer the question—Why would I work for your company? Did the applicants in the region place a different priority on work-life balance than he did? How do you present that in a recruitment process or interaction?
- How would success be measured? Was it about driving down costs or raising effectiveness? Did it have to be an either/or question? Who gets to decide?

- Clearly the regional culture would be different. And within the region, he was already picking up on the notion that Asia may be a unified geographical area but it was by no means a unified culture. How would he get insights into that? How would he understand the differences and similarities? Did they matter? What about the generational shifts?
- He knew there are differences from the UK view of diversity—but what are the differences between the various countries? And what about intracountry differences? Do they matter?
- Back to consistency versus adaptation—where to start?

Roger recorded some key points to summarize his thinking—really his questions—to try to make sense of the situation (these notes are shown in Figure 4.5). He now truly appreciated the benefits of leveraging the quantitative data and analytics, and using evidence to help answer some of these questions. Looking away from his notes, he wondered, "Can I hold to values AND adapt? Can I be a "chameleon with a spine?"

Questions:

1) Generic or specific talent?
2) Internal or external?
3) Where are they? Local or foreign
4) What do they want?
5) Do we have what they want?
6) How do we know if we've done this right?

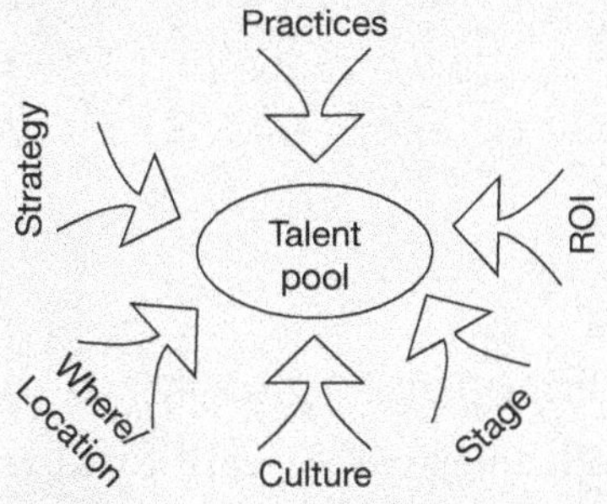

Figure 4.5
Roger's notes on global recruitment.

RECRUITING PRACTICES

We now turn to the topic of recruiting practices—that is, the methods and approaches actually used to source and attract candidates. Although we have hinted at recruiting practices throughout this chapter, we have purposely saved this topic for this later section so that these practices could be understood within the broader recruitment context and process.

Technology, and particularly the Internet, smart phones, and social media, have revolutionized the way recruiting is done in most countries and cultures. Companies that receive thousands of applications from candidates can be screened using predetermined search words. Algorithms can be used to scan for specific degree concentrations and/or GPAs and assess automatically against predefined standards (e.g., less than 2.5 GPA = automatic rejection), a feature that can defend against lawsuits based on bias if the objective standard is job-related and uniformly applied. Candidates who are located in different time zones or even continents can be interviewed over SKYPE or Hirevue rather than traveling long distances incurring costs and time. Although technology-driven recruiting practices offer many benefits, they are not a solution for every organization. Further, there is little beyond anecdotal evidence to show they are more effective than alternative (traditional) approaches. For example, empirical evidence on the success of social media in recruiting remains to be seen. Nevertheless, they are widely used because most believe these approaches can offer great potential for sourcing talent. We discuss the most important and popular recruiting practices below.

Referrals

Outplacement firms have long taught job seekers to network and build relationships with an extensive set of contacts as a way to secure a job. In reverse, organizations are increasingly promoting employee referrals as a way of driving applicant flow. Not only are costs associated with employee referrals typically lower than those associated with most other sources, but there is the assumed benefit associated with a candidate referred by someone known to the organization. This poses an interesting dilemma of the traditional view of nepotism in the West and the obligations of family inherent in highly collective cultures. Recruitment based on referrals usually has a better conversion rate than other approaches, but at the same time can result in a "sameness" and lack of diversity. Use of referrals is particularly common in some cultures. In European/North American cultures, companies often offer referral bonuses to staff who recommend someone who is hired and remains for a defined period of time.

Conversely, in highly collectivist cultures helping friends and relatives get jobs is considered part of the obligations of the friendship or extended family. The danger here is the tension between maintaining the quality of the standards against the backdrop of the culture. In parts of the Middle East, tribal influences can play heavily. What those from an individualistic culture may fail to understand is the pressure and tensions of the local staff between their family and friends versus adherence to the governance of a Western organization. One of us (Julian) saw an example of this in the Philippines when interviewing for civil engineers: "At the end of the three days, the list showed that I was due to interview—let's say—Jorge Gonzalez. I held out my hand and said, 'Mr. Gonzalez?' He said 'Yes, I am but I am Juan. My brother Jorge could not make it but suggested I interview for the job instead.' I was so taken aback and interviewed him anyway."

Job Boards

Job boards, such as Monster, CareerBuilder, and LinkedIn, have become the first stop for many new job seekers. A confluence of the referrals and technology can be seen in the presence of discussion boards in LinkedIn groups, where open jobs in a company will be advertised by members of that group who are (presumably) connections. Technology's effect on candidate sourcing was dramatic in the late 1990s and early 2000s, as Internet job boards all but put print advertising for jobs out of business. Indeed, job boards remain a popular source for many jobs. The AIRS 2011 *Job Board and Recruiting Technology Directory* lists thousands of job boards organized by geography, industry, and business function. The landscape has continued to evolve with the emergence of aggregators, such as SimplyHired and Indeed. Aggregators are job search sites that accumulate listings from a variety of sources (such job boards, newspaper and classified listings, associations, social networks, content sites, and company career sites) into a single website. While social media remains a hot topic in recruiting, one survey found that 62% of candidates reported using job boards versus 12% for Facebook (Charney, 2014). This is consistent with the movement of recruiting from a funnel to more open network, as suggested in Figure 4.4.

Organizational Career Sites

The use of job boards as a source of candidates appears to be under some pressure for a number of reasons. Job boards can be expensive, at least relative to other sources. Most companies have upgraded their own websites to make it easier for candidates to find and apply for open jobs. This is the current equivalent of a "walk-in" candidate; access to these applicants is almost free and they have the advantage of having demonstrated both interest in the company (reducing the need to sell the candidate) and some degree of initiative. Part of the reason candidates may be steered to a firm's career site is through marketing or experience with the company's product or service. For example, "liking" a company on Facebook leads to customers getting product and company information, which may direct them to the firm's website. Once there, and lacking more specific information or experience about a firm, the information organizations present on their websites becomes the primary signal and impression that potential applicants will have of the firm. It is not surprising, then, that organizations large and small have "career" websites devoted exclusively to presenting job opportunities, RJP information, organizational vision statements, employee video testimonials, and other organizational images. Career sites are also viewed by candidates as far more helpful then review sites like Glassdoor (Charney, 2014).

Because of the obvious cost advantages associated with recruits who "find the company," organizations have devoted significant resources toward enhancing their career sites. Companies have gone beyond text-based descriptions of jobs and culture to incorporate video (e.g., Ritz Carlton), games (e.g., Marriott), blogs from employees (e.g., Cadbury Schwepps), and links to Facebook, LinkedIn, and Twitter that are relevant to the company and job openings. Further, organizations often use firms offering search engine optimization and other services intended to drive candidates searching the Web to an employer's career site (e.g., EnticeLabs and Jobs2Web). The focus on company career sites extends beyond cost control, however. Candidates usually conduct online searches to learn more about potential employers and jobs, so controlling the content of career and job information helps the firm control its image and enhance application attraction and fit (e.g., RJPs).

Web logs (blogs) can be powerful ways to source candidates. Some companies encourage employees to keep regular blogs about their job duties, challenges, and life inside the company. These blogs can become very useful as both advertisements and RJPs. On the other hand, organizations should monitor the Web to check whether employees are keeping blogs presenting negative information about the firm. Employees who routinely vent or complain on Facebook, for example, are creating the firm's RJP whether the firm likes it or not!

Organizational career websites are thus powerful recruiting tools; so much so that they have begun to attract awards (top career site awards include the Webby and the People's Voice, and are handed out by organizations such as The International Academy of Digital Arts and Sciences and ERE.net). Although the content and purpose of these websites is constantly evolving, Box 4.2 provides some fundamental suggestions that should prove helpful for making them more effective.

Box 4.2 SUGGESTIONS FOR IMPROVING ORGANIZATIONAL WEBSITES FOR RECRUITING

1. Infuse the website with your organization's unique brand.
2. Put a link to a *careers page* visibly on the homepage.
3. Ensure that relevant career and job information can be found quickly.
4. Minimize the need to scroll.
5. Post a phone number or e-mail address where applicants can contact an organizational representative—and ensure the applicant receives a timely response.
6. Use realistic job previews. Simple text descriptions can be effective, but they work best when current employees provide video testimonials.
7. The visual display of the website should be appropriate and consistent with the brand.
8. The information on the website should be current.
9. Appealing websites are those that repeat key visual cues (e.g., borders, colors, links), align links and icons consistently within and across pages, and group similar topics in close proximity.
10. Ensure the website can accommodate those with disabilities.
11. Unless you compete for technically oriented talent, do not be *quite* at the cutting edge. Requiring a website to have the newest version of Flash, Java, or related software will probably mean the website will not work for many applicants. It is better to use a slightly older—but more common—version of the Web software (or at least make the option available).
12. If you use the Internet to take applications, send applicants an e-mail acknowledging receipt of the application and a timeline for next steps.
13. Offer a link where applicants can provide feedback and suggestions.
14. The website should be easily viewable from a smartphone or mobile device.

Source: Suggestions adapted from Cober Brown, Keeping, and Levy (2004); Cober, Brown, and Levy (2004); Ployhart, Schneider, and Schmitt (2006).

Gamification

One particularly interesting approach is to steer traffic to a company's website through the use of games. This approach falls under the broader umbrella of *gamification*, which is loosely defined as using games to change behavior and attitudes to achieve social or organizational goals. Gamification in recruitment typically involves games to proactively source talent. The logic is that, as the players engage in the game, the organization can monitor their performance and follow-up with recruiting media on desirable candidates. For example, Proctor and Gamble offered an online business simulation called Just-In Case. In the simulation, participants were put into teams with members representing different functions. The teams needed to put together a business plan and provide a solution. Team members worked virtually through the Internet. Those teams that performed well were sometimes invited to learn more about the company's careers.

One of Shell's more innovative recruitment approaches is called the Gourami business challenge (see http://www.youtube.com/watch?v=9hNk1_xLtF8). This is a successful approach because it engages potential employees in experiencing the challenges faced by international companies. It is a team format, emphasizing one of the core success factors to be effective in that company. The game is intense and requires long days and short nights. It is supervised by voluntary, high-quality Shell staff across the disciplines and ends with a presentation to 3 or 4 of the 200 most senior leaders who give direct, high-quality feedback to the participating teams. It was also held in attractive locations in high-end hotels. The process is particularly successful in attracting high-quality recruits who have already done internships with other companies and are emotionally committed to them. Indeed, how could you turn down an all-expenses-paid, unconditional offer to spend five days with your peers, potential peers, and senior leaders? It contains all the right messages about teamwork, low hierarchy, high-quality talent, and develops a cohesiveness with others in similar positions that you could be attracted to . . . it becomes yet another potential reason to join the company.

Third-Party Career Websites

There are also third-party websites that allow users to provide positive or negative information about people that include supervisors, employees, or coworkers. For example, websites such Glassdoor enable employees or applicants to anonymously post information about firms, including salary levels and so forth. Applicants may post interview questions, and employees may talk about the firm's culture or things happening within the firm—good or bad. One can also see tips on how to answer specific questions used by an employer, such as IBM, as shown on this UK website: http://targetjobs.co.uk/employer-hubs/ibm/343024-how-to-answer-ibm-interview-questions. Other websites allow people to anonymously discuss information about their bosses, coworkers, companies, and so on. The information may be either positive or negative, and given the popularity of such websites, it is important for organizations to proactively respond to negative feedback. These websites are growing rapidly, and they are often renamed, acquired, or merged with other similar talent websites. This can make it difficult for firms to use or respond to information posted on these websites for recruitment purposes.

Social Media

In a global survey of over 18,000 professionals, 25% reported actively looking or casually looking (a few times a week) for another job (Srinivasan, Gager, & Ignatova, 2014). Another 60% reported being open to hearing about new opportunities (the other 15% were completely satisfied and uninterested in a move). As a result, many recruiters express frustration at the "post and pray" option of job boards and career sites. These leave them with only the minority of actively looking candidates, without providing access to the majority of passive candidates (as discussed and shown in Figure 4.4). Because of its enormous potential to reach passive candidates at an economical cost, the use of social media in recruitment has become very popular. Social media sites, such as the "big three" of LinkedIn, Twitter, and Facebook, are often used by recruiters to identify and source active *and* passive job candidates. There was a time when social media sites were used only for people to network with each other. Today, organizations have links to these sites on their career pages, have postings on these sites, and allow users to "friend them" for advanced notice on jobs, "friend-only" sales, and product launches. LinkedIn is the main professional online service, and according to the company's website, has approximately 400 million members in over 200 countries. It is becoming so widely used that the LinkedIn profile may become the standard resume (it can be easily copied and offers some degree of standardization). Yet somewhat surprisingly, one US survey (Jobvite, 2014) found that while 94% of recruiters were on LinkedIn, only 36% of job seekers were (the same survey found that more than three-fourths of social job seekers found their current job via Facebook). Clearly, both applicants and recruiters would be well advised to use multiple social networking platforms. Almost all applicant tracking systems offer some degree of integration with LinkedIn, Twitter, and Facebook, enabling the recruiter to post jobs and target messages to particular groups and even individuals.

Social media offers an incredible opportunity for sourcing candidates. Job postings can be passed along to friends or members who share relevant characteristics on their profiles. Recruiters can maintain links and networks of potential candidates, searching their text for certain keywords (e.g., "10 years IT experience") as a means to identify candidates not actively looking. Or, a recruiter can send "tweets" to a limited set of subscribers informing them of job openings relevant to their skill set.

The ability to source candidates *not actively looking* marks a critical point of departure from old recruiting methods. In the past, organizations had to provide information and incentives to attract candidates; now the organizations can search more aggressively for those who may be willing to jump jobs but are not actively looking. This increases the pool of qualified talent and escalates competition for top talent through poaching.

Another critical point of departure from old recruiting methods is the ability to source candidates from other cultures. An organization wishing to start or strengthen its operations in a foreign country may "push" a recruiting message using social media to a network of contacts and customers in that region. They can ask for assistance in finding talent, making new job postings accessible to a larger group of people, and learning tacit information about the region.

Perhaps one of the most innovative approaches to recruiting involves crowdsourcing. *Crowdsourcing* is an approach that relies on groups of people to solve a particular problem (the groups can range from only a few to several thousand). People who are members of the network are tasked to solve the problem, and whoever wishes to contribute and collaborate offers an opinion. For example, suppose a financial consulting service headquartered in Japan is going to open a field office in Brazil. It is likely that most of the field office managers will come from Japan, but the firm could use crowdsourcing as a way to hire additional managers and staff locally. The company could post, via its social networks, information about the jobs to its contacts in Brazil or South America. They could also ask for advice on how to set up the operation more efficiently. By pushing the information about the jobs to the "crowd" located in Brazil, the firm can draw on the collective expertise of people within the culture. These people will have a greater social network of indigenous colleagues and greater information about working in Brazil. In this manner, the "crowd" becomes the recruiter, and could produce hiring leads that are more accurate and faster.

Crowdsourcing for recruiting can be a valuable tool. A recruiter with a social network of 1,000 people can post a job opening for particular expertise in a manner of seconds, and instantly tap the collective networks of all 1,000 people. For such approaches to work, however, there must be incentives, rewards, and trust among members. Simply crowdsourcing a job opening may be helpful, but effectiveness will be increased if there is compensation for a successful hire, and if there is a social media policy to guide appropriate conduct.

However, one should keep in mind that use of social media is not a simple solution for sourcing challenges. First, the hype about the success of recruiting through social media is not backed up with consistent data showing actual returns. One should be skeptical of purely anecdotal accounts of success (see metrics section). Second, keep in mind that not all social media sites are equally effective tools for sourcing. One report by Shaker consulting shows that LinkedIn is perceived to be the most effective media outlet for sourcing, relative to Facebook, Yahoo Groups, Jobster, and Google Groups (http://www.shakercg.com/quality-of-hire-blog/Social-Media-and-Quality-of-Candidate-|-Candidate-Competencies-Vary-by-Source-%28Part-2%29). Not all social media sites are likely to be equally effective, and the choice of which social media forum should be used needs to reflect the organization's sourcing strategy and talent market. Thus, different websites or social media platforms are likely to be differentially useful for reaching talent, suggesting that talent pools exist in the digital world as much as in the physical world.

Finally, it is important to recognize some potential problems with using social media. First, the legality of using social media as a recruiting tool may differ by country. For example, employers in some cultures (e.g., China, Germany) expect to see pictures of job candidates while such practices are to be avoided (or are illegal) in other cultures (e.g., UK, United States) (http://blogs.hbr.org/2012/03/photos-of-attractive-female-jo/; http://www.redstarresume.com/how-resumes-differ-from-country-to-country/#.UyzbAIUVCgI). Clearly, searching social media postings will be less of a concern in the former cultures because there is less of a legal concern that online pictures may bias impressions. Second, keep in mind that not everyone uses smartphones or has Internet access, particularly in developing economies. Social media approaches may work well in many countries, but traditional recruiting methods will be necessary in other cultures. Countries that are defined as "high-context" countries, where forming relationships is essential, often use drawn-out processes and may find the impersonality of online interviews less attractive than the face-to-face interview. Finally, even in developed economies, there will not be equal access to the Internet and technology. It is often the case that those with lower incomes will be less likely to have computers or smartphones, and hence it is possible to unintentionally discriminate against protected groups (e.g., minorities) by relying solely on these technologies. Alternatively, younger generations may be more reliant on technology, and attend to different apps and websites than older individuals. Box 4.3 gives a checklist of best practices for smartly using social media (an executive briefing on social media policies can be found here: http://www.shrm.org/about/foundation/products/Documents/Social%20Media%20Briefing-%20FINAL.pdf).

Applicant Tracking Systems

The use of technology to manage applicant flow has evolved from simple personal-computer-based systems to fully integrated Web-based tools capable of handling almost all tasks related to talent acquisition. In the United States, they have become particularly advanced in companies that are subject to equal-employment-related policies because

Box 4.3 CHECKLIST FOR USING SOCIAL MEDIA IN RECRUITING

1. _____ Consult legal counsel to develop a policy for using social media. Communicate and train employees on how to interpret and implement the policy.
2. _____ Use social media only for sourcing talent. Do *not* use social media for making hiring decisions, and keep any information collected during the sourcing distinct from the selection process.
3. _____ Identify and focus on the job-related competencies for a position. Limit attention to only those competencies.
4. _____ Focus on professional networking sites (e.g., LinkedIn) for targeted sourcing; use all networking sites for crowdsourcing.
5. _____ Develop a standardized scorecard or checklist for managers tasked with reviewing social media postings.
6. _____ Have recruiters and hiring managers appropriately document the process they used to source candidates. Accurate notes will help ensure the policy is being followed.

the stringent requirements to track and report applicant flow in affirmative action plans. Applicant tracking systems (ATSs) or talent acquisition systems have expanded dramatically over the past decade such that most large organizations have an ATS that want one. The typical ATS enables recruiting professionals to manage the requisition, from approval to fulfillment, including creation of the job specification, linking it to job boards and other candidate sources, candidate screening (including automatically administered assessments and interview scheduling and management), and candidate communications (e.g., offer and regrets letters).

More recently, providers of ATSs have begun offering mobile options. As smartphones and mobile devices proliferate, organizations are adapting by enabling both applicants and recruiters to access their recruitment systems via their handheld devices. Part of the "enabling capability" of social media recruiting is the increase in smartphones, or wireless telephones that have Web access and can run networking applications (e.g., iPhone, Droid). Smartphones have spawned an entire industry developing "apps" (short for "applications"), which are mini software programs designed to provide a single service.

Another innovation in recruitment has been the increased popularity of compartmentalized recruiting processes. Essentially, the tasks associated with recruitment, such as sourcing, qualifying, screening, and administration (e.g., record keeping, communications) are divided among different groups of people who specialize in the one activity. Although realistically suitable only to very high volume recruitment efforts, compartmentalization offers a number of advantages. Work flow can be continually monitored and resources shifted from one compartment to another as bottlenecks begin to appear, portions can be "outsourced" to other countries offering lower costs and around-the-clock processing, and Total Quality Management principles applied to each step reducing variability and enhancing overall productivity. For example, Kenexa's RPO division uses such an arrangement, with much of the work done in centers of excellence located in India and Argentina.

Conclusion

The trend in recruiting continues to be one of leveraging technology to improve efficiency and effectiveness for the organization, while enhancing the applicant experience. Search engine optimization to better drive candidates to career sites, video-enabled employment branding, smartphone apps to support mobile recruiting, and social networking to source passive candidates and build brand awareness, are just a few of the trends prevalent in today's talent acquisition market. But keep in mind there will be a limit to what these technologies can accomplish. Competitors may easily copy recruitment practices. Thus, leverage the technologies as much as possible, but realize that the content on the media—the symbolic and (to a lesser extent) instrumental characteristics—will be much more important for differentiating the firm in the long run.

WHO DOES RECRUITMENT?

Recruitment thus far has been discussed with the assumption that it is being conducted entirely within the firm's HR department. However, there are many options available to organizations when it comes to sourcing candidates. While it is often the case that a recruiting specialist within the HR department handles the various activities associated with recruiting and selection, it is not always so. In many (particularly small) organizations the hiring manager may also be responsible for stimulating applicant flow. There are some distinct advantages to a partnership in this space. For example the weakness of an entirely HR-led activity is the lack of knowledge of what the job really requires and how to test for it. On the other hand, there are line managers who are convinced that they will spot talent when they see it, but who routinely fail to do so because they are relying on their gut feelings as opposed to professional interviewing and assessment techniques (discussed in chapter 5).

There are many instances, however, where recruiting is handled by persons working for the organization but not as an employee of it. In broad strokes, outsourced recruitment is of three types. First, there are many staffing agencies that offer a variety of services such as temporary staffing or temp-to-perm, and do so across a broad spectrum (e.g., Kelly Services, Adecco, and Manpower) or within a specialized niche, such as accounting, legal, or information technology. Possibly due to the challenge associated with shedding excess workers, Europe represents the largest worldwide market, while the Asia-Pacific region represents the fastest growing (http://www.strategyr.com/Employment_Services_Market_Report.asp). In the United States alone, temporary staffing generates over $100 billion in revenue (American Staffing Association; https://d2m21dzi54s7kp.cloudfront.net/wp-content/uploads/2015/09/asa-fact-sheet_staffing-recruiting-sales-july2015.pdf). While it is difficult to say exactly how many jobs were sourced in this manner, it is clearly a considerable number. As organizations continually seek flexibility in their workforce commitments, the use of contingent labor such as that provided by staffing agencies is likely to continue.

A second form of outsourced recruitment is provided by third-party search firms, both retained and contingency. Retained firms (e.g., Spencer-Stuart, Korn Ferry, Russell Reynolds, and Heidrick & Struggles) are paid a fee plus expenses regardless of whether or not the client hires their candidate and are therefore used primarily for executive level and hard-to-find specialist jobs. Contingency firms, who are paid only if their candidate is hired by the client, tend to fill lower-level professional jobs. In both cases, the fee is typically negotiated as a percentage of first-year salary, sometimes including the bonus. The *Directory of Executive and Professional Recruiters* provides a list of firms providing both retained and contingent search services organized by function, industry, and geography.

Finally, recruitment process outsourcing (RPO), as the name implies, entails outsourcing all or part of the recruitment function to a third-party vendor. The RPO provider typically offers services such as sourcing, prescreening, background checks, interviewing, record keeping, and assessments, but may also help the organization with employment branding, requisition and offer management, and workforce planning. Most users of RPO services appear to begin with a portion of their workforce, such as administrative and exempt-level professional positions, while continuing

Box 4.4 SUMMARY OF NOVEL RECRUITING METHODS

- Online chats with real or virtual personnel.
- Online games or contests to draw traffic to the website and search for talent.
- Use of social media to proactively source potential applicants currently not looking for a job.
- Tracking passive candidates based on corporate website footprint.
- Phone apps to inform applicants and customers of job openings.
- Recruiting by offering short-term experiences to prepare applicants to find more permanent jobs (e.g., McDonalds, military).
- Free ringtones and backgrounds to draw traffic to the website.
- Career websites that provide links for spouses and families.
- Employee blogs as a means to provide realistic job information.
- Crowdsourcing for job targeted searches.

to recruit higher level managerial or priority positions internally. The RPO arrangements vary substantially in terms/costs, services, and jobs, but almost all have a mutual service level agreement with the organization retaining the ultimate hiring decision.

If talent is so vital for firms' competitive advantage, then why are so many organizations willing to outsource the management of talent? There are many reasons, but efficiency is top among them. Service-level agreements typically include standards for cost-per-hire and time-to-fill, and reports estimate an average 20%–30% savings off of a pre-RPO recruitment spend (see talent acquisition reports at Bersin.com). Another key attraction to RPO is flexibility. Many organizations cut their recruiting staffs following economic downturns, and RPO offers a way to resume the critical activity without adding headcount to their own payrolls. Further, as hiring increases, it is the provider's problem to scale services. Finally, many RPO providers have a global footprint (e.g., Kenexa or SHL), enabling organizations to staff up in distant geographies without having to first build the infrastructure.

Conclusion

There are lots of options when it comes to who will conduct the recruitment activities. Technology has enabled many creative recruiting practices to be used, and some of the most important ones are summarized in Box 4.4. Given that an outcome of recruitment is the acquisition of talent that is strategically valuable and contributes to performance and competitive advantage, firms should give careful thought to whether to outsource it, keep in in house, or a mix of the two. These issues and questions need to be considered more broadly than just in terms of cost and efficiency.

Roger enjoyed his time at Spooner Electronics' Singapore office, which was situated in the Central Business District opposite the Marina Bay Sands Hotel, a magnificent building of three 55-story towers that was built entirely on reclaimed land. By now, Roger was starting to understand the difficulties in translating the Spooner strategy and the S.E. Asian talent strategy [see Chapter 2] into a regional recruitment strategy that would be tied to meaningful business metrics [Chapter 3]. He was also getting crisper in his thinking about the recruitment challenges that lay ahead—not just in scale, which was pretty formidable—but in the sense of the dynamics of the Asian market. There were differences between markets, but also some emerging similarities:

1. Talent acquisition was at the heart of organizational success, both current and future. It was not a function that should be left to chance, and the involvement of skilled, professionally competent HR staff in partnership with line managers was essential. Most of the company's regional leadership understood this, but expectedly, it was seen as an HR problem.
2. The availability of overseas education, coupled with the information revolution, meant that the new generation, the ones that were attractive to Spooner and other companies looking for high

talent with technical skills, were in a position to shape the market. The economies in which they lived were growing rapidly, there was a huge influx of foreign capital, and many candidates were extremely well informed. Companies that attracted, retained, *and* engaged the best talent would win, and that might mean adapting some of Spooner's systems to compete. For example, he had already picked up that visible progression, signified by a change of job title, was a significant motivator, almost a given, in retaining talent in these countries. That trend struggled to gel with pay systems that had wide job bands and infrequent promotions. So recruiting messages had to be careful to not overpromise and underdeliver; the RJP would need to be carefully crafted.

3. He believed that the balance of local adaptation and adherence to the regional strategy was a delicate but critical one. His view was that the pressures of reducing recruitment costs would mean that some of the local systems needed to be harmonized or eliminated. So he would develop some solutions that spread across borders, even if that meant some short-term inconvenience.
4. There would always be a case to be made for some level of expatriation both into areas of high growth and from the subsidiaries to HQ and other subsidiaries. This would help ensure the corporate culture was shared and understood, enable knowledge acquisition of key markets, and represent an attractive career development option for staff.
5. Then came the twist: of course the regional teams needed strong talent, but there should be a passion for getting the very best local talent to run country organizations, because they represented continuity and local knowledge, and they were more cost-effective than expatriates, by a wide margin.

He was beginning to sense an important tension between these last two thoughts as well as the very real pressure on costs. Although Roger was increasingly excited about a multilocated regional team, he knew that without leveraging technology and all the tools of virtual working that were available, it could lead to increased costs. This left him with questions:

- How could he balance recruitment for the strategic job families (which were the priority in the regional talent strategy) relative to the less strategic job families?
- For each job family, what elements should Spooner focus on? What was the value proposition for top talent in each strategic job family? How might this value proposition change across countries?
- What would differentiate Spooner from competitors, locally and internationally?
- Why would a talented person want to work for Spooner?
- How to balance regional efficiencies with local customization?
- Talent analytics and data were key to implementing the regional talent strategy. What recruitment metrics would be used to track success on the recruitment strategy and demonstrate return on recruitment investment?

One last thought struck him as he put down his pen. He had heard a lot about outsourcing recruitment, and knew that HQ was interested in looking at that option. But how could he develop that view further without tipping his hand and risking the alienation of the team he was about to lead? And did the potentially impersonal nature of that process fit well with the "high-context" countries in which he was now working?

He left the hotel and headed out in a taxi to a seafood restaurant located overlooking the bay, which had come with high recommendations. Jumbo Seafood beckoned.

5

Selecting Talent Globally

THIRTY-SECOND PREVIEW

- Selection refers to the strategies, processes, methods, and practices used to assess applicants on job- and organizationally related knowledge, skills, abilities, and other characteristics (KSAOs)—that is, talent.
- The goal of global selection is to identify those applicants that will *fit* with the job, organization, and national culture.
- Selection is strategically valuable because it shapes the nature of talent resources, implements the firm's strategy to drive competitive advantage, changes or reinforces the company culture, and enhances analytics and quantitative insights.
- Selection methods are the assessments and predictors used to obtain scores about talent. Methods may be signs of behavior

(e.g., ability, personality, values) or samples of behavior (e.g., interviews, assessment centers, simulations, work samples).

- There are large differences in the types of selection methods used across countries, and these differences are due primarily to legal and political, rather than cultural, factors.
- However, talent scores based on different predictor methods are highly similar across cultures and countries (assuming appropriate language translations have occurred).
- A selection process or system refers to how scores obtained from different methods are combined to enhance effectiveness, efficiency, engagement, or diversity. Selection methods vary in terms of their features and benefits, so the methods used should reflect the firm's priorities.
- The Bottom Line: Using global selection practices and methods that best identify the desired talent and are unique to one's firm and culture can create talent resources that generate competitive advantage.

Roger doodled idly on a scrap of paper from his pocket as he waited outside KLCC, the massive shopping mall underneath the Petronas towers, for the doors to open. It was Saturday, and he needed to purchase some gifts for the family back home. It had been an eventful nine months in which he had succeeded in assembling a diverse, talented team that was helping him build some bridges across the region. The team had been particularly helpful in creating a recruiting strategy and process that maintained a strong regional focus, but allowed local variations to reflect local cultural and labor conditions. Roger was very proud of the recruitment strategy, and so far it had been well received. It was time for a short break before he and his team tackled the next major challenge: developing a similarly balanced selection process.

He knew this was going to be tricky. Designing the recruitment plan required his travel to the major countries in the region to give him some firsthand sense of the similarities and differences that existed.

This was perhaps the most profound learning. He knew that he would see differences between the UK and Asia, but he was unprepared for the subtle, yet distinct, differences between the various Asian countries.

One the major commonalities of the region was its strong sense of community and family values and a strong preference for a collective outlook rooted in belief systems which hold that enhancing societal well-being depends on everyone benefiting—a focus on "we" as opposed to "me." There were variations between countries, but taken together they are in marked contrast to most of western Europe and North America. In the context of HR practices, these disparities came into sharp focus in two areas—rewards and selection. It was this second area that caused Roger the most concern. He was under significant pressure from HQ to introduce selection assessments similar to what were used successfully in Europe. In particular, group assessment centers were viewed as the prime methodology for selecting graduate-level staff. He himself had been hired using this method, and the research he had done on the strengths and weaknesses of the technique had helped persuade him of the benefits. But when he mentioned this casually to Chien-Fu Wang, the country HR manager in Taiwan, it met obvious resistance. Wang noted:

> In your culture, you value different behaviors than we do. Let's look at a couple of proverbs. Yours would be "The early bird catches the worm." This applauds initiative, aggressively pursuing your individual goals over those of others—you see life as an individual competition. Maintaining connection, relationships, consensus is less important to you. In my culture there is a saying—"the first bird to rise gets shot." Your proverb places a premium on being noticed, standing out from the crowd in the types of evaluations you are considering. Ours focuses on harmony, seeking agreement to move forward as one, and so on. So who do think will be successful if there is a mixture of Western and Asian candidates and they are assessed by a Westerner? What about the reverse scenario? Do you see how the cultural bias

> of your proposal could lead to really undesirable outcomes? Do you really want the assessment of a global workforce to be so culturally dependent?

This conversation resonated strongly with Roger, because he had tripped over unexpected consequences earlier when he had tried to adapt. Spooner was interested in improving their talent inflow of Chinese students. So they went to INSEAD, a premier business school that had a branch in Singapore, to interview some promising Chinese students. The interviews were conducted in Mandarin to try and neutralize the language effect. One of the students had reacted to that:

> Why would you interview me in Mandarin? I take all my classes in English and the reason I applied to Spooner was their international operational footprint. If I wanted to work in Mandarin I would have applied to a Chinese company.

It was so obvious in hindsight—even if the thinking behind it was well intentioned, it was clumsy. What would be a solution to the apparently intractable dilemma of local relevance and global consistency in the selection of talent? The high-stakes nature of selection and assessment would leave no room for clumsiness.

Selection refers to the strategies, processes, methods, and practices used to assess and hire applicants on job- and organization-related KSAOs. Recruiting is the process that generates a pool of talented applicants; selection is the process that assesses those applicants in order to determine who meets or exceeds qualification standards and thus should receive job offers. Selection is difficult, because it is fundamentally about making a bet about future behavior, based on the incomplete and imperfect information available today. However, this difficulty is also why selection can contribute to competitive advantage. A firm that assesses and selects in the most accurate and efficient manner is building a high-quality stock of collective talent that will be difficult or costly for competitors to copy. If selection is done right, it will help reduce costs and increase revenues by ensuring the people who receive job offers are willing to accept them, will perform well, fit the company and national culture, and contribute to strategy execution. Because selection is the "gatekeeping" function, it has arguably the strongest influence on the composition of talent resources—and by extension, competitive advantage.

Selection is based on the premise that people differ in predictable ways. Different people will behave differently in the same situation, but the same person will behave similarly across very different situations. This is because people vary in their physical and mental abilities, personalities and interests, the cultures to which they have been exposed, and the skills they have learned over the course of their lives—all of which guide their behavior. Jobs also reliably differ in terms of what it takes to do them well, as discussed in chapter 2. Matching individuals to the requirements of the job is the essence of selection. It therefore follows that the first step in successful selection is a complete and accurate description of the job and an analysis of the talent required to execute it satisfactorily. In this chapter we explore various ways of identifying which candidates possess the attributes required to perform well (i.e., demands-abilities fit; chapter 4); primarily for the near term but also for longer-term potential. It is important to note that in most cases few candidates will possess high levels of all required KSAOs. The goal of selection is usually one of ensuring that the candidate actually hired has enough of the relevant talent to perform the job successfully. It is also important to note that no selection system is perfect and hiring errors will occur. However, the right selection system will greatly increase the odds of success by systematically targeting the talent (e.g., mental and physical abilities, personality and interests, knowledge and skills) that predispose one to behave in the desired fashion.

Selection is probably the most researched component of talent acquisition, because it is universal to businesses everywhere. Selection has likely been around as long as there were organizations. For example, early Chinese dynasties used selection procedures for civil service positions (Ployhart, Schneider, & Schmitt, 2006). But it was the combination of industrialization in the Western world, along with large numbers of people living in urban areas, that created the field of selection as we know it today. Industrialization created large numbers of jobs that required specific skills, and urbanization created large numbers of people who could apply for those jobs. Two World Wars, and the resulting demands of selecting and placing large numbers of people into the appropriate positions, further stimulated selection practices. Since those early beginnings, thousands of research studies on selection-related topics have been conducted, and there is now considerable evidence to show the consequences of doing it right—and wrong (Guion, 2011; Ployhart et al., 2006). Assessment of job-related KSAOs has, in turn, become a large industry. The global market for assessment services is estimated at two billion dollars, with thousands of consultancies large and small offering a variety of assessment options (keep in mind that a large number of firms do not purchase any kind of assessment, suggesting the potential market is actually much larger). This growth in assessment services has also been spurred by the now familiar influences of technology, globalization, and Big Data. Almost any assessment service is available via the Internet. Offering a fast, reliable, and inexpensive means of capturing, scoring, storing, and feeding back results, the Internet has revolutionized the assessment industry while removing barriers to entry.

Globalization has contributed to this trend, as organizations use assessments when expanding overseas and as a means of comparing individuals across cultural and geographic boundaries. Assessments can offer a way for an organization to control the "quality of hire" in locales far away from the home office or to compare multiple candidates for the same job who come from very different parts of the world. Yet as Roger has found, this approach is not without its dangers, given language and cultural differences across regions and how those can potentially influence performance in assessment processes. Thus, selection in global organizations requires additional consideration of the cultural context within which selection occurs.

Roger knew he must start developing the selection plan for the strategic job families by aligning it with the firm's broader strategy [described in chapter 2]. The plan obviously needed to support the regiocentric recruitment strategy and balance the tension between a single global perspective (strongly endorsed by HQ) and local customization (strongly endorsed by managers within each country). Changing the opinions of this latter group was going to be his biggest challenge. The existing selection process was locally customized and presented a confusing patchwork of different methods and practices. Each country ran selection in its own unique way, even when the jobs were the same across countries. This was not only inefficient and costly, but it also interfered with building a Spooner brand and culture in the region. Indeed, employees in some countries felt selection was too easy in other countries, which contributed to some regional managers becoming arrogant and disrespectful to those from other countries. Everyone at the regional executive level and HQ knew this was a serious problem that was keeping Spooner from moving forward in the region. Of course, nobody knew how to solve it or apparently was even willing to try to solve it. Conversations with Roger's regional HR team indicated that a favorite tactic employed by local HR leaders was to argue that any "new" selection practices were illegal in their countries. There was some truth to this. Several years earlier, Spooner had come under legal and political scrutiny in the Philippines for using a group-case simulation that was very inconsistent with that country's collectivistic orientation. Candidates and managers complained, the local legal counsel pushed to shut the entire process down, the company's reputation in the region suffered, and regional managers looked incompetent—or clumsy, to Roger's eye.

But this was his job, and Roger was determined not to make the same mistakes as his predecessors. He knew that persuading his colleagues of a new approach would not be easy. He also knew that many of them were technically inclined for the words "Let the data do the talking," which rang in his ears—yet again. Although in times of change, the past might be an imperfect predictor of the future, he hatched a plan that would address the concerns of the technically minded population with whom he worked, develop and present data in a way that would at least build a platform for informed debate, and be rooted not in theory but in the outcomes present with Spooner.

Roger realized that this data would require significant time and effort to analyze but firmly believed that without an analytical approach his efforts to persuade the line and his HR colleagues of the need for a strategic and data-based approach would be ineffective. And all this work would inform the strategy for successfully selecting a workforce that would represent the customer and operational base of Spooner and that was diverse—defined across multiple dimensions. And so he began to build a story relating selection to the firm's strategy . . . but what would be the talking points to this story?

SELECTION IS STRATEGICALLY VALUABLE

As discussed in chapters 2 and 3, hiring the right person for the job can have substantial positive financial impact on the performance of the individual, the unit, and the organization. As organizations mine data on employees for insights related to performance, retention, fit, and potential, assessment data are becoming increasingly important. Understanding how talent and job experiences interact to accelerate (or slow) development is a key issue on the minds of managers everywhere.

- **Selection shapes the nature of talent resources**. Chapter 2 described the talent supply chain, and chapter 4 noted how recruitment shapes the nature of talent resources due to demands-abilities fit. Selection is thus the mechanism that managers can most directly use to shape the quality and quantity of talent resources, both individually and collectively. For example, raising the qualifications needed for selection will increase quality but reduce quantity (in at least the short term).

Selection controls the flow characteristics of the talent supply chain, and it controls the talent quality threshold (chapter 2). Just as a faucet controls the amount and temperature of water, selection controls the amount and quality of talent resources.

- **Selection helps implement the firm's strategy to drive competitive advantage**. Implementing firm strategy to achieve competitive advantage is one critical role of HR. In turn, selection is one of the key ways a firm can implement its talent strategy in the service of competitive advantage. Indeed, scientific evidence suggests that use of more effective selection programs can generate greater firm performance and growth (Huselid, 1995; Kim & Ployhart, 2014). Further, if a firm adopts a new strategy that requires a new employee skill set, then selection should change to reflect this shift in strategy. A good example is offered by Dell Computers. As it faced stiffer competition from other computer manufacturers, Dell changed its focus from direct-business sales to consumer sales. As a resulting change in strategy and operations, it needed to change how it staffed and hired existing positions and how it staffed new positions (e.g., customer support).
- **Selection helps change organizational culture**. When a company is going through a major transition, be it a restructuring or a bold new strategy, the talent needed to make the change is often lacking. Furthermore, the culture that is in place may stifle any attempts to change, even if the talent is ready to support the change. In such cases, it is sometimes more effective to hire people with a new skill set than it is to change employees to develop the new skill set. In this manner, selection can be the mechanism through which a company begins to change its culture, because it brings in new hires with a different perspective and who are not entrenched in history and culture.
- **The effects of selection are difficult and costly to duplicate**. Because selection strongly shapes the nature of talent resources, and supports the implementation of firm strategy, it is extremely difficult and costly for competitors to duplicate a firm's talent

resources. Building a critical mass of generic talent is time consuming and requires synergies with other resource and HR practices. Even if competitors adopt the same selection practices, it will take a considerable amount of time to reproduce the effects of those practices on talent resources.

- **Selection enables analytics.** As we shall see, selection should be based on evidence. Assessments are often used in selection because they provide quantitative data about talent (chapter 3). Consequently, selection data based on assessments may become useful data for making strategic decisions about talent. For example, selection data about individuals can be used to customize training (e.g., giving more training only to those who score below a certain threshold), in placement and classification (e.g., determining job rotations), and in career counseling. Selection data can also be aggregated to business unit levels to identify talent shortages or surpluses. Aggregate analytics can also be used to model the business unit consequences of collective talent resources, by drawing empirical relationships between talent and performance (chapter 3 provides many examples of such analytic models).

Conclusion

A chief HR officer (CHRO) with whom we frequently work has noted, "The selection function owns responsibility for the people-related successes and failures of the firm." This is perhaps a bit overstated, but not by much. There is no escaping the fact that the processes and practices used in selection will dramatically shape the nature of the firm's talent resources. Recruitment is obviously important, but a firm has less control over its recruitment outcomes because those outcomes are ultimately based on *attracting* qualified applicants—the applicants play a "more equal" part in the outcome. In contrast, selection starts with an applicant pool, so the discretion is more with the firm than the candidates. Thus, the firm can more immediately distinguish itself in terms of its talent and strategy execution by its selection practices. Selection is often not considered very strategic, because it can be limited to microlinkages between individual types of KSAOs and job requirements. However, in the aggregate, the consequences of selection are extremely strategic, as our CHRO colleague implies. The challenge then becomes one of how a firm appropriately exercises its discretion to build a capability for talent selection.

Roger knew that he had two main goals from selection in the region. The first was to ensure the capability to staff the strategic jobs with skilled employees. Having a solid production line, capable of operating in a manner that was consistent with global quality and safety norms, while developing a sustainable supply of local talent was an urgent need, and one that would lay the foundation for the desired regional expansion. While urgent and necessary, there was a bigger goal to be pursued. The top management at Spooner realized that the somewhat Eurocentric model of staff development had served them well for many years but potentially handicapped their ability to truly take advantage of the growth offered by the Asian market. Indeed the problem arguably went beyond that based on the quantity and quality of the supply of technically talented staff that would be required in the forward years. Several of the countries in the region significantly exceeded the percentage of degrees awarded for science and engineering compared to Western countries.

So if Spooner was to surf the leading edge of the talent curve, it had to find a way to tap into this pool of growing technical talent to stay ahead of its competitors. It also had to grow its regional commercial capability to unleash the potential inherent in these markets of growth and finally have a global leadership team that was reflective of the customer base it desired to serve—and that was most certainly one that included more non-UK talent. So these were critical times to build the foundation of that long-term supply.

"What were the options available?" Roger pondered. "What steps can I take?"

His first step was to realize the solution to the problem would not come from his own insight, but that from his team. "Leverage their combined expertise," he thought. Thus he pulled them together and tasked them with solving this strategic challenge.

In the next few weeks the regional HR team had many lengthy discussions about how to draw the connections between selection strategy and Spooner's global and regional business strategies. Roger knew building the business case for a more integrated regional selection strategy was vital for getting it implemented. The team developed the following points:

- A more regionally integrated selection plan would help implement Spooner's strategic focus on expansion within Southeast Asia. Ensuring the selection of high-quality talent would enable Spooner to develop new products faster, get them to market more quickly, and provide higher levels of customer service.
- A more regionally integrated plan would both enhance and even out the quality of talent throughout the region. This would in turn enable greater "cross-pollination" across cultures and countries, which would make succession planning and workforce planning easier.
- A more regionally integrated plan would help reinforce the Spooner culture within the region, contributing to differentiating the company even further against competitors.
- The selection strategy would rely primarily on quantitative data. Evidence and data would be used to inform hiring and talent decisions. Manager input would still be expected, but data should be used to make better decisions and influence judgments. Such data would help determine whether cultural differences in talent were real or imagined. Quantitative evidence would also be used to counter resistance from local managers. In this manner, the data would be used to reduce cultural biases and mistrust.
- More broadly, much of the global technical talent Spooner needed would be coming from Southeast Asia. Using a quantified approach to selection would enable the firm to better know who

the top talent is, and help prepare them for assignments in other parts of the world. If Spooner could be the employer of choice for the highly skilled talent in the region, it would be able to create a global competitive advantage.

Roger and the team were very excited with this approach, and their enthusiasm was not lost on the executives who later listened to the plan. Although a number of criticisms were leveled, the team addressed them and ultimately gained the support of the regional managers. As the team celebrated over drinks and dinner that evening, Roger noticed that Chakthep Senniwongs (from Thailand) looked "less than jovial." "What's wrong?" Roger asked. "As excited as I am about our getting the approval to proceed, my thoughts turn to how we will implement this strategy." Chakthep said. Roger smiled, but he suddenly felt a little less jovial as well. "We can discuss it tomorrow, but tonight we should enjoy our success," Roger said. "Another round!" he shouted.

THE PROCESS OF GLOBAL SELECTION

There are many different types of selection practices and methods (e.g., interviews), but regardless of approach, there is a fundamental process that should be used to design the selection system. Following this process will ensure people with the right types of talent are selected, in the most efficient, cost-effective, fair, and legally compliant manner. Figure 5.1 provides an overview of this process.

Figure 5.1
Global selection framework.

Step 1: Identify the Desired Talent

Chapter 2 described how one goes from a firm's strategy to identifying the needed types of talent used to implement that strategy. It should go without saying that knowing how to best select is entirely dependent on what you are trying to accomplish. The value of different types of talent is dependent on how the firm is trying to differentiate itself, operationalized via different performance metrics. Thus, the firm starts with strategy and differentiation, moves backward to operational performance, and only then into the kinds of talent in different job families that contribute to those performance metrics. Within this broader strategic context, every job also has a set of core tasks that are critical. Understanding the talent requirements of these tasks is the purpose of the task-talent matrix (chapter 2; Table 2.5). Hence, the task-talent matrix becomes the blueprint for the selection process.

However, the specifics of the job, company, and culture will determine the details of the task-talent matrix, as well as the performance metrics used to further establish the job-relatedness of the selection process. The performance metrics should be directly aligned to operational performance, external performance, and ultimately competitive advantage; from the individual to the collective or business unit levels (see Figure 2.2 in chapter 2). Yet some performance metrics are so basic that they are included in most selection systems: overall job performance, turnover, counterproductive work behaviors, citizenship performance, and potential (adaptability).

First, some form of *overall job performance* will always be of interest. Overall job performance refers to the behaviors an employee engages in on the job; they are part of the specified tasks an employee is expected to perform (Campbell, 1990). However, even job performance has a contextual component based on the culture and norms of the organization, and the influences of national culture as well (Campbell & Wiernik, 2015). Different cultures will place greater or lesser importance on "getting along" versus "getting ahead," for example. One may therefore draw a distinction between *technical* aspects of performance, which tend to be more individually focused on quantitative or analytical information, and *interpersonal* aspects of performance, which tend to be more focused on working with coworkers in team-based environments. Most jobs in today's economy have elements of both technical performance and interpersonal performance.

Second, *turnover* (or retention) is often a goal for selection, particularly at lower level jobs involving high-volume hiring (e.g., entry-level retail). Turnover may be voluntary (i.e., the employee chooses to leave) or involuntary (i.e., the employee is terminated due to poor performance or fit). The costs of either form of turnover are quite high, in terms of both the hiring and training costs associated with replacing a separated employee and the opportunity costs associated with having less experienced and cohesive groups of employees in the business (Cascio & Boudreau, 2008). Unfortunately, the logic used to develop a selection system to improve performance faces significant challenges when trying to predict turnover. It is not as straightforward to analyze a candidate's "fit" for a job and determine those characteristics that will lead someone to leave. Consider, for example, trying to predict turnover due to causes such as "went back to school" or "relocated with spouse," two common causes of voluntary turnover. None of these events is likely to be predicted by one's level of ability or personality traits. In many ways, recruitment is likely to have the bigger effect on turnover, at least when turnover is voluntary (e.g., poor needs-supplies fit; chapter 4).

Third, *counterproductive work behaviors* (CWBs) are negative behaviors that organizations seek to avoid. These include theft, sabotage, privacy breeches, threats, and violence. Costs due to employee theft can be enormous. Leaking private information can produce disastrous effects for public relations. Workplace violence is fortunately a rare event, but bullying and related forms of intimidation are surprisingly more common (Hoel, Rayner, & Cooper, 1999). And CWBs can destroy a company's culture and contribute to lower performance.

Fourth, *citizenship performance* refers to helpful performance behaviors that are more discretionary and not directly part of one's prescribed job.

These behaviors include promoting the company to outsiders, engaging in discretionary tasks (e.g., cleaning trash lying around), helping coworkers who need it, and defending the company from negative attacks, as just some examples. Contributing to a firm's environmental sustainability is yet another form of citizenship. Although citizenship performance is discretionary, it has been linked to firm performance (Sun, Aryee, & Law, 2007).

Finally, *potential* (also described as promotability, adaptability, agility, or flexibility), is often an outcome for managerial or professional jobs (e.g., research and development). Realizing that they do not have the resources to devote to the development of every single employee, most organizations have programs and processes designed to identify high-potential employees or "Hi-Po's." Assessments play a key part in these programs. While the characteristics identified by leading companies vary, in an exhaustive review of the literature on identifying Hi-Po's, Silzer and Church (2009) identified a set of core or foundational attributes related to most definitions of potential: cognitive ability (breadth of thinking, intellect, comfort with ambiguity), personality (interpersonal skills, resilience, and dominance), learning (interest in feedback, openness to feedback, and flexibility), leadership (developing others, leading change, influencing), and motivation (drive, ambition, and risk taking). Anyone possessing high levels of all these characteristics would be hard pressed to not excel, but most Hi-Po's have high levels of most of these characteristics.

Different types of talent may be related to performance in different cultures, even when performance exists within the same firm and is evaluated according to standardized performance management practices. Extensive scientific research has established broad generalizations between different types of talent and the different performance metrics. Chapter 1 described different types of talent that range from cognitive (cognitive ability, knowledge, skill) to personality, values, interests, and dispositions. Figure 5.2 illustrates relationships between these types of talent and performance. Technical aspects of performance are, not surprisingly, more

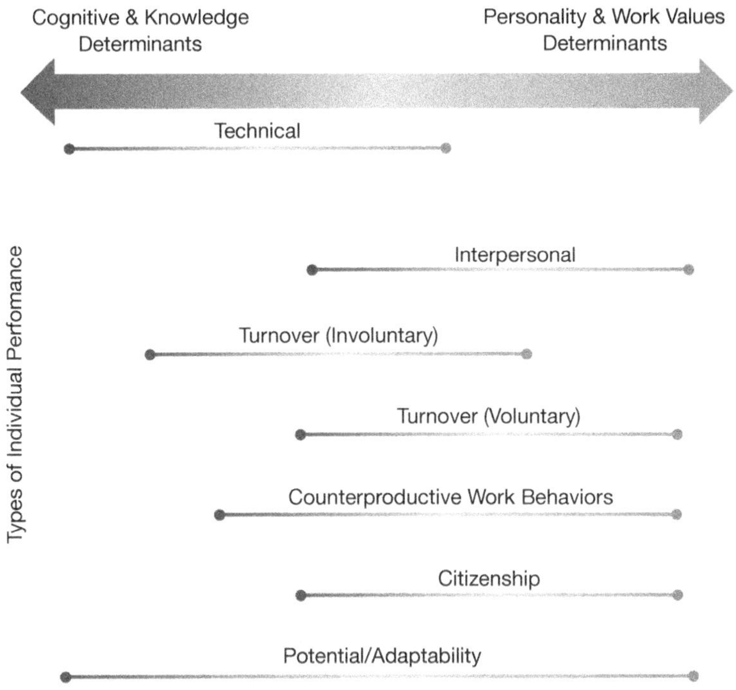

Figure 5.2
The talent determinants of different types of individual performance. The horizontal lines indicate the extent to which a particular performance dimension is determined by cognitive talent (on left), personality talent (on right), or some combination of the two. For example, technical performance is primarily determined by cognitive ability, while interpersonal performance is primarily determined by personality.

affected by cognitive talent. Interpersonal and citizenship performance dimensions are more affected by personality-based talent. Potential and CWB are affected by a combination of cognitive and personality talent types. Involuntary turnover is slightly more affected by cognitive talent, while voluntary turnover is slightly more affected by personality-based talent. Although exceptions certainly exist, these relationships generally hold across organizational settings. Across jobs, companies, and national cultures, cognitive ability and the personality trait conscientiousness are the two forms of talent most consistently related to overall job performance (Schmidt & Hunter, 1998; 2004).

Step 2: Ensure Compliance with Legal and Political Guidelines

The talent demands of different jobs may be fairly—but not completely—similar across cultures. But the specific practices and methods used to measure those forms of talent are not necessarily appropriate in every country. It is vital to ensure a selection system is compliant with legal and political guidelines. This is challenging enough within a single country, but it is far more challenging in a global context. As each country sets its own rules regarding the employment practices it deems acceptable and unacceptable, there is no simple way to ensure universal compliance. Myors et al. (2008) conducted a survey across 22 countries and found both important similarities and differences. In terms of similarities, every country identified disadvantaged groups and offered them some legal protections (note that those from disadvantaged groups are often referred to as the *minority* group and those from the advantaged groups are known as the *majority* group). All 22 countries prohibit discrimination based on sex or race/ethnicity. In terms of differences, the basis for recognizing disadvantaged status varied significantly. Race, color, religion, gender, national origin, age, disability status, political opinion, sexual orientation, and marital/family are covered in most countries. The protections offered are typically broad and almost always specifically include selection (Chile being the lone exception).

The United States was one of the first to offer applicants protection from unfair discrimination and, as a result, has influenced the laws of many other countries. Title VII of the Civil Rights Act in the United States, for example, prohibits discrimination on the basis of race, color, gender, religion, and national origin. Like the United States, many countries also make a distinction between two types of discrimination. *Disparate treatment* occurs when an individual is intentionally and overtly being discriminated against because of his/her status. Disparate treatment occurs on a person-by-person basis. For example, an interviewer who asks female candidates questions such as, "Are you married?" or "Do you have kids?," but does not ask these questions of men, illustrates disparate treatment. Women are being assumed to have greater child-rearing responsibilities, and the interviewer is treating them differently in a potentially aversive and discriminatory manner. *Disparate impact* (also frequently known as *adverse impact*) occurs when all applicants are being treated identically, but the consequences or results of selection disadvantage a protected group. Here the discrimination may be intentional or unintentional. For example, men score higher than women on most tests of physical strength. Selection based on physical strength will result in adverse impact against women, even when physical strength is required for the job.

There are a variety of ways to demonstrate disparate impact (Murphy & Jacobs, 2012). Common examples include standardized differences between groups on the test scores (the *d* statistic), group differences greater than two standard deviations, or statistically significant mean differences (see chapter 3). One simple approximation used widely in the United States is known as the four-fifth's rule. A potential flag for adverse impact (but not proof) occurs when the hiring rate for the lower scoring group (which is usually the minority group) is four-fifths or 80% of the majority group's hiring rate. For example, if the selection ratio for the majority group is 30% (meaning 30% of those who apply are hired), then the selection ratio for the minority group should be at least 24% ($30 \times .80$). If it is lower than 24%, one needs to take a closer look to see if discrimination is present. The formula for the four-fifths rule is shown here:

$$\frac{\dfrac{\text{\# minority applicants hired}}{\text{total \# minority applicants}}}{\dfrac{\text{\# majority applicants hired}}{\text{total \# majority applicants}}}$$

While adverse impact should be avoided where possible, its presence may not preclude the use of an assessment. It does require the employer to demonstrate that the assessment is job related and that no less discriminatory alternatives exists (at least in the United States). As an example, strength tests are often used to hire firefighters because of the high physical strength requirements of that job. Females, on average, perform less well on such tests, and their use inevitably creates adverse impact against women. Given the importance of strength for the job and the lack of alternatives, most fire departments have minimum strength requirements for which they test. When adverse impact occurs for reasons that might be changeable, firms should take steps to reduce it as much as reasonable. For example, one approach Shell used to proactively address the issue of women's relative lack of physical strength when applying for offshore operator positions was to provide health club access for candidates who were satisfactory on other criteria but just failed the meet the strength test requirements. They were then given another opportunity to retest and, if successful, were hired.

Importantly, there are relatively few restrictions placed on specific selection methods (e.g., polygraph in the United States, personality assessments in Switzerland). To illustrate one of the exceptions, South Africa prohibits any form of psychological testing unless the assessment is shown to be reliable, valid, and fair in South Africa. This means that employers generally have some latitude in the types of assessments they will use to hire, and that they can use many of the same assessments across multiple countries. At the same time, it is incumbent on the organization to ensure the selection methods that are used are legally appropriate. There are so many nuances to local laws, which themselves are constantly evolving, that we do not give specific guidance on what to do in this book. This is simply something that each firm must do on a case-by-case basis in close coordination with legal counsel. However, after determining which practices are appropriate, one needs to ensure the practices are consistent with the cultural expectations, as discussed next.

Step 3: Determine Cultural Influences

Of course, efforts to identify the desired talent within any given location need to balance the tension that exists between centralization within the multinational corporation relative to regional/local customization within subsidiaries. Political and legal environments vary globally, and they will create corresponding differences in selection practices and procedures. Several research studies have examined the effects of national culture and country-level influences on selection practices and validity (Salgado & Anderson, 2002; 2003; Salgado, Anderson, Moscoso, Bertua, & Fruyt, 2003; Ryan & Tippins, 2009). The most comprehensive study is by Ryan, McFarland, Baron, and Page (1999), who surveyed over 959 companies in 20 countries to evaluate the extent to which variations in staffing practices were due to country and/or culture. The following broad conclusions are supported by their research.

- **There are large differences in selection practices and methods used in different countries.** The Ryan et al. (1999) study found considerable variability across countries in the types of selection practices and methods used. About the only constants across countries are the use of educational credentials (e.g., degrees, grade point average) and one-on-one interviews. In contrast, use of family connections and life history/background information are used less frequently (or more variably). Personality assessments and medical screens are used in many countries, but not consistently within any of them. Usage of all assessments, including ability and simulations, varies considerably across countries, but none with consistently high frequency.
- **National culture does *not* strongly influence the validity of scores obtained from different assessments.** So long as the appropriate translations have been performed and the assessments are professionally developed and administered (see chapter 3), the validity of many assessments will generalize across cultures. This means that assessments used effectively in one country will likely produce scores with good predictive validity in other countries—so long as such practices are consistent with national laws and applicant expectations. MNCs can take some comfort in knowing that extensive selection

validation work conducted in one country should support use of that assessment in other countries. This helps save costs, ensures consistent standards across cultures and countries, and helps create a more uniform culture. Of course, one would need to perform the appropriate translations and establish the cultural equivalence of these scores first (see chapter 3). However, not all assessments will show high cultural generalizability. In our own research, we have found that simulations may work well in different countries, but the scoring and design of those simulations need to be adapted to the local culture.

- **Perceptions of selection practices and methods vary across cultures.** Chapter 4 showed how culture has a strong influence on recruitment and applicant perceptions. What are considered acceptable selection practices may also differ by culture (Ryan, Boyce, Ghumman, Jundt, & Schmidt, 2009). Although not a lot is known about which cultural values explain specific preferences for different practices, the suggestions provided in Table 4.3 can be used to form reasonable inferences. For example, those from a culture with a greater long-term orientation may be more willing to tolerate a lengthy selection process.
- **National laws and political systems have a stronger influence on use of selection practices and methods than national culture.** National differences account for a greater amount of the variability in selection practices than national cultural values. In fact, Ryan et al. (1999) found only uncertainty avoidance and power distance related to use of selection practices. The effects were fairly small and variable, suggesting that the influence of culture is nuanced and subtle. It is not surprising that legal and political influences constrain the effects of culture. First, selection carries high-stakes consequences for applicants and organizations, and hence falls under a number of legal and regulatory guidelines (Myors et al., 2008). Second, culture

contributes to the formation of laws and political systems, so culture is a more indirect influence on selection.

Conclusion

Global firms should realize that many (perhaps most) selection practices can be used across countries and cultures, at least so long as the practices are professionally developed and care is taken to ensure equivalence across cultures (see chapter 3). But the acceptability and legality of such practices need to be considered on a country-by-country basis. Many leading firms will strive to assess the same competencies globally (e.g., a firm's core values), but allow some variation and discretion in terms of how those competencies are assessed (Wiechmann, Ryan, & Hemingway, 2003; Ryan, Wiechmann, & Hemingway, 2003). Regional perspectives enter into this discussion as well, such that countries sharing a common language may be administered the same set of selection practices. The key point is that from a technical perspective, a firm can use a variety of different selection practices across countries and cultures; the question becomes which of those practices are acceptable from a legal and cultural perspective.

For example, interviews are used in most countries. Yet how they will be conducted will be affected by culture. In "high context" cultures (e.g., much of the Middle East and Asia), people attach significant importance to developing relationships before getting down to business. Initial meetings there may be purely about developing the relationship, with little, if any business content until an atmosphere of trust has been created. That requires significant face time. In "low context" cultures like those in western Europe and North America, the focus can be on brevity and matter-of-fact approaches to business meetings. So while it is becoming common to use Skype or video conference to conduct interviews to save money and travel, such approaches may need to be avoided in S.E. Asia—where it could be seen as cold and impersonal and nonresponsive to the culture.

Roger's team arose early—painfully early to those who celebrated a little too much—the morning after their celebration. The party was over, and there was much new work to be done in terms of implementing their plan. After a long day of discussion (and some heated arguments), the team settled on the following implementation plan:

- The regional plan would first be implemented on a small scale. They would focus on the strategic job families, starting with assembly team associates and R&D team leaders. The thinking here was similar to "rapid prototyping" in manufacturing—try the process out in a small scale, with managers who were supportive and willing to champion the changes.
- Assuming all went well, the transition would be successful and the managers would be used to promote the program to their peers in different functional areas. Any mistakes could be corrected quickly before large-scale implementation. If the entire strategy failed, then the consequences would be localized.
- The regional selection plan would attempt to leverage synergies as much as possible when the jobs were the same.
- Large volume hiring in the assembly team associate manufacturing jobs would be outsourced to a vendor who could administer systems more efficiently and cost-effectively than Spooner. As a general principle, local customization of a selection practice would be allowed only if the customized practice was significantly more:
    - predictive,
    - cost-effective,
    - culturally appropriate, and
    - legal.
- Small-volume hiring in the more strategic, and more competitive, R&D team leader jobs would be conducted internally with a collaboration of regional and local HR staff, and local managers. The selection process for these jobs would be more customized to the local R&D group.

"So far so good," he thought. But a heck of a lot of persuasion and development work was needed, and there were some nagging details about the actual process that needed to be finalized. For example, was the heart of the of the assessment process itself one founded on bias in that it sought to identify individuals separate from group performance rather than focusing on the group as a combination of individuals? There was some analogy here with the impact of attrition being greater than the sum of the individual talents. These reflections lead Roger to believe that:

- Selection methods needed to incorporate both individual and team performance.
- Preselection using online assessments for "fit" with the Spooner culture, described again in regional terms, and job-related competencies, had an important place in the screening process and could be taken with less human (biased) interactions.
- Managers, both local and expatriate, needed significant training in structured, behavioral-based interviewing and on the local culture from which the desired talent would be drawn. And an element of this training needed to minimize the effect of the cultural impact described by Geert Hofstede as power distance. This focuses on the impact of different views of how power distribution is viewed and the consequent impact on behavior and roles.

These thoughts stayed in Roger's mind as he said goodbye to the other team members.

Step 4: Determine Selection Methods

Selection methods are the actual assessments or measures used to obtain scores about talent KSAOs (see chapter 1). Selection methods are also called *predictor methods* (or simply *predictors*) because they provide scores about latent talent used for predicting performance outcomes (see chapter 3). Selection methods include interviews, tests of cognitive ability or personality, simulations, and resumes. There are multiple kinds of selection methods that can be used to assess all kinds of different talent KSAOs. One useful way to compare and contrast selection methods is by making a distinction between signs of behavior and samples of behavior. *Signs* are measures that are intended to be indicators of a particular construct believed to be related to behavior (constructs are the unobservable talent KSAOs, such as intelligence or personality traits). *Samples* are measures of the actual behavior. For example, suppose we want to measure extraversion and use it as a basis for selecting salespeople. Those who are more extraverted tend to be more outgoing, energetic, and assertive, and research has shown that those more extraverted perform better in sales jobs (Barrick & Mount, 1991). No one hires a sales person to be extraverted, but they do hire extraverts because they enjoy being around other people and hence are more likely to sell. Extraversion can be measured using either signs or samples. Predictor methods that are signs of extraversion include self-report tests administered by paper or the Internet/computer (e.g., a 20-item survey that asks questions such as, "I enjoy working with other people" or "My preference is to work with other people"). Predictor methods that are samples of extraversion may include interviews or simulations (e.g., role-play where the candidate pretends to sell to the interviewer).

One may further understand predictor methods by considering whether they are homogeneous or heterogeneous measures of talent KSAOs. Homogeneous measures are intended to assess only one type of talent at the same time. Common examples of homogeneous methods include self-report assessments like tests of personality, ability, or knowledge. Heterogeneous measures are intended to assess multiple types of talent at the same time. Samples, including simulations, interviews, role-plays, and related methods, are usually heterogeneous because they assess multiple talent KSAOs simultaneously. Table 5.1 provides an overview of different predictor method signs and samples.

TABLE 5.1 PREDICTOR METHODS: SIGNS AND SAMPLES

Sign (usually homogeneous behavioral indicators of talent)	Sample (usually heterogeneous behavioral examples of talent)
▪ Cognitive Ability	▪ Situational Judgment Test (SJT)
▪ Knowledge and Skill	▪ Interviews
▪ Personality	▪ Assessment Centers
▪ Work Interests, Styles, and Values	▪ Work samples
▪ Biographical Data (Biodata) *	▪ Simulations
▪ Fit	
▪ Situational Judgement Test (SJT)	

NOTE: * May be either homogeneous or heterogeneous. SJT may be either a sign or a sample.

Predictor Methods That Are Signs

Signs of KSAOs frequently include assessments of intelligence, cognitive ability, knowledge, skill, personality, values, and interests. The common theme among these methods is that they are usually self-report surveys administered on paper or Internet/computer platforms. Therefore, they are signs and predictor methods that assess homogeneous types of talent. The most common methods include:

- **Cognitive Ability**. As noted in chapter 1, cognitive ability refers to how well individuals can manipulate, process, and remember information. Cognitive ability is hierarchical, such that overall cognitive ability subsumes verbal, numerical, and reasoning abilities, which in turn subsume even more specific abilities (Carroll, 1993). Tests of cognitive ability vary considerably, ranging from general mental ability (GMA) to facets of intelligence like deductive and inductive reasoning, numerical and verbal ability, short-term memory, visualization, and spatial orientation. The most common assessments of cognitive ability involve taking some form of a timed test that includes some combination of numerical, verbal, and reasoning content. A considerable amount of research (e.g., Jensen, 1998; Schmidt & Hunter, 2004) has concluded that:

- Cognitive ability is a valid predictor of job performance for essentially all jobs, in essentially all parts of the developed world.
- The predictiveness of cognitive ability scores increases with the information-processing demands of the job (i.e., cognitive ability is a better predictor of performance for more complex jobs than it is for entry-level positions).
- While there are many different facets or specific cognitive abilities (e.g., processing speed, numerical ability, verbal ability), the facet-level measures usually do not predict performance better than overall cognitive ability.
- Cognitive ability has the advantages of being a universally strong predictor, widely available, easy to administer, and relatively inexpensive.
- Cognitive ability has the disadvantage of producing large racial mean differences, such that blacks and Hispanics score lower than whites and Asians. These differences can be so large that they create racial adverse impact at most selection ratios and are disliked by many job applicants. Because concerns about adverse impact are more pronounced in the United States, the use of cognitive ability as a selection approach is far more prevalent in other countries (Ryan et al., 1999).

- **Knowledge and Skill**. Job knowledge assessments are widely used and may range from very broad types of knowledge (e.g., programming languages like C++) to very specific types of knowledge (e.g., the features and benefits of a firm's product line) (see Figure 1.4 for an example). Job knowledge measures can be of either procedural knowledge (e.g., the proper sequence of steps needed to complete a task) or declarative knowledge (e.g., the exact parts constituting a diesel engine). Skills assessments measure the ability to perform all or part of a task (Ployhart, Nyberg, Reilly, & Maltarich, 2014; http://www.onetonline.org/find/descriptor/browse/Skills/). Whereas a knowledge test answers the question "Does the candidate know it?," a skills test answers the question "Can s/he do it?" There are many different kinds of skills, including problem-solving skills, interpersonal skills, persuasive skills, and presentation skills.

  - Tests of knowledge and skill have been shown to be among the best predictors of performance in many jobs and have the additional advantage of being highly face valid.
  - On the downside, effective measures of knowledge are often unique to a particular job, sometimes making them expensive to develop, and they may contribute to adverse impact. Skill assessments are often more generalizable across organizations.

- **Personality**. Personality refers to reasonably stable characteristics and behaviors that people manifest across different situations. As noted in chapter 1, personality traits are usually operationalized in terms of adjectives, and today it is widely recognized that these adjectives can be summarized into five broad trait categories (known as the Five Factor Model, or Big Five): (1) conscientiousness, or the extent to which one is dependable, persistent, planful, and achievement oriented; (2) agreeableness, or the extent to which one is friendly, good-natured, concerned for others, trusting, and pleasant; (3) extraversion, or the extent to which one is outgoing, assertive, energetic, and dominant; (4) emotional stability, or the extent to which one is stress tolerant, free from anxiety, depression, and hostility, and in control of moods and emotions; (5) openness, or the extent to which one is intellectual, interested in new experiences, analytical, and innovative (Barrick & Mount, 1991; 2012).
  - Research has shown that conscientiousness is a nearly universal predictor, meaning that more conscientious

people tend to perform better in any job (Barrick & Mount, 1991). Similarly, extraversion has been found to be a fair predictor of performance in sales and management jobs. The predictiveness of the other personality traits is less universal.

- Combining different types of personality traits relevant for specific types of performance can enhance prediction. For example, service orientation has been found to be a function of conscientiousness, agreeableness, and stability (Frei & McDaniel, 1998). In fact, it has been suggested that the Big Five personality traits can be further reduced to two—the "get along" traits of conscientiousness, agreeableness, and stability and the "get ahead" traits of openness and extraversion (Hogan, 2007).
- Personality measures have several advantages contributing to their sustained popularity: (a) they are quick and easy to administer and complete; (b) they can provide feedback to hiring managers about a candidate's likely behaviors, (c) they typically demonstrate little if any adverse impact; and (d) they can be good predictors of a variety of outcomes.
- The main disadvantage of personality tests is that they are fairly susceptible to impression management, response distortion, and faking. They also suffer from a lack of face validity in some cases. While the predictive validity of personality tests may be affected to some degree by faking, they are still practically useful.

- **Work Interests, Styles, and Values**. Work interests, styles, and values are generally considered to be important in terms of applicant attraction to different occupations and jobs (Holland, 1997; see chapters 1 and 4). However, recent research suggests that work values and interests may also be important to the prediction of job performance and turnover (Van Iddekinge, Putka, & Campbell, 2011). Not surprisingly, those whose interests better match the job tend to perform better and are less likely to quit. Interests, styles, and values show correlations with performance similar to those found with personality, but the extent to which personality predicts more strongly than interests/values is currently unclear.

- **Biographical Data (Biodata).** When it comes to predicting human behavior, nothing in psychology is as close to being a "law" as the consistency of a person's behavior over time (insurance companies have known this for years!). Consequently, measures of biographical data (abbreviated as *biodata*) are based on the premise that the best predictor of future behavior is past behavior (e.g., previous driving history being a great predictor of future accident probability). There are a variety of different types of biodata measures, that range from simple measure of job experience or organizational tenure, to measures focused on specific types of talent (e.g., past experiences related to extraversion). Thus, while most biodata measures are more like samples of behavior, they can be either homogeneous or heterogeneous. Questions about one's experiences while growing up, in school, at work on previous jobs, and so forth, have proven to be robust predictors of performance. Most biodata is empirically keyed, meaning that relationships between items and the performance outcome are deduced from the data. As a result, biodata has been criticized as being blindly empirical and not helping to justify why behavior is related to performance. Biodata has the advantage of being a good predictor if well developed, but may lack psychological interpretability and often face validity (the most valid item in one biodata assessment developed by the authors to predict sales performance was "How many times have you purchased real estate?"). Newer approaches create biodata scores that represent specific types of talent and hence have more construct validity. As described in chapter 3, the application of Big Data analytic principles to HR data captured from applicants is likely to accelerate greatly, making the expanded use of biodata as a selection method a safe bet (it is also likely that these new versions of biodata will be much more sophisticated analytically).

- **Fit.** Whereas most talent assessment methods for selection differ across jobs, tests of organizational fit will theoretically apply to every job in an organization. There are a number of ways of measuring fit (see chapter 4), but the most common approach appears to rely on the assessment of values congruence (discussed in chapter 4). O'Reilly, Chatman, and Caldwell (1991), for example, captured ratings across a number of organizations of the extent to which a number of value statements described the dominant culture. They found that organizational values can be differentiated on the basis of eight factors: (1) innovation and creativity, (2) attention to detail, (3) outcome emphasis, (4) aggressiveness/competitiveness, (5) supportiveness, (6) growth orientation, (7) team orientation, and (8) decisiveness. Organizations can be profiled on this or any other framework by compiling input from key stakeholders. Individuals can be asked to rate or rank the importance to them of the same values statements and a measure of "fit" can be created by comparing the individual's values importance ratings to the organization's profile. Individuals who share values with the organization are deemed a better fit. While we focus on company fit, fit can also be measured against a job, a team, and even a manager.
    - Although not a good predictor of job performance, measures of culture fit have been to be related to important outcomes such as engagement and retention (Arthur et al., 2006). It makes intuitive sense that employees who do not share the values of their coworkers will find their work less satisfying and be more likely to leave. Organizational fit offers the unique advantage of being applicable to every job with an organization and measures of it tend to generate scores with little adverse impact.
    - Because their link to performance is small at best, culture fit is often used as a first hurdle in the selection process to eliminate a small proportion of the candidate pool. Hence,

organizational fit is usually emphasized more as part of recruiting activities than selection activities (see chapter 4).

- **Situational Judgment Tests (SJTs).** Another approach that has seen increased use globally is that of the situational judgment test (SJT). The SJTs do not typically capture a single construct and are considered heterogeneous measurement methods rather than measures of a specific trait or ability. In the typical SJT, participants are presented with descriptions of challenging situations they are likely to encounter on the job and four to five possible ways of handling each situation. Box 5.1 provides an example SJT item intended to assess cross-cultural competence.

Box 5.1 SAMPLE SITUATIONAL JUDGMENT TEST ITEM TO MEASURE CROSS-CULTURAL COMPETENCE

You have just been assigned to lead an eight-person cross-functional project team. The team has been formed to evaluate the benefits and risks of establishing a small manufacturing facility in a new country. Because this is an international expansion, the project team is composed of members from three different countries (Japan, Norway, and Brazil). Each team member has different expertise (such as finance, HR, operations). Team members are located within their home countries. All team members speak some English, but language and time zone barriers are significant. How will you coordinate the team to accomplish its mission?

a. Conduct all communication using e-mail, copying all members on all communications.
b. Break the team into country subgroups (e.g., Norwegian team) to minimize language and time zone problems.
c. Schedule regular conference calls using a different time zone for each call.
d. Rotate a different person to serve as assistant team leader each week to handle all communication and coordination challenges.

Applicants select the best course of action, the best and worst options, or simply rate/rank the effectiveness of each option. The closer the participants' selections come to matching those of a reference group (e.g., subject matter experts such as high-performing incumbents or supervisors) the better their "judgment" is deemed to be. The SJTs have the advantage of being predictive of performance, incrementally so over most other measures, and having great face validity—some have argued that SJTs give applicants a realistic job preview by putting them "in the job" across numerous typical situations. They also tend to have smaller racial subgroup differences than other predictor methods. The potential disadvantage with SJTs is their development cost, where even the same situations may be scored differently in different cultures or organizations. For example, responses that favor individual achievement over group harmony tend to be seen as more correct in individualistic than collectivistic cultures.

Predictor Methods That Are Samples

Some of the most popular predictor methods are samples of behavior, the most notable being the selection *interview*. However, other approaches known as *assessment centers* and *work samples* are also widely used. In today's world, these latter methods are more simply known as behavioral simulations. Behavioral simulations have higher fidelity than signs, meaning that the psychological thought processes and physical actions needed to perform the assessments are more similar (if not identical) to those performed on the job. In general, the higher the fidelity the better the prediction of job performance. We start by discussing the lowest fidelity predictor, the interview, followed by assessment centers and then work samples (the highest fidelity predictor).

- **Interviews.** Interviews are by far the most commonly used method of selecting candidates (for this reason, we devote more discussion to interviews than to other predictor methods).

Despite their global ubiquity (see Ryan et al., 1999), the types and formats of the interviews can actually vary substantially from employer to employer (and even from hiring manager to hiring manager). A good number of employers use "unstructured" interviews, where the questions asked and the evaluation of the responses are left completely up to the interviewer. The questions may be drawn from the candidate's resume and related to previous work or educational experiences, they may be based on the tasks and anticipated challenges associated with the open job, or they may reflect the hiring manager's beliefs about what makes for a good employee (no matter how accurate such beliefs may be). Research has shown that such unstructured interviews to have relatively low validity (McDaniel, Whetzel, Schmidt, & Maurer, 1994). There are many reasons. Because they are unstructured, different candidates may be asked very different questions, thereby making comparisons between candidates difficult. Depending on the interviewer, the questions asked may or may not be related to the requirements of the job. Even when job-related questions are asked, the interviewer may not combine the information gathered in a way to facilitate effective decision-making. Interviewers may focus too much on one response and ignore other data points (halo error), compare the candidate to the last interviewee (contrast error), or prefer candidates that look like themselves (similarity error). Fortunately, there are better ways of conducting interviews:

- **Use structured interviews.** Research has shown that structured interviews are an effective means of identifying candidates who will perform better post-hire (Campion, Palmer, & Campion, 1997). Structured interviews predict performance two to three times better than unstructured interviews, and produce smaller racial and gender subgroup differences (so structured interviews are less likely to be discriminatory). The reason is because structured interviews employ a number of best practices that cumulatively enhance reliability and validity (Schmidt & Zimmerman, 2004; see chapter 3). When based on an unstructured interview, two interviewers are far less likely to agree on their assessment of a candidate than when the interviews are structured. This same research indicates that approximately four unstructured interviews, when their evaluations are combined, will be as valid as one structured interview. Our experience is that many employers do exactly that—deploy multiple unstructured interviews when screening candidates for professional level jobs. Obviously, the best results and most accurate predictions will occur when multiple interviewers use structured interview guides to evaluate the same candidates. Table 5.2 summarizes the key features of structured interviews, and Table 5.3 provides an example of a structured interview rating form.

- **Use both behavioral and situational questions.** Structured interviews come in two primary formats: behavioral questions and situational questions. In a *behavioral interview*, an applicant is asked to describe how she or he handled a situation in the past that is similar or relevant to the job's requirements. For example, the question "Tell me about a time when you got someone at work to do something he/she didn't want to, by using your persuasion or influence, rather than your position of authority," might be used to assess the candidate's ability to lead others. Clearly, behavioral interviews are based on the adage that past behavior predicts future behavior (see the earlier section on biodata). In a *situational interview*, an applicant is asked to describe how he or she would handle the types of situations they would likely encounter on the job: "As a manager, one of the people who reports to you doesn't think he has anywhere near the resources (budget, equipment, etc.) that are required to complete a special task you've assigned. How will you

TABLE 5.2 BEST PRACTICES IN STRUCTURED INTERVIEWS

Best Practice	Explanation
Base interview questions on a job analysis	A job analysis can identify the talent related to high performance, enabling the interviewer to ask questions to gauge the applicant's level of competence on job-related characteristics. This maximizes the amount of job-relevant information captured and reduces the time spent on less relevant questions. Job analysis is discussed in chapter 2.
Ask the same questions of all applicants	The same interview questions, be they past behavior or situational, are asked of every candidate for the same job. This ensures that the same information is collected from each candidate, making comparisons of candidate qualifications easier. In instances where the same questions cannot be asked, ask similar questions that measure the same construct or aspect of the job.
Train interviewers	Training interviewers on what to do/not to do in an interview and on the process or steps to follow before, during, and after the interview, and providing them an opportunity to practice. This enhances consistency within and between interviewers and reduces the likelihood that they will commit common judgment errors (e.g., similarity to me).
Use systematic and behavioral scoring	Ideally, each dimension is rated on some scale of behavioral effectiveness and combined into an overall score (simply adding the ratings up is far more effective than make a single summary evaluation at the end of the interview). Systematically scoring and combining items prevents biases (an interviewer overweighting a single item) and ensures consistency (each candidate is treated identically)
Use multiple interviews and interviewers	Although it is not always feasible, hiring decisions should be made on the basis of input from more than one interview and interviewer. Having multiple interviews/interviewers increases the amount of job-relevant information collected while decreasing the biases associated with any one interviewer (errors get averaged out across multiple interviewers).

respond?" Because the interviewee is asked to describe how s/he thinks s/he would respond given the context, situational interviews are based on the premise that intentions are a good predictor of behavior, that is, people generally do what they say they will. Although behavioral interviews seem to be the more commonly used type, situational interviews can be quite useful in some scenarios (e.g., when interviewing a young candidate with little or no previous work experience). Employers will often use both types of interview questions. For example, in Table 5.3, the first and third questions are situational and the second is behavioral.

- Technology has changed the way interviews are being conducted. The Internet has spawned a plethora of services offering "face-to-face" interviewing without requiring the interviewer and candidate to actually be in the same location. *Greenjobinterview, Hirevue, Jobvite, Interviewmaster*, and dozens of others enable the recruiter or hiring manager to conduct an interview with candidates remotely (not to mention *Skype*). While they save the time and costs associated with getting the candidate and interviewer in the same room, these newer approaches offer the benefit over telephone interviews of allowing the interviewer to see the candidate, read body language, and purportedly get a more accurate evaluation of the candidate's qualifications. As most hiring managers are reluctant to move forward with a candidate they have never seen, it is likely that the cost savings associated with Internet-based interviewing will spur continued growth.

- **Assessment Centers**. Assessment centers, in which participants engage in a variety of exercises designed to simulate the essential features of a job, are more commonly seen in higher-level roles such as management and sales-related occupations. The term "assessment center" actually refers to a variety of different exercises that are essentially different

TABLE 5.3 SAMPLE STRUCTURED INTERVIEW RATING FORM (FOR THE JOB R&D TEAM LEADER)

Spooner Electronics Interview Form

INTERVIEWER: ______________________________

DATE: ____________

CANDIDATE NAME: ___________________________

INTERVIEW QUESTIONS	**Behavioral Rating Examples (higher numbers are better)**
1. Cultural Agility: "Imagine you are a member of a project team. There is a lot of cultural diversity on the team. How will you work with this team to achieve a common goal?"	1 = Fails to recognize or adapt to cultural differences. 3 = Recognizes cultural differences, but does not adapt to them. 5 = Embraces cultural differences and adapts behavior to facilitate interactions.

Comments:

2. Teamwork: "Tell me about a time when you were on a very successful team. What was it about this team that made it so successful?"	1 = Cannot answer or gives extremely vague examples. 3 = Gives general answers or emphasizes own accomplishments over the team's accomplishments. 5 = Provides specific, behavioral examples of what the team did well.

Comments:

3. Performance Management: "If you were in charge of a R&D team, how would you ensure the team reaches its deliverables in a sustainable manner?"	1 = Fails to set goals, timelines, or accountability. 3 = Sets vague goals. 5 = Sets deliverables, timelines, and individual/ team accountability.

Comments:

types of simulations. Table 5.4 summarizes commonly used simulation exercises that include an in-basket, leaderless group discussion, role-play, and presentation. Most assessment centers use a combination of exercises. As participants go through the exercises, they are evaluated by trained assessors on competencies such as planning, influence/persuasion, delegation, and verbal communications. Assessment centers have the advantages of being a highly valid predictor of performance, having great face validity, and providing data useful for both selection and development (Thornton, Rupp, & Hoffman, 2015). Unfortunately, because of the labor costs associated with the assessors, the time required by multiple exercises, and the "in person" nature of most assessment center processes (i.e., time and travel), they are both expensive and cumbersome to administer. Assessment centers are used around the world, but primarily in Europe and the United States. There is an international organization that sets standards and best practices for assessment centers that can be obtained here: http://assessmentcenters.org/Assessmentcenters/media/2014/International-AC-Guidelines-6th-Edition-2014.pdf.

- **Work Samples**. Work samples duplicate, and thus assess, behaviors actually performed on the job. Work samples are typically found in lower-level jobs and require demonstration of a specific skill. While the ubiquitous typing test may be the most common work sample test, work samples have also been used to hire employees into a wide range of jobs such as mechanics (troubleshooting a malfunctioning engine), pilots (a check ride with the chief of pilots), and welders (correctly welding pipe fittings). Work samples, because they match so closely the requirements of the job, have been found to be excellent predictors of job performance. Work samples have essentially the same advantages and disadvantages as assessment centers.

TABLE 5.4 DESCRIPTION OF ASSESSMENT CENTER EXERCISES

Exercise Name	Description	Types of Talent Assessed
In-Basket	Participants are presented with a large set of tasks and activities. They need to review these materials and prioritize them, delegating some activities to others and handling some activities themselves. The name "in-basket" comes from the old days, when managers had a basket on their desk and new tasks requiring their attention would be dropped in the basket (hence, an in-basket).	Organization and planning, problem-solving, job knowledge, delegation, leadership, cognitive ability, project management, multitasking.
Leaderless Group Discussion	Participants work on a small group activity (the group size is usually 5–8 people). The activity may be a business case, problem, or creative exercise. The group needs to reach a consensus decision.	Leadership, teamwork, consensus-building, persuasion and influence, communication skills, interpersonal skills.
Role-Play	Participants need to assume a particular role (e.g., manager) of a person who is dealing with a challenging situation (e.g., complaining customer).	Persuasion and influence, communication skills, interpersonal skills, problem-solving.
Presentation	Participants are given a topic and purpose, and then asked to give a presentation on the topic (sometimes with follow-up questions).	Communication skills, persuasion and influence, interpersonal skills.

Conclusion

There are a number of different predictor methods that can be used to provide valid scores for hiring purposes. Box 5.2 lists best practices that should be followed when developing predictor methods. Obviously, there are a lot of options to use when selecting someone to fill a job. Which is the "best" predictor method depends on a variety factors, which are considered next.

Box 5.2 BEST PRACTICES IN PREDICTOR METHOD DESIGN AND USE

1. Ensure scores on the predictor method are construct valid (i.e., you know what types of talent the predictor method assesses) and reliable.
2. Ensure the predictor method will generate scores that are job related (i.e., based on a job analysis).
3. Use multiple questions/items to measure each type of talent, and ensure all applicants receive the same questions/items.
4. Ensure the questions and items are understandable, written at the lowest reading level needed for the job, and the instructions are clear.
5. For assessments that require assessors (e.g., interviewers, assessment centers):
    a. Train the assessors.
    b. Use rating scales to record behaviors.
    c. Use multiple assessors.
    d. Avoid the use of prior information (e.g., reviewing resumes before the interview).
    e. If using multiple raters, establish a process for reaching consensus.
    f. Hold assessors accountable for their ratings.
6. Continually monitor the scores of the assessments (e.g., reliability, validity).

Roger knew the key talent competencies needed for each of the strategic job families [the process used to identify talent needs was discussed in chapter 2]. Working with industrial and organizational psychologists with expertise in selection and assessment, the following selection methods were used to assess each type of talent:

- Assembly team associates. Selection for these jobs would be outsourced to a testing vendor because of the high-volume hiring and similarity across jobs. Other than language translations, the same assessments would be used in all countries because the jobs were essentially the same, and the types of talent were similar across the countries. The following assessment methods would be used:
  - Conscientiousness—biodata; Internet, or paper personality test
  - Attention to detail—Internet or paper personality test
  - Teamwork—Internet or paper personality test; situational judgment test; structured interview
- R&D team leaders. Selection for these jobs would be a mix of regional content and localized content. Internet tests would be used the same for each region (except for language translations). Assessment centers and interviews would be tailored to the job tasks at each facility, and because it was very much a team environment, be customized to the local country. However, while the content varied, the structure and nature of the exercises and interviews did not. For example, in contrast to how it was scored in Europe, the leaderless group discussion exercise would be designed to evaluate collective harmony, versus individual achievement, and thus remain consistent with local cultural norms. The following assessment methods would be used:
  - Cultural agility—Internet test; assessment center (role-play)
  - Teamwork—Internet SJT; assessment center (role-play, leaderless group discussion)
  - Performance management—assessment center (leaderless group discussion); interview

Beyond the methods noted here, both jobs would also be screened in terms of experience (using biodata) because experience was related to technical knowledge, and could serve as a proxy for knowledge and thus allow Spooner to avoid having to purchase job knowledge tests for each position. This process would ensure multiple methods were used to triangulate around the core competencies.

Roger saw a mountain of work ahead of him, but he also perceived glimpses of light between the somewhat foreboding forest of complexity ahead of him. Roger was highly self-critical and one thought continued to burn at the back of his mind. What was it that he did NOT know? Perhaps he was overlooking something that would be blindingly obvious with hindsight but right now he could not see. So he decided to test his ideas on one or two trusted regional staff as well as bouncing ideas of Gordon Baker, his old mentor back in the UK. Although Gordon had no familiarity with S.E. Asia—and indeed little with recruiting and staffing—he had a keen and insightful mind and the ability to see logical flaws in arguments, processes, and systems, born of many years of labor relations, where those abilities were tested daily.

Roger went back to his apartment, reflected on the day's work and posed himself the questions in a way that he could maximize the time for his Skype call with Gordon:

- Was he making this selection plan overly complex?
- What were the likely counterpressures and how would he overcome them?
- What was he missing because of his cultural biases?
- Were his assumptions about the cross-cultural validity of assessments accurate? And how would he ascertain that?
- What was his backup plan if the various pieces of analysis produced either no statistically relevant patterns or results that posed more problems than answers? For example how would he deal with an outcome that suggested that countries in which Spooner had aspirations were structurally challenged in terms of talent

quality? Spooner might be an influential employer but they could not singlehandedly solve a poor education system or politically designed systems (for example one that mandated for indigenous-language science and math education when textbooks in that language were unavailable).

And for his regional team, Roger had more specific questions:

- The assembly team associates selection process was a high-volume operation. How in the world could they conduct interviews on so many people? Surely they would need a way to reduce the total number of applicants to something more manageable.
- R&D team leader applicants were in high demand. Would the assessment center nature of the process conflict with their cultural values and expectations? Could they structure the process to balance leadership with collective orientation?
- What about the process—what would that look like? Test everyone at the same time? Should there be stages to the process? How would they manage applicant flow?
- Every process involves trade-offs. What were they with the selection process for either job? How would they balance benefits against constraints?

Step 5: Structure the Selection Process

A selection process is not a static activity but a continual flow of talent through the system. Rarely will all candidates be tested on the exact same assessments at the exact same time. Usually, there is a sequence of steps to the selection process, and there is a flow of candidates through these steps. The major decision to make regarding the selection process is whether it will be compensatory or noncompensatory (Gatewood, Feild, & Barrick, 2011). In a *compensatory system*, all applicants are tested on all predictors, and low scores on one assessment can be offset by higher scores on another assessment. In a *noncompensatory system*, candidates scoring too low on any specific assessment are removed from the selection system. There are two types of noncompensatory systems. A *cutoff* approach is used when all candidates are tested on *all* parts of the selection system, but they must pass minimum thresholds on the predictor methods. For example, getting a driver's license might require passing a written test and a driving test, and high scores on one does not compensate for a poor score on the other. In a *multiple hurdle* approach, the candidate must pass a sequential series of cutoffs, but will only continue through the process until he or she fails to pass a hurdle. For example, getting a driver's license might require one to pass a vision test as a first hurdle. Only those candidates passing the vision test would be offered the written test, and then only those passing the written test would be given the driving assessment.

Multiple hurdle approaches are used to balance a number of practical concerns, with cost, applicant flow numbers, and the importance of the job being the most common. To minimize cost, a firm might first screen applicants using fast and inexpensive methods such as minimum qualifications (e.g., educational degree, years of experience, or certifications). Those meeting these minimum qualifications might next complete online assessments of ability, personality, or situational judgment. Personality and situational judgment measures are often used as a "sift" or early reduction of the candidate pool because they are cost-effective and have only small to moderate subgroup differences (e.g., sex, race). Only those clearing these first two hurdles would experience more involved and expensive assessments such as interviews, assessment centers, or work samples.

The issue of applicant flow (or the number of candidates to be processed) is closely related to cost. Retailers, call center operators, and hospitality companies typically experience relatively high turnover at lower levels and, as a result, must sort through a large number of candidates. Cost and efficiency become paramount considerations, and selection methods are often quite brief and inexpensive. However, as noted here and in chapter 3, selection methods with even very modest validities can have significant payoffs when applied to large numbers of hiring decisions.

The strategic importance of the job is usually related to the type and duration of the selection process. At higher levels, where the strategic impact and consequences of error are greater, it is common to find a more extensive and time-consuming assessment process. Final-slate candidates for executive-level jobs are often asked to participate in an assessment center or an in-depth individual assessment. The latter typically involves a comprehensive assessment battery and lengthy interview with a consulting psychologist and concludes with a feedback call to the hiring manager. The higher the level, the more an organization is willing to spend on selection and the more exhaustive the process. This is logical simply because the potential consequences of a wrong selection are greater organizationally and financially.

Step 6: Make Selection Decisions

Regardless of which predictor methods and structure are used, once information is collected it must be used in some way to make a hiring decision. At the most basic level it is critical to determine whether the information is merely advisory or whether it is compulsory. Organizations adopting the former philosophy leave it up to the hiring manager as to how best to combine information to reach a decision, and organizations relying exclusively on interviews tend to default toward that approach. Even when data are available from other sources (e.g., tests), it is often up to the hiring manager as to how best to use these data. In the UK, for example, users of assessments are trained on proper use of assessments, but the combination of tests scores to reach a decision is not specified. Compulsory information does not leave any discretion in the hands of the hiring manager, but is rather done in a more mechanical manner.

There is an extensive body of research illustrating that mechanical combinations of data (e.g., combining it via an algorithm or a regression model; see chapter 3) consistently predict better than human-based judgmental combinations (Kuncel, Klieger, Connelly, & Ones, 2013). When information is combined mechanically, there are a number of options available.

One of the more common is to use an algorithm to "add up" the various inputs into a total or overall score for decision-making. The algorithm can be based on a regression model or one of the various Big Data methodologies. These types of models will weight each predictor score according to a statistically optimal composite. It is for this reason that human prediction cannot be greater than the algorithm—because the best a human can do is reach the same optimal conclusion as the regression model does every time. These mechanical models tend to be compensatory, allowing high scores on one or more measures to compensate for a low score on another measure. Given an overall score, the hiring manager can simply select from the top down until openings are filled, or focus interview time on those scoring above an established threshold. Besides simplicity and consistency, this approach has the advantage of maximizing utility or the expected benefit from use of the selection tools (Table 3.6 in chapter 3 illustrated this approach).

Step 7: Evaluate the Selection Process

As we have seen, there are many different methods and processes for selecting applicants. The system that is "best" is the system that best meets the strategic and tactical goals (see Step 1 and chapter 2) in a way that is sustainable and produces balanced results. However, there are a number of desirable features that most employers want. These desirable features become the goals for the selection system and determine the nature of performance metrics used to evaluate the selection system (see Table 5.5).

First, *effectiveness* goals (operationalized via accuracy or validity) emphasize the fact that selection systems are often designed to enhance performance. The focus may be on individual performance, unit performance, retention, and even employee engagement, but the system will be gauged a success to the extent it results in improvements in whatever metric is of importance to the firm. Chapter 3 discussed a number of different ways of demonstrating an "effect," and that is exactly what one is trying to

TABLE 5.5 COMMON SELECTION METRICS

Type of Metric	Examples
Effectiveness Metrics (Validity)	
Manager Reactions	Perceptions of whether the candidates hired are of appropriate quality; manager support for the selection system.
Applicant Experience	Perceptions of the selection system's job relatedness, fairness, invasiveness, timeliness, etc.
Job Performance	Job performance metrics such as supervisor and/or peer evaluations, sales, productivity, citizenship, accidents, and so on. Common performance dimensions include technical, interpersonal, counterproductive, citizenship, and potential/adaptability.
Unit Performance	Branch, division, subsidiary, etc., operational and financial metrics (e.g., controllable profit). Compare within and across units.
Diversity	Representation of hires in terms of racial, gender, cultural, functional, and related forms of diversity. Often indexed with adverse impact statistics.
Engagement	Employee satisfaction, motivation, commitment, identification, etc.
Individual Turnover	Individual level includes length of employment and reasons for turnover. Estimates for voluntary, involuntary, and overall turnover. Consequences/costs of individual turnover.
Unit Turnover	Collective turnover rates within a unit or organization, along with consequences/costs of turnover rates on unit metrics. Estimates for voluntary, involuntary, and overall turnover rates.
Quality of Hire	Subjective or objective assessments of the quality of the general and/or specific talent. Often based on test scores, performance, engagement, promotability, etc.

(continued)

TABLE 5.5 CONTINUED

Type of Metric	Examples
Efficiency Metrics	
Cost	Overall selection function cost, cost per applicant, cost per hire, cost per assessment.
Time-to-hire	Time between start of selection process and hiring decision.
Conversion Rates	Ratio of the number of applicants who pass assessments relative to those who accept; ratio of the number of applicants who pass assessments relative to those who perform well.
Return on Investment	Monetary estimates of financial return relative to selection costs. Often expressed in terms of utility analysis estimates or direct linkages to unit financial outcomes.

do here—demonstrate the effect of selection. For example, when evaluating individual performance, organizations often compare the performance of people selected with the system to those selected without it. This very simple method has the benefit of being easy to compute and explain (e.g., to doubting managers), but does not control for potential alternative explanations for any differences observed. The Taylor-Russell tables (Taylor & Russell, 1939) enable one to easily compute the increase in "hit rate" associated with use of a selection method. For example, assuming half of current employees are considered successful (a 50% hit rate), an assessment of moderate validity used to hire only candidates scoring at or above average on it will increase the hit rate to 60%. While this may sound very unimpressive, a 10% increase in above-average performers would transform most organizations.

Second, most employers also want the system to be *efficient*, be *cost-effective*, and minimize the amount of "seat time" it takes a candidate to complete the process. This is particularly true in high-volume hiring scenarios, where an employer may be screening millions of candidates each year for a single job (e.g., high-volume retail stores). The more complicated the selection process, typically the more expensive it is to develop, deliver, and maintain. One way to evaluate the selection system is to compare its efficiency to previous approaches. Efficiency is usually operationalized in terms of time-to-hire and/or cost-per-hire, such that any system that is faster or cheaper will be more efficient (see also chapters 2 and 3). A test inserted before an interview (multiple hurdle), for example, may be a faster and cheaper way to screen candidates when the cost of the test is compared to the cost of the interviewers' time. While it is our experience that few selection systems are designed for the sole purpose of enhancing efficiency, it is often a byproduct and one of the easiest criteria to monetize. Many ways of evaluating the impact of selection in financial terms have been developed, and these vary greatly in terms of complexity (see chapter 3; Table 3.4). For example, the Brogden-Cronbach-Gleser model enables the computation of utility in monetary terms for any period of time and has been expanded (Cascio & Boudreau, 2008) to take into consideration a host of variables (e.g., taxes, variable costs, and discounted net present values). Our experience is that these methods are usually met with skepticism by managers because (1) many of the terms are difficult to understand (e.g., how much more is a high performer worth per year than an average performer) and (2) the resulting estimates are often so large as to elicit incredulous derision. The gains from effective selection have under any method of estimating utility far exceeded the associated costs, so methods that are understood and accepted by management are generally the best (see chapter 3). In our personal experience, we have found that managers are most persuaded by analytics that directly link talent scores to key business performance metrics at the unit level (e.g., store controllable profit) (see chapter 3).

Third, as employment branding has grown in importance, many also want the selection process to be *engaging*. Efforts to make assessments look like games or to embed them in a realistic job preview as part of "day-in-the-life" experience all reflect a desire to make the experience more interesting and enjoyable. As companies take a life-span perspective toward employment, the selection process that can serve multiple purposes has come into vogue. For example, some organizations have created assessments that can (1) determine which jobs for which a candidate is the best fit, (2) determine whether they are a sufficiently good fit to justify an interview, and (3) for those fortunate enough to be hired, provide *developmental feedback* for use in employee development efforts. Other selection processes can determine whether an applicant gets hired and provide an indication of future potential for use in subsequent promotion decisions.

A final category of goals focuses on employee *diversity* and *fairness.* While many organizations have embraced diversity as a potential competitive advantage, others focus on it primarily from a legal perspective. In either instance, the selection system must ensure the workforce that is hired is diverse in accordance with strategic goals and within the guidelines of the country's legal system (see Step 2). When considering the effects of selection on diversity, it is very important to use the appropriate terminology. The terms "fairness," "bias," and "adverse impact" often are used interchangeably, but they mean very different things. There is no agreed on definition for *fairness.* What is considered "fair" depends on one's perspective. Because each person determines what is fair to him/her, fairness is in the eye of the beholder; thus it is important to consider the perspectives of multiple stakeholders when designing a selection system. Conversely, *bias* has a broadly accepted definition and a means of testing for it (often called the Cleary model of test bias; Cleary, 1968). A test is biased when the relationship between the scores in the selection process and some outcome (usually job performance) are different for different groups. That is, the validity of the assessment is not the same across different protected groups (e.g., an assessment is more valid for males than females). When evidence of bias is found, it is usually the result of a poorly designed selection method and can often be corrected by modifying its content. Somewhat between these two, adverse impact has a general definition, referring to differences in the hiring rates of demographic groups, and several different ways of testing for its presence (as was discussed in Step 2).

Designing any selection process will require that trade-offs be made (Ployhart & Holtz, 2008). For example, if maximizing validity is the objective, then including in the process a measure of mental ability would be indicated. Inclusion of such measures though is likely to result in adverse impact against some groups and will most certainly increase testing time. As another example, assessment centers produce large amounts of data that can be used for both selection and development. They are also among the most expensive, labor-intensive options, and therefore unsuitable for most high-volume situations. Among the least costly types of selection tools are tests that can be administered via the Internet (the applicant reads the question and selects an answer from a closed-ended set of options, enabling the server to score it instantly and forward the results to the hiring manager). Applicants universally find these types of assessments the least engaging. See Tables 5.6 and 5.7 for a summary of trade-offs.

It can be particularly challenging to balance effectiveness, efficiency, and diversity. Historically, the rule of thumb has been to pick two out of the three goals to maximize. To maximize effectiveness and diversity,

TABLE 5.6 SELECTION GOALS, BENEFITS, POTENTIAL TRADE-OFFS, AND METRICS

Goal	Benefit	Potential Trade-Off
Effectiveness (Accuracy/ Validity)	Maximizing validity means improving the accuracy of predictions. Maximizing validity usually means including a measure of cognitive ability in the selection process.	While cognitive ability is among the best predictors of performance, it also produces adverse impact, usually by race. Depending on the measure, including cognitive ability can also significantly increase "seat time."
Efficiency/ Cost-Effective	Especially in high-volume contexts, where an employer may assess millions of candidates per year for the same job, cost-per-person becomes a key concern.	Minimizing costs per applicant eliminates any method reliant on humans for gathering or scoring information (and the associated labor costs), such as assessment centers and interviews.

(continued)

TABLE 5.6 CONTINUED

Goal	Benefit	Potential Trade-Off
Seat Time	Employers are typically concerned with minimizing the amount of time it takes a candidate to complete an assessment. To enhance the applicant experience, they seek to minimize seat time.	To reduce the time required of applicants means measuring only the most essential things in the simplest fashion. This in turn makes branding and engaging interactive experience, developmental feedback, and maximal validity, less likely.
Engaging Applicant Experience	The use of "day-in-the-life" assessments and other employer-branded and interactive experiences are all intended to improve the applicant experience and the employer's reputation.	Interactivity, realistic job previews, multimedia presentation of assessment information, and the like all increase cost and seat time. The fastest and cheapest assessments tend to be common text-based measures.
Developmental Feedback	Employers often want assessments to provide information not only for selection or promotion purposes but also for purposes of employee development.	Assessment centers, individual assessments, and assessment batteries often yield information on a candidate's strengths and weaknesses and may even include a development plan to exploit strengths and minimize weaknesses. This often adds time and cost to the process.
Diversity/ Fairness	Many firms seek to increase diversity. Although adverse impact does not preclude the use of a selection technique, most organizations will go out of their way to avoid it.	Minimizing the potential for adverse impact means reducing cognitive ability measures and those related to it (e.g., job knowledge measures, some situational judgment measures) or including noncognitive measures (which increase seat time).

TABLE 5.7 SELECTION GOALS, BENEFITS, AND POTENTIAL TRADE-OFFS, BY PREDICTOR

	Selection Goal					
Predictor	**Effectiveness (Accuracy/ Validity)**	**Efficiency/ Cost Effective**	**Seat Time (Slower is Worse)**	**Engaging Applicant Experience**	**Developmental Feedback**	**Diversity/ Fairness**
Cognitive Ability	High	High	Fast	Low	Low	Low
Knowledge and Skill	Moderate	High	Fast	Low	Moderate	Moderate
Personality	Low	High	Fast	Low	Moderate	High
Work Interests, Styles, and Values	Low	High	Fast	Low	Moderate-High	Moderate
Biographical Data	Moderate	High	Fast	Low	Low	Moderate-Low
Fit	Low	High	Fast	Low	Moderate-Low	Moderate
Situational Judgment	Moderate-High	High	Fast	Moderate	Moderate-High	Moderate
Structured Interviews	Moderate-High	Moderate-Low	Moderate-Slow	Moderate	Moderate	Moderate-High
Assessment Centers	High	Low	Slow	High	High	High
Work Samples	High	Moderate-Low	Moderate-Slow	High	High	High

NOTE: This table is adapted from Ployhart and Holtz (2008); Ployhart, Schneider, and Schmitt (2006); Schmidt and Hunter (1998).

the selection system will be costly and less efficient. To maximize effectiveness and efficiency, the selection system will likely hurt diversity. To maximize efficiency and diversity, the selection system will likely sacrifice effectiveness. However, technology, increases in diversity and educational opportunities, and scientific research are helping to make balancing these three goals more likely. Technology in particular has offered new ways of administering assessments, so we discuss these in some detail.

The regiocentric selection plan was unfolding. The high-volume hiring in the assembly team associate jobs would leverage its large volume to save money and time by outsourcing to a vendor. The R&D team leader hiring would focus on quality and differentiation. Cumulatively, value would be added by being efficient and effective. But how to maximize both? Here the team struggled, until they came to the conclusion that the selection system for each job could be based on slightly different processes, each intended to maximize the specific goals:

- Assembly team associates. High-volume hiring. Cost of hiring errors small to moderate. Focus on efficiency and cost-savings.
  - Hurdle 1: Basic screen on biodata, experience, eligibility for employment. Internet (or paper) personality assessment (conscientiousness and attention to detail) and SJT (teamwork)
  - Hurdle 2: Structured interview
- R&D team leaders. Low-volume hiring; highly strategic and competitive. Cost of hiring errors significant. Focus on quality and diversity.
  - Hurdle 1: Basic screen on biodata, experience, eligibility for employment.
  - Hurdle 2: Internet (or paper) personality assessment and SJT (cultural agility and teamwork)

- Hurdle 3: Assessment center (tailored to each location). Structured panel interview (tailored to each location; multiple interviewers from the team and peer group)
- Hurdle 4: Final interviews with immediate supervisor, country-specific general manager and HR manager, and regional HR.

In the assembly team selection process, applicants would have to pass minimum thresholds in each hurdle to move onto the next hurdle. In the R&D team leader selection process, applications would have a cutoff in the first and second hurdles. In hurdle three, the testing would be done onsite over a 2-day period. Those candidates in hurdle three would then be rank ordered, and the top candidates would be brought back onsite for a final round of interviews. The minimum thresholds for setting the pass points in any hurdle would be established based on a job analysis.

Validity [see chapter 3] of the scores would be established via content validity and, later, criterion-related validity by correlating the assessment scores with performance (probably 6 months to a year later). Efficiency would be gauged in terms of cost-per-hire and time-to-hire. Diversity would be evaluated in a manner consistent with local regulations. Surveys of the candidates' experience with the process would be used to track engagement. Surveys of key stakeholders, including peers and supervisors, would be used to further build a data-based evaluation of the selection process.

TECHNOLOGICAL DEVELOPMENTS IN SELECTION

We now turn to consideration of the role technology plays in selection and assessment. There have been many rapid technical developments in the area of selection that are radically changing the way it is conducted. In most cases these developments have been pulled to the market by customer demand and technological advancements, rather than pushed to it by scientific research.

Virtual Branding

One example alluded to earlier has been the melding of assessments and employment branding. Increasingly, assessments are being made to be far more interactive and engaging. Combining features of gaming simulation technology, a realistic job preview, and traditional tests, assessments can be presented as a "day-in-the-life" experience or a virtual "job tryout." Personality measures, SJTs, and cognitive ability tests can be intertwined with information about the job/company and presented in a highly contextualized format that feels less like taking a test and more like doing the job. In addition to the employment branding benefits of such approaches, the candidate experience is also often enhanced (although the time required to complete the test may also go up).

Gamification

The same technology that has made branding assessments viable has seen its way into the development of simulations and game-oriented tests. Situational judgment tests have long been presented via video and increasingly using very realistic avatars. Not only are the latter cheaper to develop than video vignettes, but also minor changes in "actor" clothing, scenario background, and voice-over accommodate use in different countries and cultures much more cheaply. Similarly, creating simulations to capture measures of key abilities for jobs performed largely at a keyboard are very straightforward. Bank tellers, call center operators, and many other jobs can be similarly replicated and measures of proficiency at the key tasks measured. Assessments are also being developed to look and feel like games. Using common game features such as likes, points, badges, and leader boards, game-based assessments are intended to be more engaging and entertaining and to potentially draw people to apply. *Happy Hour*, for example, is a game in which participants play the job of bartender but actually measures their ability to multitask. *My Marriott Hotel* enables players to open a restaurant, buy equipment and food stock, hire and train employees, and ultimately serve guests. Much like *Farmville*, the game enables players to earn points for good customer service and potentially earn a profit. While the use of game-like assessments for selection is currently limited, the popularity of these on social sites makes expanded use highly likely.

Computer Adaptive Testing

Another advancement is the growing use of computer adaptive testing (CAT), particularly for the assessment of ability. In the typical cognitive ability test, the number of items a candidate gets correct is compared to the number answered correctly by a comparison sample and conclusions drawn about the candidate's relative ability (e.g., if you get more questions right than 75% of the comparison, you are deemed above average on that ability). In a CAT-enabled test, each candidate begins the test by answering an item of average difficulty. Those getting the item correct receive a more difficult item, while those getting it wrong are presented an easier second item. The testing engine continually updates the estimate of the test-taker's ability and serves up successive items designed to maximize the information gleaned. Such CAT-enabled tests have two significant advantages. First, they are typically shorter, often taking less than half as many items to reach an equally accurate ability estimate. Second, because the test adapts to each person, different applicants will see different sets of items. While CAT tests are quite expensive to develop, for at least two reasons they are likely to see ever increasing use. First, because not everyone sees the same set of questions, test security can be enhanced, which is an important consideration in the world of unproctored testing. Second, the growing popularity of testing on mobile devices will make test brevity more important.

Internet-Delivered Assessments

Assessments delivered over the Internet have been around for a long time, but they create special challenges, for which solutions are still evolving. The biggest issue is whether the assessment process will be proctored or monitored directly by the company. In a proctored environment, the candidate completing the assessment is known and his or her access to other information typically controlled. With the advent of Internet testing, many candidates now complete an assessment in an unproctored environment. Unproctored Internet testing (UIT) offers many advantages that explain its rapid adoption: it is cheaper for the company, more convenient for the candidate, and may increase the number of candidates that consider an opening. The downsides, of course, are that the company cannot be sure who is completing the assessment, nor can it control the testing environment (e.g., the candidate could be using the same Internet to help answer questions, solicit the input of friends and family, copy the content for future dissemination, or simply be interrupted frequently by other life matters).

Research on UIT is growing. We do know, for example, that tests taken online are essentially the same as those taken with paper and pencil (Ployhart, Weekley, Holtz, & Kemp, 2003). Self-report measures, like personality and biodata, are also apparently unaffected by lack of proctoring (Tippins et al., 2006). Intuitively this makes sense; these types of measures can be faked in the presence of a proctor as easily as not. For tests with right/wrong answers, the evidence is mixed. One study showed a moderate decline in scores when participants took a proctored cognitive ability test following an unproctored one, while another showed small differences between conditions (Beaty & Shepard, 2002). In response to the concern that UIT results in cheating, a number of potential solutions have been offered (see Table 5.8). Our advice is that, in high-stakes settings, the

TABLE 5.8 ALTERNATIVES TO PROCTORED OR UNPROCTORED TESTING

Options	Pros	Cons
Unproctored Internet testing (UIT) only	More convenient for candidates Faster results = faster offers Most cost-effective	Does not address security concerns Unknown testing environment Some individuals may not have access to a computer
Proctored testing only	Addresses all security concerns	Forgoes all advantages of UIT Is the most expensive option
UIT followed by brief, proctored confirmatory test	Benefits of UIT realized Confirmation address some of the security concerns	Requires development of two tests Candidate must travel to testing site for short testing session Some time is required of test administrator(s)
Noncognitive tests administered unproctored and cognitive tests administered proctored	Addresses test security for ability and knowledge tests. Candidates may take noncognitive portion at his/her convenience	Overall testing process may be increased Combining tests scores more difficult
Randomized items from a pool	Addresses content security to a degree Can be used in conjunction with other options	Does not address issues of cheating or identity Items need to be calibrated

employer should consider an unproctored assessment score to be necessary but insufficient evidence of a candidate's qualifications. You know you don't want those who do not pass the UIT, but some confirmation of the qualifications of those who do should be sought (e.g., retesting, in-depth interviewing, etc.).

Mobile Assessments

The administration of assessments on mobile devices, such as smartphones and tablets, represents the further evolution of Internet testing. In some countries, use of hand-helds to access the Internet is already higher than the PC, and growth trends clearly favor the smartphone. The fact that assessment is moving to the smartphone is inevitable; the form this will take is still being hotly debated. Already vendors are offering "responsive design," wherein the item format changes to fit the screen on which it is being displayed. Nonetheless, there are some types of items and assessments that will simply not be amenable to presentation on a phone (most tablets are analogous to a PC and do not present the same quandaries that a phone does). This is important because organizations do not want to use assessments where the scores can be dramatically impacted by the type of device on which the assessment is completed (there is evidence that blacks and Hispanics go online using hand-helds at a much higher rate than whites). Two recent studies have shown that personality measures are largely invariant (i.e., the same) across smartphones and PCs, while cognitive ability scores were significantly lower when completed on a hand-held device (Arthur, Doverspike, Munoz, Taylor, & Carr, 2014; Morelli, Illingworth, Scott, & Lance, 2012). We believe that mobile deployment will place a premium on items that require very little screen real estate and on short tests. Elaborate simulations, branded tests, case studies, drag-and-drop items, and the like will simply be very difficult for a candidate to complete on a hand-held. Instead, organizations may turn increasingly to shorter, smartphone-friendly screening assessments and administer longer, more comprehensive assessments on-site for those clearing this first hurdle.

Passive Profiling

One final trend worth discussing is the use of "passive profiling." This term reflects any broad attempt to learn about the job candidate without soliciting information directly from the candidate. This can be as simple as checking the candidate's Facebook page for embarrassing photos or more elaborate counts of "recommended by's" on LinkedIn. One survey found that more than three-fourths of recruiters used search engines to find information about candidates and more than a third reported eliminating candidates. Even more sophisticated than an online search for "digital dirt" are the attempts to analyze digital footprints to measure things like personality and cognitive ability. Several studies have reported on attempts to use blogs, tweets, and other public domain written text to derive a profile of the individual's personality. Some of these approaches are similar to biodata we noted earlier. While use of derived measures of personality to predict performance are increasingly being used in practice, there is not much evidence of their effectiveness or legality.

Conclusion

There are not only many different types of predictor measures and methods but also increasingly diverse types of media through which to deliver them. While the classic types of assessments still predominate (e.g., multiple choice measures of personality, cognitive ability, and skills), the immutable forces of mobile, social, and Big Data are increasingly impacting what is assessed and how it is measured. Those responsible for selection systems must stay current on both technological and legal developments, as technology is truly revolutionizing the manner in which selection is conducted—usually for the better, but not always.

Several months into the implementation of the regional selection strategy, it became apparent that the high-volume hiring process for assembly team associates needed some refinement. The vendor's assessments appeared to work reasonably well, but turnover remained high and line managers were frustrated that the new system did not increase retention. Results of the candidate experience survey also suggested that the process was not very engaging. Candidates found the Internet assessments to be rather time consuming and out of touch with their mobile lives. In hindsight this made sense, given that many of the candidates for these positions were young, single, and heavily into technology. They were also highly mobile because of the intense competition for generic talent in Asia. So Roger worked with the vendor to enhance the candidate experience by leveraging more advanced technology.

The vendor's belief was that technical/manufacturing skills are less culturally dependent, and hence lessons they had learned on building online simulations for manufacturing jobs in other countries could be used to create an engaging assessment for Spooner. Roger was skeptical, but his regional HR team strongly encouraged him to give it a try. Through various focus group meetings, it was determined that an online simulation could be developed. The simulation would be a fusion of situational judgment, work sample, and in-basket predictor methods. Gamification principles would be used to present candidates with a fictional manufacturing facility not unlike the one used in Spooner plants. Participants would need to attend to changing production details and schedules, and manage relationships with other team members. The idea was to create a simulation that was fast, engaging, provided a realistic job preview (to let candidates gauge their own fit), and would represent the Spooner brand. This was great in theory and operationally it looked very impressive. But it was also very expensive relative to the existing approach.

Hence Roger and his team struggled with the all-too-common question: "What is the ROI of this new method?" In close coordination with the vendor, they worked through a number of different scenarios, coming to the conclusion that if this new assessment could reduce turnover even 10%, the short-term costs would be offset by long-term gains in retention. The decision was made to try the new assessment experimentally in two countries: the Philippines and Malaysia. The simulation would be added to the existing selection system, and applicants would be expected to take the new and old assessments. However, only the old assessments would be used to make hiring decisions until sufficient validity evidence and related data could be collected on the simulation. If the simulation was related to turnover of the magnitude needed to justify its expense, was related to performance at least as well as the old system, and performed similarly or better with respect to diversity, the new simulation would be made operational across all of the regional hiring in manufacturing. This had the potential to be a highly visible victory for the regional talent strategy.

Roger was cautiously optimistic, but he did allow himself a brief moment to imagine the friendly jabbing he would be giving to his friend Andrew back in the UK—"Hey Andrew, this is Roger. You'll never believe what I pulled off."

6

Global Staffing and Talent Management

THIRTY-SECOND PREVIEW:

- *Global talent management* refers to how a firm chooses to create, combine, deploy, and divest of its talent resources within an international context.
- The talent management elements that are most affected by talent acquisition include onboarding and socialization, promotion and internal mobility, succession planning, development, engagement, and retention. These elements are all interrelated with staffing.
- Staffing fundamentally shapes the nature and effectiveness of other talent management elements. Improving staffing generally makes all subsequent aspects of talent management more effective, efficient, cost-effective, and provides greater return on investment.

- The Bottom Line: Staffing is one piece of the global talent management puzzle, but it is the piece that affects all other talent management practices, so it is critical to ensure that global staffing is aligned with other talent management practices in culturally appropriate ways.

Roger waited quietly outside the office of Ibrahim Abdul Hassan, the VPHR for S.E. Asia and the country HR manager for Malaysia. In simpler terms, Hassan was Roger's boss, and he had called a meeting to review the regional talent acquisition strategy.

It had been nearly a year since the S.E. Asian Regional Talent Acquisition Strategy had been implemented. Everyone agreed Roger had made incredible progress. There were some missteps, to be sure, and some mistakes created by cultural misunderstandings, but most people knowledgeable about the project agreed that the new strategy was considerably better than the patchwork of practices that was in place before. In fact, news of the success traveled as far back to HQ in Slough, England.

Roger did not know the purpose of the meeting, other than they were to discuss the regional talent acquisition strategy. Always the pragmatist, Roger knew there were many areas that needed additional tweaking and refinement, and he suspected the purpose of this meeting was to identify those areas and consider the operational ramifications and associated budgetary needs. Yet he couldn't help feeling excited with the expectation that part of their meeting would be Hassan congratulating Roger on a job well done. This expectation turned out to be quite wrong.

As the door to Hassan's office opened, Roger immediately saw Alias Ismail, the VP of HR Asia-Pacific and Middle East, and Patrick Chieng, the VP of S.E. Asia. Roger felt sick to his stomach, and any thoughts of rousing celebrations and compliments quickly disappeared. Ibrahim Abdul Hassan had not asked him to discuss details about the talent acquisition strategy. Rather, after exchanging pleasantries consistent with the local cultural norms, Alias Ismail said:

"Roger, we are all impressed with how you developed and implemented the regional talent acquisition strategy so quickly. It's remarkable, really. But this success has also come with a cost in terms of local relationships and, as a result, to your reputation in this region. Other HR leaders, and even some of the country HR leaders, are concerned. Culturally, it is not surprising that they haven't expressed these concerns directly to you, but I have seen several signs that they view your success as an implied criticism of their competence and operational effectiveness. Some of these leaders are very proud of what has been built in the past, but don't seek individual recognition for those achievements. But your success and the accolades that have been awarded to you have left some raw emotions and hurt feelings. They have been good citizens and team players for many years and yet you have come into this region and are not seen to have played by their rules nor given due respect for their achievements. Your success has gained strong recognition for you but not the team, and so they feel their values and trust have been violated."

Of course, Roger had no such intentions and he explained this as best he could. The three executives were widely experienced in both business and corporate politics, and Ibrahim in particular had a keen sense for navigating those tensions. They discussed this matter for the better part of the afternoon, deciding to reconvene and discuss what to do about it over dinner. Later that evening, Ibrahim had developed a possible solution. He waited for the right moment and then introduced it to the group:

"Roger, there is no benefit in undoing what you have done. HQ clearly does not want that, and the local functional leaders will likely find it offensive if we tried to suggest that. They would see it as indecisive and perhaps even manipulative. But you might gain their acceptance by seeking their guidance. These are experienced and highly committed leaders. They will listen to a business case to link your efforts to theirs. I suggest you approach them to seek their guidance and assistance. Ask how you can tie the regional staffing strategy to their strategic goals. Keep your focus on the region's talent management plan; promote your staffing strategy to be in the service of their strategies. Make their success your success. This puts you in a vulnerable position, to be sure, but it likely will help these leaders save face and help you gain their acceptance. Remember, HQ may be impressed, but your long-term viability in this region is dependent on collective support."

So Roger left realizing he would need to demonstrate how the benefits of the regional staffing and talent acquisition strategy were also benefits for other talent management activities throughout the region. Roger was privately frustrated and casually contemplated whether he should simply ride his success for a transfer back to the UK. But he was not a person who backed down from a challenge and he realized that winning one battle does not mean the war is over. If he were to look for that option, there would always be question marks about his ability to operate in an international arena and, as an aspiring global leader that would be a death knell. "No," he said to himself. "This is a challenge that I must take on and conquer."

But this next battle was going to require tact, political savvy, and an ability to connect across different functional areas. "Help me help you" would be his motto, and, although he accepted the concept, he wondered whether he had the strength of character to see it through.

To shed these doubts he started to develop a list of questions:

1. What are the major functional elements and activities in the regional talent management process?
2. How do these elements tie together?
3. Do regional and cultural differences affect these elements?

4. Who is in charge of each element? What challenges do we share; what challenges are unique?
5. How do the talent management elements connect?
6. How does talent flow through this system?

Does success in one area of talent management cause success in other areas? He assumed so, but began to wonder if it was a valid assumption.

Global talent management refers to how a firm chooses to create, combine, deploy, and divest of its talent resources within an international context—that is, how it enacts its talent strategy. We have already touched on global talent management many times in earlier chapters. Chapter 1 noted the role of high-performance work systems (HPWS) that shape the nature of collective talent resources. Chapter 2 discussed the role of talent strategy and the global talent supply chain. Chapter 3 introduced analytical approaches for providing a rigorous evaluation of staffing processes and gave some empirical examples of how staffing relates to training. Chapter 4 noted how recruitment relates to both attraction and retention. Chapter 5 considered how selection practices create the quality and quantity of collective talent resources. Each chapter considered talent acquisition from a strategic and global perspective, but these perspectives were primarily focused on external talent acquisition. In this final chapter we consider how talent acquisition can also occur from within and explore how these touchpoints shape other key elements of global talent management. By providing a "big picture" view of how talent acquisition relates to other important talent management activities and functions, the focus will be on what talent acquisition does for these other activities and functions and less about the functions themselves. Therefore, we do not go into great detail around each of the specific talent management activities and functions (readers interested in more detail may consult Cappelli, 2008; Scullion & Collings, 2011).

A FRAMEWORK FOR GLOBAL TALENT MANAGEMENT

At this point it should be clear that effective global talent acquisition can directly contribute to reducing costs, growing revenues, and enhancing strategy to differentiate the firm from competitors to achieve a competitive advantage. But talent acquisition also contributes to these outcomes indirectly via a number of other talent management activities and functions. As noted throughout this book, global talent acquisition sets the foundation for all subsequent HR and talent management activities and functions.

Figure 6.1 provides a framework for global talent management, and shows the foundational role that talent acquisition plays in ensuring the effectiveness of these other activities. It also shows that global talent management is influenced by tensions between multinational corporation (MNC) versus subsidiary pressures, and between universalistic versus local customization pressures. There are six major talent management

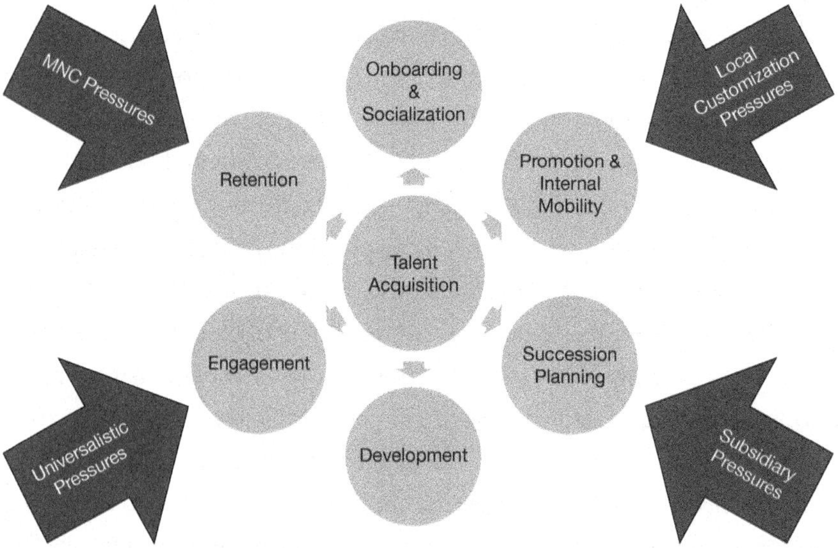

Figure 6.1
The fundamental role of talent acquisition in shaping global talent management.

activities that talent acquisition shapes (these activities are discussed in more detail shortly):

- Onboarding and socialization: how employees experience their transition from a candidate to an employee; their first few months on the job and how they are socialized to the organization's values, culture, climate, and strategy.
- Promotion and internal mobility: filling open jobs from within by promoting current employees, and the benefits and disadvantages of doing so.
- Succession planning: workforce planning to operationalize the firm's strategy and talent strategy; focused on ensuring there is the sufficient quality and quantity of talent to meet current and future needs.
- Development: a longer-term approach to create among existing employees the types of talent (e.g., knowledge, abilities) needed to support the promotion and succession plan in pursuit of business needs.
- Engagement: the extent to which employees are satisfied, motivated, committed, and supportive of the organization's strategy and goals.
- Retention: ensuring employees who are performing sufficiently will stay with the firm; whereas those who are underperforming will either be developed or let go.

Talent acquisition has the most immediate effect on onboarding and socialization, but influences all of the activities in profound ways. Recall from chapter 2 that talent acquisition shapes the quality and quantity of collective talent resources, and that the capacity of the talent resources influences subsequent performance. A similar concept applies here. If the talent resources are below requirements, then all subsequent talent management activities will likewise suffer unless firms invest additional resources to address the deficiencies. For example, hiring candidates who lack basic skills will result in greater costs in onboarding, training, and development. Employee engagement and retention may suffer because new hires do not fit or are unable to perform the job. Succession planning will in turn be difficult because poor employee quality and high turnover leave a shallow pool from which to draw. In this manner, effective talent acquisition is a tide that raises all ships; every talent management function benefits from the more effective acquisition of talent. This is why talent acquisition is presented in the center of the talent management framework.

What may be less apparent in Figure 6.1 is how talent acquisition both shapes and is shaped by *global* talent management. The tensions between the MNC and subsidiaries, and the tensions between universalistic (ethnocentric) and locally customized (polycentric) talent management approaches, influence the manner in which talent acquisition is employed. Difficulty attracting sufficient talent in some locations may require the local customization of development practices in that region. Large companies with strong brand awareness may apply ethnocentric staffing practices, but then use the cost savings by enabling local customization in onboarding and retention. Firms operating in countries with laws making it difficult to terminate underperforming employees will require much more rigorous selection practices than firms in countries with lower separation constraints. That is, if it is difficult to terminate a poorly performing employee, then extra time and expense is justified in pre-hire assessment to minimize the chances of being in that position. Global succession planning impacts talent acquisition, as gaps in succession plans may need to be addressed through external hiring.

There is a fine balance that companies with strong ethical frameworks strive to operate within. When Shell was conducting broad staff reduction programs—for example in the face of a protracted downturn in crude oil prices as occurred at the turn of the century—the actual execution would be constrained and designed to be done within local legal frameworks. But there were very clear guidelines that staff should be treated in accordance with the three values espoused by the company—honesty, integrity and respect for people. As would be described to individuals who questioned this rigorous process: "Just because you *can* do something does not mean you *should* do it."

Thus, the talent management framework presented in Figure 6.1 is very much a system of components that must be aligned to balance global-local tensions. The components are interrelated, which means that changes in one of the components will produce changes in the other components. In this manner all of the components are important. But there are never enough resources to improve all components, so the question is, which component will produce the greatest overall benefit (e.g., return on improved performance; Boudreau, 2010)? In our experience, most of these questions cannot be answered without first considering the role of talent acquisition. And often it is improvements in talent acquisition that generate the greatest overall benefit, because the benefits of all other talent management activities are necessarily constrained by the quality of the talent acquired.

Roger had been working with his team to learn much more about other countries (e.g., their histories, cultures, values, etc.) and the HR leaders within those countries, as well as different functional HR leaders. "How stupid of me to not have thought of this before!" he thought over and over. Roger's cultural orientation was very Western and individualistic, and, despite the initial research he had done on the cultures that he was about to encounter before embarking from England, in the hectic frenzy of activity in which he found himself he had forgotten much of what he had read and not fully appreciated the critical importance of building consensus with HR leaders outside of his own functional area. "How ironic that my success means collective failure" he thought. But he also realized that understanding, respecting, and working with the local cultures was central to achieving business goals and that his own UK culture was not the only valid culture—all cultures have legitimacy. So he continued his efforts and sought ways to connect his functional focus on regional talent acquisition to the other functional areas of HR. He posed a series of questions to his team:

1. How can success in staffing not contribute to success in other HR functions? On the surface this didn't make sense. So he either didn't understand the system, or he needed to better explain how success in staffing created success in other functional areas of HR.
2. How can he get buy-in from other HR leaders? The business case for his staffing plan was already accepted by line managers; but he had not considered making an "HR case" for other HR leaders. How might he do that?
3. How do cultural differences in the countries and region shape the way staffing connects to the other talent management activities? He knew the S.E. Asian countries were very different both from the UK and between themselves—many mistakes had taught him that lesson the hard way—but surely there must be some common talent management principles—what were they?
4. What data do these other leaders have that is different, greater or less, than mine? What assumptions are they making based on their data that are leading toward a behavior that, from Roger's view, was wrong?

TALENT ACQUISITION'S ROLE IN GLOBAL TALENT MANAGEMENT

We now examine how talent acquisition shapes the nature and implementation of related global talent management activities.

Onboarding and Socialization

Onboarding and socialization refer to how new employees experience the organization, job, and coworkers, and the practices that organizations can implement to enhance this experience (Bauer & Erdogan, 2014). Such practices may include new employee orientation programs, employee handbooks, mentoring/buddy systems, goal setting, training, and career counseling (Bauer, 2011; Bauer, Bodner, Erdogan, Truxillo, & Tucker, 2007). Onboarding is all about shrinking the time to productivity and engagement. Firms that better onboard and socialize employees often see higher engagement, retention, performance, and even profit growth (Bauer, 2011; Laurano, 2013). Onboarding and socialization take on even greater significance in global firms when employees are not only joining the firm but also coming from a country and culture different from the parent company. Note this is restricted to those employees who actually expatriate, but other types of culture shock can happen even when a person stays within one's own culture but works for an MNC from a different culture. For example, if you are from Australia and working in Australia for a company that is headquartered in Holland, you will probably experience some aspects of HR policy that are rooted in either Dutch law or at least the Dutch culture. Your culture shock may not be as extreme as someone actually moving to Holland, but certain aspects of the that culture will impact you and make take you unawares. Thus, talent acquisition contributes to onboarding and socialization in a number of important ways:

- **Greater fit**. Chapter 4 explained how greater cultural, company, job, and coworker fit contribute to greater performance, satisfaction, and retention. Effective recruitment and selection enhances fit by ensuring the right people are hired for the right place and the right time. Use of realistic job previews (RJPs), for example, often helps reduce early turnover. Onboarding supports this by furthering indoctrination into the company's culture and values, its policies and procedures, and how things get done.
- **Better performance and less turnover**. Effective talent acquisition ensures new employees will have the necessary talent to perform the job on Day 1. An inability to perform is linked to turnover. Given that new hires have less attachment to the organization, it stands to reason that turnover is often higher among new external hires, especially when there are few

barriers to mobility (e.g., as is often the case with lower-level hires) (Cappelli & Keller, 2014; Holtom, Mitchell, Lee, & Eberly, 2008). Effective onboarding ensures that new hires know what is expected of them (e.g., that they have clear goals), have access to whatever training may be needed, and receive regular feedback early on in their tenure.

- **Faster assimilation**. Many jobs require collaboration and teamwork. Such work often also requires knowledge sharing and a great deal of trust and common understanding. Hence, it is vital for employees to become assimilated into the workgroup as quickly as possible. Effective onboarding increases the chances that new employees will be accepted and assimilated into the group by ensuring early and frequent contact with key stakeholders and others impacting and impacted by the new hire.
- **Greater returns**. Because effective talent acquisition ensures greater fit, performance, and assimilation, firms can often spend less on onboarding and socialization, and recover greater returns on the resources that are spent. Fewer training courses may be required, assimilation may be faster, and better performance may happen more quickly.

Promotion and Internal Mobility

To a great extent the topics covered thus far have been focused on external hiring. But not all jobs are filled externally, as in many cases a current employee is promoted or moved laterally to fill an open job. How many jobs are filled internally? It is a surprisingly difficult question to answer. Surveys of employers have shown internal moves to fill between 26% (Krell, 2015) and 42% (Crispin & Mehler, 2013) of openings. Regardless of the percentage, current employees are a significant source of candidates for most open jobs. Many organizations have a "promote from within" policy. These policies take many forms but usually require that a job be posted and made available exclusively to interested internal candidates for a period of time (e.g., 2 weeks). Only after the exclusivity period is over and no qualified internal candidates have applied can the search be extended to include external candidates. There are a number of potential benefits to promoting from within.

- **Greater familiarity**. Internal candidates are already familiar and comfortable with the company's culture, know its policies and procedures, and have an established support network. It is also usually much cheaper and faster to promote from within than to source candidates externally. The information on internal candidates is more comprehensive than that available on external candidates (e.g., the organization has direct evidence of past performance and accomplishments), such that fewer hiring mistakes should be made.
- **Internal hires perform better and cost less**. Promotion from within is part of HPWSs, which have been to be related to measures of organizational performance (Huselid, 1995). In a recent study, Bidwell (2011) compared employees hired from the outside with those promoted from within and found that, while the external hires were paid 18% more than the internal promotees, their performance was significantly lower and they were 61% more likely to get fired. Even though the external hires had more experience and education, it took 2 years after hire before their performance caught up to the internal promotees. Clearly, in this study there was a significant period of time required for external hires to learn how to get things done within the company.
- **Enhances engagement**. Promotion from within signals career prospects to the promoted person and is generally a boost to engagement throughout the organization. High-performing employees are likely to leave an organization if they do not feel their performance will lead to desired advancement opportunities (Cappelli, 2008).

- **Potential risks.** There are, of course, some potential downsides to sourcing candidates internally. For example, filling jobs internally promotes a degree of "sameness," which may choke off innovation and perpetuate demographic imbalances. The need for rapid change (e.g., new skill sets) or rapid growth, of course, are ill-suited to a promote-from-within strategy. In such cases, the organization is almost always compelled to hire from the outside. Furthermore, unless the organization is shrinking, promotion from within creates a ripple effect. For example, if person A is promoted, and person B is promoted to fill A's job, and person C is promoted to fill B's job, eventually someone is hired from the outside to backfill C's job. Such ripple effects can result in several people learning a new job and temporarily lower productivity in several places within the organization.

Succession Planning

Succession planning is a form of workforce planning, and is primarily concerned with ensuring there is a sufficient pipeline of talent to implement the firm's strategy, now and in the future. In the past, succession planning often involved identifying and grooming multiple back-ups for key jobs (e.g., resulting in replacement charts). Greater competition and economic uncertainty have made long-range succession planning all but impossible (Cappelli, 2008). Indeed, Cappelli and Keller (2014) note that such uncertainty is one reason why external hiring has grown so much in the last few decades. This should not be taken as evidence for eliminating succession planning. Rather, it means that firms must think strategically and flexibly about talent. It means they must consider talent as a resource that can be leveraged, bundled, or rebundled for different purposes. It means firms should strive to achieve a reasonable balance of external hiring and internal promotion and development. It often requires firms to hire as much for potential as to hire for specific job competencies. This, of course, is really what campus hiring is focused on primarily. Even though an individual may have a degree in chemical engineering, which forms the basic set of knowledge necessary to work in a continuous process plant, a company will have specific pieces of equipment, processes, and procedures that the new graduate will need to learn. And with a graduate you are not intending them to stay at the entry level for their entire career—you are betting on the future development and upward mobility. The best way to prepare for an uncertain world is to ensure you have competent and engaged employees who are willing and able to grow and adapt to a variety of roles. Talent acquisition thus contributes to succession planning in multiple ways:

- **Hiring for potential.** Hiring for potential means looking beyond immediate job responsibilities and thinking about future jobs the employee may be able to fill. Effective talent acquisition can ensure candidates will fit the company and culture, which facilitates retention and engagement. New employees will already have the mindset that they will be deployed into different assignments, and hence selection can ensure they have the dispositional qualities to be open to different assignments and the abilities needed to perform well in different contexts.
- **Greater willingness to work with people from other cultures.** In global firms, there is a need to ensure new hires will fit the country culture as well as the company culture. This will often require working with people who are from different cultures, who speak different languages, and hold different values. Effective talent acquisition increases the odds that new hires will fit and assimilate to the new cultures.
- **Creating global talent pools.** As mentioned, uncertainty makes long-range planning difficult, and elaborate succession plans with carefully developed promotional ladders are largely a thing of the past (Cappelli & Keller, 2014). In response, firms may often use talent pools, which are groups of people within

job families or geographic regions that have been identified as having growth potential. Chapter 4 talked about talent hubs in finance and technology, and the benefit of such hubs is that they allow a firm to draw from them when the talent is in demand. Because talent is unequally distributed around the globe (chapters 1 and 2), firms that develop talent pools should be more adaptive than those that do not (Boudreau & Ramstad, 2007). Effective talent acquisition contributes to such pools by ensuring a sufficient quality and quantity of generic talent that can quickly be deployed for different purposes. Firms can use selection information (e.g., assessment scores) as a basis for assessing the depth of their talent pools. Although external hiring has increased considerably, it is clear that most firms seek to strike a reasonable balance between external and internal hiring (see Cappelli, 2008, for much greater depth around this topic).

- **More accurate placement and classification**. Talent acquisition is often used not only for hiring but also for placement and classification (the latter term often being reserved for military settings). That is, talent acquisition is as much about finding the right job and place for an external applicant or internal promotee as it is about deciding whom to hire or promote in the first place. For example, firms often use a single test battery for a wide variety of jobs. By weighting the traits differently to reflect differences in the requirements across the various jobs, the same assessment can be used to predict success in several jobs. A single assessment can be used as a means to determine whom to hire and which job to place them in. Sometimes candidates may be assessed in one of three classifications—Ready Now, Ready 3–5 years, and Candidate Long-Term. Shell has, since the 1960s, used a formalized assessment of current estimate of potential (CEP), which is discussed on an annual or biannual basis with individuals and is a cornerstone of their succession and development planning activities.

Development

Employee development is about creating the talent needed in the future by training existing employees and exposing them to different job challenges (Noe, Clarke, & Klein, 2014). One of the most-often-cited studies on the topic (Lombardo & Eichinger, 1996) found that employee development occurs primarily via on-the-job experiences (70%), interactions with others (20%), and formal training (10%). As such, development is necessarily tied closely to succession planning, and many of the benefits of talent acquisition described in the prior section apply here as well. Yet in contrast to succession planning, development is often seen as an alternative to external hiring, probably because it is exclusively focused on current employees (i.e., build versus buy). In our experience, this is a false dichotomy, as most organizations use a blend of external hires ("buy") and internal promotees ("build"). Talent acquisition and development should be closely linked, as occurs, for example, when an organization acquires skills from the external labor market it does not have the time or expertise to build. Some additional benefits include:

- **Greater willingness for development.** Most employees want their organization to provide opportunities for development, and this is particularly true for high-performing employees (Noe et al., 2014). Development activities involve internal training, job rotations, stretch and special assignments, and mentoring programs as well as external education and coursework. But not all employees desire such forms of development, and not all employees will benefit equally from the different development programs. Hence, the same talent acquisition activities that drive selection can also be used to both identify who has a desire for development and who will benefit the most from different types of development programs.
- **Greater willingness for international assignments.** One of the greatest challenges in global firms is having talent willing to relocate to different countries and cultures. These assignments

are frequently short, perhaps spanning only a couple months to a couple of years. They present incredible opportunities, but can also be quite disruptive from the employee's perspective (e.g., buying and selling a house; children and school) and are incredibly expensive for the organization. The cost of an expatriate (an employee sent from his/her home country to live and work in another) is usually much greater than what the same employee would cost in his or her home country. Yet in some global companies these may represent a substantial element of development, even a rite of passage, for employees who are seen as having the potential to lead large segments of a company. Talent acquisition can be used to better identify those willing and able to accept such international assignments (e.g., recruitment to ensure those who accept are willing to relocate, and selection can be used to ensure candidates have the cultural flexibility to successfully adapt to new cultures).

- **Increase the return on development investments.** Talent acquisition can help identify who wants development, the types of development needed, and the timing of development. This in turn enables more efficiency in terms of the resources devoted to development and the ROI that will come from such investments. For example, development programs can be customized to the individual employee based on his or her assessment scores captured during the selection process. Similarly, firms can use these same scores to identify who most needs development so that developmental costs and activities can be sequenced to optimize efficiency.

Engagement

Engagement refers to the constellation of traits, psychological states, and behaviors that employees have toward their work and organization (Macey & Schneider, 2008). Engagement is multifaceted and is expressed in multiple ways that include satisfaction, affect, commitment, and support. Engagement is also closely related to motivation, involvement, proactivity, and adaptability. Hence engagement is a very useful concept because it captures the extent to which employees will exert discretionary effort toward achieving the organization's goals and strategy. In fact, research suggests there is considerable variability in the extent to which employees are engaged, and firms with more engaged employees significantly outperform those with less engaged employees (Harter, Schmidt, Asplund, Killham, & Agrawal, 2010). Effective selection can enhance employee engagement in multiple ways:

- **Increasing likelihood of engagement**. Employee engagement has a dispositional component, which means that some people are naturally more likely to be engaged than others. Recruitment and selection practices can be used to identify those individuals who not only may have the ability to do the work but also are likely to have the personality characteristics that will predispose them to be more satisfied and committed. Such dispositionally engaged employees may be particularly important in global firms that deploy personnel in many different types of international and national assignments.
- **Greater motivation and citizenship**. Employees who are more engaged tend to be more motivated. More motivated employees exhibit better job performance, of course, but they are also more likely to exhibit greater discretionary effort (e.g., citizenship performance; chapters 3 and 5), support the organization both on and off the job, and collaborate and share knowledge. Thus, hiring those disposed to be engaged directly and indirectly contributes to individual and group performance.
- **Greater innovation and creativity**. More satisfied, happier, and motivated employees tend to generate more innovative and creative ideas (Zhou & Hoever, 2014). Innovation requires having the persistence to overcome disappointments and mistakes, and those with greater engagement are likely to

demonstrate such perseverance. Consequently, recruitment can be used to attract those with a preference for less-structured or certain work contexts, and selection can be used to identify people with the characteristics linked to innovation and creativity.

Retention

It has long been known that turnover is costly, but recent research suggests that the costs are even greater than prior studies estimated (Shaw, 2011). One reason is because when people work in groups, a turnover event disrupts the entire group's functioning. Further, turnover can deplete the quantity and quality of collective talent resources (Call, Nyberg, Ployhart, & Weekley, 2015). Although some turnover is necessary and perhaps even desirable, most turnover is negative and should be avoided. As the function that attracts and selects candidates, talent acquisition should bear a considerable amount of accountability for turnover. Stated differently, effective talent acquisition should contribute to lower levels of undesirable turnover. We covered turnover briefly in Chapters 2, 4, and 5, and here consider other features of turnover that may be influenced by talent acquisition.

- **Increases embeddedness.** Embeddedness refers to why people stay in their jobs, and is largely based on three elements: *links* to people and places, *fit* with companies and regions, and *sacrifices* that may occur by leaving (Lee, Burch, & Mitchell, 2014). For example, people become embedded because they like where they live; their spouse would not find similar employment elsewhere, and their children want to finish school with their friends. Clearly, more effective recruitment will contribute to embeddedness by enhancing fit, ensuring the region satisfies the person's preferences (wants), and providing a realistic preview of the job. Alternatively, turnover can be reduced if selection processes identify candidates already embedded to some degree (e.g., strong ties to the community).

- **Enhances functional turnover**. Not all turnover is dysfunctional (Dalton, Todor, & Krackhardt, 1982). Functional turnover occurs when those who do not fit, or cannot perform the work, decide to leave of their own volition. Often, the replacement employee turns out to be better than the one who left. Some amount of turnover is healthy and even desirable if it contributes to new skills and fresh perspectives. For example, GE's former CEO Jack Welch believed such an approach was necessary and sought to remove the bottom-performing employees from the firm—and this process was mandated every year. Talent acquisition contributes to functional turnover by ensuring that replacement hires are of sufficient quality and quantity to enhance firm performance over the long run.
- **Reduces voluntary turnover**. There are two types of turnover (see also chapters 4 and 5). *Voluntary turnover* occurs when the employee decides to leave; *involuntary turnover* occurs when the employee must be fired or let go (and hence was unwilling to quit or leave). Involuntary turnover is problematic because it requires a termination, but may ultimately be functional if the replacement performs better. Voluntary turnover is usually more costly because it is undesirable and often dysfunctional. Both types of turnover can represent a failure in talent acquisition if they occur because of unmet expectations, poor fit, or lack of ability. By ensuring these do not happen, effective talent acquisition should reduce the possibility of involuntary turnover. Voluntary turnover can be difficult to impact, because people may leave for many reasons that have nothing to do with the job or their ability to perform it (e.g., better pay, desire to live in a warmer climate, going back to school, spouse gets relocated, etc.). Because voluntary turnover involves a choice by the employee, talent acquisition impacts it indirectly (e.g., use of

realistic job previews in recruitment or culture-fit assessments in selection).

Conclusion

Talent acquisition is fundamental to talent management and all subsequent HR practices, because it is the function responsible for the attraction and selection of talent. That is, talent acquisition is in many ways the input to all subsequent talent management outputs. This means that if talent acquisition is done more effectively, the firm will have employees who are more able to do their jobs, are more flexible and adaptive, are more willing to accept and benefit from developmental and promotional opportunities, and are motivated, engaged, and likely to stay. Talent acquisition is the function that will often generate the greatest return on investment and result in the largest improvements in performance. Given that talent resources are unevenly distributed geographically, these effects are even greater in global organizations.

The regional talent acquisition team had worked hard and done their due diligence. They identified the leader of each HR function, responsibilities and backgrounds of the country HR leaders, and informal leaders who had deep respect and understand HR throughout the region. Roger learned a lot in this process, and he realized this was something he should have done much earlier.

Their approach would be to meet with each relevant leader and try to understand his or her challenges and opportunities. Roger would then search for ways in which he could help them through linking talent acquisition to their efforts. "Help me help you" was the theme that would run through each meeting. But this was extremely time consuming. Roger vetted the key points with his team and supporters from different parts of the organization. He took Ibrahim Abdul

Hassan up on his offer to help, and made several trips to his office for guidance and counsel.

It was through these meetings that Ibrahim began to develop a deep respect for Roger. He initially thought Roger was competent but clumsy; typical of many Westerners who accepted assignments to S.E. Asia. But there was something different about Roger. He seemed to truly want to understand the cultural nuances of the region, and he truly valued cultural diversity and the multitude of perspectives that it provides. Roger did not take the easy route; he took the slower and more effortful route but the one that was right. Ibrahim slowly came to be a huge supporter of Roger—indeed a real friendship began to develop between the two of them, which was to survive Roger's assignment in Asia.

Roger sensed this shift in Ibrahim's thinking, but he knew the focus must remain on understanding the other HR leaders. And so after their last meeting, he left Ibrahim's office with quiet resolve to meet the first of many HR leaders over the coming weeks. He was going to make things right. And the advice of Ibrahim rang in his ears:

1. Greet each leader according the conventions of his/her local culture.
2. Understand the challenges each leader faces, from his or her perspective. Don't assume; ask. Don't tell; question. Listen and learn.
3. Search for mutual understanding. Look for shared challenges and concerns. But always frame your suggestions from their perspective.
4. Ask them how you can help. Don't boast about the success of the talent acquisition strategy, but instead offer facts about how talent acquisition can address their concerns. Let the data do the talking.
5. Restate the issues and the steps moving forward.
6. Come to consensus on how you will define success for your collaboration. Define deliverables, timelines, and resources.

7. Follow up the conversation with an e-mail that restates these points and ask for assurance they are correct. Treat this e-mail as your contract.

This sequence of questions flowed through Roger's head as he looked out the window of the taxi taking him to the airport. Soon a confidence began to grow within him. "I can do this," he thought. And so his mind turned to other matters and he began to draft a quick e-mail to his friend Andrew back in the UK. "Hey Andrew, guess what? I have some interesting news."

REFERENCES

CHAPTER 1

Ackerman, P. L., & Heggestad, E. D. (1997). Intelligence, personality, and interests: Evidence for overlapping traits. *Psychological Bulletin, 121,* 219–245.

Altbach, P. G., Reisberg, L., & Rumbley, L. E. (2009). Trends in global higher education: Tracking an academic revolution: A report prepared for the UNESCO 2009 World Conference on Higher Education. In *Executive summary*. Paris, France: United Nations Educational, Scientific and Culture Organization.

Barney, J. B. (1991). Firm resources and sustained competitive advantage. *Journal of Management, 17,* 99–120.

Barrick, M. R., & Mount, M. K. (2012). Nature and use of personality in selection. In N. Schmitt (Ed.), *The Oxford Handbook of Personnel Assessment and Selection* (pp. 225–251). New York, NY: Oxford University Press.

Becker, G. S. (1964). *Human capital.* New York, NY: Columbia University Press.

Bhattacharya, M., Gibson, D. E., & Doty, D. D. (2005). The effects of flexibility in employee skills, employee behaviors, and human resource practices on firm performance. *Journal of Management, 31,* 622–640.

Boudreau, J. W., & Ramstad, P. M. (2007). *Beyond HR: The new science of human capital.* Boston, MA: Harvard Business School Press.

Brand channel: Always branding. Always on. (n.d.). Retrieved from www.brandchannel.com

Boxall, P. F. (1996). The strategic HRM debate and the resource-based view of the firm. *Human Resource Management Journal, 6,* 59–75.

Campbell, B. A., Coff, R., & Kryscynski, D. (2012). Rethinking sustained competitive advantage from human capital. *Academy of Management Review, 37,* 376–395.

Cappelli, P. (2008). *Talent on Demand.* Boston, MA: Harvard Business School Press.

Cappelli, P., & Crocker-Hefter, A. (1996). Distinctive human resources are firms' core competencies. *Organizational Dynamics, 24,* 7–21.

Cappelli, P., & Keller, J. R. (2014). Talent management: Conceptual approaches and practical challenges. *Annual Review of Organizational Psychology and Organizational Behavior, 1,* 305–331.Chaloff, J., & Lemaitre, G. (2009). *Managing highly skilled migration: A comparative analysis of migration policies and challenges in OECD countries.* OECD Social, Employment and Migration Working Papers No. 79.

Coff, R. W. (1997). Human assets and management dilemmas: Coping with hazards on the road to resource-based theory. *Academy of Management Review, 22,* 374–402.

Coff, R. W. (1999). When competitive advantage doesn't lead to performance: The resource-based view and stakeholder bargaining power. *Organization Science, 10,* 119–133.

Crook, T. R., Ketchen, D. J., Jr., Combs, J. G., & Todd, S. Y. (2008). Strategic resources and performance: A meta-analysis. *Strategic Management Journal, 29,* 1141–1154.

Deloitte. (2014). *Global human capital trends 2014: Engaging the 21st-century workforce.* Retrieved from http://www2.deloitte.com/global/en/pages/human-capital/articles/human-capital-trends-2014.html

Deloitte. (2014). *The talent paradox: A 21st century talent and leadership agenda.* Retrieved from http://www.deloitte.com/assets/Dcom-UnitedStates/Local%20Content/Articles/Deloitte%20University/ONtalent_Selected_Articles_From_DeloitteReview.pdf

Gorney, C. (2014, January). Far from home. *National Geographic,* 70.

Hagel, J. I., Brown, J. S., & Davidson, L. (2009). *Measuring the forces of long-term change: The 2009 shift index.* Deloitte Center for the Edge.

Hancock, J. I., Allen, D. G., Bosco, F. A., McDaniel, K. M., & Pierce, C. A. (2013). Meta-analytic review of employee turnover as a predictor of firm performance. *Journal of Management, 39,* 573–603.

Harter, J. K., Schmidt, F. L., Asplund, J. W., Killham, E. A., & Agrawal, S. (2010). Causal impact of employee work perceptions on the bottom line of organizations. *Perspectives on Psychological Science, 5,* 378–389.

Hofstede, G. (2001). *Culture's consequences: Comparing values, behaviors, institutions and organizations across nations* (2nd ed.). Thousand Oaks, CA: SAGE.

Holland, J. L. (1997). *Making vocational choices: A theory of vocational personalities and work environments* (3rd ed.). Odessa, FL: PAR.

House, R. J., Hanges, P. J., Javidan, M., Dorfman, P. W., & Gupta, V. (Eds.). (2004). *Culture, leadership, and organizations: The GLOBE study of 62 societies.* Thousand Oaks, CA: Sage.

Huselid, M. A. (1995). The impact of human resource management practices on turnover, productivity, and corporate financial performance. *Academy of Management Journal, 38,* 635–672.

Huselid, M. A., Beatty, R. W., & Becker, B. E. (2005, December). "A players" or "A positions?": The strategic logic of workforce management. *Harvard Business Review, 38,* 110–117.

International comparison of math, reading, and science skills among 15-year-olds. (n.d.). Retrieved from http://www.infoplease.com/ipa/A0923110.html#ixzz3S2uP9wxV

Jiang, K., Lepak, D. P., Hu, J., & Baer, J. (2012). How does human resource management influence organizational outcomes? A meta-analytic investigation of mediating mechanisms. *Academy of Management Journal, 55,* 1264–1294.

Kim, Y., & Ployhart, R. E. (2014). The effects of staffing and training on firm productivity and profit growth before, during, and after the great recession. *Journal of Applied Psychology, 99,* 361–389.

Kimes, M. (2009). *Keeping your senior staffers.* Retrieved from http://archive.fortune.com/2009/07/10/news/companies/basf_retaining_senior_engineers.fortune/index.htm?section=money_latest

Lepak, D. P., Liao, H., Chung, Y., & Harden, E. E. (2006). A conceptual review of human resource management systems in strategic human resource management research. *Research in Personnel and Human Resources Management, 25,* 217–271.

Manpower Group. (2011). *Talentism.* Retrieved from http://www.manpowergroup.com/wps/wcm/connect/manpowergroup-en/home/thought-leadership/human-age/Talentism+_new/#.VOPTK0go4y8

Manpower Group. (2015). *Talent shortage survey.* Retrieved from http://www.manpowergroup.com/wps/wcm/connect/408f7067-ba9c-4c98-b0ec-dca74403a802/2015_Talent_Shortage_Survey-lo_res.pdf?MOD=AJPERES&ContentCache=NONE

Michaels, E., Handfield-Jones, H., & Axelrod, B. (2001). *The war for talent.* Boston, MA: Harvard Business School Press.

Myors, B., Lievens, F., Schollaert, E., Van Hoye, G., Cronshaw, S. F., Mladinic, A., . . . Sackett, P. R. (2008). International perspectives on the legal environment for selection. *Industrial and Organizational Psychology, 1,* 206–246.

Park, T. Y., & Shaw, J. D. (2013). Turnover rates and organizational performance: A meta-analysis. *Journal of Applied Psychology, 2,* 268–309.

Penrose, E. T. (1959). *The theory of the growth of the firm.* Oxford, England: Basil Blackwell.

Ployhart, R. E., Nyberg, A. J., Reilly, G., & Maltarich, M. A. (2014). Human capital is dead; long live human capital resources! *Journal of Management, 40,* 371–398.

Ployhart, R. E., Van Iddekinge, C. H., & MacKenzie, W. I. (2011). Acquiring and developing human capital in service contexts: The interconnectedness of human capital resources. *Academy of Management Journal, 54,* 353–368.

PwC. (2014). *Key findings from 18th annual global CEO survey.* Retrieved from http://www.pwc.com/gx/en/ceo-survey/2015/key-findings/index.jhtml

Rabl, T., Jayasinghe, M., Gerhart, B., & Kühlmann, T. M. (2014). A meta-analysis of country differences in the high-performance work system–business performance relationship: The roles of national culture and managerial discretion. *Journal of Applied Psychology, 99,* 1011–1041.

Schmidt, F., & Hunter, J. (2004). General mental ability in the world of work: Occupational attainment and job performance. *Journal of Personality and Social Psychology, 86,* 162–173.

Stephan, M., Vahdat, H., Walkinshaw, H., & Walsh, B. (2014). *Global Human Capital Trends 2014: Engaging the 21st-century workforce.* Deloitte University Press.

Wernerfelt, B. (1984). A resource-based view of the firm. *Strategic Management Journal, 5,* 171–180.

CHAPTER 2

Banerjee, N. (2002). Shell Oil to acquire Pennzoil. *New York Times.* Retrieved from http://www.nytimes.com/2002/03/26/business/shell-oil-to-acquire-pennzoil.html

Barney, J. B. (1991). Firm resources and sustained competitive advantage. *Journal of Management, 17,* 99–120.

Bartlett, C. A., & Ghoshal, S. (1989). *Managing across borders: The transnational solution.* Boston, MA: Harvard Business School Press.

Becker, B. E., Huselid, M. A. & Ulrich, D. (2001). *The HR scorecard: Linking people, strategy, and performance.* Boston, MA: Harvard Business School Press.

Brown, D., & Fersht, P. (2014). *Executive report: The State of Services and Outsourcing in 2014.* KPMG and HfS Research.

Caligiuri, P., & Paul, K. B. (2010). *Selection in multinational organizations.* New York, NY: Routledge/Taylor & Francis.

Cappelli, P., & Keller, J. R. (2013). Classifying work in the new economy. *Academy of Management Review, 38,* 575–596.

Colakoglu, S., Tarique, I., & Caligiuri, P. M. (2009). Towards a conceptual framework for the relationship between subsidiary staffing strategy and subsidiary performance. *International Journal of Human Resource Management, 6,* 1288–1305.

Collings, D., & Scullion, H. (2006). Approaches to international staffing. In H. Scullion & D. G. Collings (Eds.), *Global Staffing* (pp. 17–38). London: Routledge.

Dierickx, I., & Cool, K. (1989). Asset stock accumulation and sustainability of competitive advantage. *Management Science, 35,* 1504–1511.

Freelancers Union & Elance-oDesk. (2014). *Freelancing in America: A national survey of the new workforce.* http://fu-web-storage-prod.s3.amazonaws.com/content/filer_public/c2/06/c2065a8a-7f00-46db-915a-2122965df7d9/fu_freelancinginamericareport_v3-rgb.pdf

Guion, R. M. (1961). Criterion measurement and personnel judgments. *Personnel Psychology, 14,* 141–149.

Guthridge, M., Lawson, E., & Komm, A. (2008). Making talent a strategic priority. *The McKinsey Quarterly, 1,* 49–59.

Heenan, D. A., & Perlmutter, H. V. (1979). *Multinational organization development.* Reading, MA: Pearson Addison Wesley.

Horowitz, S. (2013). The freelance movement has gone global. *Freelancers Union.* https://www.freelancersunion.org/blog/ dispatches/2013/12/19/q-joel-dullroy-freelancers-mvmnt-germany/

Hunter, J. E. (1983). A causal analysis of cognitive ability, job knowledge, job performance, and supervisory ratings. In F. Landy, S. Zedeck, & J. Cleveland (Eds.), *Performance measurement and theory* (pp. 257–266). Hillsdale, NJ: Erlbaum.

Kaplan, R. S., & Norton, D. P. (2004). *Strategy maps: Converting intangible assets into tangible outcomes.* Boston, MA: Harvard Business School Press.

Kim, Y., & Ployhart, R. E. (2014). The effects of staffing and training on firm productivity and profit growth before, during, and after the great recession. *Journal of Applied Psychology, 99,* 361–389.

Pan, K. Y. (2012). *Why strategic alliances are now in fashion.* http://www.kpmg.de/Topics/32199.htm

Lepak, D. P., Liao, H., Chung, Y., & Harden, E. E. (2006). A conceptual review of human resource management systems in strategic human resource management research. *Research in Personnel and Human Resources Management, 25,* 217–271.

Lepak, D. P., & Snell, S. A. (1999). The human resource architecture: Toward a theory of human capital allocation and development. *Academy of Management Review, 24,* 31–48.

Nyberg, A. J., & Ployhart, R. E. (2013). Context-emergent turnover theory: A theory of collective turnover. *Academy of Management Review, 38,* 109–131.

Perlmutter, H. (1969). The tortuous evolution of the multinational corporation. *Columbia World Journal of Business, 4,* 9–18.

Peteraf, M. A., & Barney, J. B. (2003). Unraveling the resource-based tangle. *Managerial and Decision Economics, 24,* 309–323.

Prahalad, C. K., & Doz., Y. L. (1987). *The multinational mission: Balancing local demands and global vision.* New York, NY: The Free Press.

Ployhart, R. E., & Moliterno, T. P. (2011). Emergence of the human capital resource: a multilevel model. *Academy of Management Review, 36,* 127–150.

Ployhart, R. E., Nyberg, A. J., Reilly, G., & Maltarich, M. A. (2014). Human capital is dead; long live human capital resources! *Journal of Management, 40,* 371–398.

Ployhart, R. E., Schneider, B., & Schmitt, N. (2006). *Staffing organizations: Contemporary practice and research.* Mahwah, NJ: Erlbaum.

Ployhart, R. E., & Weekley, J. A. (2014). Recruitment and selection in the global organization. In D. G. Collings, G. Wood, & P. Caligiuri (Eds.), *The Routledge Companion to International Human Resource Management.* Routledge.

Rasch, R. (2014). *Your best workers may not be your employees: A global study of independent workers.* Somers, NY: IBM Corporation.

Statista. (2014). http://www.statista.com/statistics/189788/global-outsourcing-market-size/

Ulrich, D., Younger, J., Brockbank, W., & Ulrich, M. (2012). *HR from the outside in: Six competencies for the future of human resources.* New York, NY: McGraw-Hill.

Wall Street Journal. (2015, June 1). Next comes the on-demand sales force.

CHAPTER 3

Bliese, P. D. (2000). Within-group agreement, non-independence, and reliability: Implications for data aggregation and analysis. In K. Klein & S. W. J. Kozlowski (Eds.), *Multilevel theory, research, and methods in organizations: Foundations, extensions, and new directions* (pp. 349–381). San Francisco, CA: Jossey-Bass.

Boudreau, J. W., & Ramstad, P. M. (2007). *Beyond HR: The new science of human capital.* Boston, MA: Harvard Business School Press.

Brislin, R. W. (1970). Back-translation for cross-cultural research. *Journal of Cross-Cultural Psychology, 1,* 185–216.

Cascio, W., & Boudreau, J. (2008). *Investing in people: Financial impact of human resource initiatives.* Upper Saddle River, NJ: Pearson Education.

Fulmer, I. S., & Ployhart, R. E. (2014). "Our most important asset": A multidisciplinary/multilevel review of human capital valuation for research and practice. *Journal of Management, 40,* 161–192.

Guzzo, R. A., Fink, A., King, E., Tonidandel, S., & Landis, R. (2015). Big data recommendations for industrial-organizational psychology. *Industrial and Organizational Psychology, 8,* 491–508.

Kim, Y., & Ployhart, R. E. (2014). The effects of staffing and training on firm productivity and profit growth before, during, and after the great recession. *Journal of Applied Psychology, 99,* 361–389.

Kuncel, N. R., Klieger, D. M., Connelly, B. S., & Ones, D. S. (2013). Mechanical versus clinical data combination in selection and admissions decisions: A meta-analysis. *Journal of Applied Psychology, 98,* 1060–1072.

LeBreton, J. M., & Senter, J. L. (2008). Answers to twenty questions about interrater reliability and interrater agreement. *Organizational Research Methods, 11,* 815–852.

Little, T. D. (1997). Mean and covariance structures (MACS) analyses of cross-cultural data: Practical and theoretical issues. *Multivariate Behavioral Research, 32,* 53–76.

Ployhart, R. E. (2015). The reluctant HR Champion. In D. Ulrich, W. A. Schiemann, & L. Sartain (Eds.), *The Rise of HR: Wisdom from 73 Thought Leaders.* HRCI. E-Book distributed by HRCI.

Ployhart, R. E., Van Iddekinge, C. H., & MacKenzie, W. I. (2011). Acquiring and developing human capital in service contexts: The interconnectedness of human capital resources. *Academy of Management Journal, 54,* 353–368.

Ployhart, R. E., Weekley, J. A., & Baughman, K. (2006). The structure and function of human capital emergence: A multilevel examination of the attraction-selection-attrition model. *Academy of Management Journal, 49,* 661–677.

Ployhart, R. E., Weekley, J. A., Holtz, B. C., & Kemp, C. F. (2003). Web-based and paper-and-pencil testing of applicants in a proctored setting: Are personality, biodata, and situational judgment tests comparable? *Personnel Psychology, 56,* 733–752.

Ployhart, R. E., Weekley, J. A., & Ramsey, J. (2009). Consequences of human resource stocks and flows: A longitudinal examination of unit service orientation and unit effectiveness. *Academy of Management Journal, 52,* 996–1015.

CHAPTER 4

Barber, A. E. (1998). *Recruiting employees: Individual and organizational perspectives.* Thousand Oaks, CA: Sage.

Barrick, M. R., Mount, M. K., & Gupta, R. (2003). Meta-analysis of the relationship between the five-factor model of personality and Holland's occupational types. *Personnel Psychology, 56,* 45–74.

Baumeister, R. F., Bratslavsky, E., Finkenauer, C., & Vohs, K. D. (2001). Bad is stronger than good. *Review of General Psychology, 5,* 323–370.

Bidwell, M. (2011). Paying more to get less: The effects of external hiring versus internal mobility. *Administrative Science Quarterly, 56,* 369–407.

Cable, D. M., & Judge, T. A. (1996). Person–organization fit, job choice decisions, and organizational entry. *Organizational Behavior and Human Decision Processes, 67,* 294–311.

Caligiuri, P., & Paul, K. B. (2010). *Selection in multinational organizations.* New York, NY: Routledge/Taylor & Francis Group.

CEB. (2015). Employment value proposition (EVP). Retrieved from http://www.executiveboard.com/exbd/human-resources/evp/index.page

Chambers, E. G., Foulon, M., Handfield-Jones, H., Hankin, S., & Michaels, E. (1998). The war for talent. *McKinsey Quarterly, 3*, 1–8.

Chapman, D. S., Uggerslev, K. L., Carroll, S. A., Piasentin, K. A., & Jones, D. A. (2005). Applicant attraction to organizations and job choice: A meta-analytic review of the correlates of recruiting outcomes. *Journal of Applied Psychology, 90*, 928–944.

Charney, M. (2014). The state of candidate experience in 10 statistics. Retrieved from http://recruitingdaily.com/the-state-of-candidate-experience-in-10-statistics

Cober, R. T., Brown, D. J., Keeping, L. M., & Levy, P. E. (2004). Recruitment on the net: How do organizational Web site characteristics influence applicant attraction? *Journal of Management, 30*, 623–646.

Cober, R. T., Brown, D. J., & Levy, P. E. (2004). Form, content, and function: An evaluative methodology for corporate employment Web sites. *Human Resource Management, 43*, 201–218.

Collings, D., & Scullion, H. (2012). *Global Staffing*. London, UK: Routledge.

Economist Intelligence Unit. (2013). *Hot spots 2025: Benchmarking the future competitiveness of cities*. Retrieved from http://www.citigroup.com/citi/citiforcities/pdfs/hotspots2025.pdf

Evans, P., & Lanvin, B. (2013). *The world's most talent competitive countries*. Retrieved from http://knowledge.insead.edu/talent-management/the-worlds-most-talent-competitive-countries-3006

Guo, C., & Miller, J. K. (2010). Guanxi dynamics and entrepreneurial firm creation and development in China. *Management and Organization Review, 6*, 267–291.

Harold, C. M., & Ployhart, R. E. (2008). What do applicants want? Examining changes in attribute judgments over time. *Journal of Occupational and Organizational Psychology, 81*, 191–218.

Hartnell, C. A., Ou, A. Y., & Kinicki, A. (2011). Organizational culture and organizational effectiveness: A meta-analytic investigation of the competing values framework's theoretical suppositions. *Journal of Applied Psychology, 96*, 677–694.

Holland, J. L. (1997). *Making vocational choices: A theory of vocational personalities and work environments* (3rd ed.). Odessa, FL: PAR.

Jobvite. (2014). *Jobvite job seekers nation study: An authoritative survey of the social, mobile job seeker*. Retrieved from http://web.jobvite.com/rs/jobvite/images/2014%20Job%20Seeker%20Survey.pdf

Koncept Analytics. (2015). *Global recruitment market report: 2015 edition*. Retrieved from http://www.konceptanalytics.com/Researchreport/global-recruitment-market-report-2015-edition-185.aspx

Kops, D., & Lyon, J. (2013). The future of finance talent. *Association of Chartered Certified Accountants Report*. Retrieved from http://www.accaglobal.com/us/en/technical-activities/technical-resources-search/2013/april/the-future-of-finance-talent.html

Kristof, A. L. (1996). Person-organization fit: An integrative review of its conceptualizations, measurement, and implications. *Personal Psychology, 49*, 1–48.

Lievens, F., & Highhouse, S. (2003). The relation of instrumental and symbolic attributes to a company's attractiveness as an employer. *Personnel Psychology, 56*, 75–102.

Ma, R., & Allen, D. G. (2009). Recruiting across cultures: A value-based model of recruitment. *Human Resource Management Review, 19,* 334–346.

McPherson, M., Smith-Lovin, L., & Cook, J. (2001). Birds of a feather: Homophily in social networks. *Annual Review of Sociology, 27,* 415–444.

Mitchell, T., Holtom, B., Lee, T., Sablynski, C., Erez, M. (2001). Why people stay: Using job embeddedness to prdeict voluntary turnover. *Academy of Management Journal, 44,* 1102–1121.

Muchinsky, P. M., & Monahan, C. J. (1987). What is person-environment congruence? Supplementary versus complementary models of fit. *Journal of Vocational Behavior, 31,* 268–277.

O'Reilly, C. A., Chatman, J., & Caldwell, D. F. (1991). People and organizational culture: A profile comparison approach to assessing person-organization fit. *Academy of Management Journal, 34,* 487–516.

Ployhart, R. E., Schneider, B., & Schmitt, N. (2006). *Staffing organizations: Contemporary practice and research.* Mahwah, NJ: Erlbaum.

Quinn, R. E., & Rohrbaugh, J. (1983). A spatial model of effectiveness criteria: Towards a competing values approach to organizational analysis. *Management Science, 29,* 363–377.

Ramesh, A., & Gelfand, M. J. (2010). Will they stay or will they go? The role of job embeddedness in predicting turnover in individualistic and collectivistic cultures. *Journal of Applied Psychology, 95,* 807–824.

Schneider, B. (1987). The people make the place. *Personnel Psychology, 40,* 437–453.

Stone, D., Stone-Romero, E., & Lukaszewski, M. (2007). The impact of cultural values on the acceptance and effectiveness of human resource management policies and practices. *Human Resource Management Review, 17,* 152–165.

Srinivasan, L., Gager, S., & Ignatova, M. (2014). *Talent trends 2014: What's on the minds of the professional workforce.* Retrieved from https://business.linkedin.com/content/dam/business/talent-solutions/global/en_US/c/pdfs/linkedin-talent-trends-2014-en-us.pdf

Tarique, I., Schuler, R. S., & Gong, Y. (2006). A model of multinational enterprise subsidiary staffing composition. *International Journal of Human Resource Management, 17,* 207–224.

Yates, K., & Sejen, L. (2011). Employee value proposition: Creating alignment, engagement, and stronger business results. *Society for Human Resources Management.* Webcast.

CHAPTER 5

Arthur, W., Jr., Bell, S. T., Villado, A. J., & Doverspike, D. (2006). The use of person-organization fit in employment decision making: An assessment of its criterion-related validity. *Journal of Applied Psychology, 91,* 786–801.

Arthur, W., Doverspike, D., Muñoz, G. J., Taylor, J. E., & Carr, A. E. (2014), The use of mobile devices in high-stakes remotely delivered assessments and testing. *International Journal of Selection and Assessment, 22,* 113–123.

Barrick, M. R., & Mount, M. K. (1991). The Big Five personality dimensions and job performance: A meta-analysis. *Personnel Psychology, 44,* 1–26.

Barrick, M. R., & Mount, M. K. (2012). Nature and use of personality in selection. In N. Schmitt (Ed.), *The Oxford Handbook of Personnel Assessment and Selection* (pp. 225–251). New York, NY: Oxford University Press.

Beaty, J., Fallon, J., & Shepherd, W. J. (2002). *Proctored versus unproctored Web-based administration of a cognitive ability test.* Symposium at the Society for Industrial & Organizational Psychology conference, Toronto, Canada.

Campbell, J. P. (1990). Modeling the performance prediction problem in industrial and organizational psychology. In M. Dunnette & L. M. Hough (Eds.), *Handbook of industrial and organizational psychology* (2nd ed., Vol. 1, pp. 687–732). Palo Alto, CA: Consulting Psychologists Press.

Campbell, J. P., & Wiernik, B. M. 2015. The modeling and assessment of work performance. *Annual Review of Organizational Psychology and Organizational Behavior, 2,* 47–74.

Campion, M. A., Palmer, D. K., & Campion, J. E. (1997). A review of structure in the selection interview. *Personnel Psychology, 50,* 655–702.

Cascio, W., & Boudreau, J. (2008). *Investing in people: Financial impact of human resource initiatives.* Upper Saddle River, NJ: Pearson Education.

Cleary, T. A. (1968). Test bias: Prediction of grades of Negro and white students in integrated colleges. *Journal of Educational Measurement, 5,* 115–124.

Frei, R. L., & McDaniel, M. A. (1998). Validity of customer service measures in personnel selection: A review of criterion and construct evidence. *Human Performance, 11,* 1–27.

Gatewood, R., Feild, H., & Barrick, M. (2011). *Human resource selection.* Mason, OH: Cengage Learning.

Guion, R. M. (2011). *Assessment, measurement, and prediction for personnel decisions* (2nd ed.). New York, NY: Routledge.

Hoel, H., Rayner, C., & Cooper, C. L. (1999). Workplace bullying. In C. L. Cooper & I. T. Robertson (Eds.), *International review of industrial and organizational psychology* (pp. 195–230). New York, NY: Wiley.

Hogan, R. (2007). *Personality and the fate of organizations.* Mahwah, NJ: Erlbaum.

Holland, J. L. (1997). *Making vocational choices: A theory of vocational personalities and work environments* (3rd ed.). Odessa, FL: PAR.

Huselid, M. A. (1995). The impact of human resource management practices on turnover, productivity, and corporate financial performance. *Academy of Management Journal, 38,* 635–672.

Jensen, A. R. (1998). *The g factor.* Westport, CT: Praeger.

Kim, Y., & Ployhart, R. E. (2014). The effects of staffing and training on firm productivity and profit growth before, during, and after the great recession. *Journal of Applied Psychology, 99,* 361–389.

Kuncel, N. R., Klieger, D. M., Connelly, B. S., & Ones, D. S. (2013). Mechanical versus clinical data combination in selection and admissions decisions: A meta-analysis *Journal of Applied Psychology, 98,* 1060–1072.

McDaniel, M. A., Whetzel, D., Schmidt, F. L., & Maurer, S. (1994). The validity of the employment interview: A comprehensive review and meta-analysis. *Journal of Applied Psychology, 79,* 599–616.

Morelli, N. A., Illingworth, A. J., Scott, J. C., & Lance, C. E. (2012). Are Internet-based, unproctored assessments on mobile and non-mobile devices equivalent? In J. C. Scott (Chair), *Chasing the tortoise: Zeno's paradox in technology-based assessment.* Symposium presented at the 27th Annual Conference of the Society for Industrial and Organizational Psychology, San Diego, CA.

Murphy, K. R., & Jacobs, R. R. (2012). Using effect size measures to reform the determination of adverse impact in equal employment litigation. *Psychology, Public Policy, and Law, 18,* 477–499.

Myors, B., Lievens, F., Schollaert, E., Van Hoye, G., Cronshaw, S. F., Mladinic, A., . . . Sackett, P. R. (2008). International perspectives on the legal environment for selection. *Industrial and Organizational Psychology, 1,* 206–246.

Sun, L. Y., Aryee, S., & Law, K. S. (2007). High-performance human resource practices, citizenship behavior, and organizational performance: A relational perspective. *Academy of Management Journal, 50,* 558–577.

O'Reilly, C. A., Chatman, J., & Caldwell, D. F. (1991). People and organizational culture: A profile comparison approach to assessing person-organization fit. *Academy of Management Journal, 34,* 487–516.

Ployhart, R. E., & Holtz, B. C. (2008). The diversity–validity dilemma: Strategies for reducing racio-ethnic and sex subgroup differences and adverse impact in selection. *Personnel Psychology, 61,* 153–172.

Ployhart, R. E., Nyberg, A. J., Reilly, G., & Maltarich, M. A. 2014. Human capital is dead. Long live human capital resources! *Journal of Management, 40,* 371–398.

Ployhart, R. E., Schneider, B., & Schmitt, N. (2006). *Staffing organizations: Contemporary practice and research.* Mahwah, NJ: Erlbaum.

Ployhart, R. E., Weekley, J. A., Holtz, B. C., & Kemp, C. F. (2003). Web-based and paper-and pencil testing of applicants in a proctored setting: Are personality, biodata, and situational judgment tests comparable? *Personnel Psychology, 56,* 733–752.

Ryan, A. M., Boyce, A. S., Jundt, D., Ghumman, S., Schmidt, G., & Gibby, R. (2009). Going global: Cultural values perceptions of selection procedures. *Applied Psychology: An International Review, 58,* 520–556.

Ryan, A. M., McFarland, L. A., Baron, H., & Page, R. (1999). An international look at selection practices: Nation and culture as explanations for variability in practice. *Personnel Psychology, 52,* 359–391.

Ryan, A. M., & Tippins, N. (2009). *Designing and implementing global selection systems.* West Sussex, UK: Wiley-Blackwell.

Ryan, A. M., Wiechmann, D., & Hemingway, M. (2003). Designing and implementing global staffing systems: Part II—Best practices. *Human Resource Management, 42,* 85–94.

Salgado, J. F., & Anderson, N. (2002). Cognitive and GMA testing in the European community: Issues and evidence. *Human Performance, 15,* 75–96.

Salgado, J. F., Anderson, N., Moscoso, S., Bertua, C., & Fruyt, F. D. (2003). International validity generalization of GMA and cognitive abilities: A European Community meta-analysis. *Personnel Psychology, 56,* 573–605.

Schmidt, F. L., & Hunter, J. E. 1998. The validity and utility of selection methods in personnel psychology: Practical and theoretical implications of 85 years of research findings. *Psychological Bulletin, 124,* 262–274.

Schmidt, F., & Hunter, J. (2004). General mental ability in the world of work: Occupational attainment and job performance. *Journal of Personality and Social Psychology, 86,* 162–173.

Schmidt, F. L., & Zimmerman, R. D. (2004). A counterintuitive hypothesis about employment interview validity and some supporting evidence. *Journal of Applied Psychology, 89,* 553–561.

Silzer, R., & Church, A. H. (2009). The pearls and perils of identifying potential. *Industrial and Organizational Psychology, 2,* 377–412.

Taylor, H. C., & Russell, J. T. (1939). The relationship of validity coefficients to the practical effectiveness of tests in selection. *Journal of Applied Psychology, 23,* 565–578.

Thornton, G. C., Rupp, D. E., & Hoffman, B. (2015). *Assessment center perspectives for talent management strategies.* New York, NY: Routledge.

Tippins, N. T., Beaty, J., Drasgow, F. D., Gibson, W. M., Pearlman, K., Segall, D., & Shepherd, W. J. (2006). Unproctored, Internet testing in employment settings. *Personnel Psychology, 59,* 189–225.

Van Iddekinge, C. H., Putka, D. J., & Campbell, J. P. (2011). Reconsidering vocational interests for personnel selection: The validity of an interest-based selection test in relation to job knowledge, job performance, and continuance intentions. *Journal of Applied Psychology, 96,* 13–33.

Wiechmann, D., Ryan, A. M., & Hemingway, M. (2003), Designing and implementing global staffing systems: Part I—Leaders in global staffing. *Human Resource Management, 42,* 71–83.

CHAPTER 6

Bauer, T. N. (2011). *Onboarding new employees: Maximizing success.* SHRM Foundation's Effective Practice Guideline Series. Arlington, VA: Society for Human Resource Management.

Bauer, T. N., Bodner, T., Erdogan, B., Truxillo, D. M., & Tucker, J. S. (2007). Newcomer adjustment during organizational socialization: A meta-analytic review of antecedents, outcomes, and methods. *Journal of Applied Psychology, 92,* 707–721.

Bauer, T. N., & Erdogan, B. (2014). Delineating and reviewing the role of newcomer capital on organizational socialization. In F. Morgeson & S. Ashford (Eds.), *Annual Review of Organizational Psychology and Organizational Behavior, 1,* 439–457.

Bidwell, M. (2011). Paying more to get less: SPECIFIC skills, matching, and the effects of external hiring versus internal promotion. *Administrative Science Quarterly, 56,* 369–407.

Boudreau, J. W. (2010). *Retooling HR.* Boston, MA: Harvard Business School Publishing.

Boudreau, J. W., & Ramstad, P. M. (2007). *Beyond HR: The new science of human capital.* Boston, MA: Harvard Business School Press.

Call, M., Nyberg, A. J., Ployhart, R. E., Weekley, J. A. (2015). The dynamic nature of collective turnover and unit performance: The impact of time, quality, and replacements. *Academy of Management Journal, 58,* 1208–1232.

Cappelli, P. (2008). *Talent on demand.* Boston, MA: Harvard Business School Press.

Cappelli, P., & Crocker-Hefter, A. (1996). Distinctive human resources are firms' core competencies. *Organizational Dynamics, 24,* 7–21.Cappelli, P., & Keller, J. R. (2014). Talent management: Conceptual approaches and practical challenges. *Annual Review of Organizational Psychology and Organizational Behavior, 1,* 305–331.

Crispin, G., & Mehler, M. (2013). Sources of hire 2013: Perception is reality. *CareerXRoads.* http://www.careerxroads.com/news/SourcesOfHire2013.pdf

Dalton, D. R., Todor, W. D., & Krackhardt, D. M. (1982). Turnover overstated: The functional taxonomy. *Academy of Management Review, 7,* 212–218.

Harter, J. K., Schmidt, F. L., Asplund, J. W., Killham, E. A., & Agrawal, S. (2010). Causal impact of employee work perceptions on the bottom line of organizations. *Perspectives on Psychological Science, 5,* 378–389.

Holtom, B. C., Mitchell, T. R., Lee, T. W., & Eberly, M. B. 2008. Turnover and retention research: A glance at the past, a closer review of the present, and a venture into the future. *Academy of Management Annals, 2,* 231–274.

Huselid, M. A. (1995). The impact of human resource management practices on turnover, productivity, and corporate financial performance. *Academy of Management Journal, 38,* 635–672.

Krell, E. (2015). *Weighing internal vs. external hires.* http://www.shrm.org/publications/hrmagazine/editorialcontent/2015/010215/pages/010215-hiring.aspx

Laurano, M. (2013). *Strategic onboarding 2013: A new look at new hires.* Boston, MA: Aberdeen Group.

Lee, T. W., Burch, T., & Mitchell, T. R. (2014). The story of why we stay: A review of job embeddedness. *Annual Review of Organizational Psychology and Organizational Behavior, 1,* 199–216.

Lombardo, M. M., & Eichinger, R. W. (1996). *The career architect development planner.* Minneapolis: Lominger.

Noe, R. A., Clarke, A. D. M., & Klein, H. K. (2014). Learning in the twenty-first-century workplace. *Annual Review of Organizational Psychology and Organizational Behavior, 1,* 245–275.

Macey, W. H., & Schneider, B. (2008). The meaning of employee engagement. *Industrial and Organizational Psychology, 1,* 3–30.

Scullion, H., & Collings, D. (2011). *Global talent management.* London, UK: Routledge.

Shaw, J. D. (2011). Turnover rates and organizational performance. *Organizational Psychology Review, 1,* 187–213.

Zhou, J., & Hoever, I. J. (2014). Research on workplace creativity: A review and redirection. *Annual Review of Organizational Psychology and Organizational Behavior, 1,* 333–359.

INDEX